Density Functional Methods in Physics

NATO ASI Series

Advanced Science Institutes Series

A series presenting the results of activities sponsored by the NATO Science Committee, which aims at the dissemination of advanced scientific and technological knowledge, with a view to strengthening links between scientific communities.

The series is published by an international board of publishers in conjunction with the NATO Scientific Affairs Division

A Life Sciences **B Physics**	Plenum Publishing Corporation New York and London
C Mathematical and Physical Sciences	D. Reidel Publishing Company Dordrecht, Boston, and Lancaster
D Behavioral and Social Sciences **E Engineering and Materials Sciences**	Martinus Nijhoff Publishers The Hague, Boston, and Lancaster
F Computer and Systems Sciences **G Ecological Sciences**	Springer-Verlag Berlin, Heidelberg, New York, and Tokyo

Recent Volumes in this Series

Volume 118—Regular and Chaotic Motions in Dynamic Systems
edited by G. Velo and A. S. Wightman

Volume 119—Analytical Laser Spectroscopy
edited by S. Martellucci and A. N. Chester

Volume 120—Chaotic Behavior in Quantum Systems: Theory and Applications
edited by Giulio Casati

Volume 121—Electronic Structure, Dynamics, and Quantum Structural Properties of Condensed Matter
edited by Jozef T. Devreese and Piet Van Camp

Volume 122—Quarks, Leptons, and Beyond
edited by H. Fritzsch, R. D. Peccei, H. Saller, and F. Wagner

Volume 123—Density Functional Methods in Physics
edited by Reiner M. Dreizler and João da Providência

Series B: Physics

Density Functional Methods In Physics

Edited by

Reiner M. Dreizler
Institute for Theoretical Physics
Johann Wolfgang Goethe University
Frankfurt/Main, Federal Republic of Germany

and

João da Providência
Department of Physics
University of Coimbra
Coimbra, Portugal

Plenum Press
New York and London
Published in cooperation with NATO Scientific Affairs Division

Proceedings of a NATO ASI on
Density Functional Methods in Physics,
held September 5-16, 1983,
in Alcabideche, Portugal

Library of Congress Cataloging in Publication Data

Main entry under title:

Density functional methods in physics.

(NATO ASI series. Series B, Physics; v. 123)
"Proceedings of a NATO ASI on density functional methods in physics, held September 5-16, 1983, in Alcabideche, Portugal"—T.p. verso.
Bibliography: p.
Includes index.
1. Density functionals—Congresses. 2. Mathematical physics—Congresses. 3. Quantum theory—Congresses. I. Dreizler, Reiner M. II. Providência, João da. III. North Atlantic Treaty Organization. Scientific Affairs Division. IV. Series.
QC20.7.D43D46 1985 530.1′5 85-3563
ISBN 978-1-4757-0820-2 ISBN 978-1-4757-0818-9 (eBook)
DOI 10.1007/978-1-4757-0818-9

Softcover reprint of the hardcover 1st edition 1985
A Division of Plenum Publishing Corporation
233 Spring Street, New York, N.Y. 10013

PREFACE

With the Nato Summer School on "Density Functional Methods in Physics" we planned to offer a forum, where the advances in this field could be discussed and where the exchange of ideas between the different groups responsible for this advancement could be stimulated.

Density functional methods emerged historically in connection with the discussion of atomic and molecular systems in the early days of quantum mechanics. The foundation of the method and a more systematic insight into its intricate structures was developed only in the mid sixties. In the meantime we have learned to accept density functional methods as a suitable tool for tackling a variety of aspects of quantum mechanical many body systems.

At the School substantial progress in the development of the general principles and the basic formalism as well as an impressive number of applications, including the fields of nuclear physics and solid state physics in addition to the traditional domain of applications, were reported.

We intended to provide a bridge over the gaps that unfortunately divide physics of today into isolated areas. We are glad that to a large degree this aim was achieved. We have to thank all the participants for their efforts towards this aim. In particular we would like to thank the lecturers, who willingly faced an " inhomogeneous " audience and the resulting pattern of insistent questions.

The School was held in Alcabideche, Portugal during the period from September 5 to September 16, 1983. The proceedings contain all the main lectures presented at the School as well as two of the four short contributions offered. The first three lectures address mainly the mathematical foundations. This is continued in the fourth, where an existence theorem for time dependent density functional methods is presented. In the fourth lecture we also find the transition to the discussion of atomic and molecular systems, which is continued in the subsequent three lectures. Solid state applications as well as further insight into the structure of the theory are found in the following three lectures. After an interlude on hadronic systems, the concluding lectures deal with a variety of aspects from the world of nuclear physics and the corresponding development of the general formalism. We have decided to include a list of the posters presented or, where possible, an abstract of the posters.

Our thanks are due to the North Atlantic Treaty organization for sponsoring this meeting and to the Calouste Gulbenkian Foundation, the Instituto Nacional de Investigacão Científica and the Portuguese Physical Society for financial support. We are grateful to the Town Hall of Cascais for hospitality.

We thank the advisory committee J. Eichler, Berlin, N. March, Oxford and J.R. Schrieffer, Santa Barbara, for their help and last but not least Mrs. M. Schwarz, who provided the secretarial assistance without which this meeting would not have been possible.

J. da Providencia and
R. M. Dreizler

CONTENTS

DENSITY FUNCTIONAL THEORY: BASIC RESULTS AND SOME OBSERVATIONS 1
Walter Kohn, Santa Barbara

THE CONSTRAINED SEARCH FORMULATION OF DENSITY FUNCTIONAL THEORY 11
Mel Levy and John P. Perdew, New Orleans

DENSITY FUNCTIONALS FOR COULOMB SYSTEMS 31
Elliott H.Lieb, Princeton

DENSITY FUNCTIONAL APPROACH TO TIME DEPENDENT AND TO RELATIVISTIC SYSTEMS 81
Eberhard K.U. Gross and Reiner M. Dreizler, Frankfurt

DENSITY FUNCTIONAL THEORY IN CHEMISTRY 141
Robert G. Parr, Chapel Hill

A DENSITY FUNCTIONAL FORMALISM FOR CONDENSED MATTER SYSTEMS 159
A.K. Rajagopal, Baton Rouge

DENSITY FUNCTIONALS FOR CORRELATION ENERGIES OF ATOMS AND MOLECULES 177
Hermann Stoll and Andreas Savin, Stuttgart

DENSITY-FUNCTIONAL THEORY AND EXCITATION ENERGIES 209
Carl O. Almbladh and Ulf von Barth, Lund

DENSITY FUNCTIONALS AND THE
DESCRIPTION OF METAL SURFACES 233
Norton D. Lang, Yorktown Heights

WHAT DO THE KOHN-SHAM ORBITAL ENERGIES MEAN ?
HOW DO ATOMS DISSOCIATE ? 265
John P. Perdew, New Orleans

HADRONIC DENSITY OF STATES 309
Rajat K. Bhaduri, Hamilton

SEMICLASSICAL DESCRIPTION OF
NUCLEAR BULK PROPERTIES 331
Matthias Brack, Regensburg

THE SCALING APPROACH TO NUCLEAR
GIANT MULTIPOLE RESONANCES 381
Gottfried Holzwarth, Siegen

DENSITY FUNCTIONALS IN HIGH-ENERGY
HEAVY-ION COLLISIONS 401
Joachim A. Maruhn, Frankfurt

ON THE SEMICLASSICAL DESCRIPTION
OF NUCLEAR FERMI LIQUID DROPS 417
Peter Schuck, Grenoble

AVERAGE NUCLEAR PROPERTIES FROM THE
NUCLEAR EFFECTIVE INTERACTION 457
Jacques Treiner, Orsay

ON CHARGE SHARING IN DIATOMIC
QUASIMOLECULES 497
Jörg Eichler, Berlin

GRADIENT EXPANSIONS AND QUANTUM MECHANICAL
EXTENSIONS OF THE CLASSICAL PHASE SPACE 503
Byron K. Jennings, Vancouver

POSTER SESSION 509

INDEX 529

DENSITY FUNCTIONAL THEORY: BASIC RESULTS AND SOME OBSERVATIONS

Walter Kohn†

Institute for Theoretical Physics
University of California
Santa Barbara, CA 93106

ABSTRACT

We review first the basic results of density functional theory: the lemma of Hohenberg and Kohn establishing the density n(r) as a sufficient variable for a description of a non-degenerate or degenerate ground state; the energy variational principle, including recent reformulations by Levy and Lieb; the self-consistent Kohn-Sham equations and the most recent improved approximations for the exchange-correlation functional. This is followed by a number of observations: the issue of v-representability is discussed, including, depending on the context, its significance and non-significance in the light of recent work by Levy, Lieb, and Kohn; a discussion is presented of two different approaches to the basic functionals F[n(r)] and $E_{xc}[n(r)]$ -- by means of many-body techniques or by using the definition of F[n(r)] as a minimum over a constrained set of antisymmetric functions; other interesting recent developments of density functionals are briefly mentioned; some remarks are offered concerning a proposed concept of generalized local approximations; and finally a few other promising directions for future research are listed.

In these lectures, which come at the beginning of a workshop dedicated to density functional theory (DFT), I would like to attempt two objectives: to refresh your memory about the basic facts of DFT, and to present some remarks about the present situation and the outlook for the future.

---

†Supported in part by the National Science Foundation under Grant No. PHY77-27084, supplemented by funds from the National Aeronautics and Space Administration.

BASIC RESULTS

Density functional theory in its simplest form is the venerable Thomas-Fermi theory,[1] a simple, very useful though highly approximate description of interacting Fermi systems in their ground state.

The basis of the modern, rigorous DFT is a simple lemma due to Hohenberg and Kohn (HK).[2]

Lemma

Let n(r) be the density of a ground state (which may be degenerate) of an interacting electron system in an external potential v(r), and let n'(r) have the same relationship to v'(r). Then, n'(r) = n(r) implies v'(r) = v(r) + C where C is a constant.*

In other words, a knowledge of a ground state density n(r) implicitly determines (to within a trivial constant) the external potential of the system. Obviously n(r) also determines the total number of particles N. Thus, rather remarkably, a knowledge of n(r) determines the entire Hamiltonian operator, H, since the other parts of H, kinetic energy T and interaction energy U, are given by the fact that we are dealing with electrons.

Once the Hamiltonian is known from n(r), all properties of the system are implicitly determined via the Schrödinger equation, e.g. all stationary states and energies, Green's functions, etc.

Variational Principle

The second element of DFT is the variational principle for the ground state energy. A key ingredient in this principle is the functional F[n(r)], defined as follows. If n(r) is a ground state density associated with some external potential (or V-representable, VR) then

$$F[n(r)] = \langle T + U \rangle \qquad (1)$$

where < > denotes expectation value in the ground state defined by n(r). Recently the definition of F[n(r)] has been extended by Levy[3,4] and Lieb[5] to include also non-VR densities n(r). This broader definition is

$$F[n(r)] = \min\left(\Psi,\ (T+U)\ \Psi\right) \qquad (2)$$

---

*This is the generalization of the original HK lemma which was limited to non-degenerate ground states. See W. Kohn in Highlights in Condensed Matter Theory, International School of Physics "Enrico Fermi," F. Bassani and F. Fermi, Eds. (1983), to be published.

where the Ψ's range over all antisymmetric N-particle wave functions giving rise to the density n(r).** With the aid of this functional we can now state the variational principle. For a given v(r), we can define the functional of n(r) (with $\int n(r)dr = N$ = integer)

$$E_v[n(r)] \equiv \int v(r)n(r)dr + F[n(r)] \; . \tag{3}$$

Using the conventional Rayleigh-Ritz variational principle it is easy to show that this functional attains its minimum when n(r) is the (or, in degenerate situations, a) ground state density, and that its minimum value is the ground state energy.

If F[n] is, somehow, known (we shall come back to this issue later) the determination of ground state densities and energies becomes extremely simple compared to the problem of solving the 3N-dimensional Schrödinger equation: we just vary the density n(r), a function of only three variables, regardless of the number of particles involved, until we find the minimum (or one of the minima) of $E_v[n]$.

Self-consistent Equations

In practice the most accurate results have generally been found not by direct minimization of $E_v[n(r)]$ but by solving the self-consistent Kohn-Sham equations,[6] similar to the Hartree equations, which have been derived from the basic variational principle. To obtain these equations we limit ourselves to non-degenerate ground states and decompose F[n] as follows:

$$F[n] = T_s[n] + \frac{1}{2}\int \frac{n(r)n(r')}{(r-r')}\, drdr' + E_{xc}[n(r)] \, , \tag{4}$$

where $T_s[n]$ is the kinetic energy of a non-interacting electron system of density distribution n(r) in some external potential $v_s(r)$; the next term is the classical Coulomb repulsion energy and $E_{xc}[n(r)]$, the exchange-correlation energy, is the remainder. This decomposition evidently presupposes that the given density n(r) can be reproduced as the ground state density of non-interacting electrons in some external potential, $v_s(r)$, a question to which we shall return. If the decomposition (4) is possible one can show straightforwardly that the variational principle for $E_v[n]$ leads to the set of self-consistent equations

---

**All mathematically well-behaved densities can be obtained from an antisymmetric wave function. See J.E. Harriman, Phys. Rev. A 24:680 (1981).

$$\left.\begin{aligned}
&\left\{-\frac{1}{2}\nabla^2 + v_{eff}(r) - E_i\right\}\psi_i = 0 \\
&v_{eff}(r) = v(r) + \frac{1}{2}\int\frac{n(r')}{(r-r')}\,dr' + \frac{\delta E_{xc}[n]}{n(r)} \\
&n(r) = \sum_i^N |\psi_i(r)|^2 .
\end{aligned}\right\} \qquad (5)$$

Here the sum goes over the lowest N eigenstates. The ground state energy is given by

$$E = \left(\sum_1^N \varepsilon_i - \int v_{eff}(r)n(r)dr\right) + \int v(r)n(r)dr$$

$$+ \frac{1}{2}\int\frac{n(r)n(r')}{(r-r')}\,drdr' + E_{xc}[n(r)] . \qquad (6)$$

In this formulation of DFT, the essential quantity is the exchange-correlation energy $E_{xc}[n]$. If an accurate and tractable form of E_{xc} is available the solution of the full many-body problem, limited only by the accuracy of $E_{xc}[n]$, is only minimally more difficult than the solution of the Hartree equations.

Since the original Kohn-Sham paper a great deal of research has been done on the functional $E_{xc}[n]$. We mention here only the first and the most recent work. The simplest and yet surprisingly effective approximation is the local density approximation (LDA),[6]

$$E_{xc}^{LDA}[n(r)] = \int \varepsilon_{xc}(n(r))\; n(r)dr , \qquad (7)$$

where $\varepsilon_{xc}(n)$ is the exchange-correlation energy of a uniform electron gas of density n. Evidently this approximation will be the more accurate, the slower the spatial variation of n(r). The same kind of approximation can also be made for systems with density n(r) and magnetization density m(r).[6,7]

The most recent work to improve on the LDA by a functional which depends both on n(r) and $|\nabla n(r)|$ is due to Langreth and Mehl.[8] In the atomic systems to which it has been applied, it reduces the errors of the LDA by a factor of 2-6.

OBSERVATIONS

v-Representability

Twice in the foregoing pages we have encountered the question of v-representability (VR): Can a given density n(r) be reproduced by a ground state density in some external potential v(r)? (Depending on the context one may also need to specify whether the ground state is non-degenerate or degenerate, and whether the electrons are interacting or non-interacting.)

In this area there have been several recent developments. Levy[4] and Lieb[5] have exhibited explicitly examples of mathematically very well-behaved functions n(r) which are not VR by non-interacting ground states, either non-degenerate or degenerate. The same result can also be easily established for interacting systems. This means that there exist many well-behaved functions n(r) for which the original definition (1) of the basic HK functional F[n(r)] does not apply. On the other hand, the recent, more general definition of F[n(r)] applies to all densities which are representable by an antisymmetric wave function, and this includes all smooth density distributions, such as the examples of Levy and Lieb.

Another result bearing on the question of v-representability has recently been established by Kohn.[9] For a lattice version of the Schrödinger equation it was shown that if n(r) is the density of a non-degenerate ground state in some external potential v(r), then all "neighboring" density distributions, n(r) + δn(r), are also obtainable from a "neighboring" potential v(r) + δv(r). (Here δn(r) is arbitrary, except that it must be small enough and obey the particle-conserving condition $\Sigma\delta n(r) = 0$.) This means that all densities n(r), "near" to non-degenerate ground state densities, are VR. In particular, the density distributions of Levy's and Lieb's counter-examples, although smooth and well-behaved since they are not VR, are in fact very unphysical in the sense of being "far" from any density of a non-degenerate ground state. Since, in any practical application to a physical non-degenerate ground state problem, one has to work with densities, n(r), which are "near" to the physical ground state, the non-v-representability of Levy and Lieb becomes irrelevant as soon as one is sufficiently close to the physical density.

We now turn to the relevance of v-representability for the KS self-consistent equations. The derivation of these equations requires that the physical density n(r) of the interacting non-degenerate ground state and neighboring densities n(r) + δn(r) be VR by non-degenerate ground states of systems of non-interacting electrons, moving in the effective potential, v_{eff}, of the KS equation. How can we know if this condition is satisfied? Suppose that in some way the physical non-degenerate density, n(r), is known to some accuracy (e.g. in molecules as superposition of atomic densities). One can then solve the KS equation using, for example, the LDA for $v_{eff}(r)$

$$v_{eff}^{LDA}(r) = v(r) + \int \frac{n(r')}{(r-r')} dr' + \frac{d}{dn} [\varepsilon_{xc}(n)n]_{n=n(r)} \quad . \tag{8}$$

If the resulting density, $n^{LDA}(r)$, which is by construction VR for non-interacting electrons, is "close" to the physical n(r) then all is well, since the VR result of Kohn implies that the physical n(r) itself will then also be VR by a non-interacting, non-degenerate ground state. (Of course, unfortunately we do not know, at present, what is meant by "close" in quantitative terms.)

Thus we see that the question of v-representability is important for density functional theory, especially the KS equations which, in practice, have been used in most applications. We now know some examples of mathematically well-behaved but, in fact, very unphysical functions, n(r), which are not VR; and we know that every density, n(r), "near" a physical density is VR, at least for a lattice version of the Schrödinger equation. Even if a complete characterization of all those densities which are VR and all those which are not VR may not be within reach, one may hope that in the near future the classes of known VR and those of known non-VR density distribution will both be considerably broadened.

Many-Body Approach vs. Wave Function Variation Approach

As one of the authors of the original HK and KS papers I would like to express our general orientation concerning the use of DFT. We observed that the theory of the ground state energy and the ground state density could be formulated completely in terms of the functionals F[n(r)] and $E_{xc}[n(r)]$ respectively. The derivation of these results depended on the familiar Rayleigh-Ritz variational principle for the ground state energy. But once the general density functional theory was established we hoped that useful approximations could be found without, for each system, returning to the original Rayleigh-Ritz principle. This has, in fact, been the case. Beginning with the original local density approximation[6] up to the recent much more sophisticated work of Langreth and Mehl,[8] useful approximations for F[n] and $E_{xc}[n]$ have been developed with the aid of many-body theory and without further recourse to the Rayleigh-Ritz principle for each system. This is important especially for extensive systems, for there the Rayleigh-Ritz approach with a trial function

$$\Psi = \sum_{1}^{M} a_m \Psi_m \quad , \tag{9}$$

with M finite is entirely useless, leading to improvements tending to zero as the size of the system increases.

The recent developments of Levy[3] and Lieb,[5] leading to the definition

$$F[n(r)] = \min \left(\Psi, (T + U) \Psi\right) , \tag{2}$$

where the Ψ's range over all antisymmetric functions giving rise to n(r), has considerable value in allowing a simpler and cleaner derivation of the HK variational principle. However, in dealing with a specific system (e.g. the He atom) it would not make sense to (I) choose a trial n(r), (II) use the definition (2) to find and approximate F[n], (III) compute

$$E_v[n] = \int v(r)n(r)dv + F[n] , \tag{3}$$

and (IV) vary the original n(r) to minimize $E_v[n]$. Through the introduction of the intermediate quantity n(r), this procedure is considerably more complicated than the direct Rayleigh-Ritz minimization of $(\Psi, H\Psi)$, without reference to n(r).

However, it may be hoped that the new definition (2) of F[n] may in the future lead to improved approximations for F[n] expressed in terms of n(r) itself. The same would, of course, be hoped for $T_s[n]$ which is the F[n] for non-interacting particles. The quantity $E_{xc}[n]$ can then be computed from Eq. (4). Possibly the work of Maschke and Zumbach,[10] reported in this meeting, may point the way to such developments.

OTHER RECENT DEVELOPMENTS

In the last few years DFT has seen a number of other very interesting theoretical developments. Here we shall only list a few of these because they are the subject of other papers given at this meeting: the recognition of the highest KS eigenvalue as the exact ionization energy; the discontinuity of v_{xc} (in the KS equation) as function of N, the total number of electrons (regarded as a continuous variable); the implications of this last fact for the density functional theory of the energy gap of insulators; the generalizations of DFT in several new directions, especially time-dependent phenomena; and new, mathematically rigorous results for Thomas-Fermi theory which is an especially simple form of DFT.

CONCLUDING REMARKS

This writer has the sense that the present conference is being held at a time when important progress in DFT is occurring, both in the further development of the general theory and in more accurate approximations for the density functional $E_{xc}[n]$.

I would like to make some remarks about the nature of approximations. The local density approximation is currently defined as

$$E_{xc}^{LDA}[n(R)] = \int \varepsilon_{xc}(n(r))n(r)dr \ , \tag{7}$$

where $\varepsilon_{xc}(n(r))$ is the exchange-correlation energy per particle of a uniform electron gas at density $n(r)$. This requires the knowledge of a function of one variable, $\varepsilon_{xc}(n)$. Simple gradient expansions, as well as more sophisticated work like that of Langreth and Mehl, make a replacement of the form

$$\varepsilon_{xc}(n(r)) \to \varepsilon_{xc}(n(r), |\nabla n(r)|) \ . \tag{10}$$

I suggest that these may usefully be regarded as generalized local approximations (LA) since the integrand in Eq. (10) depends only on the local density and the local density gradient. These theories require the knowledge of a function of <u>two</u> variables $\varepsilon_{xc}(n, |\nabla n|)$. Clearly, in principle one can continue in this way to higher order in the partial derivatives of n, $\partial^2 n/\partial x_i \partial x_j$ etc. which requires the knowledge of functions of 6 variables, etc. If this were a widely applicable and rapidly convergent procedure it would be very attractive because it would characterize a wide class of many-body systems with increasing accuracy, by functions of 1,2,6,... variables. Clearly this approach represents a partial summation of the gradient expansion since, in each order beyond the lowest, arbitrary powers of the derivative operator $\partial/\partial x_i$ occur. The work of Langreth and Mehl[8] shows that nonlinear gradient approximations, at least for the atomic systems studied, lead to significantly better results than simple, low-order gradient expansions. Further theoretical studies of the range of validity and rate of convergence of this sequence of local approximations, as well as the practical search for the "best" universal functions of 1,2,6,... variables ($\varepsilon_{xc}(n)$, $\varepsilon_{xc}(n,|\nabla n|)$,...) might be fruitful.

Proceeding in the opposite direction, it may be interesting to examine cases or features which elude even the best LA in the above generalized sense. For example, one can verify directly that no finite order LA which is a differentiable function of its arguments can account for the discontinuity of v_{xc} noted in certain cases by Perdew and Levy[11] and Sham and Schlüter.[12]

Many other areas appear ready for further progress. These include nonlocal theories, effective descriptions of surface and other nonclassical regions (like metal surfaces, the edges of atoms, etc.), further development of DFT for time-dependent processes, degenerate ground states, excited states, and v-representability by temperature ensembles.

REFERENCES

1. L.H. Thomas, Proc. Camb. Phil. Soc. 23:542 (1927); E. Fermi, Rend. Acad. Naz. Linzei 6:602 (1927).
2. P. Hohenberg and W. Kohn, Phys. Rev. 136:B864 (1964).
3. M. Levy, Proc. Natl. Acad. Sci. (USA) 76:6062 (1979).
4. M. Levy, Phys. Rev. A 26:1200 (1982).
5. E.H. Lieb, in "Physics as Natural Philosophy: Essays in Honor of Laszlo Tisza on his 75th Birthday," H. Feshbach and A. Shimoni, eds., MIT Press, Cambridge (1982).
6. W. Kohn and L.J. Sham, Phys. Rev. 140:A1133 (1965).
7. U. von Barth and L. Hedin, J. Phys. C 5:1629 (1972).
8. D.C. Langreth and M.J. Mehl, Phys. Rev. B 28:1809 (1983).
9. W. Kohn, Phys. Rev. Lett. 51:1596 (1983).
10. G. Zumbach and K. Maschke, Phys. Rev. A 28:544 (1983).
11. J.P. Perdew and M. Levy, Phys. Rev. Lett. 51:1884 (1983).
12. L.J. Sham and M. Schlüter, Phys. Rev. Lett. 51:1888 (1983).

THE CONSTRAINED SEARCH FORMULATION OF DENSITY FUNCTIONAL THEORY

Mel Levy and John P. Perdew

Departments of Chemistry and Physics
Tulane University
New Orleans, Louisiana 70118

THE CHARM OF THE DENSITY

Consider N interacting electrons in a local spin-independent external potential v. The Hamiltonian is

$$H = T + Vee + \sum_{i=1}^{N} v(\vec{r}_i) , \tag{1}$$

where T and Vee are, respectively, the kinetic and electron-electron repulsion operators. For the above kind of Hamiltonian, perhaps no theorem has brought forth the charm and power of the density more than the Hohenberg-Kohn theorem.[1] According to the theorem, a given ground-state electron density $n_o(\vec{r})$ determines its ground-state wavefunction Ψ_o, and consequently the properties of the system, even though there are an infinite number of antisymmetric functions that give the density. Accordingly, it is our purpose to describe here how the density contains all this information, and by means of the "constrained-search" formulation[2,3] We shall emphasize how simple it is to see the $n_o \rightarrow \Psi_o$ mapping.

Of all those antisymmetric functions that yield n_o, what is so special about the ground-state Ψ_o? Well, Ψ_o is simply that antisymmetric function which yields n_o and simultaneously minimizes $\langle T + Vee \rangle$. So, we see how n_o determines Ψ_o. Moreover, since Ψ_o cannot be an eigenfunction of more than one H, it follows immediately that n_o determines H; change H by more than an additive constant in v and n_o must change. In short, $n_o \rightarrow \Psi_o \rightarrow H$. Also, it should be clear that if the minimum in $\langle T + Vee \rangle$ is achieved by more than one antisymmetric function, then each minimizing function must give the same expectation value with respect to any H, so it follows that each minimizing function is a ground-state of the same H. Thus, again $n_o \rightarrow H$, and degeneracies present no problems.

THE CONSTRAINED-SEARCH UNIVERSAL FUNCTIONALS

In the spirit of the preceding section, we define the following universal variational functional[2,3] for the determination of the ground-state density, n_o, and energy, E_o, of H:

$$Q[n] = \text{Min}\langle\Psi_n|T + Vee|\Psi_n\rangle \tag{2}$$

or

$$Q[n] = \langle\Psi_n^{min}|T + Vee|\Psi_n^{min}\rangle \tag{3}$$

Q[n] searches each and every antisymmetric function, Ψ_n, which yields the trial n, and delivers the minimum in $\langle T + Vee\rangle$. (That a minimum always exists has recently been proven by Lieb.[4])

With Q[n] defined by equations (2) or (3) it follows that

$$\int d\vec{r}v(\vec{r})n(\vec{r}) + Q[n]$$

$$= \langle\Psi_n^{min}|\sum_i v(\vec{r}_i) + T + Vee|\Psi_n^{min}\rangle \tag{4}$$

$$\geq E_o \ .$$

Hence, Q[n] is the desired universal variational functional, in that

$$E_o \leq \int d\vec{r}v(\vec{r})n(\vec{r}) + Q[n] \ , \tag{5}$$

with the equality if and only if $n = n_o$. Or,

$$E_o = \underset{n}{\text{Inf}}\{\int d\vec{r}v(\vec{r})n(\vec{r}) + Q[n]\} \ . \tag{6}$$

Also, note that Q[n] is equal to the original Hohenberg-Kohn F[n] if n is interacting v-representable[2,3]--that is, a ground-state density for <u>some</u> H. However, unlike F[n], Q[n] does not require v-representability.

The expectation value $\langle T + Vee\rangle$ with any Ψ_n provides an upper bound to Q[n]. With this in mind, Nyden and Parr[5] and Zumbach and Maschke[6] have already provided specific algorithms for the implementation of equation (2). As a simple example for spherical n, for four electrons, let's employ the Harriman[7] construction and form a single determinant from two doubly-occupied orthonormal space orbitals given by $\phi_k(r) = N^{-\frac{1}{2}}n^{\frac{1}{2}}(r)\exp[ikf(r)]$, with k an integer. Minimization of $\langle T + Vee\rangle$ by optimization of the integer k, within an orthogonalized Hartree expression,[8,9] gives rigorously

$$Q[n] \leq \int \frac{(\nabla n)^2}{8n} d\vec{r} + 3\int [\nabla f(r)]^2 n(r) d\vec{r}$$

$$+ \tfrac{1}{2}\left(\frac{4-1}{4}\right) \iint n(r_1) n(r_2) r_{12}^{-1} d\vec{r}_1 d\vec{r}_2 \tag{7}$$

Note, incidentally, how the Fermi-Amaldi correction naturally arises in the above equation. More work has to be done along these lines; but with the upper bounds embodied in Q[n], systematic approximations are now possible. In contrast, the original Hohenberg-Kohn formulation is less explicit and provides no mechanism for the assertion of upper bounds.

In closing this section, it should be emphasized that one should not look upon the ground-state existence theorems, for the original Hohenberg-Kohn F[n] or for the newer Q[n], as mere tautologies. If this were the case, then we would be able to fabricate analogous definitions and existence theorems for individual excited states. On the contrary, as also noted by Lieb in this book, it is impossible to define a universal functional for <T + Vee> that would always generate both the exact energy and density for an individual excited state. In short, the substance and potential of ground-state density-functional theory is underscored by the inherent incapabilities of excited-state density functional theory; what we can do for ground states is made more meaningful by the fact that we can't do the very same for individual excited states.

EXCHANGE-CORRELATION ENERGY

Positive Kinetic Contribution

In the Kohn-Sham theory,[10] the exchange-correlation energy, E_{xc}, is defined by

$$E_{xc}[n] = T[n] - T_s[n] + Vee[n]$$

$$- \tfrac{1}{2}\iint d\vec{r}_1 d\vec{r}_2 n(\vec{r}_1) n(\vec{r}_2) r_{12}^{-1} \ , \tag{8}$$

where, in line with our constrained-search discussion

$$T[n] = \langle \Psi_n^{min} | T | \Psi_n^{min} \rangle \ , \tag{9}$$

and

$$Vee[n] = \langle \Psi_n^{min} | Vee | \Psi_n^{min} \rangle \ . \tag{10}$$

But, what about $T_s[n]$? To answer this for now, assume as Kohn-Sham did (although the assumption is not really necessary) that n is non-interacting v-representable. That is, assume that n is a ground-state density for some H in equation (1) with Vee absent. Then, in accordance with our theme it follows that[2,14]

$$T_s[n] = \langle\Phi_n^{min}|T|\Phi_n^{min}\rangle \ , \tag{11}$$

where Φ_n^{min} is that antisymmetric function which yields n and simultaneously minimizes $\langle T\rangle$. It follows immediately that[3]

$$T[n] > T_s[n] \ , \tag{12}$$

so the conclusion is that the exchange-correlation energy must contain a positive kinetic energy contribution. Moreover, the magnitude of $T[n] - T_s[n]$ is often about the order of the correlation energy itself. Since $E_{xc}[n]$ has to be negative, and its kinetic contribution has to be positive, this all means that the correction $|E_{xc}[n]|$ is less than it would otherwise be and therefore easier to obtain. In short, one has to recover less negative energy than otherwise, a fact which is partially responsible for the success of many of the density-functional calculations. On the other hand, the neglect, in part, of this positive kinetic contribution may be responsible, in part, for the fact that the local density approximation to $E_{xc}[n]$ tends to overestimate the magnitude of the correlation energy.

Virial Theorem

Consider the equilibrium nuclear orientation, so that the virial theorem takes the form

$$E_o^{tot} = -T[n_o] \ , \tag{13}$$

where E_o^{tot} is the total energy, E_o plus the nuclear-nuclear repulsions. But, by equation (12)

$$E_o^{tot} < -T_s[n_o] \ . \tag{14}$$

Hence, it is not necessarily desirable for

$$E_o^{tot} = -T_s[n_o] \ . \tag{15}$$

In other words, equation (15) cannot be simultaneously satisfied by the exact ground-state energy and density. A small deficiency in X-α theory[11] then is reflected by the fact that the equality in equation (15) is given by that popular theory.

ELECTRON DENSITIES IN SEARCH OF HAMILTONIANS

The original Hohenberg-Kohn-Sham[10] formulation requires that a given trial n be associated simultaneously with some interacting ground state (is interacting v-representable)[2,3] and some non-interacting ground state (is non-interacting v-representable).[2,3] Not all densities, however, are pure-state v-representable.[3,4,12]

Theorem: Unless prevented by certain linear dependencies, n is not pure-state v-representable if it can be expanded as a convex sum of degenerate ground-state densities:[3,4]

Proof: Start with the convex sum

$$n = \sum_{i=1}^{q} d_i n_{ii}, \quad 0 < d_i < 1, \quad \sum_{i=1}^{q} d_i = 1 \; ; \tag{16}$$

where $\Psi_i \rightarrow n_{ii}$, where Ψ_i is the ith degenerate ground-state wavefunction of Hamiltonian H (interacting or non-interacting) with external potential v, and where q is the order of the ground-level degeneracy.

Let's sketch the spirit of the proof.[3,4] First, for n to be a ground-state density of H, Ψ_n^{min} would have to be expandable in terms of the Ψ_i. But, when we take

$$\Psi = \sum_i C_i \Psi_i \; , \tag{17}$$

the density for Ψ will contain cross-terms of the form n_{ij}, arising from

$$n_{ij} = \int \Psi_i^* \Psi_j \; . \tag{18}$$

But, the n in equation (16) contains no such cross-terms, unless there are certain linear dependencies. (Consult refs. 3 and 4 for a more detailed presentation of this point.) Thus Ψ_n^{min} is not a ground state of H, so n can't be a ground-state density of H.

All that remains for completion of the proof is the demonstration that n is not a ground-state density of any other Hamiltonian, H', where H' differs from H by more than an additive constant. To show this, we can now utilize the conclusion of the first part of our proof, namely

$$\langle \Psi_n^{min} | H | \Psi_n^{min} \rangle > \langle \Psi_i | H | \Psi_i \rangle \; . \tag{19}$$

Or, since $\sum_i d_i = 1$,

$$\langle \Psi_n^{min} | H | \Psi_n^{min} \rangle > \sum_{i=1}^{q} d_i \langle \Psi_i | H | \Psi_i \rangle \; . \tag{20}$$

Now, the density is the same on both sides of equation (20), so we can replace v by v', or H by H', without changing the inequality. Obtain

$$\langle\Psi_n^{min}|H'|\Psi_n^{min}\rangle > \sum_{i=1}^{q} d_i \langle\Psi_i|H'|\Psi_i\rangle \ . \tag{21}$$

Since the right-hand-side of equation (21) is an upper bound to the ground-state energy of H', it follows that the strict inequality in equation (21) completes the proof.

A simple example of an n which is not non-interacting v-representable is the one formed, say, from the sum of the densities of the ten lowest states of Na^{+10} plus 1/9 times the sum of the densities of the next nine different states of Na^{+10}. For the n so described, there is no effective one-body Hamiltonian, h_{eff}, such that the total density of the <u>lowest</u> 11 states of h_{eff}, with the Pauli Principle obeyed, add to n. That is, there is <u>no</u>

$$h_{eff}(\vec{r}) = -\tfrac{1}{2}\nabla^2 + v_{eff}(\vec{r}) \tag{22}$$

such that

$$h_{eff}\phi_i = \varepsilon_i\phi_i \ ; \ i = 0,1,2,\cdots,10 \tag{23}$$

where

$$n = \sum_{i=0}^{10} \phi_i^*\phi_i \ ; \quad \varepsilon_0 \leq \varepsilon_1 \leq \varepsilon_2 \cdots \leq \varepsilon_{10} \ . \tag{24}$$

Significant is the fact that the original Kohn-Sham theory[10] requires the existence of an equation (23) for an auxiliary non-interacting system of the true interacting ground-state density of interest, where $\phi_0 = \phi_1$, $\phi_2 = \phi_3$, , $\phi_8 = \phi_9$.

In closing this section, it is important to note that Englisch and Englisch[12] have exhibited non-v-representability with densities which do not arise from degeneracies.

GENERALIZED KOHN-SHAM THEORY

In this section, within the constrained search formulation,[2,3] a most general Kohn-Sham theory is presented[13]--one which obviates the need for v-representability. Expanded upon and extended is the fractional occupation number formulation of Perdew and Zunger,[14] the work of Lieb,[4] and the non-interacting boson formulation of Levy, Perdew, and Sahni.[15]

The <u>general</u> energy expression is[13]

$$E_o \leq -\tfrac{1}{2}\int\nabla^2\gamma + \int vn + w[n] \tag{25}$$

$$\gamma = \sum_i d_i b_i^*(\vec{r}\,')b_i(\vec{r}) \;;\; \gamma \rightarrow n \;, \tag{26}$$

$$w[n] = Q[n] - g[n] \;, \tag{27}$$

$$g[n] = \min \int -\tfrac{1}{2}\nabla^2\bar{\gamma} \;;\; \bar{\gamma} \rightarrow n \tag{28}$$

$$\bar{\gamma} = \sum_i c_i \phi_i^*(\vec{r}\,')\phi_i(\vec{r}) \;, \tag{29}$$

where in the definition of g[n] it is only required that each "searched" one-matrix $\bar{\gamma} \rightarrow n$ be of the same form as the one-matrix $\gamma \rightarrow n$. In other words, if γ is restricted to be idempotent, then each searched $\bar{\gamma}$ is restricted to be idempotent. On the other hand, if fractional occupation numbers are allowed in γ, with $\langle b_i|b_j\rangle = \delta_{ij}$, then fractional occupation numbers are allowed in each searched $\bar{\gamma}$, with $\langle\phi_i|\phi_j\rangle = \delta_{ij}$. Or, with g[n] defined accordingly, γ could even be the one-matrix for a system of non-interacting bosons, where all $d_i = 1$ and all b_i are identical. Equation (25) would then take the form[15]

$$E_0 \leq \int n^{\frac{1}{2}}(-\tfrac{1}{2}\nabla^2)n^{\frac{1}{2}} + \int vn + w[n] \tag{30}$$

The variational principle for the true interacting ground-state energy E_0 and density n_0 would always hold, with respect to equation (25), for all the above cases because by construction and definition

$$-\tfrac{1}{2}\int\nabla^2\gamma \geq g[n] \;. \tag{31}$$

In sum, there is nothing which restricts γ to be a one matrix associated with a non-interacting fermion ground-state. One may add and subtract <u>any</u> kinetic energy to and from the right-hand-side of equation <u>(25)</u>, and neither interacting nor non-interacting v-representabilities are required.

SELF-CONSISTENT ORBITAL EQUATIONS FOR NON-V-REPRESENTABILITY

In this section, we follow Levy and Perdew.[13] We are concerned here with situations where the interacting ground-level energy E_0 is achieved by a n_0 which is <u>not</u> non-interacting v-representable.

What happens when there is no non-interacting v-representability? Can one still use the traditional Kohn-Sham equations of the form of equations (23) without fractional occupation numbers, as Harris[16] argues is desirable when the local density approximation is employed? Let's consider primarily, although not necessarily, the case when the absence of non-interacting v-representability is due to ground-level degeneracies.[3,4] Degeneracy situations are not uncommon. Indeed these situations tend to arise when "nice" symmetries are imposed upon densities in high-symmetry systems (spherical symmetry in atoms and cylindrical symmetry in diatomic molecules, etc.).

The answer to the question is that equations of the form of (23) still arise, but the optimum orbitals are not the first N one-body states of an effective Hamiltonian with the Pauli Principle imposed. To see this, form equation (25) with the one-matrices always restricted to be idempotent; all $d_i = 1$ and $\langle b_i | b_j \rangle = \delta_{ij}$. Now, in terms of our constrained-search theme, it should be clear that the orbitals which minimize the right-hand-side of equation (25) are those which yield n_0 and simultaneously minimize $-\frac{1}{2}\sum_i \int b_i \nabla^2 b_i$. Hence, optimum orbitals, called a_i, are generated by taking[13]

$$\delta\{\sum_i [\langle b_i | -\tfrac{1}{2}\nabla^2 | b_i \rangle + \int d\vec{r}\, b_i^*(\vec{r}) b_i(\vec{r}) \lambda(\vec{r})\} = 0 \tag{32}$$

where $\lambda(\vec{r})$ is a functional Lagrangian multiplier[17-19] associated with the constraint that $\sum_i b_i^*(\vec{r}) b_i(\vec{r})$ yield $n_0(\vec{r})$. Unless prevented perhaps by the existence of the kinds of non-v-representable situations described by Englisch and Englisch,[12] the result of the implementation of (32) is

$$-\tfrac{1}{2}\nabla^2 a_i(\vec{r}) + \lambda(\vec{r}) a_i(\vec{r}) = f_i a_i(\vec{r}), \tag{33}$$

where

$$\lambda(\vec{r}) = v(\vec{r}) + (\delta w/\delta n)_{n=n_0} \; . \tag{34}$$

In other words, $\lambda(\vec{r})$ turns out to be a density-functional effective potential, so that the optimum orbitals, a_i, do in fact obey Kohn-Sham-like equations of the form (23), with all the a_i as eigenfunctions of the same one-body Hamiltonian, and with the f_i as the eigenvalues. But, the a_i are not the first N non-interacting fermion states of $\lambda(\vec{r})$. Instead, there are now holes below the Fermi level. (That is, there exists at least one partially filled or unoccupied level below the highest occupied level. A simple one-electron example is provided by the density of the 2s orbital in the H-atom. There, $\lambda(\vec{r})$ would turn out to be the Coulomb potential; the 2s-level would be occupied and the 1s-level would be empty.)

What does this all mean in practice for the few or many-electron problem? It means that there are significant non-v-representable situations where one "is allowed" to carry out the usual Kohn-Sham-like procedure, but self-consistency at the interacting ground-level E_0 and n_0 would only be achieved with the presence of the holes below the Fermi level. If at the end of each trial iteration, on the other hand, one were to only extract the lowest

N fermion states of $-\frac{1}{2}\nabla^2 + v(\vec{r}) + \delta w/\delta n$, then one would not be able to achieve simultaneously self-consistency, E_o, and n_o.

UNIVERSAL FUNCTIONALS FOR EXCITED STATES AND MIXED SYMMETRY STATES

In this section we follow Englisch and Englisch,[12] Levy,[20] Lieb,[21] and von Barth.[22,23] The existence of universal functionals for excited states (without restriction to lowest states of given symmetries) may be immediately established by defining the functional $Q[n_B,n_A]$ and showing that it is compatible with the variational principle. Let $Q[n_B,n_A]$ first find that antisymmetric wavefunction, Ψ_A^{min}, which yields n_A and simultaneously minimizes $<T + Vee>$. Then let $Q[n_B,n_A]$ find that antisymmetric wavefunction which yields n_B, is orthogonal to Ψ_A^{min}, and minimizes $<T + Vee>$. This minimum in $<T + Vee>$ is then identified as the value of $Q[n_B,n_A]$.[20]

Note that $Q[n_B,n_A]$ is truly universal. Also, no symmetry requirements are placed upon n_B or n_A, other than the fact that they must correspond to fermions. The definition of $Q[n_B,n_A]$ leads directly to the desired bound which is in accordance with the variational principle (with nondegeneracies assumed for simplicity):

$$E_1^v \leq \int d\vec{r}v(\vec{r})n(\vec{r}) + Q[n,n_o^v] \tag{35}$$

where n_o^v signifies the ground-state density of external potential v and E_1^v signifies the first-excited state energy. The equality applies when $n = n_1^v$. Extensions to higher excited states are straightforward. For example, if one were interested in the second excited state, then the appropriate universal functional would be $Q[n_C,n_B,n_A]$, and one would evaluate $Q[n,n_1^v,n_o^v]$ for each trial n. (Again, nondegeneracies are assumed.)

Let us now discuss the excited state formulation of Theophilou[24] in terms of our general theme which stresses that practically all universal density and density matrix functionals simply constitute constrained searches and minimizations. Accordingly, in spite of the fact that Theophilou's development is somewhat involved and employs the usual proof by contradiction, we assert,[20] as also done by Englisch and Englisch,[12] that the formal display of his implied universal functional may be made in a surprisingly easy manner through the constrained-search formulation.

Although we don't really have to, for simplicity, let us assume that the M lowest lying states of the Hamiltonian H, with a local external potential, are nondegenerate. Given this nondegeneracy condition, we assert that Theophilou's universal functional of the density, for $<T + Vee>$, is simply[12,20]

$$Q_M[n]=\mathrm{Min}\int\cdots\int[T(1,2,\cdots N) + Vee(1,2,\cdots N)]D_n^M(1',2',\cdots N'|1,2,\cdots N) \quad (36)$$

The search in Eq. (36) involves all those statistical ensemble N-particle density matrices, D_n^M, which integrate to the composite trial density n. In addition, all occupation numbers are restricted to be equal. Namely,

$$D_n^M(1',2',\cdots N'|1,2\cdots N) = \sum_{j=0}^{M-1} \Psi_j^*(1',2',\cdots N')\Psi_j(1,2,\cdots N) \quad (37)$$

where the Ψ_j's are orthonormal and where D_n^M has been normalized to M. According to familiar theorems of the linear variational method (see Quantum Mechanics by Kemble, for instance), with D_n^M defined by Eq. (37), our previous theorems dictate that Eq. (36) assures adherence to the variational principle for the sum of the first M eigenvalues of H when the external potential term is added to $Q_M[n]$. Moreover, when the total composite energy achieves a minimum, the optimized trial composite n must be the sum of the first M eigenstate densities of H, with the total composite energy equal to the sum of the M lowest eigenvalues of H.

The universal functional $Q[n_B,n_A]$ is related to Eq. (36). On the other hand, the recent presentation of Valone and Capitani[25] features McDonald's theorem and thus presents a somewhat different orientation, although constrained searches and minimizations are emphasized. Valone and Capitani do not, however, isolate a truly universal functional for <T + Vee>.

Gunnarsson and Lundqvist[26] proved that the Hohenberg-Kohn theorem for ground-state variational calculations can be generalized to the lowest energy state of a given symmetry, thus encompassing first excited states with symmetries which differ from the ground-states. Symmetry specifications could include, for instance, total orbital and spin angular momenta. The Gunnarsson-Lundqvist functional requires that the density be realized by some external potential. It is our purpose now to identify the Gunnarsson-Lundqvist functional and to simultaneously expand its domain so as to not require that the density be associated with some external potential. We have[12,20,22]

$$E_\lambda \leq \int d\vec{r}v(\vec{r})n(\vec{r}) + G_\lambda[n], \quad (38)$$

where $G_\lambda[n]$ is identified as

$$G_\lambda[n] = \mathrm{Min}\langle\Psi_n\lambda|T+Vee|\Psi_n\lambda\rangle \quad . \quad (39)$$

The search on the right-hand-side of Eq. (39) is restricted to all antisymmetric wavefunctions $\Psi_n\lambda$ which yield the density n and which possess the specified symmetry indicated by λ. The equality in Eq. (38) holds when n is obtainable from the lowest energy eigenstate of symmetry λ.

The general recipe may be directly extended to density functional theory for mixed symmetry states as discussed by von Barth.[22,23] Simply restrict the search in Eq. (39) to all wavefunctions which possess the desired well-defined mixture of symmetries. We emphasize that a proof by reduction to a contradiction involving external potentials is not required. Also, degeneracy considerations need not be explored.

Finally, analogues of $G_\lambda[n]$ may be defined for Hartree-Fock, unrestricted Hartree-Fock, and other approximate wavefunction formulations. For example, for the restricted Hartree-Fock case, the search on the right-hand-side of Eq. (39) is restricted to all single determinant wavefunctions which yield the trial n and which possess the specified symmetry indicated by λ .

The ground-state universal Legendre transform functional[4] is now extended to the sum of the M lowest eigenstate energies for N electrons. (This is also done by Lieb in this book.) Define

$$\bar{F}_M[n] = \sup_w\left[\sum_{j=0}^{M-1} E_j(w) - \int d\vec{r}w(\vec{r})n(\vec{r})\right] \tag{40}$$

where $n(\vec{r})$ is a trial sum of densities such that

$$\int d\vec{r}n(\vec{r}) = (M)(N) \ . \tag{41}$$

Each $w(\vec{r})$ is the local external potential associated with

$$H(w) = T + Vee + \sum_{i=1}^{N} w(\vec{r}_i) \ . \tag{42}$$

Also,

$$E_0(w) \leq E_2(w) \leq \cdots \leq E_{M-1}(w) \ , \tag{43}$$

where $E_j(w)$ is the jth eigenstate of $H(w)$. The equality in equation (43) applies whenever a degeneracy exists.

Assume that one is interested in the sum of the M lowest eigenstates for the Hamiltonian in equation (1). Call the $E_j(v)$ the eigenvalues of interest. Now, by the definition of "supremum", it follows immediately that

$$\sum_{j=0}^{M-1} E_j(v) - \int d\vec{r}v(\vec{r})n(\vec{r})$$
$$\leq \sup_w \left[\sum_{j=0}^{M-1} E_j(w) - \int d\vec{r}w(\vec{r})n(\vec{r})\right] \tag{44}$$

But, the right-hand-side of equation (44) is simply $\bar{F}_M[n]$. Hence,

$$\sum_{j=0}^{M-1} E_j(v) - \int d\vec{r} v(\vec{r}) n(\vec{r}) \leq \bar{F}_M[n] \tag{45}$$

and with rearrangement obtain

$$\sum_{j=0}^{M-1} E_j(v) \leq \int d\vec{r} v(\vec{r}) n(\vec{r}) + \bar{F}_M[n] , \tag{46}$$

which is the desired result. The equality holds when n is the sum of the first M eigenstate densities of the H(v) given in equation (1), and follows directly from the variational principle involving $\bar{F}_M[n]$ and wavefunctions.

The Individual Excited-State Limitations

As stated at the beginning of this chapter, density functional theory for individual excited states is limited by the fact that the exact energy and corresponding density are not guaranteed for each individual excited state. The problem arises from the existence, at times, of the following situation:

$$\langle \Psi_j^v | H(w) | \Psi_j^v \rangle < E_j(w) \ ; \ \ w \neq v \ ; \ j>0 , \tag{47}$$

where Ψ_j^v is the jth eigenstate of Hamiltonian H(v). The strict inequality in equation (47) is the source of trouble. In contrast, the inequality never appears, of course, in ground-state theory, a fact which is the source of the very foundation and success of ground-state theory. But, in individual excited-state density-functional theory, within, say, the universal Legendre transform representation $F_j[n]$ defined by[21]

$$F_j[n] = \operatorname*{Sup}_w [E_j(w) - \int d\vec{r} w(\vec{r}) n(\vec{r})] , \tag{48}$$

it would unfortunately follow, when equation (47) applies, that

$$F_j[n] > \langle \Psi_j^v | T + Vee | \Psi_j^v \rangle \ ; \ \Psi_j^v \rightarrow n , \tag{49}$$

which leads to the inequality

$$E_j(v) < \int d\vec{r} v(\vec{r}) n(\vec{r}) + F_j[n] \tag{50}$$

when $n(\vec{r})$ is the exact density for the jth eigenstate of H(v). (The inequality in equation (47), when in effect, follows from the fact that $E_j(w)$ is not concave, an observation made in Lieb's chapter.)

There are meaningful exact bounds, nevertheless, for individual excited states. For instance, Lieb[21] presents universal functionals which give, upon density optimization, upper bounds to

$E_j(v)$. Following is a universal functional which we present as a lower bound to $E_j(v)$ upon density optimization. Define $B_j[n]$ by

$$B_j[n] = \underset{w}{\mathrm{Inf}}\{<\Psi_j^w|T + Vee|\Psi_j^w>|\Psi_j^w \rightarrow n\}, \tag{51}$$

where Ψ_j^w is the jth state of Hamiltonian $H(w)$ with external potential $w(\vec{r})$. Or, if no $\Psi_j^w \rightarrow n$, then set

$$B_j[n] = F_j[n] \ . \tag{52}$$

With $B_j[n]$ defined by equations (51) and (52) it follows that

$$E_j(v) \geq \underset{n}{\mathrm{Inf}}\{\int d\vec{r}v(\vec{r})n(\vec{r}) + B_j[n]\} \ . \tag{53}$$

So, if $j = 1$, then the right-hand-side of equation (53) is bounded below by $E_o(v)$ and above by $E_1(v)$--that is, bounded above by the first excited-state energy and below by the ground-state energy, and probably much closer to the excited state than the ground state. In any case, equation (53) is our lower bound counterpart to the upper bound formula[21]

$$E_j(v) \leq \underset{n}{\mathrm{Inf}}\{\int d\vec{r}v(\vec{r})n(\vec{r}) + F_j[n]\} \tag{54}$$

Excited-Ground Connections and Bounds

We know, of course, a fair amount about explicit computational approximations to the ground-state functionals, but very little about explicit computational approximations to the excited-state functionals. With this in mind, it would be helpful to at least list some bounds that connect the formal excited-state functionals with the formal ground-state functionals.

First, if $n(\vec{r})$ is pure-state v-representable, then

$$Q_M[n] \geq M^{-1}Q[n] \ , \tag{55}$$

$$\bar{F}_M[n] \geq M^{-1}Q[n] \ , \tag{56}$$

$$F_j[n] \geq B_j[n] \geq Q[n] \tag{57}$$

Further, regardless of the v-representability status of n, we assert that

$$Q_M[n] \geq M^{-1}F_{DM}[n] \ , \tag{58}$$

$$\bar{F}_M[n] \geq M^{-1}F_{DM}[n] \ , \tag{59}$$

$$F_j[n] \geq B_j[n] \geq F_{DM}[n] \ . \tag{60}$$

The constrained-search $F_{DM}[n]$ is given by[4,27]

$$F_{DM}[n] = \text{Min} \sum_i d_i <\Psi_i|T + V_{ee}|\Psi_i> ; \tag{61}$$

$$<\Psi_i|\Psi_i> = 1, \ 0 \leq d_i \leq 1, \ \sum_i d_i = 1 , \tag{62}$$

where each searched ensemble in equation (61) is constrained to yield n. The Ψ_i are antisymmetric functions.

OPEN QUESTIONS

V-Representability for Ground-State Theory

As has been emphasized with specific examples in this chapter, there exist an infinite number of very "nice" densities which, by virtue of degeneracies, are not pure-state v-representable.[3,4] These densities, however, are nevertheless non-interacting ensemble v-representable,[3,12] which means that in the implementation of modified[13] Kohn-Sham theory,[10] one would have to invoke only limited fractional orbital occupancies,[14] or alternatively, one would be able to achieve energy minimization and orbital self-consistency with, perhaps, only limited need of empty orbitals[13] below the Fermi level when fractional occupancies are not employed.

Englisch and Englisch[12] have given examples of densities which are not even ensemble v-representable. On the other hand, Kohn[28] has very recently proven that, on a lattice, all densities in the neighborhood of a non-degenerate ground-state are pure-state v-representable and thus automatically satisfy the weaker condition of ensemble v-representability. So, should ensemble v-representability normally be expected for densities which possess reasonable physical characteristics for ground states? Well, for finite dimensional quantum systems with strictly positive densities, Englisch and Englisch[29] have answered affirmatively; for classical systems, Chayes, et al.[30] have answered affirmatively; but for infinite dimensional quantum systems of Fermions, the question is still open (the 2s orbital-density of the H-atom is not ensemble v-representable, but this density has a node which is not a reasonable characteristic for a ground-state density).

Towards an answer, let's start[13] with equation (4.5) in Lieb:[4]

$$F[\rho] = F_{DM}[\rho] \tag{63}$$

In our notation, equation (63) reads

$$\underset{w}{\mathrm{Sup}}[E(w) - \int d\vec{r}w(\vec{r})n(\vec{r})]$$

$$= \mathrm{Inf}[\sum_i <\Psi_i|G|\Psi_i>] \tag{64}$$

$E(w)$ is the ground-level energy associated with external potential $w(\vec{r})$, where as in equation (61), it is understood that each searched ensemble[27] is constrained to yield n, and where $G = T + \mu V_{ee}$; $\mu = 1$ for interacting considerations and $\mu = 0$ for non-interacting considerations. The μ is a coupling constant.[16,26,31,32]

Now, Inf may always be replaced by Min[4] on the right-hand-side of equation (64). Further, suppose the supremum on the left-hand-side is a maximum, given by say $w = \bar{w}$. Then,

$$E(\bar{w}) - \int d\vec{r}\bar{w}(\vec{r})n(\vec{r}) = \mathrm{Min}[\sum_i d_i <\Psi_i|G|\Psi_i>] \ . \tag{65}$$

Rearrangement of equation (65) is the key step and results in

$$E(\bar{w}) = \int d\vec{r}\bar{w}(\vec{r})n(\vec{r}) + \mathrm{Min}[\sum_i d_i <\Psi_i|G|\Psi_i>] \ , \tag{66}$$

or

$$E(\bar{w}) = \mathrm{Min}[\sum_i d_i <\Psi_i|\sum_j \bar{w}(\vec{r}_j) + G|\Psi_i>] \ . \tag{67}$$

Therefore, by the equality in equation (67), the conclusion[13] is that n has to be ensemble v-representable when, in the search over potentials in equation (64), the supremum turns out to be a maximum.

Assume that n exhibits interacting ensemble v-representability. Now, at fixed n, let μ very gradually decrease to zero. For all but perhaps a very weird n, it seems most reasonable that at no point along the path $\mu = 1$ to $\mu = 0$ would the presence of a maximum suddenly vanish, especially since we are turning the repulsion off and not on. Consequently, we hypothesize[13] that if a density is interacting ensemble v-representable then it is <u>extremely likely</u> to be simultaneously non-interacting ensemble v-representable. Similarly, we conjecture that it is extremely likely that an interacting pure-state v-representable density is also simultaneously non-interacting ensemble v-representable, if not non-interacting pure-state v-representable. The significance of all this should be clear with respect to Kohn-Sham and generalized Kohn-Sham theories, as discussed above.

The Individual Excited State

How tight are the bounds in equations (53) and (54)?

RECENT CONSTRAINED SEARCHES

This chapter has been concerned primarily with energies and densities, at zero temperature, for spin-independent Hamiltonians of the form of equation (1). With these Hamiltonians, constrained-search functionals have recently been identified by von Barth[22] for spin-density functional theory,[33,34] where the search involves all those antisymmetric functions which yield a given trial spin-density.[33,34] The constrained-search formulation is thereby applicable to spin-dependent Hamiltonians as well as to spin-independent Hamiltonians. Capitani, Nalewajski, and Parr[35] have formulated constrained-search functionals for non-Born-Oppenheimer cases. Rajagopal[36] has generalized the constrained-search approach to temperature-dependent relativistic situations for electron-nuclei systems. Bartolotti[37] and Gross[38] have formulated time-dependent constrained-search functionals. Stoll and Savin[39] have defined density functionals for correlation energies with the Hartree-Fock density functional as the starting point. Henderson[40] has modified Q[n] for momentum space,[40,41] and Bauer[42] has very recently put down corresponding Kohn-Sham-like equations. Valone has extended Q[n] to ensemble searches.[27]

According to Perdew, Parr, Levy, and Balduz, Jr., the highest occupied Kohn-Sham eigenvalue becomes well-defined and is identified as the <u>exact ionization energy</u>[43] when density-functional theory is extended to fractional electron number.[43] (See references 14, 22 and 44-51 for significant related work.) In this case, the constrained search involves statistical mixtures of M and M + 1 electrons, where M is an integer.[43] The search is carried out over all those statistical mixtures which yield the given density.[43]

REFERENCES

1. P. Hohenberg and W. Kohn, Inhomogeneous electron gas, <u>Phys. Rev.</u> 136:B864 (1964).
2. M. Levy, Universal variational functionals of electron densities, first-order density matrices, and natural spin-orbitals and solution of the v-representability problem, <u>Proc. Natl. Acad. Sci. (USA)</u> 76:6062 (1979); M. Levy, Universal functionals of the density and first-order density matrix, <u>Bull. Amer. Phys. Soc.</u> 24:626 (1979).
3. M. Levy, Electron densities in search of hamiltonians, <u>Phys. Rev. A</u> 26:1200 (1982).

4. E. H. Lieb, Density functionals for coulomb systems, in "Physics as Natural Philosophy: Essays in Honor of Laszlo Tisza on his 75th Birthday", H. Feshbach and A. Shimony, eds., M.I.T. Press, Cambridge (1982); E. H. Lieb, Density functionals for coulomb systems, Int. J. Quantum Chem. 24:243 (1983).
5. M. R. Nyden and R. G. Parr, Restatement of conventional electronic wavefunction determination as a density functional procedure, J. Chem. Phys. 78:4044 (1983); M. R. Nyden, An orthogonality constrained generalization of the Weizacker density functional method, J. Chem. Phys. 78:4048 (1983).
6. G. Zumbach and K. Maschke, New approach to the calculation of density functionals, Phys. Rev. A 28:544 (1983).
7. J. E. Harriman, Orthonormal orbitals for the representation of an arbitrary density, Phys. Rev. A 24:680 (1981).
8. M. Levy, T.-S. Nee, and R. G. Parr, Method for direct determination of localized orbitals, J. Chem. Phys. 63:316 (1975).
9. P. A. Christiansen and W. E. Palke, A study of the ethane internal rotation barrier, Chem. Phys. Lett. 31:462 (1975).
10. W. Kohn and L. J. Sham, Self-consistent equations including exchange and correlation effects, Phys. Rev. 140:A1133 (1965).
11. J. C. Slater, "The Self-Consistent Field for Molecules and Solids", McGraw-Hill, New York (1974); J. W. D. Connally, The $X\alpha$ Method, in Modern Theoretical Chemistry 7, G. A. Segal, ed., Plenum, New York (1977).
12. H. Englisch and R. Englisch, Hohenberg-Kohn theorem and non-v-representable densities, Physica 121A:253 (1983).
13. M. Levy and J. P. Perdew, Generalized density-functional orbital theories and v-representability, unpublished.
14. J. P. Perdew and A. Zunger, Self-interaction correction to density-functional approximations for many-electron systems, Phys. Rev. B 23:5048 (1981). They have also proved that regardless of the v-representability status of a given orbital it must be free of self interaction in the exact theory.
15. M. Levy, J. P. Perdew, and V. Sahni, Exact differential equation for the density of a many-particle system, unpublished.
16. J. Harris, The role of occupation numbers in HKS theory, Int. J. Quantum Chem. S13:189 (1980); J. Harris, The adiabatic-connection approach to Kohn-Sham, unpublished.
17. J. K. Percus, The role of model systems in the few-body reduction of the N-Fermion problem, Int. J. Quantum Chem. 13:89 (1978).
18. P. W. Payne, Density functionals in unrestricted Hartree-Fock theory, J. Chem. Phys. 71:490 (1979).
19. K. F. Freed and M. Levy, Direct first principles algorithm for the universal electron density functional, J. Chem. Phys. 77:396 (1982).

20. This section, equations (35) to (46), was sent to Englisch and Englisch in January, 1983. See the addendum which they kindly include in their article (reference 12). Englisch and Englisch independently assert these equations in reference 12. We thank them for having sent us a copy of their manuscript before publication.
21. E. H. Lieb, excited-state section of his chapter in this book. We thank him for kindly having sent us a copy of his excited-state section before publication.
22. U. von Barth, Density functional theory for solids, NATO Advanced Study Institute, Gent, Summer 1982, Plenum, in press.
23. U. von Barth, Local-density theory of multiplet structure, Phys. Rev. A 20:1693 (1979).
24. A. K. Theophilou, The energy density functional formalism for excited states, J. Phys. C 12:5419 (1979); J. Katriel, An alternative interpretation of Theophilou's extension of the Hohenberg-Kohn theorem to excited states, J. Phys. C 13:L375 (1980).
25. S. M. Valone and J. F. Capitani, Bound excited states in density-functional theory, Phys. Rev. A, 23:2127 (1981).
26. O. Gunnarsson and B. I. Lundqvist, Exchange and correlation in atoms, molecules, and solids by the spin-density functional formalism, Phys. Rev. B 13:4274 (1976).
27. S. M. Valone, A one-to-one mapping between one-particle densities and some n-particle ensembles, J. Chem. Phys. 73:4653 (1980); S. M. Valone, Consequences of extending 1 matrix energy functions from pure-state representable to all ensemble representable 1 matrices, J. Chem. Phys. 73:1344 (1980).
28. W. Kohn, "V-representability and density functional theory", Phys. Rev. Lett. 51:1596 (1983).
29. H. Englisch and R. Englisch, V-representability in finite-dimensional space, unpublished. See also S. T. Epstein and C. M. Rosenthal, The Hohenberg-Kohn theorem, J. Chem. Phys. 64:247 (1976); J. Katriel, C. J. Appelof, and E. R. Davidson, Mapping between local potentials and ground state densities, Int. J. Quantum Chem. 19:293 (1981).
30. J. T. Chayes, L. Chayes, and E. H. Lieb, The inverse problem in classical statistical mechanics, unpublished manuscript.
31. D. C. Langreth and J. P. Perdew, The exchange-correlation energy of a metallic surface, Solid State Commun. 17:1425 (1975); Exchange-correlation energy of a metallic surface: wave-vector analysis, Phys. Rev. B 15:2884 (1977).
32. C. O. Almbladh, Technical Report, University of Lund (1972).
33. U. von Barth and L. Hedin, A local exchange-correlation potential for the spin-polarized case I, J. Phys. C 5:1629 (1972).

34. A. K. Rajagopal and J. Callaway, Inhomogeneous electron gas, Phys. Rev. B 7:1912 (1973). See also A. K. Rajagopal, Adv. in Chem. Phys. 41:59 (1980).
35. J. F. Capitani, R. F. Nalewajski, and R. G. Parr, Non-Born-Oppenheimer density functional theory of molecular systems, J. Chem. Phys. 76:568 (1982).
36. A. K. Rajagopal, A density functional formalism for condensed matter systems, chapter in this book.
37. L. J. Bartolotti, Time-dependent extension of the Hohenberg-Kohn-Levy energy-density functional, Phys. Rev. A 24:1661 (1982).
38. Erich Runge and E.K.U. Gross, Phys. Rev. Lett. 52:997 (1984).
39. H. Stoll and A. Savin, Density functionals for correlation energies of atoms and molecules, chapter in this book, and references within.
40. G. A. Henderson, Variational theorems for the single-particle probability density and density-matrix in momentum space, Phys. Rev. A 23:19 (1981).
41. R. N. Pathak, P. V. Panat, and S. R. Gadre, Local-density functional model for atoms in momentum space, Phys. Rev. A 26:3073 (1982).
42. G. E. W. Bauer, unpublished. See also G. E. W. Bauer, General operator ground-state expectation values in the Hohenberg-Kohn-Sham density-functional formalism, Phys. Rev. B 27:5912 (1983).
43. J. P. Perdew, R. G. Parr, M. Levy, and J. L. Balduz, Jr., Density-functional theory for fractional particle number: Derivative discontinuities of the energy, Phys. Rev. Lett. 49:1691 (1982).
44. R. G. Parr, R. A. Donnelly, M. Levy, and W. E. Palke, Electronegativity: The density functional viewpoint, J. Chem. Phys. 68:3801 (1978).
45. R. G Parr and L. J. Bartolotti, Some remarks on the density functional theory of few electron systems, J. Phys. Chem. 87: 2810 (1983).
46. C.-O. Almbladh and U. von Barth, Exact results for the charge and spin densities, exchange-correlation potentials, and density-functional eigenvalues, unpublished manuscript (1983). See also chapter in this book; C.-O. Almbladh and A. C. Pedroza, unpublished manuscript (1983).
47. J. P. Perdew and M. Levy, Density functional theory for open systems, in "Many-Body Phenomena at Surfaces", D. C. Langreth and H. Suhl, eds., Academic, in press.
48. J. P. Perdew and M. Levy, Physical content of the exact Kohn-Sham orbital energies: Band gaps and derivative discontinuities, Phys. Rev. Lett. 51: 1884 (1983); L. S. Sham and M. Schluter, Density functional theory of the energy gap, Phys. Rev. Lett. 51: 1888 (1983).

49. M. Levy, On long-range behavior and ionization potentials, technical report, University of North Carolina, Chapel Hill (1975).
50. J. P. Perdew, What do the Kohn-Sham orbital energies mean? How do atoms dissociate?, chapter in this book.
51. M. Levy, J. P. Perdew, and V. Sahni, Exact differential equation for the density of a many-particle system, unpublished manuscript (1983). This manuscript contains an extensive discussion and a convincing theorem which states that the Kohn-Sham effective potential tends asymptotically to zero. See also references 14, 22, 43, and 46-50.

DENSITY FUNCTIONALS FOR COULOMB SYSTEMS

Elliott H. Lieb

Departments of Mathematics and Physics
Princeton University
Princeton, New Jersey 08544, U.S.A.

INTRODUCTION

The idea of trying to represent the ground state (and perhaps some of the excited states as well) of atomic, molecular, and solid state systems in terms of the diagonal part of the one-body reduced density matrix $\rho(x)$ is an old one. It goes back at least to the work of Thomas [1] and Fermi [2] in 1927. In 1964 the idea was conceptually extended by Hohenberg and Kohn (HK) [3]. Since then many variations on the theme have been introduced. As the present article is not meant to be a review, I shall not attempt to list the papers in the field. Some recent examples of applications are Refs. 4 and 5. Some recent examples of theoretical papers which will play a role here are Refs. 6-12. A bibliography can be found in the recent review article of Bamzai and Deb [13].

This article has three aims:

(i) To discuss and prove some of the mathematical relations between N-particle functions ψ and their corresponding single particle densities ρ.

(ii) To discuss the mathematical underpinnings of general density functional theory along the lines initiated by HK. In that theory a universal energy functional $F(\rho)$ is introduced. Despite the hopes of HK, $F(\rho)$ is not defined for all ρ because it is not true (see Theorem 3.4) that every ρ (even a "nice" ρ) comes from the ground state of some single-particle potential $v(x)$. This problem can be remedied by replacing the HK functional by the Legendre transform of the energy,as is done here. However, the new theory is also not free of difficulties, and these can be traced to the fact that the connection between v and ρ is extremely complicated and poorly understood.

(iii) To present briefly another approach to the ground state energy problem by means of functionals that, while not exact, are explicitly computable and yield upper and lower bounds to the energy.

The analysis in this paper gives rise to many interesting open problems. It is my hope that the incompleteness of the results presented here will be partly compensated if others are encouraged to pursue some of the questions raised by them.

It is not my intention to present a brief for HK theory. However, it deserves to be analyzed for at least two reasons: The HK theory is used by many workers and it gives rise to some deep problems in analysis. While it is my opinion that density functionals are a useful way to approach Coulomb systems, there are other approaches besides the HK approach [e.g., see (iii) above]. Apart from the difficulties mentioned above, the HK approach may be too general because *all* potentials have to be considered. Coulomb potentials are special and do lend themselves to a density functional approach; for example, Thomas-Fermy theory is asymptotically exact as $Z \to \infty$ (see Sect. 5E and Ref. 14). In addition to this question of generality there is also the crucial point that the "universal functional" is very complicated and essentially uncomputable. If one is going to make uncontrolled approximations for this functional, then the general theory is not very helpful.

It is a pleasure to thank Barry Simon for some helpful conversations and the proofs of Theorems 4.4 and 4.8. I also thank Haim Brezis for the proof of Theorem 1.3.

1. SINGE-PARTICLE DENSITIES

The first order of business is to describe the single-particle densities of interest. For simplicity we confine our attention to three dimensions whenever dimensionality is important. $z = (x,\sigma)$ will denote a space-spin variable, that is, $x \in R^3$ and $\sigma \in \{1,\ldots,q\}$. $q = 2$ for electrons, of course, but one might wish to consider $q = 1$, which would mean that a ferromagnetic state is under consideration. We use the notation

$$\int dz = \sum_{\sigma=1}^{q} \int dx. \tag{1.1}$$

Let $\psi = \psi(z_1,\ldots,z_N)$ be an N-particle function (which may be complex valued). To simplify notation we will not indicate N explicitly except where needed. However, the condition of fixed N is crucial and frequently glossed over. The density functionals that will be introduced later are explicitly N dependent in a highly nontrivial way (see Sect. 4A). ψ is assumed to be *normalized:*

$$\int |\psi|^2 = 1 \tag{1.2}$$

(with $\int = \int dz_1, \ldots, dz_N$) and to have *finite kinetic energy*, that is

$$T(\psi) = \sum_{i=1}^{N} \int |\nabla_i \psi|^2 < \infty . \tag{1.3}$$

Notes. $f \in L^p$ means that f is a function satisfying $\| f \|_p \equiv \{\int |f|^p\}^{1/p} < \infty$. $f \in H^1$ means that f and each component of ∇f are in L^2. Thus, the above $\psi \in H^1$. If f is differentiable everywhere, then ∇f and $T(f) = \int |\nabla f|^2$ are well defined. Otherwise, the correct definition of $T(f)$ for functions in $L^2(R^m)$ is

$$T(f) = (2\pi)^{-3} \int k^2 |\hat{f}(k)|^2 \, dk, \tag{1.4}$$

where $\hat{f}$ is the Fourier transform of f. Since $f \in L^2$, H^1 is a Hilbert space with inner product $(f,g) = \int f^* g + \int \nabla f^* . \nabla g$.

In most of the following it will be assumed that ψ satisfies the Pauli principle, that is, ψ is antisymmetric. However, some of the theorems are easier for symmetric (i.e., bosonic) ψ with $q=1$, and occasionally this will be mentioned explicitly. In either case, the symmetry implies that

$$T(\psi) = N \int |\nabla_1 \psi|^2 . \tag{1.5}$$

We define the *single-particle density* to be [see eq. (A.1)]

$$\rho(x) = N \sum_{\sigma} \int |\psi((x,\sigma_1), \ldots, (x_N,\sigma_N))|^2 \, dx_2 \ldots dx_N . \tag{1.6}$$

Notice that $\int \rho(x) \, dx = N$, not 1.

Determinants. If $\phi_1(z), \ldots, \phi_N(z)$ are orthonormal functions, we can form the determinantal

$$\psi(z_1, \ldots z_N) = (N!)^{-1/2} \det\{\phi_i(z_j)\}, \tag{1.7}$$

which is normalized. Then

$$\rho(x) = \sum_{i=1}^{N} \sum_{\sigma=1}^{q} |\phi_i(x,\sigma)|^2 . \tag{1.8}$$

$$T(\psi) = \sum_{i=1}^{N} \sum_{\sigma=1}^{q} \int |\nabla\phi_i(x,\sigma)|^2 \, dx. \tag{1.9}$$

Returning to the general case, the finiteness of $T(\psi)$ implies the following [15].

Theorem 1.1. $\rho(x)^{1/2} \epsilon L^2(\mathbb{R}^3)$ and $\nabla\rho(x)^{1/2} \epsilon L^2(\mathbb{R}^3)$, that is, $\rho(x)^{1/2} \epsilon H^1(\mathbb{R}^3)$. Moreover, $\int (\nabla\rho^{1/2})^2 \leq T(\psi)$.

Proof. $\rho^{1/2} \epsilon L^2(\mathbb{R}^3)$ because $\int\rho = N$. Now $\nabla\rho(x) = N\int^{\dagger}(\nabla_1\psi)^*\psi + N\int^{\dagger}\psi^*\nabla_1\psi$, where $\int^{\dagger}$ means the integral in (1.6). By the Schwarz inequality,

$$[\nabla\rho(x)]^2 \leq 4N\rho(x) \int^{\dagger} |\nabla_1\psi|^2 .$$

Thus $\int (\nabla\rho^{1/2})^2 \, dx = \frac{1}{4}\int(\nabla\rho)^2\rho^{-1} dx \leq T(\psi)$. □

We know $\rho^{1/2} \epsilon H^1(\mathbb{R}^3) = \{f \mid f \epsilon L^2, \nabla f \epsilon L^2\}$. (Here we use the standard convention that $\{A|C\}$ means the set of A such that condition C holds). To discuss the converse of Theorem 1.1 some definitions are useful.

Definition. $J_N = \{\rho \mid \rho(x) \geq 0, \rho^{1/2} \epsilon H^1(\mathbb{R}^3), \int\rho(x)\, dx = N\}$.

Definition. $R_N = \{\rho \mid \rho(x) \geq 0, \int\rho(x)\, dx = N, \rho \epsilon L^3(\mathbb{R}^3)\}$.

R_N contains J_N by the Sobolev inequality (see Ref. 16) because if $f \epsilon H^1(\mathbb{R}^3)$, then

$$\int |\nabla f(x)|^2 \, dx \geq 3(\pi/2)^{4/3} \left[\int |f(x)|^6 \, dx\right]^{1/3}. \tag{1.10}$$

Equation (1.10) is true only in three dimensions, but analogous inequalities hold in other dimensions. By Theorem 1.1,

$$T(\psi) \geq 3(\pi/2)^{4/3} \, \|\rho\|_3 .$$

R_N is clearly a *convex set*; that is, if ρ_1 and $\rho_2 \epsilon R_N$, then $\rho \equiv \lambda\rho_1 + (1-\lambda)\rho_2 \epsilon R_N$ for all $0 \leq \lambda \leq 1$. J_N is also *convex* by the same proof as in Theorem 1.1; that is, by the Schwarz inequality

$$(\nabla\rho)^2 \leq 4\rho[\lambda(\nabla\rho_1^{1/2})^2 + (1-\lambda)(\nabla\rho_2^{1/2})^2].$$

In particular, the functional $\int[\nabla\rho^{1/2}]$ is *convex*. The convexity of J_N will be important in Sect. 3.

Definition. A function (or functional) f is convex if

$$f(\lambda x+(1-\lambda)y) \leqq \lambda f(x)+(1-\lambda)f(y)$$

for all $0\leqq\lambda\leqq1$ and all x and y in the domain of f.

Theorem 1.2. Suppose $\rho\varepsilon J_N$. Then for either Bose or Fermi statistics there exists a ψ(which is a determinant in the fermion case) such that (1.6) holds and, moreover,

$$T(\psi) \leqq \int [\nabla\rho^{1/2}(x)]^2\, dx \text{ (bosons)}, \tag{1.11}$$

$$T(\psi) \leqq (4\pi)^2 N^2 \int [\nabla\rho^{1/2}(x)]^2\, dx \text{ (fermions)}. \tag{1.12}$$

Proof. For bosons the proof is easy; simply take

$$\psi(x_1,\ldots,x_N) = \prod_{i=1}^{N} \left(\frac{\rho(x_i)}{N}\right)^{1/2} .$$

For fermions the construction is much more complicated. Some ideas from Ref. 17 will be used in the following. Write $x = (x^1,x^2,x^3)$ and define

$$f(x^1) = \left(\frac{2\pi}{N}\right) \int_{-\infty}^{x^1} ds \int_{-\infty}^{\infty} dt \int_{-\infty}^{\infty} du\, \rho(s,t,u).$$

Then f is monotone increasing from 0 to 2π. For $k = 0,\ldots,N-1$ define

$$\phi^k(x) = [\rho(x)/N]^{1/2}\exp[ikf(x^1)].$$

It is easy to check that ϕ^k are orthonormal functions in $L^2(\mathbb{R}^3)$. [First do the x^2 and x^3 integrations and then note that the overlap integral is of the form $\int_{-\infty}^{\infty} (df/dx^1)\exp[i(\Delta k)f(x^1)]dx^1 =$ $\{\exp[i(\Delta k)f(\infty)] - \exp[i(\Delta k)f(-\infty)]\}/i(\Delta k) = 0$. Furthermore,

$$N\int |\nabla\phi^k|^2 = \int (\nabla_\rho{}^{1/2})^2 + \left(\frac{2\pi k}{N}\right)^2 \int_{-\infty}^{\infty} g(s)^6\, ds, \tag{1.13}$$

with

$$g(s)^2 \equiv \int_{-\infty}^{\infty}\int dt\, du\, \rho(s,t,u).$$

As in Theorem 1.1, we conclude that

$$g \in H^1(\mathbb{R}^1) \quad \text{and} \int \left(\frac{dg}{ds}\right)^2 ds \leq \int (\nabla\rho^{1/2})^2 \equiv A.$$

Since

$$g(s)^2 = 2\int_{-\infty}^{s} g(y)\left(\frac{dg(y)}{dy}\right) dy,$$

we conclude by the Schwarz inequality that $g(s)^4 \leq 4[\int g^2][\int (dg/dy)^2]$. Thus, the last term in (1.13) is less than $4(2\pi k/N)^2 N^2 A$. Finally, we take ψ to be a determinant as in (1.7) using the functions $\phi^k(x) \times$ (spin up). Equation (1.12) follows by summing on k. □

Theorem 1.2 is closely related to the results of Gilbert [8] and Harriman [9].

For fermions, the extra factor N^2 in (1.12) is noticeably different from the factor N^0 in Theorem 1.1. Although (1.12) can be improved, it is not easy to do so. In any case, the conclusion is that the map from ψ to $\rho^{1/2}$ given by (1.6) is a map from $H^1(\mathbb{R}^{3N})$ onto $H^1(\mathbb{R}^3)$. But the map is clearly not 1:1; different ψ's can give the same ρ.

Question 1. Is this map continuous as a map from $H^1(\mathbb{R}^{3N})$ to $H^1(\mathbb{R}^3)$? That is, if ψ is fixed and ψ_j is a sequence (with corresponding ρ and ρ_j) such that $\int|\psi-\psi_j|^2 \to 0$ and $\int|\nabla\psi-\nabla\psi_j|^2 \to 0$, does it follow that $\int|\rho^{1/2}-\rho_j^{1/2}|^2 \to 0$ and $\int|\nabla\rho^{1/2}-\nabla\rho_j^{1/2}|^2 \to 0$?

Question 2. Although the map is not invertible (since it is not 1:1), we can ask the following: Given a sequence $\rho_j^{1/2}$ that converges to $\rho^{1/2}$ in the above $H^1(\mathbb{R}^3)$ sense, and given some ψ satisfying (1.6) for ρ, does there exist a sequence ψ_j [related to ρ_j by (1.6)] that converges to ψ in the above $H^1(\mathbb{R}^3)$ sense? [This is equivalent to the statement that the map $\psi \to \rho^{1/2}$ is "open", that is, the map takes open sets in $H^1(\mathbb{R}^{3N})$ into open sets $H^1(\mathbb{R}^3)$.]

Intuitively, the answer to both questions should be affirmative. The continuity can indeed be proved, but the proof is not entirely elementary. A proof of Theorem 1.3, due to H. Brezis, is given in the appendix.

Theorem 1.3. The map $\psi \to \rho^{1/2}$ given by (1.6) is continuous as a map from $H^1(\mathbb{R}^{3N})$ to $H^1(\mathbb{R}^3)$.

I cannot offer any proof of the openness of the map, however. The fact that these questions do not have simple answers should serve as a warning that the connection between ψ and ρ is not as obvious as one might intuitively think.

2. SINGLE-PARTICLE DENSITY MATRICES

If ψ is given as before, we can define the *single-particle density matrix*

$$\gamma(x,x') = N \sum_{\sigma} \int \psi((x,\sigma_1),\ldots,(x_N,\sigma_N)) \times \psi((x',\sigma_1),\ldots,(x_N,\sigma_N))^* \, dx_2 \ldots dx_N. \tag{2.1}$$

This definition is different from the usual one because we sum on σ_1 in (2.1). Usually one defines the quantity $\tilde{\gamma}\,(x,\sigma;x',\sigma)$. Clearly, $\rho(x) = \gamma(x,x)$.

Theorem 2.1. γ satisfies

(i) $\mathrm{Tr}\,\gamma = \int\gamma(x,x)\,dx = N$.
(II) As an operator, $0 \leq \gamma \leq qI$, for fermions; that is, $0 \leq (f,\gamma f) \leq q(f,f)$. For bosons, $0 \leq \gamma \leq NI$.

Proof. (i) is "obvious" but not trivial. The point is that if an operator K is given, then its kernel $K(x,y)$ is defined only almost everywhere. In particular, $K(x,x)$ can be anything. Thus, $\mathrm{Tr}K$ need not be $\int K(x,x)\,dx$. However, (i) can be proved from (2.1). This is left as an exercise. To prove (ii), let $M(x,x') = f(x)f(x')^*$ be a one-particle operator with $(f,f) = 1$. Then $A = \Sigma_{i=1}^{N} M(x_i,x'_i)$ has as its largest eigenvalue on the antisymmteric space the value q. Moreover, A is clearly positive semidefinite. Thus, $0 \leq (f,\gamma f) = \mathrm{Tr}\gamma M = (\psi,A\psi) \leq q$. □

Definition. Let $\gamma(x,y)$ be any kernel. γ is said to be *admissible* if $\mathrm{Tr}\,\gamma = N$ and $0 \leq \gamma \leq qI$ (fermions) or $0 \leq \gamma \leq NI$ (bosons). The set of admissible γ is clearly convex. That is, if γ and δ are admissible, then so is $\alpha\gamma + (1-\alpha)\delta$ for $0 \leq \alpha \leq 1$.

Now we come to a subtle point. If γ is an admissible operator, we can ask two questions:

Question 3. Does an N-particle *density matrix* Γ always exist, where $\Gamma = \Gamma(z_1,\ldots,z_N;z'_1,\ldots,z'_N)$, so that γ is given by (2.1) with $\psi\psi^*$ replaced by Γ ? (Γ is a density matrix if $0 \leq \Gamma$ and $\mathrm{Tr}\Gamma = 1$. Γ must also satisfy the appropriate symmetry).

Question 4. Does a ψ always exist so that (2.1) holds; that is, can Γ be chosen to be a *pure state*, namely, $\Gamma = \psi\rangle\langle\psi$?

The answer to question 4 is No!(for fermions). For bosons, the answer is yes.

The proof of question 3 (which we now call Theorem 2.2) has been known for a long time. An explicit construction is given in Ref. 26. An example in which Γ fails to be of the form $\psi\rangle\langle\psi$, for $N = 2$ and $q = 1$, is the case in which γ has three nonzero eigenvalues $1,\frac{1}{2},\frac{1}{2}$. To see this, let the normalized eigenvectors of γ be $f(x), g(x)$, and $h(x)$, respectively; that is

$$\gamma(x,x') = f(x)f(x')^* + \tfrac{1}{2}g(x)g(x')^* + \tfrac{1}{2}h(x)h(x')^*.$$

Let $A = -\gamma(x_1,x'_1) - \gamma(x_2,x'_2)$ be an operator on the antisymmetric states. Its lowest eigenvalue is $-1 -1/2 = -3/2$, which is doubly degenerate. If $\Gamma = \psi\rangle\langle\psi$, then ψ must be a ground state since $\mathrm{Tr}\Gamma A = -\mathrm{Tr}\gamma^2 = -1 -1/4 -1/4 = -3/2$. But every ground state is of the form $\psi = 2^{-\frac{1}{2}}\det(f,\rho)$ where $\rho = ag + bh$, $|a|^2+|b|^2 = 1$. But then $\gamma = f\rangle\langle f + \rho\rangle\langle\rho$, and this is never of the form $f\rangle\langle f + \frac{1}{2}g\rangle\langle g + \frac{1}{2}h\rangle\langle h$.

The moral of all this is the following: on the one-particle level we can study density matrices $\gamma(x,x')$ or densities, $\rho(x) = \gamma(x,x)$. The former do not always come from pure states $\psi\rangle\langle\psi$. The latter do, as Theorem 1.2 shows. While γ is more complicated than ρ (it has two variables), it has the distinct advantage that the map $\Gamma\to\gamma$ is *linear*! The map $\psi\to\rho$ is *nonlinear*, and this, as will be seen, is the source of some difficulty.

The relation among ψ, Γ, γ, and ρ can be summarized by the following diagram:

$$\psi \to \Gamma \to \gamma \to \rho, \qquad (2.2)$$

by which we mean (i) the map $\psi\to\Gamma = \psi\rangle\langle\psi$, (ii) $\Gamma\to\gamma$ by (2.1) with $\psi\psi^*$ replaced by Γ,(iii) $\gamma\to\gamma(x,x) = \rho(x)$. (ii) and (iii) are linear while (i) is nonlinear.

Notation. We shall use the symbol $\psi \to \rho$ (or any other combination such as $\gamma \to \rho$) to indicate that ψ and ρ are related by the above maps.

Technical remarks. Since γ is self-adjoint and trace class, it can always be written in the form

$$\gamma(x,x') = \sum_{j=1}^{\infty} \gamma_j f_j(x) f_j(x')^*, \tag{2.3}$$

where the f_j are orthonormal and $0 \leq \gamma_j \leq q$ (fermions), $0 \leq \gamma_j \leq N$ (bosons). If $\Gamma \epsilon H^1(\mathbb{R}^{3N})$, by which we mean $\mathrm{Tr}-\Sigma_{i=1}^{N} \Delta_i \Gamma < \infty$, then each $f \epsilon H^1(\mathbb{R}^3)$ (see Theorem 1.1). Although $\gamma(x,x)$ is not *a priori* well defined, as stated before, it is well defined almost everywhere (in $\mathbb{R}^3$) by (2.3) and

$$N = \mathrm{Tr}\gamma = \sum_{j=1}^{\infty} \gamma_j . \tag{2.4}$$

3. GENERAL DENSITY FUNCTIONAL THEORY

The problem that will concern us in calculating the ground state energy for N electrons interacting with each other via a repulsive Coulomb potential $|x_i - x_j|^{-1}$ and also interacting with a single-particle potential $v(x)$. If $v = 0$, the Hamiltonian is

$$H_0 = K + \sum_{1 \leq i < j \leq N} |x_i - x_j|^{-1} , \tag{3.1}$$

where K is the kinetic energy operator

$$K = - \sum_{i=1}^{N} \Delta_i \tag{3.2}$$

in units in which $\hbar^2/2m = 1$. Also of interest is the case where $H_0 = K$ alone (see Sec. 4C). Recall that N is fixed and will not be mentioned unless necessary. Also, to simplify matters we shall confine our attention in the following to fermions. However, many of the following results have obvious analogs for bosons.

The total Hamiltonian is

$$H_v = H_0 + V, \tag{3.3}$$

where

$$V = \sum_{i=1}^{N} v(x_i). \tag{3.4}$$

The ground state energy E(v) is defined to be

$$E(v) = \inf\{(\psi, H_v\psi) | \psi \varepsilon W_N\}, \tag{3.5}$$

where

$$W_N = \{\psi | \, \|\psi\| = 1, \, T(\psi) < \infty\}. \tag{3.6}$$

Technical remark. Something should be said about the meaning of $(\psi, H_v\psi)$ and about the class of v's under consideration. We shall always interpret $(\psi, H_v\psi)$ in the sense of a quadratic form; in particular, this means that $(\psi, K\psi) \equiv T(\psi)$. It is not assumed that $\Delta\psi \varepsilon L^2$. Since $\psi \varepsilon H^1$, it is easy to prove that $(\psi, |x_i - x_j|^{-1}\psi)$ is finite for all $i \neq j$. The part containing v is $\int \rho(x)v(x)dx$. As $\rho \varepsilon L^1$ and $\rho \varepsilon L^3$ (since $\nabla\rho^{1/2} \varepsilon L^2$), $\rho \varepsilon L^p$ for all $1 \leq p \leq 3$. The integral is then well defined if $v \varepsilon L^{3/2} + L^\infty$. This means that *we consider v's that can be written as* $v = v_{3/2} + v_\infty$ with $v_{3/2} \varepsilon L^{3/2}$ and with $|v_\infty|$ *a bounded function*. This choice precludes v's that go to ∞ as $|x| \to \infty$, such as the harmonic oscillator potential. Unbounded potentials can also be handled by the methods given here, but then we have to place additional restrictions on ρ so that $\int v\rho$ makes sense. We restrict ourselves here to $L^{3/2} + L^\infty$ for simplicity of exposition. The class includes Coulomb potentials because $|x|^{-1} = \theta(x)|x|^{-1} + [1-\theta(x)]|x|^{-1}$ with $\theta(x) = 1$ if $|x| \leq 1$, $\theta(x) = 0$, $|x| > 1$. The two terms on the right are in $L^{3/2}$ and in L^∞, respectively.

$L^{3/2} + L^\infty$ is a Banach space with the norm

$$\|v\| = \inf\{\|g\|_{3/2} + \|h\|_\infty | g + h = v\}. \tag{3.7}$$

Technical remark. R_N is a subset of the Banach space $X = L^3 \cap L^1$. X^*, the dual of X, is $Y = L^{3/2} + L^\infty$. However, the dual of Y is not X because while L^∞ is the dual of L^1, L^1 is not the dual of L^∞. However, $X \subset Y^*$. The duality will be useful.

There may or may not be a minimizing ψ for (3.5), and if there is one it may not be unique (for bosons it is unique because it is a positive function). Any minimizing ψ (called a *ground state*) would satisfy

$$H_v\psi = E(v)\psi \tag{3.8}$$

in the distributional sense. The proof of this assertion is not difficult. For example, a minimizing ψ will not exist if v is an attractive square well and if N is too large; the extra, unbound electrons will simply "leak away" to infinity. In such a case, E(v) would still have physical signifiance. It would be the ground state energy for *fewer* than N particles.

There are three simple, but important, properties of E(v):

Theorem 3.1.(i) E(v) is concave in v: that is,

$$E(v) \geqq \alpha E(v_1) + (1-\alpha)E(v_2), \tag{3.9}$$

for all v_1, v_2, $0 \leqq \alpha \leqq 1$ and $v = \alpha v_1 + (1-\alpha)v_2$.

(ii) E(v) is monotone decreasing: that is, if $v_1(x) \leqq v_2(x)$ for all x, then $E(v_1) \leqq E(v_2)$.

(iii) E(v) is continuous in the $L^{3/2} + L^\infty$ norm and is, moreover, locally Lipschitz. In particular, E(v) is finite.

Proof. (i) If $\psi \in W_N$, then

$$(\psi, H_v\psi) = \alpha(\psi, H_{v_1}\psi) + (1-\alpha)(\psi, H_{v_2}\psi) \geqq \alpha E(v_1) + (1-\alpha)E(v_2).$$

(ii) $(\psi, H_{v_2}\psi) \geqq (\psi, H_{v_1}\psi) \geqq E(v_1)$.

(iii) Fix v_o and let $\delta = v - v_o$. We want to show that when $\|\delta\| \leqq L/3$, $|E(v)-E(v_o)| < C\|\delta\|$ for some C, independent of v.[Here, L is the constant in (1.10).] Since E(v) is concave, it is sufficient to show that for some fixed D, $E(v)-E(v_o) \geqq D$ whenever $\|\delta\| = L/3$; because if $0 \leqq \gamma \leqq 1$,

$$\gamma[E(v_o+\delta)-E(v_o)] \leqq E(v_o+\gamma\delta)-E(v_o) \leqq \gamma[E(v_o)-E(v_o-\delta)].$$

Let $E(v,\frac{1}{2})$ denote(3.5)with K replaced by K/2. Then

$$E(v) \geqq E(v_o,\tfrac{1}{2}) + \inf(\psi, [\tfrac{1}{2}K + \Sigma\delta(x_i)]\psi).$$

The last term is bounded by -LN/2 bacause $\delta = g+h$ with $\|g\|_{3/2} < L/2$ and $\|h\|_\infty < L/2$. Thus,

$$\int \delta\rho > -(L/2)[|\rho|_3 + N].$$

But$(\psi,K\psi)/2 \geq (L/2)|\rho|_3$ by (1.10) and Theorem 1.1. Finally, note that $(v_o,\frac{1}{2})-E(v_o)$ is a constant, D', independent of v.

Now we begin the study of density functional theory in the manner of Hohenberg and Kohn. Their work is based on the following theorem[3]:

Theorem 3.2. Suppose ψ_1(respectively,ψ_2) is a ground state for v_1 (respectively v_2) and $v_1 \neq v_2$ + constant. Then $\rho_1 \neq \rho_2$.

Proof. Suppose $\rho_1 = \rho_2 = \rho$. $\psi_1 \neq \psi_2$ because they satisfy different Schrödinger equations, (3.8). [Note. To prove this we must know that $v_1\psi = v_2\psi$ implies that $v_1 = v_2$. This, in turn, requires that $\psi(x)$ does not vanish on a set of positive measure. This technical point is discussed in rermark (ii) preceding Theorem 3.5.] Moreover, ψ_2 (respectively, ψ_1) does not satisfy (3.8) for v_1 (respectively,v_2). Therefore,

$$E(v_1) < (\psi_2,H_{v_1}\psi_2) = E(v_2) + \int (v_1-v_2)\rho .$$

Likewise, $E(v_2) < E(v_1) + \int (v_2-v_1)\rho$. This is a contradiction. □

Hohenberg and Kohn assume that *every* ρ comes from some ψ that is a ground state for some v. For such ρ they *define* the functional

$$F_{HK}(\rho) = E(v) - \int v\rho , \tag{3.10}$$

and we shall retain this definition for $\rho \varepsilon A_N$, where

$$A_N = \{\rho | \rho \text{ comes from a ground state}\}. \tag{3.11}$$

$A_N \neq J_N$, as remarked earlier, and it is not convex (see Theorem 3.4)! The definition given by (3.10) requires Theorem 3.2, according to which there is a unique v (up to a constant) associated with ρ. We can also define

$$V_N = \{v | H_v \text{ has a ground state}\}. \tag{3.12}$$

It then follows easily that for $v \varepsilon V_N$

$$E(v) = \min \{ F_{HK}(\rho) + \int v\rho \,|\, \rho \varepsilon A_N \}. \tag{3.13}$$

This is the HK *variational principle*, but it is important to note that it holds only for $v \varepsilon V_N$, which is unknown, and that the variation is restricted to the unknown set A_N.

We also do not know what F_{HK} is, and that is a very serious problem. But there are also conceptual problems, which will be addressed here.

If F is to be used in a variational principle, it is clearly desirable that F be a convex functional. In particular, it should be defined everywhere on J_N, or at least on some known convex subset of J_N.

The domain of F_{HK} (i.e., $\mathcal{A}_N$) is not all of J_N and it is not convex. This last fact is closely connected with the following difficulty: One can define a functional for all ρ in J_N by *

$$\tilde{F}(\rho) = \inf\{(\psi,H_0\psi)\,|\,\psi\to\rho,\psi\varepsilon\mathcal{W}_N\}. \tag{3.14}$$

It then follows trivially that

$$E(v) = \inf\{\tilde{F}(\rho) + \int v\rho\,|\,\rho\varepsilon J_N\}, \tag{3.15}$$

$$\tilde{F}(\rho) = F_{HK}(\rho), \qquad \text{if } \rho\varepsilon\mathcal{A}_N . \tag{3.16}$$

So far, so good. The difficulty is that $\tilde{F}$ is not convex either. However, $\tilde{F}$ has one important property that is proved in the appendix.

Theorem 3.3. For each ρ in J_N there is a $\psi\varepsilon\mathcal{W}_N$ such that $\tilde{F}(\rho)=(\psi,H_0\psi)$. In other words, the infimum in (3.14) is a minimum.

The following functional F is one choice for "the density functional" that remedies the difficulties mentioned so far:

$$F(\rho) = \sup\,\{E(v) - \int v\rho\,|\,v\varepsilon L^{3/2} + L^{\infty}\}. \tag{3.17}$$

We shall explore the properties of F, but it, too will be seen to have subtle difficulties of its own.

Remarks. (i) (3.17) defines $F(\rho)$ for all $\rho\varepsilon R_N$, not just J_N, provided F is interpreted in the extended sense as a function that can have the value $+\infty$. In fact, (3.17) defines F on the much larger set $X = L^3 \cap L^1$, without the restrictions $\rho(x)\geq 0$ and $\int\rho = N$. As Theorem 3.5 shows, however, it is only necessary to consider F on the convex subset J_N of X.

---

* Levy [10] also defined $\tilde{F}(\rho)$ which he called $\mathcal{Q}$, and derived (3.15). He did not prove Theorem 3.3, but assumed the existence of a minimizing ψ. Also, he did not establish the connection between $\tilde{F}$ and the Legendre transform, F (Theorem 3.7). In Ref. 11, Levy proved Theorem 3.4(ii), independently and virtually at the same time as myself, using essentially the same construction. See Ref. 12 for additional remarks about $\mathcal{Q}$.

(ii) Recall that F depends explicitly on N through E.

(iii) Since F is the supremum of a family of linear functionals, it is convex.

(iv) Theorem 3.8 shows that $F(\rho) = +\infty$ if $\rho \notin J_N$. There is an alternative definition of F, namely, F', by which F' is finite on the set

$$J'_N \equiv \{\rho \,|\, \rho(x) \geq 0 \text{ and } \nabla\rho^{1/2} \varepsilon L^2\},$$

without requiring $\int\rho = N$. This is

$$F'(\rho) = \left(\int\rho\right) F\left(\rho\Big/\int\rho\right), \qquad \rho \not\equiv 0,$$

$$F'(0) = 0. \tag{3.18}$$

It is easy to check that the convexity and lower semicontinuity (a concept to be defined later) of F carry over to F'. This definition has the virtue that F' is finite on a dense subset of the set of nonnegative functions in X. However, this does not change the theory in any important way, so we shall continue to use the definition given by (3.17).

(v) Other characterizations of F, directly in terms of $\tilde{F}$, are given in Theorem 3.7, and in eqs.(4.5)-(4.7).

There is an obvious relation between F and $\tilde{F}$, namely

$$F(\rho) \leq \tilde{F}(\rho) \qquad \text{for all} \quad \rho\varepsilon J_N, \tag{3.19}$$

since $E(v) \leq \tilde{F}(\rho) + \int v\rho$ for all $\rho\varepsilon J_N$. Furthermore, since F is convex and $\tilde{F}$ is not convex (by Theorem 3.4), there are ρ's in J_N for which $F(\rho) < \tilde{F}(\rho)$.

First we prove that not all ρ's come from ground states. The essential ingredient is the existence of v with a degenerate ground state. (Such v's, incidentally, preclude the existence of a map $v \to \rho$).

Theorem 3.4. Let $N > \int\rho$ = number of spin states. Then

(i) $\tilde{F}(\rho)$) is not convex

(ii) There exists a $\rho\varepsilon J_N$ that does not come from a ground state ψ. Moreover this ρ is a convex combination of ρ's that do come from a ground state.

Proof. Let v be a spherically symmetric potential having a ground state and with the property that its ground state has orbital angular momentum $L \geqq 1$. We assume the degeneracy is no greater than necessary, namely $M = 2L + 1$. The orthonormal ground states are $\psi_1, \ldots, \psi_M$ and $\psi_i \to \rho_i$. Under simultaneous rotation of all N coordinates, they transform as a basis for the M-dimensional irreducible representation of O(3). The following fact is easy to prove:(a) If $\bar{\rho} = M^{-1}\Sigma\rho_i$, then $\bar{\rho}$ (x) is spherically symmetric: that is, $\bar{\rho}$ depends only on $r = |x|$.

A second fact that will be needed is (b): if ϕ is any ground state (and hence a linear combination of the ψ_i) and $\phi \to \rho$, then ρ is not spherically symmetric. This fact must follow from some group-theoretic agreement, but I have not found one. However, it is not hard to see that (b) is equivalent to (c): There exists a perturbation of $v, v(x) \to v(x) + \lambda w(x)$, with w bounded and of compact support, so that to first order in λ the M-fold degeneracy is broken. Such pairs v and w certainly exist, so we can henceforth assume that (b) holds.[A proof that a v satisfying (b) exists is the following. First, take the case that $H_o = K$, that is, independent particles. The ground states are determinants. Choose v so that the ground state has $L \geqq 1$, in which case (b) obviously holds. Next, consider $H = K + \lambda\Sigma|x_i - x_j|^{-1} + V$. Angular momemtum is still conserved and for sufficiently small λ the ground state will have the same L and, by continuity of the ground states, (b) will continue to hold for small λ. We are interested in $\lambda = 1$ but, under the scaling

$$x \to x/\lambda, \qquad v(x) \to \lambda^{-2} v(x/\lambda) = v'(x),$$

the v, λ problem is converted into the v', $\lambda = 1$ problem. Thus, v' has the desired properties. I thank B. Simon for this remark.]

Clearly, $\tilde{F}(\rho_i) = \text{constant} = D = E(v) - \int v\rho_i$. We claim $\tilde{F}(\bar{\rho}) > D$, thereby proving lack of convexity. Obviously, $\tilde{F}(\bar{\rho}) \geqq D$, for otherwise we could use $\bar{\rho}$ instead of ρ_i in (3.15). (Note: $\int v\bar{\rho} = \int v\rho_i = \text{constant} = C$,) Suppose $\tilde{F}(\bar{\rho}) = D$. Then $\bar{\rho}$ comes from some ϕ that must be a ground state for v. But (a) and (b) show this to be impossible. Thus $\tilde{F}(\bar{\rho}) > D$. Moreover, $\bar{\rho}$ cannot come from any ground state ψ for any other v'. If it did, then

$$E(v') = \tilde{F}(\bar{\rho}) + \int v'\bar{\rho} > M^{-1}\,\Sigma\left(\tilde{F}(\rho_i) + \int v'\rho_i\right).$$

This implies that for some $1 \leqq i \leqq M$, $\tilde{F}(\rho_i) + \int v'\rho_i < E(v')$, which is a contradiction.□

Remarks. (i) The foregoing proof holds just as well if $(\psi,H_0\psi)$ is replaced by $T(\psi)$ in the definition of $\tilde{F}$. [This functional will be denoted by $\tilde{T}(\rho)$ and the analog of (3.17) by $T(\rho)$.] In other words, the interelectron Coulomb repulsion plays no role in Theorem 3.4 (see Sec. 4C).

(ii) There are other ρ's that do not come from some v, namely, those $\rho\epsilon J_N$ that vanish on a nonempty open set. If $v\epsilon L^{3/2}+L^\infty$ and ψ cannot vanish in an open set by the unique continuation theorem[18]. (Strictly speaking, this theorem is only known to hold for $v\epsilon L^3_{loc}$, but it is believed to hold for $L^{3/2}+L^\infty$.) Presumably, such ρ's can, in many cases, be obtained as limits in which $v\to\infty$ on the open set. Therefore, if the set of allowed v's can be extended properly to include infinite v's, the existence of such ρ's may not have any particular importance. The question is very delicate, however, as Englisch and Englisch [7] showed recently. Even for one particle there are densities which never vanish but which do not come from any v, even if one allows density matrices (see Sect. 4B) instead of pure states. These densities have regions in which they are "small" so that the obvious v (defined by $v=\rho^{-1/2}\Delta\rho^{1/2}$) has the property that $-\Delta+v$ cannot be defined as a semibounded operator.

Theorem 3.5.

$$E(v) = \inf\ \{F(\rho) + \int \rho v \mid \rho\epsilon L^3 \cap L^1\}, \tag{3.20}$$

$$E(v) = \inf\ \{F(\rho) + \int \rho v \mid \rho\epsilon J_N\}\ . \tag{3.21}$$

Remark. The right sides of (3.20) and (3.21) are automatically concave functionals, which is a property we already proved for E.

Proof. Let $M^-(v)$ [respectively, $M^+(v)$] be the infimum in (3.20) [respectively, (3.21)]. Obviously, $M^-(v)\leq M^+(v)$. First, pick v_0. Clearly $F(\rho)\geq E(v_0) - \int v_0\rho \equiv F_1(\rho)$. Therefore,

$$M^-(v)\geq \inf\ \{F_1(\rho) + \int \rho v \mid \rho\epsilon L^3 \cap L^1\ \}.$$

and hence $M^-(\underset{\sim}{v_0})\geq E(\underset{\sim}{v_0})$. Second, by (3.19), $F(\rho)\leq\tilde{F}(\rho)$, so that $M^+(v)\leq \inf\ \{F(\rho)+\int\rho v \mid \rho\epsilon J_n\} = E(v)$. □

Let us pause briefly to review the situation. Three functionals have been defined:

F_{HK} defined on $\mathcal{A}_N \subset J_N$;

$\tilde{F}$ defined on $J_N \subset L^3 \cap L^1$;

F defined on $X = L^3 \cap L^1$.

Of these, only F is convex and only $\tilde{F}$ satisfy the variational principle for all v.

The next step is to find out something of the nature of F. It is at this point that the analysis becomes complicated and where difficulties and incompleteness arise. The basic reason is that the connection between v and ρ is anything but simple. We have $X = L^3 \cap L^1$ and its dual $X^* = L^{3/2}+L^\infty$. Although X is not the dual of X^*, it is a subset of X^{**}, the dual of X^*.

<u>Definitions</u>. (i) A sequence $\rho_n \varepsilon X$ is said to converge to $\rho \varepsilon X(\rho_n \to \rho)$ if and only if $\|\rho_n - \rho\|_3 \to 0$ and $\|\rho_n - \rho\|_1 \to 0$. This is also called *norm convergence*. ρ_n *converges weakly* to $\rho(\rho_n \rightharpoonup \rho)$ if and only if $\int v(\rho_n - \rho) \to 0$ for all $v \varepsilon Y = X^*$. Clearly, strong convergence implies weak convergence.

(ii) A functional f on X is *continuous* (or norm continuous) if and only if $\rho_n \to \rho$ implies $f(\rho_n) \to f(\rho)$. *Weak continuity* requires the concept of nets to define but, if f is weakly continuous, then whenever $\rho_n \rightharpoonup \rho, f(\rho_n) \to f(\rho)$. Weak continuity implies norm continuity.

(iii) A real functional f on X is *lower semicontinuous* (l.s.c.) if and only if $\rho_n \to \rho$ implies $f(\rho) \leq \liminf f(\rho_n)$. *Weak lower semicontinuity* requires nets to define, but if f is weakly l.s.c. then $\rho_n \rightharpoonup \rho$ implies $f(\rho) \leq \liminf f(\rho_n)$. (Weak) lower semicontinuity is equivalent to the following: $\{\rho | f(\rho) \leq \lambda\}$ is (weakly) closed for all real λ.

<u>Remarks</u>. (i) Weak lower semicontinuity always implies lower semicontinuity, but not conversely. It is a theorem of Mazur[19], however, that if f is *convex* and norm l.s.c., then it is automatically weakly l.s.c.

(ii) The function $\rho(x) \equiv 0$ is *not* in the $L^3 \cap L^1$ weak closure of F_N.

The reader may be puzzled by all these definitions, especially lower semicontinuity, because finite convex functions on $\mathbb{R}^n$ are always continuous. Unfortunately, this is not true in infinite-dimensional spaces such as the space X we are considering. Even l.s.c. cannot be taken for granted.

Theorem 3.6. $F(\rho)$ is weakly (and hence also norm) lower semicontinuous.

Proof.

$$K_\lambda \equiv \{\rho \mid F(\rho) \leqq \lambda\} = \{\rho \mid E(v) - \int v\rho \leqq \lambda \text{ for all } v\varepsilon Y\}.$$

Now if $\rho_n \to \rho$ in norm and $\rho_n \varepsilon K_\lambda$, then, for each $v\varepsilon Y$.

$$E(v) - \int v\rho = \lim \left[E(v) - \int v\rho_n \right] \leqq \lambda.$$

Therefore K_λ is norm closed, so that F is norm l.s.c. Weak l.s.c. is a consequence of Mazur's theorem. □

Next we define the *convex envelope* (CE).

Definition. Let f be a real functional defined on a subset A of X. $f(\rho)$ is allowed to be $+\infty$, but not for all $\rho\varepsilon A$. CE f is defined on all of X as follows: CE $f(\rho) = \sup \{g(\rho) \mid g$ is weakly l.s.c. is convex on X, and $g(\rho') \leqq f(\rho')$ for all $\rho'\varepsilon A\}$.

It is easy to check that CE f is convex and weakly l.s.c. and CE $f(\rho) \leqq f(\rho)$ for all $\rho\varepsilon A$. However, CE $f(\rho)$ may be $+\infty$ for some ρ.

The function of interest is CE $\tilde{F}$ with $A = J_N$. Note that is convex and that $\tilde{F}$ (and hence CE $\tilde{F}$) is finite on A by Theorem 1.2. Since CE $\tilde{F} \leqq \tilde{F}$ on A, it is obvious from (3.19) and Theorem 3.6 that $F \leqq$ CE $\tilde{F}$ on X. On the other hand, suppose we use CE $\tilde{F}$ instaed of $\tilde{F}$ in (3.15), This gives a new function, which we call E'. Clearly $E' \leqq E$. Then, if E' is used in (3.17), we get a new function F', and $F' \leqq F$. However, an infinite-dimesional generalization of Fenchel's theorem [29], [31, Theorem I.10], (which uses the Hahn-Banach theorem) states that if the original function (in our case, CE $\tilde{F}$) is convex and weakly l.s.c. in X, then its double Legendre transform (in our case F') is equal to the original function. Thus, F' = CE $\tilde{F}$ and we have

Theorem 3.7. $F(\rho)$ = CE $\tilde{F}$ for all $\rho\varepsilon L^3 \cap L^1$.

The reader may wonder what Theorem 3.7 is good for; the following is an example of the usefulness of the foregoing functional analysis (see Theorem 4.3).

Theorem 3.8. For all $\rho \in L^3 \cap L^1$ let

$$G(\rho) \equiv \int (\nabla \rho(x)^{1/2})^2 \, dx \quad \text{if } \rho \in J_N$$

$$= +\infty \qquad \text{otherwise.}$$

Then $F(\rho) \geq G(\rho)$, for all $\rho \in L^3 \cap L^1$.

Proof. G is obviously convex on X [see the remark after (1.10)]. We claim that G is norm l.s.c.(Note: The norm in question is $L^3 \cap L^1$, not the H^1 norm on $\rho^{1/2}$). If so, we are done because G is then weakly l.s.c. and, by Theorem 1.1, $G \leq \tilde{F}$ on J_N: but then $G \leq CE\, \tilde{F} = F$.

To prove norm l.s.c., let ρ_n be any sequence in X with $\rho_n \to \rho$; that is, $\|\rho_n - \rho\|_1 \to 0$ and $\|\rho_n - \rho\|_3 \to 0$, We can assume that $G = \lim (\rho_n)$ exists and is finite, and we have to show that $G \geq G(\rho)$. We can also assume $\rho(x) \geq 0$ a.e. because if $\rho < 0$ on a set S of positive measure, then, for sufficiently large n, $\rho_n < 0$ on some set of positive measure; hence $\rho_n \notin J_N$ and $G(\rho_n) = \infty$. For a similar reason we can assume $\int \rho = N$. Since $G(\rho_n) < \infty$, $\rho_n \in J_N$. Thus, if we define $g_n = \rho_n^{1/2}$ and $g = \rho^{1/2}$, we have: (a) g_n is bounded in H^1; (b) $g_n^2 \to g^2$ in L^3 and L^1. By the Banach-Alaoglu theorem there is an $f \in H^1$ such that $g_n \rightharpoonup f$ and $\nabla g_n \rightharpoonup \nabla f$ weakly in L^2. Clearly $f(x) \geq 0$. It is not hard to prove that if $g_n \rightharpoonup f$ in L^2 and $g_n^2 \to g^2$ in L^1, then $g = f$. Hence $\nabla g = \nabla f$, and thus $\nabla g_n \rightharpoonup \nabla g$. But since $\int (\nabla g)^2$ is H^1-norm continuous, it is H^1 weakly l.s.c., so that $\lim G(\rho_n) \geq G(\rho)$. □

Theorem 3.8 is certainly not obvious. Among other things it says that if $\rho \notin J_N$ (and such ρ's can be quite smooth and innocent looking), then there exists a sequence of potentials $v_n \in L^{3/2} + L^\infty$ such that $E(v_n) - \int v_n \rho \to \infty$. The reader is asked to reflect on this fact. Another interesting fact is that F is convex finite on J_N, but infinite off J_N. However, the complement of J_N (in X) is dense (in the X norm) in J_N and J_N is dense in the cone of nonnegative functions in X.

The following upper bound complements Theorem 3.8.

Theorem 3.9. If $\rho \in J_N$, then

$$F(\rho) \leq \tilde{F}(\rho) \leq (4\pi)^2 N^2 G(\rho) + \tfrac{1}{2}\iint \rho(x)\rho(y)|x-y|^{-1}\, dx\, dy. \quad (3.22)$$

Proof. Use the definition (3.14). By Theorem 1.2 there is a determinantal ψ, with $\psi \to \rho$, such that (1.12) holds. With this ψ we can calculate the Coulomb repulsion $I = (\psi, \Sigma|x_i - x_j|^{-1}\psi)$. I has a direct term, given in (3.22), plus an exchange term. The latter is negative, as is well known, since $|x-y|^{-1}$ is a positive definite kernel. Thus $\tilde{F}(\rho) \leq$ right side of (3.22). Then use (3.19). □

Remark. By one of Sobolev's inequalities,

$$D \equiv \iint \rho(x)\rho(y)|x-y|^{-1}\, dx\, dy \leq (\text{const.})\|\rho\|_{6/5}^2.$$

By Hölder's inequality $\|\rho\|_{6/5}^4 \leq \|\rho\|_1^3 \|\rho\|_3$. Again, by Sobolev's inequality, (1.10), $\|\rho\|_3 \leq (\text{const.})\, G(\rho)$. Thus,

$$D \leq (\text{const.}) N^{3/2} G(\rho)^{1/2} \leq (\text{const.})[N + N^2 G(\rho)].$$

To continue the study of F the following concept is needed.

Definition. Let f be a real functional on a subset A of a Banach space B, and let $\rho_o \in A$. A linear functional l on B is said to be a *tangent functional* (TF) *at* ρ_o if and only if for all $\rho \in A$

$$f(\rho) \leq f(\rho_o) - l(\rho - \rho_o). \quad (3.23)$$

l may not be unique. If l is continuous, then l is a *continuous tangent functional* at ρ_o.

l is a continuous linear functional on X if and only if it has the form $\int v\rho$ with $v \in X^* = Y$. If f is convex, then at every point ρ_o at which f is finite, f has at least one TF. This is guaranteed by the Hahn-Banach theorem. However, f may have no *continuous* TF at ρ_o.

The functional of interest is obviously F. In general, $F \leq \tilde{F}$, but the following says something about those ρ for which $F(\rho) = \tilde{F}(\rho)$.

Theorem 3.10. Let $\rho_0 \varepsilon J_N$. The following are equivalent:

(1) $F(\rho_0) = \tilde{F}(\rho_0)$ and F has a continuous TF at ρ_0.

(2) $\rho_0 \varepsilon A_N$.

(3) $\tilde{F}$ has a continuous TF at ρ_0; $\tilde{F}(\rho) \geqq \tilde{F}(\rho_0) - \int v(\rho-\rho_0)$ for $\rho\varepsilon J_N$.

(4) (3) and (5) hold with the same v.

(5) $E(v) = \tilde{F}(\rho_0) + \int v\rho_0$ for some v.

(6) (5) holds and, in addition, $v\varepsilon V_N$ and $v \to \rho_0$.

(7) $\tilde{F}$ has a continuous TF at ρ_0 and v is unique up to a constant. Moreover, F has the same continuous TFs at ρ_0, and no others.

Proof. (1)$\Rightarrow$(3): For $\rho\varepsilon J_N$,

$$\tilde{F}(\rho) \geqq F(\rho) \geqq F(\rho_0) - \int v(\rho-\rho_0) = \tilde{F}(\rho_0) - \int v(\rho-\rho_0).$$

(3)$\Rightarrow$(4): Let $F_1(\rho) = \tilde{F}(\rho_0) - \int v(\rho-\rho_0) \leqq \tilde{F}(\rho)$. Then

$$\tilde{F}(\rho_0) + \int v\rho_0 \geqq E(v) \geqq \inf\,\{F_1(\rho) + \int \rho v \,|\, \rho\varepsilon J_N\} = \tilde{F}(\rho_0) + \int v\rho_0.$$

(4)$\Rightarrow$(5), (7)$\Rightarrow$(3), (6)$\Rightarrow$(5): All trivial.

(5)$\Rightarrow$(1): $\tilde{F}(\rho_0) + \int\rho_0 v = E(v) \leqq F(\rho_0) + \int\rho_0 v \Rightarrow F(\rho_0) \geqq \tilde{F}(\rho_0) \Rightarrow F(\rho_0) = \tilde{F}(\rho_0)$. Then, for all $\rho\varepsilon X$, $F(\rho) + \int\rho v \geqq E(v) = F(\rho_0) + \int\rho_0 v$.

(2)$\Rightarrow$(5): By (3.16).

(5)$\Rightarrow$(2),(6): By Theorem 3.3, $\tilde{F}(\rho_0) = (\psi, H_0\psi)$ for some ψ with $\psi \to \rho_0$. Then $E(v) = (\psi, H_0\psi) + \int v\rho_0 \Rightarrow \rho_0\varepsilon A_N$, $v\varepsilon V_N$, and $v \to \rho_0$. Thus (1)-(6) are equivalent and (7)$\Rightarrow$ (3). Now we show that (1)-(6)$\Rightarrow$ (7). If v is a continuous TF for F, then v is a continuous TF for $\tilde{F}$[by the proof of (1)$\Rightarrow$(3)]. If v is a continuous TF for $\tilde{F}$, then $F(\rho) \geqq E(v) - \int v\rho$, so v is a continuous TF for F.

Suppose $\tilde{F}$ has two continuous TFs v and w with $v - w \neq$ constant. Then $E(v) = \tilde{F}(\rho_o) + \int v\rho_o$ and $E(w) = \tilde{F}(\rho_o) + \int w\rho_o$. Since $\rho_o \varepsilon \mathcal{A}_N$, this is impossible by Theorem 3.2. □

It should be noted that the only place that the HK Theorem 3.2 entered in the analysis of F was in establishing the uniqueness (modulo constants) in (7).

Now we turn to two important questions whose answers we cannot give but that are obviously important for the theory. We replaced F_{HK} by F bacause F_{HK} was not defined on all of J_N. Theorem 3.10 states that on $\mathcal{A}_N$, where F_{HK} is defined, $F = \tilde{F} = F_{HK}$ and F has an essentially unique *continuous* TF.

Question 5. For which points of J_N does F have a continuous TF?Where there is one, is it unique (modulo adding a constant to v)?

Question 6. If F has a continuous TF at $\rho_o \varepsilon J_N$ given by some $v \varepsilon L^{3/2} + L^\infty$, is this $v \varepsilon \mathcal{V}_N$?

Questions 5 and 6 have alternative formulations, given below.

Theorem 3.11. Let $\rho_o \varepsilon J_N$ and $v \varepsilon L^{3/2} + L^\infty$. v is not necessarily in $\mathcal{V}_N$. Then, for all ρ,

$$F(\rho) \geqq F(\rho_o) - \int v(\rho - \rho_o) \qquad \text{(continuous TF)} \tag{3.24}$$

if and only if

$$E(v) = F(\rho_o) + \int v\rho_o \qquad \text{[minimum in (3.21)]} \tag{3.25}$$

Proof. Assume (3.24) and let $M_v(\rho)$be its right side. Then

$$E(v) \geqq \inf \{M_v(\rho) + \int \rho v\} = F(\rho_o) + \int v\rho_o \geqq E(v).$$

For the converse,

$$F(\rho) + \int v\rho \geqq E(v) = F(\rho_o) + \int v\rho_o. \ \square$$

Question 5 is equivalent to the following: For which $\rho_o \varepsilon J_N$ is there a v such that (3.25) holds? Is this v unique (up to constants)? Question 6 is the following: If (3.25) holds, is $v \varepsilon \mathcal{V}_N$?

Some insight into the continuous TFs of F are provided by the Bishop-Phelps theorem. We refer the reader to Ref. 20 for this as well as other interesting facts about convexity. A definition is needed.

Definition. Let F be a real functional on a real Banach space B with dual B^* (the set of continuous linear functionals on B). $b^* \epsilon B^*$ is said to be F-*bounded* if there is a constant C(depending on b^* but not in b) such that $F(b) \geq b^*(b)+C$ for all $b \epsilon B$.

In our case B = X and F is our density functional.

Theorem 3.12. Every $v \epsilon X^* = L^{3/2} + L^\infty$ is F bounded.

Proof. By Theorem 3.8, $F(\rho) = \infty$ if $\rho \not\epsilon J_N$, so we only have to consider $\rho \epsilon J_N$ and prove that $G(\rho) \geqq \int v\rho + C$ for some C. The proof of this is identical to the last part of the proof of Theorem 3.1. □

The Bishop-Phelps theorem is the following.

Theorem 3.13. Let F be a l.s.c. convex functional on a real Banach space B. (Note: Norm and weak l.s.c. are identical). F can take the value $+\infty$, but not everywhere. Then

(i) The continuous tangent functionals to F (over all of B) are B^*-norm dense in the set of F-bounded functionals in B^*.

(ii) Suppose $b_o \epsilon B$ and $b_o^* \epsilon B^*$ with $F(b_o) < \infty$. For every $\epsilon > 0$ there exists $b_\epsilon \epsilon B$ and $b_\epsilon^* \epsilon B^*$ such that $\| b_\epsilon^* - b_o^* \|_{B^*} \leqq \epsilon$ and

$$\epsilon \| b_\epsilon - b_o \|_B \leqq F(b_o) + b_o^*(b_o) - \inf\{F(x) + b_o^*(x) \,|\, x \epsilon B\}.$$

Moreover, b_ϵ^* is tangent to F at b_ϵ, namely $F(b) \geqq F(b_\epsilon) - b_\epsilon^*(b - b_\epsilon)$ for all b.

The significance of Theorem 3.13(i) is the following. There are certainly many v's in Y that are not in V_N. (Example: Suppose $v \epsilon L^{3/2}$ and $\|v\|_{3/2} < L$, where L is the constant in (1.10). Then $(\psi, H_v \psi) > 0$ for all ψ, but $E(v) = 0$ because we can always take a sequence ψ_n

that "leaks away to infinity"). Let $v \notin V_N$, whence $\tilde{F}(\rho)+\int v\rho$ does not have a minimum. What Theorem 3.13 says is that there always exists a sequence v_n (not necessarily in V_N) such that

(a) $F(\rho) + \int \rho v_n$ has a minimum at some $\rho_n \varepsilon J_N$ and this minimum is $E(v_n)$.

(b) $F(\rho) \geqq F(\rho_n) - \int v_n(\rho-\rho_n)$ for all ρ.

(c) $v_n \to v$ in the $L^{3/2}+L^\infty$ norm.

Point (c) means the following: $v_n = v + g_n + h_n$ with $\|g_n\|_{3/2} \to 0$ and $\|h_n\|_\infty \to 0$. In particular, if $v \varepsilon L^{3/2}$ with $\|v\|_{3/2} < L$, then $\|v+g_n\| < L$ for large n. Hence $v+g_n \notin V_N$, then it can only be because of the (vanishingly small) L^∞ piece h_n.

One consequence of Theorem 3.13(ii) is the following.

Theorem 3.14. Let $\rho_o \varepsilon J_N$. Then there exists a sequence $\rho_n \varepsilon J_N$ such that

(i) $\rho_n \to \rho_o$ in $L^3 \cap L^1$ norm.

(ii) F has a continuous TF at ρ_n.

Proof. Given n>0, by (3.17) there exists v_n such that $E(v_n)-\int \rho_o v_n > F(\rho_o)-1/n$. Hence

$$F(\rho) \geqq E(v_n) - \int \rho v_n \geqq F(\rho_o) - \int v_n(\rho-\rho_o) - 1/n.$$

Take $\varepsilon = 1$ in Theorem 3.13. There exists $w_n \varepsilon Y$ such that w_n is a continuous TF at some ρ_n and

$$\|\rho_n - \rho_o\| \leqq F(\rho_o) + \int v_n \rho_o - Z,$$

with

$$Z = \inf\{F(\rho) + \int v_n \rho \,|\, \rho \varepsilon X\}.$$

By the above

$$Z \geqq F(\rho_o) + \int \rho_o v_n - n^{-1}. \quad \square$$

4. ADDITIONAL REMARKS ABOUT DENSITY FUNCTIONALS

A. The N-Dependence of F

As was stressed earlier, any functional F that satisfies (3.20) or (3.21) must depend explicitly on the particle number N. This fact is unavoidable and frequently overlooked. Let us denote the N dependence by $F(N,\rho)$. It might be hoped that F is jointly convex in N and ρ in the sense that for $N \geqq 2$

$$F(N+1,\rho_1)+ F(N-1,\rho_2) \geqq 2F(N,\tfrac{1}{2}\rho_1+\tfrac{1}{2}\rho_2). \tag{4.1}$$

This convexity definitely does not hold as a general feature, as will be demonstrated.

The importance of convexity is shown by the following.

Theorem 4.1. Consider the following two statements about any two functionals, F and E: (i) $F(N,\rho)$ is jointly convex in N and ρ in the sense of (4.1). (ii) $E(N,v)$ is convex in N for all fixed v; that is, for $N \geqq 2$

$$E(N+1,v) + E(N-1,v) \geqq 2E(N,v). \tag{4.2}$$

(a) If (3.17) holds, then (ii) implies (i).

(b) If either (3.20) or (3.21) holds, then (i) implies (ii).

Proof. (a) For each v, $E(N,v) - \int v\rho$ is jointly (N,ρ) convex. By (3.17), $F(N,\rho)$ is the supremum of such convex functions and hence is convex.

(b) Pick $\varepsilon > 0$. Not $N + 1$ there is a ρ_+ such that

$$A \equiv F(N+1,\rho_+) + \int \rho_+ v \leqq E(N+1,v)+ \varepsilon.$$

Likewise,

$$B \equiv F(N-1,\rho_-) + \int \rho_- v \leqq E(N-1,v)+ \varepsilon.$$

For the N problem, define $2\rho = \rho_+ + \rho_-$. Then

$$2 \{ F(N,\rho) + \int \rho v\} \leqq A + B.$$

Since this holds for all $\varepsilon > 0$, (ii) is proved. □

Equation (4.2) has a simple physical meaning. The ionization potential increases as the number of electrons is decreased. This is intuitively expected to be true, but if it is true, it must be because of some special property of the Coulomb repulsion. A non-Coulombic counterexample is given below.

The kinetic energy functional $\tilde{T}(N,\rho)$ is not even convex in ρ (Theorem 3.4), but the Legendre transform $T(N,\rho)$ is jointly convex. This is so because $E(N,v)$ is indeed convex in N for independent particles as the explicit expressen for E, as the sum of the first N eigenvalues (counted with an extra multiplicity q) shows.

What about the convexity of F when the Coulomb repulsion is included? While it has been conjectured that $E(N,v)$ is convex in N (for *all* v) in the case of Coulomb repulsion, this has never been proved. It has not even been proved that $E(3,v) + E(1,v) \geqq 2E(2,v)$.

Lest the reader think that convexity in N is a general feature, we present a counterexample. Replace $|x|^{-1}$ by the hard-core repulsion $\theta(x) = \infty$ if $|x|<1$ and $\theta(x) = 0$ otherwise. Pick four distinct points x_o, y_1, y_2, y_3 in $\mathbb{R}^3$ such that $|y_i - y_j|>1$ for all $i \neq j$ but $|x_o - y_i|<1$ for all i. Let $v(x) = -2\lambda<0$ in small balls about the y_i, $v(x) = -3\lambda$ in a small ball about x_o, and $v(x) = 0$, otherwise. If the kinetic energy be neglected, then $E(1,v) = -3\lambda$, $E(2,v) = -4\lambda$, and $E(3,v) = -6\lambda$. Convexity does not hold. This can be turned into a proper example by letting λ be sufficiently large so that the kinetic energy can effectively be neglected; it is also possible to replace the hard core by a soft core.

Remark. The foregoing example is not applicable if θ is replaced by $|x|^{-1}$, thereby keeping alive the hope that convexity holds in the Coulomb case. The reason is the following: Given any four points x_o, y_1, y_2, y_3 let

$$|x_o - y_1| = \max_i \{|x_o - y_i|\}.$$

Then

$$|x_o - y_1|^{-1} \leqq |y_1 - y_2|^{-1} + |y_1 - y_3|^{-1}.$$

The proof of this is left as an exercise, as well as the implication that if the kinetic energy is neglected, then convexity holds in the Coulomb case.

Question 7. For the case of Coulomb repulsion, is $F(N,\rho)$ jointly convex in N and ρ?

B. Density Matrices

Another possible modification of the theory of Sect. 3 is to replace densities $\rho(x)$ by single-particle admissible density matrices $\gamma(x,x')$. (See Questions 3 and 4 in Sec. 2. We do not restrict ourselves to γ's that come from pure states $\psi\rangle\langle\psi$). This set of γ's is convex, and $\tilde{F}(\gamma)$, defined analogously to (3.14), is convex[see the proof of Theorem 4.1(b)].

Despite the attractive feature just mentioned, there are three drawbacks to the approach:

(i) The problems about continuous tangent functionals remain and may even be more complex than before.

(ii)The original aim of the theory was to express the energy in terms of $\rho(x)$ and not $\gamma(x,x')$.

(iii)While the set of admissible γ's is well defined, it is not easy to identify. Given some γ, it is easy to verify that $\mathrm{Tr}\,\gamma=N$, but it is difficult to verify that $0\leqq\gamma\leqq qI$.

Still another possible modification is to retain ρ but to consider all N-particle density matrices Γ instead of merely pure states $\psi\rangle\langle\psi$. In other words, consider $\Gamma\to\rho$ instead of $\psi\to\rho$ and define

$$F_{DM}(\rho) = \inf\{\mathrm{Tr}H_o\Gamma \mid \Gamma\to\rho\} \qquad (4.3)$$

on J_N and $F_{DM}(\rho)= +\infty$ otherwise. Because $\Gamma\to\rho$ is linear, F_{DM} is convex on J_N. (Note: The example in Theorem 3.4 does not yield nonconvexity of F_{DM}).

Obviously, the analog of (3.15) holds, namely,

$$E(v) = \inf\{F_{DM}(\rho)+\int \rho v \mid \rho\varepsilon J_N\}. \qquad (4.4)$$

Since F_{DM} is convex,(4.4) can be used directly instead of (3.20) or (3.21).

Both F and F_{DM} are convex. The amusing fact is that

$$F(\rho) = F_{DM}(\rho), \qquad \rho\varepsilon J_N. \qquad (4.5)$$

Equation (4.5) is not at all obvious, but it does say that the modification does not change the theory in any way. Equation (4.5) also yields another characterization of F. Equation (4.5) is proved in Theorem 4.3.

First, Γ is admissible if and only if

$$\Gamma(z,z') = \sum_{i=1}^{\infty} \lambda_i \psi_i(z)\psi_i(z')^*$$

with $0 \leq \lambda_i$, $\Sigma\lambda_j = 1$, and the ψ_i are orthonormal. If $\psi_i \to \rho_i$, then

$$\mathrm{Tr} H_o \Gamma = \Sigma\lambda_i(\psi_i, H_o\psi_i).$$

Thus we conclude that for all $\rho \varepsilon J_N$

$$F_{DM}(\rho) = \inf \{ \sum_{i=1}^{\infty} \lambda_i \tilde{F}(\rho_i) \,|\, \Sigma\lambda_i\rho_i = \rho, \rho_i \varepsilon J_N, \lambda_i \geq 0,\ \Sigma\lambda_i = 1\}. \quad (4.6)$$

A simpler expression (which has to be proved) is

$$F_{DM}(\rho) = \inf \{ \Sigma\lambda_i \tilde{F}(\rho_i) \,|\, \Sigma\lambda_i\rho_i = \rho, \rho_i \varepsilon J_N,\ \lambda_i \geq 0, \Sigma\lambda_i = 1\}, \quad (4.7)$$

where the sums in (4.7) are restricted to finite sums. In view of (4.5), (4.7) is an alternative characterization of $F(\rho)$ for $\rho \varepsilon J_N$.

Theorem 4.2. Equation (4.7) is true.

Proof. Pick $\varepsilon > 0$. Using (4.6), let $\{\lambda_i, \rho_i\}$ be an infinite sequence satisfying $\Sigma\lambda_i\rho_i = \rho, \rho_i \varepsilon J_N$, and $F_{DM}(\rho) \geq \Sigma\lambda_i \tilde{F}(\rho_i) - \varepsilon$. Since $\Sigma\lambda_i = 1$ and $\Sigma\lambda_i \tilde{F}(\rho_i) < \infty$, there exists K such that $A \equiv \sum_{i=K}^{\infty} \lambda_i \leq \varepsilon$ and $B \equiv \sum_{i=K}^{\infty} \lambda_i F(\rho_i) \leq \varepsilon$. Assume $A>0$ for otherwise we are done. By Theorem 1.1 and the convexity of $G(\rho) = \int (\nabla\rho^{1/2})^2$,

$$\varepsilon \geq \sum_K^{\infty} \lambda_i \tilde{F}(\rho_i) \geq \sum_K^{\infty} \lambda_i G(\rho_i) \geq AG(\rho^K)$$

with $\rho^K = \sum_K^{\infty} \lambda_i\rho_i / A \varepsilon J_N$. By Theorem 3.9 and the remark following it

$$\tilde{F}(\rho^K) \leq C[N^2 G(\rho^K) + N].$$

Therefore the finite sequence $\{\lambda_i, \rho_i\}_{i=1}^K$ with $\{\lambda_K, \rho_K\} = \{A, \rho^K\}$ satisfies $\Sigma\lambda_i \tilde{F}(\rho_i) \leqq F_{DM}(\rho) + \varepsilon CN(N+1) + \varepsilon$. □

<u>Theorem 4.3.</u> Equation (4.5) is true.

<u>Proof.</u> The easy part is that for $\rho \varepsilon J_N$, $F_{DM}(\rho) \geqq F(\rho)$. By (4.4). $E(v) \leqq F_{DM}(\rho) + \int \rho v$ for all v. Hence, by (3.17), $F_{FM}(\rho) \geqq F(\rho)$. The hard part is contained in Corollary 4.5, which will be assumed for now. Then: (i) $F_{DM}(\rho) \leqq \tilde{F}(\rho)$ by (4.6); (ii) F_{DM} is convex and l.s.c. Hence $F_{DM}(\rho) \leqq CE\ \tilde{F}(\rho) = F(\rho)$ by Theorem 3.7. □

<u>Theorem 4.4.</u> Suppose $\{\rho_n\}$ and $\rho \varepsilon J_N$ and $\rho_n \rightharpoonup \rho$ weakly on L^1. Then there exists a density matrix Γ, with $\Gamma \to \rho$, such that $\mathrm{Tr} H_o \Gamma \leqq \liminf F_{DM}(\rho_n)$.

The proof of Theorem 4.4, due to Barry Simon, is given in the appendix.

<u>Corollary 4.5.</u> (i) F_{DM} is (norm and weakly) l.s.c.

(ii) If $\rho \varepsilon J_N$, there exists a density matrix Γ with $\Gamma \to \rho$ such that $\mathrm{Tr} H_o \Gamma = F_{DM}(\rho)$ (see Theorem 3.3).

<u>Proof.</u> (i) If $\rho_n \to \rho$, $F_{DM}(\rho) \leqq \mathrm{Tr} H_o \Gamma \leqq \liminf F_{DM}(\rho_n)$. Norm l.s.c. implies weak l.s.c.

(ii) Take $\rho_n = \rho$ in Theorem 4.4. □

C. The Kinetic Energy Functional

Kohn and Sham (KS) [30] define a kinetic functional $T_{KS}(\rho)$. There are several other possible kinetic energy functionals and we shall explore their interrelations, as well as the fact that T_{KS} does *not* have a property assumed by KS. KS define the *exchange and correlation functional* $E_{xc}(\rho)$ by

$$F_{HK}(\rho) \equiv \tfrac{1}{2} \int \int \rho(x)\rho(y)|x-y|^{-1}\, dx\, dy + T_{KS}(\rho) + E_{xc}(\rho). \quad (4.8)$$

F_{HK} and T_{KS} are defined on different subsets of J_N, so E_{xc} is defined only on a third unknown subset of J_N. This difficulty can be remedied by using $\tilde{F}$ and $\tilde{T}$ in (4.8), but there is another point that should be stressed: There is no reason to believe that E_{xc} is convex on J_N.

First, let us give some definitions. These use K instead of H_o but otherwise are self-explanatory (with the aid of the equation numbers on the left):

$$\begin{array}{lll} (3.5\ ): & E'(v) & \text{on } L^{3/2} + L^{\infty}; \\ (3.10): & T_{KS}(\rho) & \text{on } A'_N \quad (3.11); \\ (3.14): & \tilde{T}(\rho) & \text{on } J_N\ ; \\ (3.17): & T(\rho) & \text{on } L^3 \cap L^1\ . \end{array} \qquad (4.9)$$

($T(\psi)=(\psi,K\psi)$ was defined in (1.3) but it is quite different from $T(\rho)$above. It is hoped that this notational lapse will not be confusing). All the previous theorems [except for 3.9, wherein the last term in (3.22) should be omitted] carry over to these quantities. The primes on $E'(v)$ and A'_N indicate that these are different from before. Since Theorem 3.4 still holds, A'_N is not J_N. It is left as an exercise to show that $A_N \neq A'_N$.

Question 8. What is $A_N \cap A'_N$?

There is one more kinetic energy functional that can be defined on J_N, namely,

$$T_{det}(\rho) = \inf\{(\psi,K\psi)\,|\psi\to\rho,\psi\in W_n,\psi \text{ is a determinant}\}. \qquad (4.10)$$

Clearly, $T_{det}(\rho) \geq \tilde{T}(\rho)$. The question to be addressed is whether $T_{det} = \tilde{T}$. The answer is NO!, *not even on all of* A'_N. KS assumed implicitly that $T_{KS}(\rho)=T_{det}(\rho)$for $\rho\varepsilon A'_N$; any such ρ_N minimizes K+V,but it is not true that such a $\rho(x)$ can always be written as $\Sigma_{i=1}^N |\psi_i(x)|^2$ with the ψ_i being orthonormal functions on $\mathbb{R}^3$. (Spin is a complication that is ignored at this point for simplicity). In other words, not every ground state of K+V is a determinant when degeneracy is present. I Thank B. Simon for drawing my attention to this subtlety and for the construction in Theorem 4.8, which is reminiscent of the construction in Theorem 3.4.

Of course, $T_{KS} = \tilde{T}$ on A'_N by definition. Also $\tilde{T}=T$ on A'_N by

Theorem 3.10. The following shows that there are cases in which $\tilde{T} = T_{det}$.

Theorem 4.6. Suppose $\rho \in A'_N$, so that K + V has a ground state. If this ground state is nondegenerate, then $T_{det}(\rho) = \tilde{T}(\rho)$.

Proof. The ψ that minimizes $(\psi, [K + V]\psi)$ is, of course, a determinant. □

The following analog of Theorem 3.3 will be needed for Theorem 4.8.

Theorem 4.7. Let $\rho \in J_N$. Then there exists a determinant that minimizes $(\psi, K\psi)$ under the condition that $\psi \to \rho$, $\psi \in W_n$, and ψ is a determinant. Thus, (4.10) is actually a minimum.

Proof. Let D_j be a sequence of determinants with

$$D_j \to \rho \qquad \text{and} \qquad \lim(D_j, KD_j) = T_{det}(\rho).$$

The proof of Theorem 3.3 shows that ψ exists such that (i) $\psi \to \rho$; (ii) $(\psi, K\psi) = T_{det}(\rho)$; (iii) $D_j \to \psi)$ strongly in L^2. It suffices to show that ψ is a determinant. Let f_j^i, $i=1,\ldots,N$, be the orthonormal single-particle functions of D_j. By the Banach-Alaoglu theorem, N functions $f^1,\ldots,f^N$ exist so that (after passing to a subsequence) $f_j^i \to f^i$ weakly. The f^i are not necessarily orthonormal. The function

$$P_j(z_1,\ldots,z_N) = \prod f_j^i(z_i)$$

then converges weakly to $P = \prod f^i$. This so because any $\psi \in L^2(\mathbb{R}^{3N})$ can be approximated in norm by sums of product functions. Therefore,

$$D_j \rightharpoonup (N!)^{1/2} \det[f^i(z_j)] \equiv D \quad \text{weakly.}$$

But $D_j \to \psi$, so $D = \psi$. □

Theorem 4.8. Let N = 7 and q = 1. Then there is a $\rho \in A'_N$ such that $T_{det}(\rho) > \tilde{T}(\rho)$.

Proof. Take $v(x) = |x|^{-1}$, the hydrogen potential. The eigenvalues of $-\Delta + v$ are $-1/4$(onefold), $-1/16$(fourfold), $-1/36$(ninefold). All other eigenvalues are greater than $-1/36$. The ground state for N=7 and q = 1 is $\binom{9}{2} = 36$-fold degenerate, and a basis for this eigenspace consists of the determinants $(7!)^{-1/2}\det(1S,2S,2P_1,2P_2,2P_3,fg)$ where f and g are any orthonormal functions in the nine-dimensional space M spanned by $S,P_1,P_2,P_3,D_1,\ldots,D_5$(an orthonormal set for the 3S,3P, and 3D waves). Let d(f,g) denote the above normalized determinant and let

$$3^{1/2}\psi = d(S,D_1) + d(D_2,D_3) + d(D_4,D_5).$$

Then $\psi \to \rho$ with $\rho = \rho_a+\rho_b$ and

$$\rho_a(x) = |1S(x)|^2+|2S(x)|^2+\sum_{i=1}^{3} |2P_i(x)|^2,$$

$$3\rho_b(x) = |S(x)|^2+\sum_{i=1}^{5} |D_i(x)|^2.$$

Clearly $\rho \epsilon A'_N$ since ψ is a ground state.

If $T_{det}(\rho)= \tilde{T}(\rho)$, then there exists a determinant ϕ with $\phi\to\rho$ and such that ϕ *must be* a ground state. Therefore $\phi = d(f,g)$ for some orthonormal $f,g\epsilon M$. Thus,

$$|f(x)|^2 + |g(x)|^2 = \rho_b(x). \qquad (4.11)$$

I claim that this is impossible. Write f = A+D and g = B+d, with A and B being linear combinations of S and the P_i while D and d are linear combinations of the D_i. Now the S,P, and D waves behave as $|x|^0,|x|^1,|x|^2$, respectively, near the origin. By examining the behavior of (4.11) near the origin we conclude that

$$|D(x)|^2 + |d(x)|^2 = \frac{1}{3}\sum_{1}^{5}|D_i(x)|^2.$$

Since all the D_i waves have the same radial wave functions, this is really an equality about spherical harmonics Y_{2m}. The right side of the last equality is spherically symmetric, so the problem is to find two linear combinations F and G of the Y_{2m} such that

$$|F(\Omega)|^2 +|G(\Omega)|^2 = \text{constant} > 0.$$

This is impossible, and the proof is left as an exercise. (It is easily carried out if the following five basis functions are used:

xyr^{-2}, xzr^{-2}, yzr^{-2}, $3x^2r^{-2}-1$, $3y^2r^{-2}-1$, with $r^2= x^2 +y^2 +z^2$). □

Remarks. (i) N = 7 is not special; it was chosen for convenience in the proof.

(ii) An alternative way of viewing Theorem 4.8 is following. Suppose K + V has a degenerate ground state, so that the ground eigenspace G is more than one-dimensional. $\psi \varepsilon G$ is a linear combination of determinants. Consider a perturbation w of v, namely, $v \to v + \lambda w$. In first-order perturbation theory, $V + \lambda W$ picks out a subspace g of G as the new ground eigenspace. If g is one dimensional, then g consists of one determinant since the ground eigenspace of $V + \lambda W$ always contains determinants (see Theorem 4.6). Now we ask, if $\psi_o \varepsilon G$ and $\psi_o \to \rho_o$, can w be chosen so that g is one dimensional and $g = \{\psi_o\}$? Alternatively, can w be chosen so that min $\{\int w\rho \,|\, \psi \to \rho$ and $\psi \varepsilon G\}$ occurs uniquely for $\rho = \rho_o$? If so, ψ_o is a determinant. Theorem 4.8 says that there can be a ρ_o such that no w can pick it out uniquely.

Even though $T_{det}(\rho) > \tilde{T}(\rho)$ for some ρ, T_{det} still satisfies the variational principle for E(v).

Theorem 4.9. For all $v \varepsilon L^{3/2} + L^\infty$

$$E'(v) = \inf \{T_{det}(\rho) + \int \rho v \,|\, \rho \varepsilon J_N\}. \tag{4.12}$$

Proof. Equation (4.12) is equivalent to the following:

$$E'(v) \equiv \inf \{(\psi, [K + V]\psi \,|\, \psi \varepsilon W_N\}$$

$$= \inf \{(\psi, [K + V]\psi \,|\, \psi \varepsilon W_N, \psi \text{ is a determinant}\}$$

$$\equiv \tilde{E}(v).$$

Clearly $E'(v) \leq \tilde{E}(v)$. Consider the operator $-\Delta + v(x)$. We *define* its "eigenvalues" $e_1 \leq e_2$,...(here, spin degeneracy is included) by the min-max principle:

$$e_{n+1} = \sup \{\varepsilon_n(\phi_1, \ldots, \phi_n)\},$$

where

$$\varepsilon_n(\phi_1,\ldots,\phi_n) = \inf\{(\phi,[-\Delta+v]\phi) \mid \phi\varepsilon H^1, \|\phi\|_2 = 1$$

$$\text{and } \phi \text{ is orthogonal to } \phi_1,\ldots,\phi_n\}.$$

From this definition, it follows by a standard argument that

$$E_N(v) \equiv \sum_{i=1}^{N} e_i = \inf\{\sum_{i=1}^{N} (\phi_i,[-\Delta+v]\phi_i) \mid \phi_1,\ldots,\phi_N \text{ are orthonormal}\}. \tag{4.13}$$

But this least infimum equals

$$\inf\{\sum_{i=1}^{\infty} \lambda_i(\phi_i,[-\Delta+v]\phi_i) \mid \phi_1,\phi_2,\ldots, \text{ are orthonormal,}$$

$$0 \leq \lambda_i \leq 1 \text{ and } \sum_{i=1}^{\infty} \lambda_i = N\}.$$

This is easy to verify.

Let $\psi\varepsilon W_N$ and let $\gamma = \Sigma\lambda_i f_i\rangle\langle f_i$ be its one-particle density matrix (including spin and with the f^i-orthonormal. $0\leq\lambda_i\leq 1$, $\Sigma\lambda_i = N$). Then

$$(\psi,[K+V]\psi) = \Sigma\lambda_i(f_i,[-\Delta+v]f_i).$$

Thus $E'(v) \geq E_N(v)$. But $E_N(v) = \tilde{E}(v)$ by inspection. □

<u>Remark.</u> This proof gives a formula for $E'(v)$, namely $E_N(v)$.

The situation is complicated, so let us summarize it. T_{KS} is defined only on A'_N, the set of ρ's that come from ground states for some v. A'_N has a smaller subset, A''_N, in which ρ comes from a determinantal ground state, A''_N includes, but is larger than, A'''_N, the set of ρ's that come from nondegenerate ground states. (Note: By Theorem 3.2 any ρ comes from a unique v (up to constants). Thus, if ρ comes from a determinant in a degenerate ground eigenspace, then $\rho\varepsilon A'''_N$). On A''_N we have

$$T_{KS}(\rho) = T_{det}(\rho) = \tilde{T}(\rho).$$

Elsewhere on $\mathcal{A}'_N$,

$$T_{det}(\rho) > T_{KS}(\rho) = \tilde{T}(\rho).$$

Thus, there are two choices for (4.8): either T_{KS} or T_{det}. On $\mathcal{B}_N$, the complement of $\mathcal{A}'_N$, T_{KS} is not defined (but $\tilde{T}$, T_{det}, and T are defined). The preferred functional here is $T(\rho)$ because it is convex and hence most manageable. On $\mathcal{B}_N$; $T(\rho)$ is probably strictly less than $\tilde{T}(\rho) \leq T_{det}(\rho)$— at least this is so when T has a continuous tangent functional(Theorem 3.10). Such points are dense (Theorem 3.14). In any case, since T, $\tilde{T}$, and T_{det} can be interchangeably used in (4.12); it makes no difference which is used as far as E'(v) is concerned.

D. Density Functionals for Excited States

Several functionals of the density ρ have been presented in section 3, that have the property that their infima over ρ yields the ground state energy E(v).It is desirable also to have functionals whose infima yield information about excited state energies. Some results in this direction will now be given.

Two functionals (in addition to F_{HK}) have been introduced earlier. These are

$$F(\rho) \equiv \sup_v \{E(v) - \int v\rho\} \tag{4.14}$$

$$= \min_\Gamma \{\mathrm{Tr}H_o\Gamma \mid \Gamma \to \rho\}, \tag{4.15}$$

where Γ is an N-particle density matrix (Theorems 4.3, 4.4).

$$\tilde{F}(\rho) = \min_\psi \{(\psi, H_o\psi) \mid \psi \to \rho\}. \tag{4.16}$$

(Theorem 3.3).

Both F and $\tilde{F}$ satisfy

$$E(v) = \inf \{F(\rho) + \int \rho v \mid \rho \in J_N\}. \tag{4.17}$$

Denote the excited state energies of H_v by

$$E(v) \equiv E_o(v) \leqq E_1(v) \leqq E_2(v) \leqq \ldots \tag{4.18}$$

The well known *max-min principle* states that

$$E_n(v) = \sup_{\phi_o,\ldots,\phi_{n-1}} \quad \inf_{\psi\perp\{\phi_1,\ldots\phi_{n-1}\}} (\psi,H_v\psi) \ , \tag{4.19}$$

where $\phi_o,\ldots\phi_{n-1}$ is an arbitrary set of orthonormal functions and $\psi|\phi$ means $\psi,\phi = 0$ in the ordinary L^2 ($\mathbb{R}^{3N}$) inner product. Eq. (4.19) may be taken as a definition of $E_n(v)$,

From (4.19) it is easy to deduce the variational principle for the *sum* $\sigma^n(v)$:

$$\sigma^n(v) \equiv \sum_{j=o}^{n} E_j(v) = \inf_{\phi_o,\ldots\phi_n} \sum_{j=o}^{n} (\phi_j,H\phi_j) \ , \tag{4.20}$$

where $\phi_o,\ldots,\phi_n$ are orthonormal in $L^2(\mathbb{R}^{3N})$. $\sigma^n(v)$ has the same properties E(v) (Theorem 3.1). By imitating the proofs of Theorems 3.3 and 4.4, one can show that the inf in (4.20 is a minimum.

If

$$\phi_j \to \rho_j \text{ and } \sum_{j=o}^{n} \rho_j = \rho\varepsilon J_{Nn}$$

we write $\{ \phi_o,\ldots,\phi_n\} \to \rho$. Note that $\int \rho = Nn$.

Likewise, if $\hat{\Gamma}$ is a positive semidefinite operator on H^1 ($\mathbb{R}^{3N}$) with eigenvalues $\leqq 1$, but with TR $\Gamma = n$ (note the nomalization) we can, using the notation of section 2, write $\Gamma \to \rho\varepsilon J_{Nn}$. In analogy with (4.14)-(4.16)(all the proofs go through essentially as before), we have for $\rho\varepsilon J_{Nn}$

$$F^n(\rho) \equiv \sup_v \{\sigma^n(v) - \int \rho v\} \tag{4.21}$$

$$= \min_{\hat{\Gamma}} \{TrH_o \ \hat{\Gamma}|\hat{\Gamma}\to \rho \ \}, \tag{4.22}$$

$$\tilde{F}^n(\rho) = \min_{\phi_o,\ldots,\phi_n} \{ \sum_{j=o}^{n} (\phi_j, H_o \phi_j) \,|\, \sigma_o,\ldots,\phi_n \text{ orthonormal and } \overset{n}{\rightarrow} \rho \} \tag{4.23}$$

$F^n(\rho)$ is convex, but $\tilde{F}^n(\rho)$ is not.

Both F^n and $\tilde{F}^n$ satisfy the variational principle

$$\sigma^n(v) = \inf \{F^n(\rho) + \int \rho v \,|\, \rho \varepsilon J_{Nn}\}. \tag{4.24}$$

Thus, if we are content with the sum $\sigma^n(v)$, the functionals F^n or $\tilde{F}^n$ do the job. This variational principle was derived earlier, but I am not sure by whom.

Next, let us turn to the question of trying to calculate $E_n(v)$ *alone*, not $\sigma^n(v)$. First note, that there *cannot exist* a functional $F_n(\rho)$ (for $n \geqq 1$) with the property that

$$E_n(v) = \inf \{F_n(\rho) + \int \rho v \,|\, \rho \varepsilon J_N\} \tag{4.25}$$

for, if such an F_n existed thus, by (4.25), $E_n(v)$ would be concave. But $E_n(v)$ is *not* concave, as simple examples show. Restricting the ρ's in (4.25) (such as ρ's that come from excited states) will not alter this conclusion.

The problem is that $E_n(v)$ is not concave. Therefore we introduce

$$\hat{E}_n(v) = \text{concave envelope of } E_n(v). \tag{4.26}$$

<u>Remark.</u> The convex envelope was defined before Theorem 3.7. The concave envelope is the inf of weakly upper semicontinuous functions $\geqq E_n$. Alternatively $E_n(v) = -CE\,(-E_n(v))$, where CE is the convex envelope.

It will now be shown that one can indeed find a variational principle for $\hat{E}_n(v)$ in the form (4.25). By what was said above, this is the best one can hope to achieve.

Fix n and, temporarily, fix a set of orthonormal functions $\phi_o,\ldots,\phi_{n-1}$, which will be denoted collectively by ϕ. Let us define the following quantities in analogy with those in section 3.

$$E_n(v\phi) \equiv \inf_{\psi\perp\phi} (\psi, H_v\phi) \tag{4.26}$$

$$= \inf_{\Gamma\perp\phi} \mathrm{Tr}\Gamma H_v \tag{4.27}$$

$$F_n(\rho,\phi) \equiv \sup_v E_n(v,\phi) - \int v\rho \tag{4.28}$$

$$\tilde{F}_n(\rho,\phi) \equiv \inf \{(\psi, H_o\psi) | \psi\perp\phi,\ \psi\to\rho \} \tag{4.29}$$

$$F_{DM,n}(\rho,\phi) \equiv \inf \{\mathrm{Tr}\Gamma H_o | \Gamma\perp\phi,\ \Gamma\to\rho \} \tag{4.30}$$

Here, ϕ is an arbitrary function in H' with $(\psi,\psi) = 1$ and Γ is an arbitrary density matrix. $\rho\varepsilon J_N$ and $v\varepsilon L^{3/2} + L^\infty$ as before. $\psi\perp\phi$ means $(\psi,\phi_j) = o$, $j=o,\ldots,n-1$. $\Gamma|\phi$ means $(\phi_j,\Gamma\phi_j) = o$, $j=o,\ldots,n-1$ (equivalently, $\Gamma\phi_j=o$).

Relative to the previous definitions of $E,F,\tilde{F},F_{DM}$ all we have done is to restrict the variations to the orthogonal complement of the finite dimensional space span $\{\phi_o,\ldots,\phi_{n-1}\}$. All the theorems go through as before (note that a condition such as $(\psi,\phi) = o$ is preserved under weak limits). In particular: the infima in (4.29) and (4.30) are minima; $E_n(v,\phi)$ is concave in v (unlike $E_n(v)$) and it is continuous;

$$F_n = \mathrm{CE}\ \tilde{F}_n \text{ (as a function of } \rho) = F_{DM,n} > \tilde{F}_n \tag{4.31}$$

$$E_n(v,\phi) = \inf_\rho F_n(\rho,\phi) + \int v\rho \tag{4.32}$$

$$= \inf_\rho \tilde{F}_n(\rho,\phi) + \int v\rho. \tag{4.33}$$

Next let us define

$$E_n(v) \equiv \sup_\phi E_n(v,\phi) \tag{4.34}$$

$$F_n(\rho) \equiv \sup_\phi F_n(\rho,\phi) \tag{4.35}$$

$$= \sup_{\phi} F_{DM,n}(\rho,\phi) \equiv F_{DM,n}(\rho) \tag{4.36}$$

$$\tilde{F}_n(\rho) \quad \sup_{\phi} \tilde{F}_n(\rho,\phi). \tag{4.37}$$

In passing from (4.26) to (4.34) concavity is lost. However, $F_n(\rho)$ is convex and l.s.c. since it is the supremum of a family of such functions [31].

Since sup sup = sup sup always,
$\quad$ A B $\quad$ B A

$$F_n(\rho) = F_{DM,n}(\rho) = \sup_{\phi} \sup_{v} E_n(v\phi\ ) - \int v\rho$$

$$= \sup_{v} \sup_{\phi} E_n(v,\rho) - \int v\rho = \sup_{v} E_n(v) - \int v\rho. \tag{4.38}$$

Thus, F_n is the Legrendre transform of E_n.

A generalization of Fenchel's theorems [31, Theorem I.10] states that the Legendre transform of F_n is the concave envelope of E_n. Thus,

$$\hat{E}_n(v) = \inf\{F_n(\rho) + \int \rho v | \rho \varepsilon C_N\}. \tag{4.39}$$

This is the desired *density functional variational principle for excited states*.

Using Fenchel's theorem again,(4.38) can be replaced by

$$F_n(\rho) = D_{DM,n}(\rho) = \sup_{v} \hat{E}_n(v) - \int v\rho. \tag{4.39}$$

5. SOME DENSITY FUNCTIONALS THAT ARE BOUNDS

In this sections we forego the abstract functional theory of the previous sections and instead expound adifferent philosophy. Rather than pursuing "the correct density functional", which seems to be uncomputable, we shall content ourselves here with finding upper and lower bounds to the various quantities of interest in terms of $\rho(x)$. This latter program can provide rigorous bounds on ground state energies that, while they may not always be extremely accurate, do have a proper place in our conceptual scheme. Some of these bounds will be briefly displayes here; the interested reader is referred to the original papers for proofs.

It should be remembered that if one has bounds on two quantities (e.g., T and I; see below) and even if these bounds are optimal, then, in general, the sum of the bounds is *not* optimal for the sum of the two quantities (e.g., T + I).

A. Kinetic Energy Lower Bound

Lieb and Thirring (LT)[21](also see Ref. 16) proved (for fermions in three dimensions) that if $\psi \to \rho$, then (for all N)

$$T(\psi) \geqq K^c (4\pi)^{-2/3} q^{-2/3} \int \rho(x)^{5/3}\, dx, \tag{5.1}$$

where K^c is the "classical" value $(3/5)(6\pi^2)^{2/3}$. LT conjectured that (5.1) holds in three dimensions with the $(4\pi)^{-2/3}$ deleted. [Note: Although an analog of (5.1) holds in all dimensions, the corresponding constant is definitely less than K^c in one and two dimensions]. In Ref. 22 (also see Ref. 16) $(4\pi)^{-2/3}$ was replaced by $1.496(4\pi)^{-2/3}$.

Incidentally, the statement $T(\psi) \geqq K q^{-2/3} \int \rho^{5/3}$ for all ψ, all N, and some K is equivalent to the following [21]. Let v be any non-positive potential in $L^{5/2}(\mathbb{R}^3)$ and let $e_1 \leqq e_2 \leqq \ldots$, be the negative eigenvalues (if any) of $-\Delta + v(x)$ counting degeneracy, but not counting the q-fold degeneracy. Then

$$\Sigma\, e_i \geqq -L \int |v(x)|^{5/2}\, dx$$

with $K = (3/5)(2/5L)^{2/3}$.

B. Kinetic Energy Upper Bound

There is, of course, no upper bound for $T(\psi)$ in terms of ρ. March and Young (MY)[17] proposed that for all $\rho \varepsilon J_N$ there is a *determinantal* ψ, with $\psi \to \rho$, such that

$$T(\psi) \leqq q^{-2/n} K^c \int \rho(x)^{(n+2)/n} dx + \int [\nabla \rho^{1/2}(x)]^2\, dx, \tag{5.2}$$

where n is the dimension and $K^c = \pi^2/3$ for n = 1. (Compare (5.2) with Theorem 1.2). They proves (5.2) for n = 1, but their proof for n>1 has an error. Equation (5.2) for n>1 is still an *open problem*. The MY construction for n = 1 motivated the construction in the proof of Theorem 1.2.

C. Lower Bound For The Indirect Part Of The Coulomb Repulsion

Let Γ be a density matrix (which may be a pure state, $\Gamma = |\psi\rangle\langle\psi|$) with $\Gamma \to \rho$. Let

$$I(\Gamma) = \mathrm{Tr}\,\{\,\Gamma \sum_{1 \leq i<j \leq N} |x_i - x_j|^{-1}\} \tag{5.3}$$

be the Coulomb repulsive energy. The *indirect part* of this energy, $E(\Gamma)$, is *defined* by

$$I(\Gamma) = D(\rho) + E(\Gamma), \tag{5.4}$$

with

$$D(\rho) = \tfrac{1}{2} \int\int \rho(x)\rho(y)|x-y|^{-1}\, dx\, dy \tag{5.5}$$

being the *direct part*.

In Ref. 23 it was shown that

$$E(\Gamma) \geq -C \int \rho(x)^{4/3}\, dx \tag{5.6}$$

with C = 8.5. In Ref. 24 this was improved to C = 1.68. The sharp (i.e., best) C in (5.6) is not known, but it is larger than 1.23.

It is well known that in any pure, *determinantal* state, $E(\Gamma) < 0$. For other states, $E(\Gamma)$ can be positive. Indeed, for any fixed ρ there is no upper bound for $E(\Gamma)$ (see Ref. 24).

There is no q-dependence in (5.6) and, indeed, (5.6) holds for all statistics (i.e., C does not depend on statistics). This is explained in Ref. 24. The Dirac approximation has $C_D q^{-1/3}$ in (5.6) with $C_D = 3(6/\pi)^{1/3}/4 = 0.93$, but this q dependence is an artifact of the particular q-dependent determiantal ψ used to evaluate E from (5.4).

It should be noted that the bound

$$I(\Gamma) \geq D(\rho) - C \int \rho^{4/3} \tag{5.7}$$

is not *convex* in ρ. It is not even positive. These two faults lead to absurd conclusions when the right side of (5.7) is used in Thomas-Fermi-Dirac theory (see Ref. 25).

Since $\Gamma \to \rho$ is linear,

$$\tilde{I}(\rho) = \inf\{I(\Gamma) | \Gamma \to \rho\} \tag{5.8}$$

is convex in ρ. In other words, an optimal positive, convex lower bound must exist. Any reader who is devoted to abstract density functional theory, in the spirit of Sec. 3 or (5.8), should try to guess a plausible form for $\tilde{I}(\rho)$. (Proving it is another matter). It will quickly be seen that $\tilde{I}(\rho)$must be extremely complicated, and to say that it is "nonlocal" is an understatement. To see this, consider $N = 2$ and ρ consisting of two "bumps", ρ_1 and ρ_2, very far apart. As long as $\int\rho_1 = \int\rho_2 = 1$, $\tilde{I}(\rho) - D(\rho) \simeq 0$, independently of ρ_1 and ρ_2. But when $\int\rho_1 > 1$, $\int\rho_2 < 1$, then $\tilde{I}(\rho) - D(\rho)$ depends heavily on ρ_1 but not on ρ_2. The reason is that in the former case the two electrons can be far apart in the two bumps; in the latter case the two electrons must partly be close together in the first bump.

A problem that is physically more relevant and that illustrates the hidden complexity of density functional theory is the following problem about induced dipolar (or Van der Waals) forces raised in Ref. 25. When two atoms are a distance R apart, and R is large, there is an attraction $-R^{-6}$ (neglecting retardation effects). This attraction comes from the Coulomb repulsion, but it is not a static effect. The atomic dipole moment is almost zero. (There are, in fact, tiny dopile moments, but these are opposite in sign by symmetry, and hence repulsive. They must exist by the Feynmann-Hellman theorem: dE/dR = electric potential at the nucleus. I thank C. Herring for this remark). There is almost no static dipole moment because to create one would cost a polarization energy αd^2. The attractive energy is $-d^2R^{-3}$ and, if $R^{-3} < \alpha$, $d = 0$ for minimum energy. The cause of the $-R^{-6}$ energy is more subtle, but it has a semiclassical basis: The electrons in each atom *move in phase* while maintaining the spherical symmetry about each atom. The energy cost is then αd^4 and the minimum energy occurs when $2\alpha d^2 = R^{-3}$. Thus, the $-R^{-6}$ attraction comes from the fact that the electron cloud cannot be thought of as a simple "fluid". This effect is somehow built into $\tilde{I}(\rho)$, but an explicit form of $\tilde{I}(\rho)$that will produce this effects has yet to be displayed.

D. A Variational Principle

E(v) given by (3.5), satisfies (by definition) the well-known variational principle

$$E(v) \leq (\psi, H_v \psi). \tag{5.9}$$

Can an upper bound for E(v) be given in terms of ρ alone? If (5.2) were true, then, for any $\rho \varepsilon J_N$,

$$E(v) \leq \text{right side of (5.2)} + D(\rho) + \int v\rho. \tag{5.10}$$

[See the remark about E(Γ) for determinants in Sec. 5C.]

An upper bound for E(v) can, indeed, be given in terms of the one-particle density *matrices* $\gamma(z,z')$ as follows [26]:

Let γ be any admissible one-particle density matrix ($0 \leq \gamma \leq I$, $\mathrm{Tr}\gamma = N$). (Note: γ includes spin. It was called $\bar{\gamma}$ in Sec. 2). Then

$$E(v) \leq \mathrm{Tr}\,\gamma(-\Delta + v(x)) + \tfrac{1}{2}\int K_2(z,z')|x-x'|^{-1}\,dz\,dz', \tag{5.11}$$

where $\int dz = \Sigma_\sigma \int dx$ and

$$K_2(z,z') = \gamma(z,z)\gamma(z',z') - |\gamma(z,z')|^2 . \tag{5.12}$$

The form (5.11) is well known *if* γ came from a pure state $|\psi\rangle\langle\psi|$ with ψ being a determinant. The point about (5.11) is that it holds for *all* admissible γ. Incidentally, the minimum of (5.11) over all admissible γ occurs when γ comes from a determinant ψ. In other words, the *best* Hartree-Fock function minimizes (5.11), but (5.11) is interesting precisely because this HF function is unknown.

E. Thomas-Fermi Theory

This theory (see Ref. 25 for an exposition) does not yield bounds and therefore does not properly belong here. However, it illustrates the usefulness of the bounds in Secs. 5A-C.

The TF functional is

$$\mathcal{E}^{TF}(\rho) = K^c q^{-2/3}\int \rho^{5/3} + D(\rho) + \int v\rho, \tag{5.13}$$

while the TF Weizaecker functional $\mathcal{E}^{TFW}(\rho)$ is the right-hand side of (5.10). If $-C\int\rho^{4/3}$ is added to the right-hand side of (5.13), the result is TF Dirac theory.

The TF energy for N particles is defined by

$$E^{TF} = \inf \{ \mathcal{E}^{TF}(\rho) \mid \int \rho = N \}, \qquad (5.14)$$

and similarly for E^{TFW} and E^{TFD}.

Now suppose that v is an atomic or molecular potential, that is

$$v(x) = - \sum_{j=1}^{k} z_j |x-R_j|^{-1} , \qquad (5.15)$$

with the $z_j > 0$. It is a fact [14]that under the scaling $z_j \to \lambda z_j$ and $N \to \lambda N$, as $\lambda \to \infty$

$$E^{TF}/E\ (v) \to 1, \qquad (5.16)$$

where E(v) is the true ground state energy. Note that (5.16) also holds if E^{TF} is replaced by E^{TFW} or E^{TFD}(see Ref. 25).

Thus we see that if the conjecture in Sec. 5A holds, then, combining (5.1) with (5.7), TFD theory is lower bound that is asymptotically exact. Similarly, if (5.2) holds, then, as remarked in Sec. 5C TFW theory is an upper bound that is asymptotically exact.

F. Two-Body Density Matrices

If one is willing to go beyond the one-body density ρ or one-body density matrix γ and consider the two-body reduced density matrix $\gamma^{(2)}$, then E(v) is directly and exactly expressible in terms of $\gamma^{(2)}$, since H_v has only one- and two-body terms. The problem is that it is very difficult to decide when a given $\gamma^{(2)}$ is, in fact, the reduction of an admissible N-body density matrix Γ. This is called the N-*representability problem* and it has not been solved. (This is to be compared with the fact that there is a simple necessary and sufficient condition for a one-body γ to be N-representable; see Sec. 2).

It is possible, however, to find some necessary conditions and some sufficient conditions for $\gamma^{(2)}$ to be N-representable. Using these, bounds on E(v) can be derived. Since this approach is outside the scope of this article, we refer the reader to the excellent review of Percus [27].

APPENDIX: PROOFS OF THEOREMS 1.3, 3.3, and 4.4

The following proof of Theorem 1.3 is due to H. Brezis (private communication).

Proof. For simplicity of presentation we take $N = 2$ and $q = 1$ (no spin). Therefore we have $\psi(x,y)$ and

$$F(x) = \left(\int |\psi(x,y)|^2 dy \right)^{1/2} = [\rho(x)/2]^{1/2}.$$

Suppose that $\psi_n \to \psi$ in $H^1(\mathbb{R}^3 \times \mathbb{R}^3)$; that is, $\psi_n \to \psi$ and $\nabla\psi_n \to \nabla\psi$ in L^2. We want to show that $F_n \to F$ in $L^2(\mathbb{R}^3)$ and $\nabla F_n \to \nabla F$ in $L^2(\mathbb{R}^3)$. The former is trivial:

By the Schwarz inequality,

$$\left(\int |\psi_n(x,y)|^2 \, dy \int |\psi(x,y)|^2 \, dy \right)^{1/2} \geq \tfrac{1}{2}\int \psi_n(x,y)^* \psi(x,y) \, dy$$

$$+ \tfrac{1}{2}\int \psi(x,y)^* \psi_n(x,y) \, dy.$$

Therefore

$$\int |F_n(x) - F(x)|^2 \, dx \leq \int dx \int |\psi_n(x,y) - \psi(x,y)|^2 \, dy,$$

and the right-hand side converges to zero.

The proof that $\nabla F_n \to \nabla F$ is the difficult one. It is sufficient to prove convergence for *some* subsequence n_j, $j = 1,2,\ldots$.(If $\nabla F_n \not\to \nabla F$, then there is some subsequence and some $\varepsilon > 0$ such that $\| \nabla F_{n_j} - \nabla F \| > \varepsilon$. But then this subsequence clearly does not have a subsequence which converges to ∇F). Now since $\psi_n \to \psi$ in H^1 there is some subsequence and some function $G \in H^1$ such that

$$|\psi_{n_j}(x,y)| \leq G(x,y) \text{ and } |\nabla\psi_{n_j}(x,y)| \leq G(x,y),$$

a.e. in $\mathbb{R}^6$. (The proof of this fact is the same as the first half of the proof of the Riesz-Fischer lemma that L^1 is complete).

Henceforth, we shall replace n_j by n. We shall also assume, for simplicity, that $F_n(x)$ and $F(x)>0$ for all x (otherwise, an approximation argument can be used).

Now

$$2\nabla F_n(x) = A_n(x)/F_n(x) + \text{c.c.}$$

with

$$A_n(x) = \int B_n(x,y)\,dy \quad \text{and} \quad B_n(x,y) = \psi_n(x,y)^* \nabla\psi_n(x,y).$$

As we saw, $F_n \to F$ in L^2, so, by passing to a subsequence, we can assume $F_n(x) \to F(x)$ a.e. Furthermore, a.e.

$$|B_n(x,y)| \leq G(x,y)^2 \varepsilon L^1(\mathbb{R}^6).$$

By passing to a subsequence we can assume $\nabla\psi_n \to \nabla\psi$ and $\psi_n \to \psi$ a.e. Thus, by dominated convergence, $B_n \to B$ in $L^1(\mathbb{R}^6)$. For this subsequence $|B_n(x,y) - B(x,y)| \to 0$, a.e. $(\mathbb{R}^6)$. Then, for a.e. x, $|B_n(x,y) - B(x,y)| \to 0$ a.e. y. Thus, by dominated convergence $\int dy\,|B_n(x,y) - B(x,y)| \to 0$, a.e. x. In other words, for some subsequence, $\nabla F_n(x) \to \nabla F(x)$, a.e.

Finally, we note that, by the Schwarz inequality,

$$|\nabla F_n(x)|^2 \leq \int |\nabla\psi_n(x,y)|^2\,dy \leq \int G(x,y)^2\,dy \equiv C(x)^2.$$

Since C is a fixed L^2 function, $\nabla F_n \to \nabla F$ in L^2 by dominated convergence.

Proof of Theorem 3.3. Let ψ_j (with $\psi_j \to \rho$) be a minimizing sequence for $\tilde{F}(\rho)$. The ψ_j are obviously bounded in $H^1(\mathbb{R}^{3N})$, so, by the Banach-Alaoglu theorem, there is a $\psi \varepsilon H^1(\mathbb{R}^{3N})$ such that $\psi_j \rightharpoonup \psi$ weakly in $H^1(\mathbb{R}^{3N})$. Obviously, ψ has the same symmetry as the ψ_j. It is well known that under weak limits positive quadratic forms decrease. Thus

$$\tilde{F}(\rho) = \lim\,(\psi_j, H_0\psi_j) \geq (\psi, H_0\psi).$$

If we can show that $\psi \rightarrow \rho$, we are done. To do so it is sufficient to prove that $\psi_j \rightarrow \psi$ strongly because if $\psi \rightarrow \tilde{\rho}$ we have, by the easy part of Theorem 1.3 that $\rho^{1/2} = \rho_j^{1/2} \rightarrow \tilde{\rho}^{1/2}$ in L^2, so that $\tilde{\rho} = \rho$.

Strong convergence will be proved by showing that $\int |\psi|^2 = 1$. Let S be the characteristic function of some bounded set in $\mathbb{R}^{3N}$. By the Rellich-Kondrachov theorem [28] there is a subsequence (which can be chosen independent of S) of the ψ_j such that $S\psi_j$ converges strongly (in L^2) to $S\psi$. Pick $\varepsilon > 0$ and let χ be the characteristic function of a bounded set in $\mathbb{R}^3$ such that

$$\varepsilon > \int \rho(1 - \chi) \equiv \int |\psi_j|^2 \sum_j [1 - \chi(x_i)].$$

But

$$\Sigma[1 - \chi(x_i)] \geqq 1 - S,$$

where $S = \Pi\chi(x_i)$. Thus, $\int |\psi_j|^2 S \geqq 1 - \varepsilon$. Since $|\psi_j|^2 S \rightarrow \int |\psi|^2 S$, we have that $\int |\psi|^2 \geqq \int |\psi|^2 S \geqq 1 - \varepsilon$ for all $\varepsilon > 0$. □

Remark. The symmetry of ψ was not needed in this proof provided one generalizes definition (1.6) to

$$\rho(x) = \sum_\sigma \sum_{i=1}^N \int |\psi(z_1, \ldots, (x,\sigma_i), \ldots, z_N|^2 dx_1 \ldots dx_{i-1} dx_{i+1} \ldots dx_N. \tag{A.1}$$

The following proof of Theorem 4.4 is due to B. Simon (private communication). It is closely related to the proof of Theorem 3.3 just given.

Proof. Without loss, replace H_o by $h^2 = H_o + 1$ in the definitions. h^{-1} is a bounded operator. We can assume that $g_n \equiv F_{DM}(\rho_n) < \infty$, $g \equiv \lim g_n$ exists, and $\mathrm{Tr}\, h\Gamma_n h = \mathrm{Tr}\Gamma_n h^2 \leqq g_n + 1/n$ with $\Gamma_n \rightarrow \rho_n$. Thus, $y_n \equiv h\Gamma_n h$ is uniformly bounded in the trace norm. The dual of the compact operators, com, is the trace class operators t, and $\gamma \varepsilon t$ takes $A \varepsilon$ com into $\mathrm{Tr}\gamma A \rightarrow \mathrm{Tr}\gamma A$ for all $A \varepsilon$ com.

The Banach-Alaoglu theorem states that a norm-closed ball of finite radius in t is compact in the weak* topology. For us this means that there exists y with Tr $y < \infty$, so that, for a subsequence,

$\mathrm{Tr}\, y_n A \to \mathrm{Tr}\, yA$ for every compact A. Clearly, $y \geq 0$ and therefore $\underline{\lim}\, \mathrm{Tr}\, y_n \geq \mathrm{Tr}\, y$. Also, y obviously has the correct (Pauli) symmetry. If we can show that $\Gamma \equiv h^{-1} y h^{-1}$ (which is in trace class) satisfies $\Gamma \to \rho$, we are done. To do this we shall show that if $\Gamma \to \rho'$, then $\int (\rho_n - \rho')f \to 0$ for any $f \in L^\infty$. This would mean that $\rho_n \rightharpoonup \rho'$ weakly in L^1. But since $\rho'_n \rightharpoonup \rho$ in L^1, $\rho' = \rho$.

As in the proof of Theorem 3.3, for any $\varepsilon > 0$ there is a χ (= characteristic function of a bounded set in $\mathbb{R}^3$) such that

$$\int \rho(1 - \chi) < \varepsilon \qquad \text{and} \qquad \int \rho'(1 - \chi) < \varepsilon .$$

Since $\rho_n \rightharpoonup \rho, \int \rho_n(1 - \chi) < \varepsilon$ for n sufficiently large. If

$$\phi_n(x_1,\ldots,x_N) = \Gamma_n(x_1,\ldots,x_N;x_1,\ldots,x_N)$$

(after summing on spins), and similarly for ϕ, we have (as in Theorem 3.3)

$$\int \phi_n(1 - S) < \varepsilon \qquad \text{and} \qquad \int \phi(1 - S) < \varepsilon$$

where $S = \Pi\chi(x_i)$. In view of this, it is sufficient to show that

$$\int_n P \to \int \phi P \qquad \text{with} \qquad P = S \sum_i f(x_i).$$

Let $P = P(x_1,\ldots,x_N)$ be any bounded functions of compact support and let M_P be the operator (in L^2) of multiplication by P. It is a fact that $A_P \equiv h^{-1} M_P h^{-1}$ is compact. (This is essentially the same as the Rellich-Kondrachov theorem used in Theorem 3.3). Therefore

$$\mathrm{Tr}\, \Gamma_n M_P = \mathrm{Tr}\, y_n A_P \to \mathrm{Tr}\, yA_P = \mathrm{Tr}\, \Gamma M_P .$$

□

ACKNOWLEDGEMENT

This work was partially supported by U.S. National Science Foundation grant no. PHY-7825390-A02. This papaer first appeared in *Physics as Natural Philosophy: Essays in Honor of Laszlo Tisza on his 75th Birthday*, H. Feshbach and A. Shimony eds. (M.I.T. Press, Cambridge, 1982), pp. 111-149. A revised version appeared in the International Journal of Quantum Chemistry 23, 243-277(1983). The present paper contains an additional section 4.D which was prompted by the question raised by M. Levy at the 1983 NATO ASI on DENSITY FUNCTIONALS, namely whether density functionals exist for excited states.

REFERENCES

1. L.H. Thomas, Proc. Camb. Phil. Soc. 23:542 (1927).
2. E. Fermi, Rend. Accad. Naz. Lincei 6:602 (1927).
3. P.Hohenberg and W. Kohn, Phys. Rev. B136:864 (1964).
4. M.M. Morell, R.G. Parr and M. Levy, J. Chem. Phys.62:549 (1975).
5. R.G. Parr, S. Gadre and L.J. Bartolotti, Proc. Natl. Acad. Sci. 76:2522 (1979).
6. R.A. Donnelly and R.G. Parr, J. Chem. Phys. 69:4431 (1978).
7. H. Englisch and R. Englisch, "Hohenberg-Kohn theorem and non-v-representable densities", Physica A, to be published.
8. T.L. Gilbert, Phys. Rev. B6:211 (1975).
9. J.E. Harriman, Phys. Rev.A6:680 (1981).
10. M. Levy, Proc. Natl. Acad. Sci. USA 76:6062 (1979).
11. M. Levy, Phys. Rev. A26:1200 (1982).
12. S.M. Valone, J. Chem. Phys. 73:1344 (1980); ibid. 73:4653 (1980).
13. A.S. Bamazai and B.M. Deb, Rev. Mod. Phys. 53:95 (1981). Erratum, 53:593(1981).
14. E.H. Lieb and B. Simon, Adv. Math. 23:22 (1977) See also Thomas-Fermi theory revisited, Phys. Lett.31:681 (1973. See also Refs. 16 and 25.
15. M. Hoffmann-Ostenhof and T. Hoffmann-Ostenhof, Phys. Rev. A16: 1782 (1977).
16. E.H. Lieb, Rev. Mod. Phys. 48:553 (1976).
17. N.H. March and W.H. Young, Proc. Phys. Soc.72:182 (1958).
18. M. Reed and B. Simon, Methods of Modern Mathematical Physics (Academic. New York,1978), Vol. 4.
19. S. Mazur, Studia Math. 4:70 (1933).
20. R.B.Israel, Convexity in the Theory of Lattice Gases(Princeton NJ. (1979).
21. E.H. Lieb and W.E. Thirring, "Inequalities for the moments of the eigenvalues of the Schrödinger hamiltonian and their relation to Sobolev inequalities", in Studies in Mathematical Physics, E.H. Lieb, B. Simon and A.S. Wightman, Eds. (Princeton U.P., Princeton, N.J. 1976). See also Phys. Rev. Lett. 687 (1975); Errata 35:1116 (1975).

22. E.H. Lieb, Am. Math. Soc. Proc. Symp. Pure Math. 36:241 (1980).
23. E.H. Lieb, Phys. Lett. A70:444 (1979).
24. E.H. Lieb and S. Oxford, Int. J. Quantum Chem. 19:427 (1981).
25. E.H. Lieb, Rev. Mod. Phys. 53:603 (1981); Errata, 54:311 (1982).
26. E.H. Lieb, Phys. Rev. Lett.46:457 (1981); Erratum, 47_69 (1981).
27. J.K. Percus, Int. J. Quantum Chem. 13:89 (1978).
28. R.A. Adams, Sobolev Spaces (Avademic Press, New York)(1975).
29. W. Fenchel, Can. J. Math. 1:23 (1949).
30. W. Kohn and L.J. Sham, Phys. Rev. A140:1133 (1965).
31. H. Brezis, Analyse Functionelle, Theorie et Applications, Masson, Paris (1983).

DENSITY FUNCTIONAL APPROACH TO TIME-DEPENDENT AND TO RELATIVISTIC SYSTEMS

Eberhard K.U. Gross and Reiner M. Dreizler

Institut für Theoretische Physik
Universität Frankfurt
D-6000 Frankfurt am Main
Federal Republic of Germany

1. Introduction

Density functional methods, in general, can be discussed on three different levels. The first is the level of fundamental existence theorems such as those of Hohenberg, Kohn and Sham[1,2] (HKS) and related mathematical topics. The second level consists of the various approaches[2-32] to a systematic construction of density functionals. Prominent among these are the different gradient expansion techniques. The approximations required on this level depend, of course, on the particular situation to be described. The third level consists of practical schemes and numerical solutions for real physical systems.

This paper will address all three levels: In section 2 we shall present a discussion of time-dependent systems[33] on level 1. It will be shown that the exact density can, in principle, be calculated either from a set of hydrodynamical equations or from a stationary action principle or from a set of single-particle orbitals fulfilling an effective time-dependent Schrödinger equation.

In the subsequent sections, stationary systems will be considered on level 2 and 3: In section 3 a gradient expansion technique which has first been advanced in a similar form by Kirzhnits[16] is described in some detail. This method is used to derive second order gradient contributions to the kinetic and the exchange energy. There will be some emphasis on nonlocal contributions to the kinetic energy. In the following section, the functionals are applied to physical systems (level 3); numerical results will be presented for atoms and for two-center scattering systems.[34-39]

In the final section, a density functional representation of relativistic inhomogeneous systems is developed.[40-44] The consistent application of the Kirzhnits technique leads to inhomogeneity corrections even in the lowest order, i.e. there will be nongradient corrections due to the inhomogeneity of the system which are not present in the nonrelativistic case. Some preliminary results will be shown for relativistic atoms.

2. Existence theorems for time-dependent systems

The successful application of density functional methods to stationary systems has recently sparked new interest in treating time-dependent (td) problems in terms of such methods: atomic[45-48] and nuclear[49-51] scattering processes, photoabsorption in atoms[52] and the dynamical response of inhomogenous metallic systems[53,54] have been successfully discussed. However, as yet, a fundamental existence theorem comparable to the HK theorem[1] could not be demonstrated for arbitrary td systems. The proof of the traditional HK theorem for the stationary ground state is based on the Rayleigh-Ritz principle. The difficulty for td systems arises from the fact that no minimum principle is available; the action integral

$$A = \int_{t_o}^{t_1} dt \, \langle\Phi(t) \mid i\, \partial/\partial t - \hat{H}(t) \mid \Phi(t)\rangle \tag{1}$$

provides only a stationary point (but, in general, no minimum) at the solution of the td Schrödinger equation

$$i \frac{\partial}{\partial t} \Phi(t) = \hat{H}(t)\, \Phi(t) \tag{2}$$

(throughout this paper, we use the units $\hbar = c = 1$). The Hamiltonian

$$\hat{H}(t) = \hat{T} + \hat{V}(t) + \hat{W}$$

is assumed to consist of the kinetic energy

$$\hat{T} = \sum_s \int d^3r \, \hat{\psi}_s^+(\vec{r}) \left(-\frac{1}{2m}\nabla^2\right) \hat{\psi}_s(\vec{r}) \, ,$$

a td, local, and spin-independent single-particle potential

$$\hat{V}(t) = \sum_s \int d^3r \, v(\vec{r}t)\, \hat{\psi}_s^+(\vec{r})\, \hat{\psi}_s(\vec{r}) \, ,$$

and some spin-independent particle-particle interaction

$$\hat{W} = \frac{1}{2} \sum_{s} \sum_{s'} \int d^3r \int d^3r' \; \hat{\psi}_s^+(\vec{r}) \; \hat{\psi}_{s'}^+(\vec{r}')w(\vec{r},\vec{r}')\hat{\psi}_{s'}(\vec{r}')\hat{\psi}_s(\vec{r}) \; .$$

By solving the td Schrödinger equation (2) with various potentials $v(\vec{r}t)$ and a fixed initial state

$$\Phi(t_o) = \Phi_o \tag{3}$$

we obtain a map

$$F : v(\vec{r}t) \mapsto \Phi(t) \; . \tag{4}$$

The range of allowable single-particle potentials will be specified later. Next we calculate the densities

$$\rho(\vec{r}t) = \langle\Phi(t) \mid \hat{n}(\vec{r}) \mid \Phi(t)\rangle \tag{5}$$

with the density operator

$$\hat{n}(\vec{r}) = \sum_{s} \hat{\psi}_s^+(\vec{r}) \; \hat{\psi}_s(\vec{r}) \tag{6}$$

for all the td wave functions resulting from F. This defines another map

$$G : v(\vec{r}t) \mapsto \rho(\vec{r}t) \; . \tag{7}$$

In order to establish a td version of the HK theorem one has to show that G is invertible. So far, a HK like theorem has been proven only for two special td cases: a) If the pontentials $v(\vec{r}t)$ are restricted to functions having a periodic dependence on time and if, in addition, the occurring frequencies are assumed to be so small that transitions to excited states are impossible then the "adiabatic" td groundstate energy possesses the usual minimum property. In that case it is possible to derive[55] a variational theorem even for non-v-representable densities following the proof given by Levy[56] for the stationary case. b) If the potentials consist of a static (internal) and a td (external) part $v(\vec{r}t) = v_o(\vec{r}) + v_{ext}(\vec{r}t)$ then, in a small neighbourhood of $v_{ext} \equiv 0$, the inverse of G can be constructed within linear response theory.[57] In both cases, the allowed potentials are "álmost static" though in a different sense. Therefore, the validity of the corresponding existence theorems is restricted to a regime very close to that of the traditional HK theorem for stationary states.

Before we proceed with a theorem for a more general td situation, it should be noted that one cannot expect an exact 1 - 1 correspondence between potentials and densities: Consider two potentials $\hat{V}(t)$ and $\hat{V}'(t)$ differing by an additive merely td scalar function C(t):

$$\hat{V}'(t) = \hat{V}(t) + C(t).$$

Then the corresponding solutions $\Phi(t)$ and $\Phi'(t)$ of the td Schrödinger equation will differ by a merely td phase factor

$$\Phi'(t) = e^{-i\alpha(t)}\Phi(t) \tag{8}$$

with

$$\dot{\alpha}(t) = C(t) \tag{9}$$

and, consequently, the resulting densities will be identical $\rho'(\vec{r}t) = \rho(\vec{r}t)$. Thus, all the potentials differing by an additive merely td function are mapped on the same density, i.e. there is no exact 1 - 1 correspondence between potentials and densities. However, if it is possible to establish the invertibility of G up to such an additive td function then the wave function is fixed by the density up to a td phase via

$$\Phi(t) = FG^{-1}\rho(\vec{r}t) \tag{10}$$

and any expectation value $<\Phi(t)|\hat{O}|\Phi(t)>$ can be regarded as a functional of the density (the ambiguity in the phase cancels out provided $\hat{O}$ contains no time derivatives).

In the following it will be shown that the map G is in fact invertible up to such a trivial td function. The only restriction on the set of admissible potentials will be a smoothness condition concerning the dependence on time.

<u>Theorem 1.</u> For all single-particle potentials $v(\vec{r}t)$ which can be expanded into a Taylor series with respect to the time coordinate around $t = t_0$, a map $G : v(\vec{r}t) \mapsto \rho(\vec{r}t)$ is defined by solving the time-dependent Schrödinger equation with a fixed initial state $\Phi(t_0) = \Phi_0$ and calculating the corresponding densities $\rho(\vec{r}t)$. This map can be inverted up to an additive, merely time-dependent function in the potential.

Proof: Let $v(\vec{r}t)$ and $v'(\vec{r}t)$ be two potentials which differ by more than a td function, i.e. $v(\vec{r}t) - v'(\vec{r}t) \neq c(t)$. This does of course not exclude that the potentials are identical at $t = t_0$. However, since the potentials can be expanded into a Taylor series around t_0 there must exist at least one time derivative in which the potentials differ by more than a constant. In other words, there must exist some minimal nonnegative integer k such that

$$\frac{\partial^k}{\partial t^k} [v(\vec{r}t) - v'(\vec{r}t)]_{t=t_0} \neq \text{const} . \tag{11}$$

This inequality is essential for the proof. If it holds for $k = 0$ the potentials differ already at $t = t_0$; if it holds for some $k > 0$ the potentials will become different infinitesimally later than t_0.

The only thing to be proven is that the densities $\rho(\vec{r}t)$ and $\rho'(\vec{r}t)$ corresponding to $v(\vec{r}t)$ and $v'(\vec{r}t)$ are different if (11) is fulfilled with some $k \geq 0$. In a first step, we show that the corresponding current densities $j(\vec{r}t)$ and $j'(\vec{r}t)$ are different. We use the representation

$$\vec{j}(\vec{r}t) = \langle\Phi(t) \mid \hat{\vec{j}}(\vec{r}) \mid \Phi(t)\rangle \tag{12}$$

with

$$\hat{\vec{j}}(\vec{r}) = \frac{1}{2mi} \sum_s ([\vec{\nabla}\hat{\psi}_s^+(\vec{r})]\hat{\psi}_s(\vec{r}) - \hat{\psi}_s^+(\vec{r}) [\vec{\nabla}\hat{\psi}_s(\vec{r})]) . \tag{13}$$

It should be noted that the particle and current densities corresponding to $v(\vec{r}t)$ and $v'(\vec{r}t)$ are of course identical at the initial time t_0 since we consider only wave functions which evolve from a fixed initial state Φ_0 :

$$\vec{j}(\vec{r}t_0) = \vec{j}'(\vec{r}t_0) = \langle \Phi_0 \mid \hat{\vec{j}}(\vec{r}) \mid \Phi_0 \rangle \equiv \vec{j}_0(\vec{r}) \tag{14}$$

$$\rho(\vec{r}t_0) = \rho'(\vec{r}t_0) = \langle \Phi_0 \mid \hat{n}(\vec{r}) \mid \Phi_0 \rangle \equiv \rho_0(\vec{r}) . \tag{15}$$

In order to discuss the time evolution of the current density we use the equation of motion

$$i \frac{d}{dt} \langle \Phi(t) \mid \hat{O}(t) \mid \Phi(t) \rangle = \langle\Phi(t) \mid i\frac{\partial}{\partial t} \hat{O}(t)+[\hat{O}(t),\hat{H}(t)] \mid \Phi(t)\rangle \tag{16}$$

which holds for arbitrary td operators $\hat{O}(t)$. For the current density (12), this yields

$$i \frac{\partial}{\partial t} \vec{j}(\vec{r}t) = <\Phi(t) \mid [\hat{\vec{j}}(\vec{r}),\hat{H}(t)] | \Phi(t)> \tag{17}$$

and

$$i \frac{\partial}{\partial t} \vec{j}'(\vec{r}t) = <\Phi'(t) \mid [\hat{\vec{j}}(\vec{r}),\hat{H}'(t)] | \Phi'(t)> \quad . \tag{18}$$

Since $\Phi(t)$ and $\Phi'(t)$ evolve from the same initial state Φ_o we obtain at $t = t_o$ by subtracting (18) from (17)

$$i \frac{\partial}{\partial t} [\vec{j}(\vec{r}t) - \vec{j}'(\vec{r}t)]_{t=t_o} = <\Phi_o | [\hat{\vec{j}}(\vec{r}),\hat{H}(t_o) - \hat{H}'(t_o)] | \Phi_o >$$

$$= \frac{i}{m} \rho_o(\vec{r})\vec{\nabla}(v(\vec{r}t_o) - v'(\vec{r}t_o)) \tag{19}$$

where the last equality follows from a straightforward calculation of the commutator. If the potentials differ at $t = t_0$ (i.e. if (11) holds for $k = 0$) then the right hand side of (19) will be different from zero and thus $\vec{j}(\vec{r}t)$ and $\vec{j}'(\vec{r}t)$ will become different infinitesimally later than t_0. If the minimum integer k for which (11)holds is greater than zero eq. (16) has to be applied k times. Derivatives of the potentials with respect to space coordinates (as far as required to calculate the commutators in (16))are assumed to exist. After some straightforward algebra we obtain

$$(i\frac{\partial}{\partial t})^{k+1} [\vec{j}(\vec{r}t)-\vec{j}'(\vec{r}t)]_{t=t_o} =$$

$$\frac{i}{m} \rho_o(\vec{r}) \; \vec{\nabla}((i\frac{\partial}{\partial t})^k [v(\vec{r}t)-v'(\vec{r}t)]_{t=t_o}) \neq 0. \tag{20}$$

Again this means that $\vec{j}(\vec{r}t)$ and $\vec{j}'(\vec{r}t)$ will become different infinitesimally later than t_o which completes the proof for the current vectors.

Next we consider the corresponding densities. By use of the continuity equation we have

$$\partial/\partial t[\rho(\vec{r}t)-\rho'(\vec{r}t)] = -\mathrm{div}[\vec{j}(\vec{r}t)-\vec{j}'(\vec{r}t)\;]. \tag{21}$$

Taking the (k+1)-th derivative of this equation we obtain by help of (20) for some $k \geq 0$

$$\frac{\partial^{k+2}}{\partial t^{k+2}} [\, \rho(\vec{r}t) - \rho'(\vec{r}t)\,]_{t=t_o} =$$

$$- \operatorname{div} [\, \rho_o(\vec{r})\vec{\nabla} \, (\frac{1}{m} \frac{\partial^k}{\partial t^k} [v(\vec{r}t)-v'(\vec{r}t)]_{t=t_o})]. \qquad (22)$$

It remains to be shown that the right hand side of (22) cannot vanish if (11) holds. The proof is by reductio ad absurdum: Assume that $\operatorname{div}[\, \rho_o(\vec{r}) \, \vec{\nabla} u(\vec{r})] = 0$ with $u(\vec{r}) \neq$ const and consider the following integral

$$\int d^3r \, \rho_o(\vec{r}) \, (\vec{\nabla} u(\vec{r}))^2 \; . \qquad (23)$$

By use of Green's theorem, we obtain

$$= - \int d^3r \, u(\vec{r}) \, \operatorname{div} [\rho_o(\vec{r})\vec{\nabla} u(\vec{r})] + \oint d\vec{S} \cdot (\rho_o(\vec{r}) u(\vec{r}) \vec{\nabla} u(\vec{r})) = 0.$$

The first term vanishes by assumption while the surface integral vanishes if the initial density $\rho_o(\vec{r})$ is assumed to fall off rapidly enough in the asymptotic region. Thus the value of the integral (23) is zero. Since the integrand functions of (23) are nonnegative, we can conclude

$$\rho_o(\vec{r}) \, (\vec{\nabla} u(\vec{r}))^2 \equiv 0.$$

This is in contradiction to $u(\vec{r}) \neq$ const provided $\rho_o(\vec{r})$ is reasonably well behaved (we merely have to exclude that the initial density vanishes in precisely those subregions of space where u = const, if such regions exist at all). Thus, the right hand side of (22) cannot vanish which proves that the densities $\rho(\vec{r}t)$ and $\rho'(\vec{r}t)$ become different infinitesimally later than t_o.

This completes the proof of theorem 1. In the following we shall use this theorem to establish a theoretical basis for practical schemes to calculate the td density.

Theorem 2. There exists a three-component density functional $\vec{P}[\rho](\vec{r}t)$ which depends parametrically on $(\vec{r}t)$ such that the exact particle and current densities can be determined from a set of "hydrodynamical" equations

$$\frac{\partial}{\partial t}\,\rho(\vec{r}t) = -\mathrm{div}\,\vec{j}(\vec{r}t) \tag{24}$$

$$\frac{\partial}{\partial t}\,\vec{j}(\vec{r}t) = \vec{P}[\rho](\vec{r}t) \tag{25}$$

with initial conditions

$$\rho(\vec{r}t_o) = \rho_o(\vec{r}) \tag{26}$$

and

$$\vec{j}(\vec{r}t_o) = \vec{j}_o(\vec{r}) \quad . \tag{27}$$

Proof: Since the exact particle and current densities always satisfy the continuity equation (24) it is sufficient to prove eq. (25). From theorem 1 we know that potential is determined by the density up to an additive td function C(t). This in turn fixes the wave function within a td phase factor

$$\Phi(t) = e^{-i\alpha(t)}\;\Psi[\rho](t) \tag{28}$$

where $\Psi[\rho](t)$ is defined as the wave function obtained for the choice $C(t) = 0$. By insertion of (28) into (17) the desired eq. (25) is immediately obtained if the functional $\vec{P}$ is chosen as

$$\vec{P}[\rho](\vec{r}t) = -i\langle\Psi[\rho](t)\;|\;[\hat{\vec{j}}(\vec{r}),\,\hat{H}(t)]\;|\;\Psi[\rho](t)\rangle \;. \tag{29}$$

If it is possible to construct the functional $\vec{P}[\rho]$within some reasonable approximation then the hydrodynamical equations (24), (25) provide a practical scheme to calculate approximate densities.

Of course, the two equations can be contracted into a second order equation for the density alone:

$$\ddot{\rho}(\vec{r}t) = -\mathrm{div}\,\vec{P}[\rho](\vec{r}t) \tag{30}$$

which has to be solved for the initial conditions

$$\rho(\vec{r}t_o) = \rho_o(\vec{r}) \tag{31}$$

and

$$\dot{\rho}(\vec{r}t_o) = -\mathrm{div}\, \vec{j}_o(\vec{r}) \ . \tag{32}$$

A second order equation of this type has been used to discuss the dynamical response of inhomogeneous electronic systems.[54]

Theorem 3. The action integral (1) can be represented as a functional of the density A[ρ]. If the potential $v(\vec{r}t)$ is chosen such that no additive time-dependent function can be split off, the total action can be written as

$$A[\rho] = B[\rho] - \int_{t_o}^{t_1} dt \int d^3r\, \rho(\vec{r}t)\, v(\vec{r}t) \tag{33}$$

where B[ρ] is a universal functional of the density in the sense that the same dependence on $\rho(\vec{r}t)$ holds for all external potentials $v(\vec{r}t)$. A[ρ] has a stationary point at the exact density of the system, i.e. the exact density can be computed from the Euler equation

$$\delta A \,/\, \delta\rho(\vec{r}t) = 0 \ . \tag{34}$$

Proof: Although the wave function Φ(t) is fixed by the density only within a td phase factor the matrix element $\langle\Phi(t)|i\,\partial/\partial t - \hat{T} - \hat{W} - \hat{V}(t)|\Phi(t)\rangle$ is uniquely determined since the function C(t) contained in the potential $\hat{V}(t)$ is precisely cancelled by the time derivative of the phase $\dot{\alpha}(t) = C(t)$ (see discussion leading to eq. (9)). Therefore, the action (1) is a unique functional of the density and can be written as (33) if B[ρ] is chosen as

$$B[\rho] = \int_{t_o}^{t_1} dt\, \langle\Psi[\rho](t)|i\,\partial/\partial t - \hat{T} - \hat{W}|\Psi[\rho](t)\rangle \ . \tag{35}$$

The universality of B follows trivially from the construction. Since the action (1) is stationary for the exact solution of the td Schrödinger equation (2), the corresponding density functional (33) must be stationary for the exact td density of the system.

The variational equation (34) can be used to calculate approximate densities provided the action functional (33) is known within some reasonable approximation. If we knew the exact action functional (33) then the variational equation (34) would of course be identical with the exact hydrodynamical equation (30).

A particularly attractive possibility of exploiting the stationary action principle is the construction of a td Kohn-Sham scheme: First of all, we define another density functional by

$$S[\rho] := \int_{t_o}^{t_1} dt \, \langle \Psi[\rho](t) | i\, \partial/\partial t - \hat{T} | \Psi[\rho](t) \rangle \tag{36}$$

which is, of course, universal in the same sense as $B[\rho]$. It should be pointed out that the particle-particle interaction has been kept fixed so far. If we compare two different interactions $\hat{W}$ and $\hat{W}'$ then the corresponding functionals $S_W[\rho]$ and $S_{W'}[\rho]$ will in general be different. Now let $S_o[\rho]$ be the particular functional (36) for the case $\hat{W} \equiv 0$, i.e. for noninteracting particles. Then, in analogy to the stationary case, the "exchange-correlation" part of the action can be defined as

$$A_{xc}[\rho] := \int_{t_o}^{t_1} dt \, \langle \Psi[\rho](t) | \hat{W} | \Psi[\rho](t) \rangle$$

$$- \frac{1}{2} \int_{t_o}^{t_1} dt \int d^3r \int d^3r' \, \rho(\vec{r}t) w(\vec{r},\vec{r}') \rho(\vec{r}'t) + S_o[\rho] - S_W[\rho]. \tag{37}$$

Theorem 4. The exact time-dependent density of the system can be computed from

$$\rho(\vec{r}t) = \sum_j \varphi_j^*(\vec{r}t) \, \varphi_j(\vec{r}t) \tag{38}$$

where the single-particle orbitals $\varphi_j(\vec{r}t)$ fulfil the time-dependent Schrödinger equation

$$\left(i\, \partial/\partial t + \frac{1}{2m} \nabla^2\right) \varphi_j(\vec{r}t) = v_{eff}[\vec{r}t, \rho(\vec{r}t)] \, \varphi_j(\vec{r}t) \tag{39}$$

with an effective one-particle potential given by

$$v_{eff}[\vec{r}t, \rho(\vec{r}t)] = v(\vec{r}t) + \int d^3r' \rho(\vec{r}'t) w(\vec{r},\vec{r}') + \delta A_{xc} / \delta\rho(\vec{r}t) \tag{40}$$

Proof: Let us first apply the stationary action principle (theorem 3) to a system of noninteracting particles. In this case, the Euler equation (34) which fixes the density looks as follows

$$0 = \delta S_o / \delta\rho(\vec{r}t) - v(\vec{r}t) \, . \tag{41}$$

On the other hand, the density for a noninteracting system can be constructed from $n(\vec{r}t) = \sum_j \varphi_j^*(\vec{r}t)\varphi_j(\vec{r}t)$ with single-particle orbitals fulfilling the Schrödinger equation $(i\,\partial/\partial t + \frac{1}{2m}\nabla^2)\varphi_j(\vec{r}t) = v(\vec{r}t)\varphi_j(\vec{r}t)$. However, if the particles interact with some potential $\hat{W}$ then, by definition of the "exchange-correlation" functional (37), the exact density can be obtained by solving the Euler equation

$$0 = \delta S_o/\delta\rho(\vec{r}t) - [v(\vec{r}t) + \int d^3r'\rho(\vec{r}'t)w(\vec{r},\vec{r}') + \delta A_{xc}/\delta\rho(\vec{r}t)]$$

Comparison with (41) shows that this is precisely the equation for noninteracting particles moving in the potential $v_{eff}[\vec{r}t,n(\vec{r}t)]$ defined by (40). Therefore, the exact density can be calculated from a set of single-particle orbitals fulfilling the effective td Schrödinger equation (39).

In practical calculations, an approximation of the "exchange-correlation" functional $A_{xc}[\rho]$ is required. If, for example, $A_{xc}[\rho]$ is approximated by a homogeneous electron gas exchange term $\sim\int dt\int d^3r\ \rho(\vec{r}t)^{4/3}$ one obtains the wellknown td Hartree-Fock-Slater scheme.

On the exact level the three schemes suggested here are, of course, completely equivalent. However, the most attractive alternative to calculate approximate densities is provided by the td Kohn-Sham scheme since it will produce a quantum mechanical (wiggle-) structure in the most natural way.

It should be emphasized that the functionals $P[\rho]$, $B[\rho]$, and $A_{xc}[\rho]$ as given by (29), (35), and (37) respectively are defined only for v-representable densities. The functionals remain undefined for those densities $\rho(\vec{r}t)$ which do not correspond to some potential $v(\vec{r}t)$. This fact may cause mathematical problems e.g. when variations $\delta A[\rho]$ with respect to arbitrary densities are required. At present, it is not clear how large the set of v-representable td densities is. Furthermore, it should be noted that the functional $S_o[\rho]$ is well defined only for densities corresponding to noninteracting systems. Thus the calculation of $A_{xc}[\rho]$ (according to eq. (37)) tacitly assumes the existence of a noninteracting system which reproduces the density of the interacting system in question.

Each pair $(\Phi_o, v(\vec{r}t))$ specifies one particular initial value problem of the type (2),(3). So far, the initial state Φ_o has been kept fixed while the potential was allowed to vary within a set specified in theorem 1. For this reason, the functionals $P[\rho]$, $A[\rho]$, etc. are defined only for td densities which all have the same initial shape $\rho(\vec{r}t_0)$. Therefore, the td theory presented above cannot be compared directly to the HKS theory of the ground state since the initial densities corresponding to stationary ground states are of course all different. In order to make the correspondence between the td and the stationary theory more transparent, we shall now discuss the case where Φ_0 is allowed to vary as well; i.e. we consider a larger class of initial value problems $(\Phi_0, v(\vec{r}t))$ and again ask the question whether this larger class can be discussed in terms of the density alone. If Φ_0 is allowed to be an arbitrary initial state the question is difficult to answer. However, if the set of admissible initial states is restricted to nondegenerate ground state wave functions* then the question can be answered in the affirmative: In this case the extended map

$$\tilde{G} : (\Phi_0, v(\vec{r}t)) \mapsto \rho(\vec{r}t) \qquad (42)$$

is again invertible up to a trivial additive td function in the potential. To prove this, we have to show that two densities $\rho(\vec{r}t)$ and $\rho'(\vec{r}t)$ are different provided the corresponding pairs $(\Phi_0, v(\vec{r}t))$ and $(\Phi_0, v(\vec{r}t))'$ are different. The case $\Phi_0 = \Phi_0'$, $v \neq v'$ is treated in theorem 1. The case $\Phi_0 \neq \Phi_0'$ (with the potentials being either identical or not) follows directly from the proof of the HK theorem for the ground state: $\Phi_0 \neq \Phi_0'$ implies $< \Phi_0|\hat{n}(\vec{r})|\Phi_0 > \neq < \Phi_0'|\hat{n}(\vec{r})|\Phi_0'>$ so that the densities $\rho(\vec{r}t)$ and $\rho'(\vec{r}t)$ are different at the initial time t_0.

For the set of densities obtained from $\tilde{G}$, the theorems 2,3 and 4 hold in precisely the form given above. Since this set of densities contains all stationary ground state densities it is now possible to investigate in which way the td theory reduces to the common HKS theory: Let us consider the subset of stationary initial value problems $\{(\Phi_v, v(\vec{r}))\}$ where for each static potential $v(\vec{r})$ only the ground state wave function

$$\Phi_v = e^{-i\varepsilon t_0} \Psi_{gs}$$

corresponding to this very potential is allowed as initial state. On this subset, the extended map

$$\tilde{F} : (\Phi_0, v(\vec{r}t)) \rightarrow \Phi(t) \qquad (43)$$

---

* It should be noted that for each potential $v(\vec{r}t)$ the ground states corresponding to arbitrary static potentials are allowed as initial states. Thus we have an infinite set of initial value problems $(\Phi_0, v(\vec{r}t))$ for each potential $v(\vec{r}t)$.

(which is derived from solving the td Schrödinger equation with potential $v(\vec{r}t)$ and initial state Φ_0) obviously yields only wave functions of the form

$$\Phi(t) = e^{-i\varepsilon t}\Psi_{gs} \quad . \tag{44}$$

Thus, the densities obtained from $\tilde{G}$ on this subset of initial value problems are precisely the ground state densities

$$\rho_{gs}(\vec{r}) = \langle \Psi_{gs}|\hat{n}(\vec{r})|\Psi_{gs}\rangle$$

corresponding to the various static potentials $v(\vec{r})$. By insertion of (44) into the definition of the functionals $P[\rho]$ and $A[\rho]$ it is readily seen that these functionals reduce to the correct static limit:

$$P[\rho_{gs}] = 0 \tag{45}$$

$$A[\rho_{gs}] = (t_1-t_0) \langle \Psi_{gs}[\rho_{gs}]|\varepsilon-\hat{H}|\Psi_{gs}[\rho_{gs}]\rangle$$

$$= (t_1-t_0) (\varepsilon \langle \Psi_{gs}[\rho_{gs}]|\Psi_{gs}[\rho_{gs}]\rangle - E_{HK}[\rho_{gs}]) \tag{46}$$

where E_{HK} is the common Hohenberg-Kohn energy functional. It follows from (46), that the stationary action principle (34) reduces to the common energy minimization principle. That way the whole HKS theory of the ground state is recovered.

3. Stationary nonrelativistic systems: the Kirzhnits gradient expansion

In this section, we shall describe in some detail how a density functional representation of stationary nonrelativistic systems can be derived in a systematic fashion. We start from the exact ground state energy of an interacting Coulomb system which is split into the kinetic energy E_{kin}, the external potential energy E_{en}, the particle-particle interaction E_{ee}, the exchange contribution E_{ex}, and the correlation part E_{cor}:

$$E = E_{kin} + E_{en} + E_{ee} + E_{ex} + E_{cor}. \tag{47}$$

In terms of the one- and two-particle density matrices, these contributions take the following form:

$$E_{kin} = \int d^3r \int d^3r' \delta(\vec{r}-\vec{r}') [- \frac{\Delta_r}{2m} \rho(\vec{r},\vec{r}')] \equiv \int d^3r \tau(\vec{r}) \tag{48}$$

$$E_{en} = \int d^3r \rho(\vec{r}) v_n(\vec{r}) . \tag{49}$$

$v_n(\vec{r})$ is the nuclear potential corresponding to either atomic, molecular or solid state situations.

$$E_{ee} = \frac{e^2}{2} \int d^3r \int d^3r' \frac{\rho(\vec{r})\rho(\vec{r}')}{|\vec{r}-\vec{r}'|} \tag{50}$$

$$E_{ex} = - \frac{e^2}{2} \int d^3r \int d^3r' \frac{\rho(\vec{r},\vec{r}')\rho(\vec{r}',\vec{r})}{|\vec{r}-\vec{r}'|} \equiv \int d^3r \; e_x(\vec{r}) \tag{51}$$

$$E_{cor} = \frac{e^2}{2} \int d^3r \int d^3r' \frac{\rho^{(2)}(\vec{r}\vec{r}', \vec{r}\vec{r}')}{|\vec{r}-\vec{r}'|} - E_{ee} - E_{ex} . \tag{52}$$

For the sake of brevity spin degrees of freedom are suppressed here. For later use we have expressed the kinetic energy as well as the exchange energy by the corresponding energy densities $\tau(\vec{r})$ and $e_x(\vec{r})$.

Within this paper we shall restrict ourselves to the Hartree-Fock(HF) limit, i.e. we neglect the correlation energy

$$E_{cor} \equiv 0.$$

Since E_{en} and E_{ee} are already functionals of the one-particle density, it only remains to derive a density functional representation of E_{kin} and E_{ex} which so far depend on the full density matrix.

In the HF limit, the one particle density matrix can be written as

$$\rho(\vec{r},\vec{r}') = \sum\!\!\!\!\!\!\int_\nu \; n_\nu < \vec{r}|\nu> <\nu|\vec{r}'>$$

where the single-particle orbitals $|\nu>$ are solutions of the HF variational equation

$$(\hat{t} + \hat{v}_{eff}) \; |\nu> = \varepsilon_\nu |\nu> \quad .$$

A representation of the one-particle density operator can be derived as follows: For the ground state, the occupation numbers n_ν (1 for occupied levels, 0 for unoccupied levels) can be expressed in terms of a step function θ involving the Fermi-level ε_F

$$< \vec{r}|\hat{\rho}|\vec{r}\,' > = \rho(\vec{r},\vec{r}\,')$$

$$= \sum\!\!\!\!\!\!\int_\nu \theta(\varepsilon_F-\varepsilon_\nu) <\vec{r}|\nu> <\nu|\vec{r}\,'>$$

$$= \sum\!\!\!\!\!\!\int_\nu <\vec{r}|\theta(\varepsilon_F-\hat{t}-\hat{v}_{eff})|\nu> <\nu|\vec{r}\,'> .$$

By use of the completeness of the HF single-particle basis we obtain

$$= < \vec{r}|\theta(\varepsilon_F-\hat{t}-\hat{v}_{eff})|\vec{r}\,' > .$$

For convenience we introduce a "Fermi energy operator" by

$$\hat{E}_F = \varepsilon_F - \hat{v}_{eff} \tag{53}$$

and extract from the simple argument above, the well known representation

$$\hat{\rho} = \theta(\hat{E}_f-\hat{t}) . \tag{54}$$

This representation of the HF density operator is the starting point of the Kirzhnits gradient expansion technique[16]. In the following, we shall first give a survey of the complete procedure; the details will be presented afterwards.

Using (54), the density matrix can be calculated by insertion of the complete set of plane waves

$$\rho(\vec{r},\vec{r}\,') = < \vec{r}|\hat{\rho}|\vec{r}\,' >$$

$$= 2 \int \frac{d^3k}{(2\pi)^3} < \vec{r}|\theta(\hat{E}_F-\hat{t})|\vec{k} > < \vec{k}|\vec{r}\,' > \tag{55}$$

Although the plane wave states are eigenfunctions of the kinetic energy operator

$$\hat{t}|\underline{k}> = \frac{k^2}{2m}|\underline{k}>$$

it will in general not be correct to write

$$\theta(\hat{E}_F-\hat{t}) \,|\, \underline{k}> = \theta(\hat{E}_F- \frac{k^2}{2m}) \,\underline{k}>$$

since the operators $\hat{E}_F$ and $\hat{t}$ do not commute. Thus, we are left with a mathematical problem of the following structure: We have to calculate a function of a sum of two noncommuting operators acting on an eigenstate of one of the operators:

$$f(\hat{a} + \hat{b})|a> = ?$$

$$\text{with } [\hat{a},\hat{b}] \neq 0$$

$$\text{and } \hat{a}|a> = a|a> \quad .$$

In order to evaluate this expression, we make use of the following representation

$$f(\hat{a} + \hat{b})|a> = \sum_{n=0}^{\infty} f^{(n)}(a + \hat{b})\,\hat{O}_n|a> . \tag{56}$$

We obtain an infinite series over all the derivatives of the function f which now contain the eigenvalue a in place of the operator $\hat{a}$. The operators $\hat{O}_n$ can be calculated via a recursion relation involving multiple commutators of $\hat{a}$ and $\hat{b}$:

$$\hat{O}_n = \frac{1}{n}([\hat{a},\hat{O}_{n-1}] + \sum_{i=1}^{n} \hat{C}_i\,\hat{O}_{n-1-i}) \tag{57}$$

$$\text{with } \hat{C}_i = \frac{(-1)^i}{i!}\,\underbrace{[\hat{b},[\hat{b},[\ldots[\hat{b}}_{i\ \text{times}},\hat{a}]\ldots]]]$$

$$\text{and } \hat{O}_o = 1 , \hat{O}_1 = 0 .$$

In our case, the multiple commutators derived from the kinetic energy operator $\hat{a} = -\hat{t}$ and from $\hat{b} = \hat{E}_F$ lead to an expansion of the density matrix (55) in terms of multiple derivatives of E_F:

$$\rho(\vec{r},\vec{r}') = \text{function of } E_F(\vec{r}),\ \vec{\nabla}E_F(\vec{r}),\ \partial_{ij}E_F(\vec{r})\ldots$$

With this relation, we can calculate the particle density, the kinetic and the exchange energy densities:

$$\rho(\vec{r}) = \rho\ [\ E_F(\vec{r})\ ] \tag{58}$$

$$\tau(\vec{r}) = \tau\ [\ E_F(\vec{r})\ ] \tag{59}$$

$$e_x(\vec{r}) = e_x[\ E_F(\vec{r})\ ]\ . \tag{60}$$

If the first relation can be inverted

$$E_F(\vec{r}) = E_F\ [\rho(\vec{r})\ ] \tag{61}$$

(in a sense to be specified later), then we can eliminate $E_F(\vec{r})$ in favour of $\rho(\vec{r})$ by insertion of (61) into (59) and (60). This yields the desired density functionals $\tau[\rho]$ and $e_x[\rho]$ for the kinetic and the exchange energy densities.

After this survey, we shall first indicate a proof of the theorem (56): We start from a formal Laplace representation

$$f(\hat{a}+\hat{b})|a\rangle = \int d\lambda c(\lambda)\exp\ [\lambda(\hat{a}+\hat{b})]\ |a\rangle$$

and use the factorisation of the exponential operator

$$\exp[\lambda(\hat{a}+\hat{b})] = \exp[\lambda\hat{b}]\hat{K}(\lambda)\exp[\lambda\hat{a}]\ . \tag{62}$$

This yields

$$f(\hat{a}+\hat{b})|a\rangle = \int d\lambda\ \exp\ [\lambda(a+\hat{b})]\ \hat{K}(\lambda)|a\rangle$$

where use has been made of the eigenvalue equation $\hat{a}|a\rangle = a|a\rangle$. If the operator $\hat{K}$ is expanded into a power series in λ

$$\hat{K}(\lambda) = \Sigma_{n=0}^{\infty}\ \lambda^n\hat{O}_n \tag{63}$$

one obtains

$$f(\hat{a}+\hat{b})|a> = \Sigma_{n=o}^{\infty} \left[\int d\lambda\lambda^n c(\lambda)\exp[\lambda(a+\hat{b})] \right] \hat{O}_n|a> .$$

Insertion of the Laplace representation of the derivatives of f

$$f^{(n)}(\hat{Q}) = \frac{d^n f(\hat{Q})}{d\hat{Q}^n} = \int d\lambda c(\lambda)\lambda^n \exp[\lambda\hat{Q}]$$

finally yields the expansion stated above

$$f(\hat{a}+\hat{b})|a> = \Sigma_{n=o}^{\infty} f^{(n)}(a+\hat{b}) \hat{O}_n|a> .$$

From eq. (62), one readily obtains a differential equation

$$\frac{d\hat{K}}{d\lambda} = [\hat{a},\hat{K}]+\exp(-\lambda\hat{b})[\hat{a},\exp(\lambda\hat{b})]\hat{K}(\lambda) \tag{64}$$

which is satisfied by the operator $\hat{K}(\lambda)$ with the initial condition $\hat{K}(o)= 1$. Insertion of the expansion (63) into the differential equation (64) yields the recurrence relation (57) for the operators $\hat{O}_n$.

To obtain a first estimate of the density matrix, it is reasonable to approximate the expression $f(\hat{a}+\hat{b})|a>$ by the lowest order term of the expansion (56) :

$$f(\hat{a}+\hat{b})|a> = f(a+\hat{b})|a>$$

i.e. we neglect the effects due to the noncommutativity of the operators $\hat{a}$ and $\hat{b}$. In the case of the density operator, we have to replace

$$f \to \theta , \qquad \hat{a} \to -\hat{t} , \qquad \hat{b} \to \hat{E}_F$$

which yields to lowest order

$$\theta(\hat{E}_F-\hat{t})|\hat{k}> = \theta(\hat{E}_F- \frac{k^2}{2m}) |\hat{k}> .$$

The operator $\hat{E}_F$ was defined as the difference of the Fermi energy and the HF potential (see eq.(53)). The latter can be split into a nonlocal (exchange) part A and a local (external plus direct) part $v_o(\vec{r})$:

$$\hat{E}_F = \varepsilon_F - \hat{v}_{eff} = \varepsilon_F - v_o(\vec{r}) + \hat{A} \quad . \tag{65}$$

Since for Coulomb systems the exchange term is always "smaller" than the direct term, it is reasonable to make a further expansion in the operator $\hat{A}$ around $\hat{A} \equiv 0$. To lowest order, the operator $\hat{A}$ is dropped completely. This yields

$$\theta(\hat{E}_F-\hat{t})|\hat{k}> = \theta(\varepsilon_F - v_o(\vec{r}) - \frac{k^2}{2m})|\hat{k}>$$

$$=:\theta(E_F(\vec{r}) - \frac{k^2}{2m})|\vec{k}>$$

where we have introduced a local Fermi energy by the definition

$$E_F(\vec{r}) := \varepsilon_F - v_o(\vec{r}) . \tag{66}$$

Higher orders in $\hat{A}$ will be discussed later.

Within this approximation (i.e. lowest order of the commutator expansion (56) <u>and</u> lowest order in $\hat{A}$), it is easy to calculate the density matrix from (55):

$$\rho_o(r,r') = \frac{2}{(2\pi)^3}\int d^3k\theta(E_F(\vec{r})- \frac{k^2}{2m}) < \vec{r}|\vec{k} > < \vec{k}|\vec{r}' >$$

The step function yields an upper boundary $k_F(\vec{r})$ for the momemtum integral which is the local Fermi momentum corresponding to $E_F(\vec{r})$:

$$\frac{k_F(\vec{r})^2}{2m} = E_F(\vec{r}) . \tag{67}$$

That way, we obtain

$$\rho_o(r,r') = \frac{2}{(2\pi)^3}\int^{k_F(\vec{r})} d^3k\, e^{i\vec{k}\cdot(\vec{r}-\vec{r}')}$$

$$= \frac{1}{\pi^2} k_F(\vec{r})^3 \frac{j_1(k_F(\vec{r})\cdot|\vec{r}-\vec{r}'|)}{k_F(\vec{r})\cdot|\vec{r}-\vec{r}'|} . \tag{68}$$

In the limit $\vec{r} \to \vec{r}'$ one obtains

$$\rho_o(\vec{r}) = \frac{1}{3\pi^2} k_F(\vec{r})^3 . \tag{69}$$

Insertion of (68) into (48) and (51) yields for the kinetic and the exchange energy densities

$$\tau_o(\vec{r}) = \frac{1}{10m\pi^2} \; k_F(\vec{r})^5 \tag{70}$$

$$e_x^{(o)}(\vec{r}) = -\frac{e^2}{4\pi^3} \, k_F(\vec{r})^4 \quad . \tag{71}$$

These expressions are formally identical with those of the homogeneous electron gas. The only difference lies in the $\vec{r}$-dependence of the Fermi momemtum. In the final step, we eliminate this unknown local Fermi momemtum in favour of the density by inversion of (69)

$$k_F(\vec{r}) = (\; 3\pi^2 \rho(\vec{r}))^{1/3} \; . \tag{72}$$

Insertion into (70) and (71) yields for the kinetic and the exchange energy densities

$$\tau_o(\vec{r}) = \frac{3}{10m} \, (3\pi^2)^{2/3} \rho(\vec{r})^{5/3} \tag{73}$$

$$e_x^{(o)}(\vec{r}) = -\, e^2 \, \frac{3}{4} \, (\frac{3}{\pi})^{1/3} \; \rho(\vec{r})^{4/3} \; . \tag{74}$$

One particular point should be noted: At the intermediate (semi-classical) level of eq. (69)- (71) we have a turning point problem. The local Fermi momemtum

$$k_F(\vec{r}) = [\; 2m(\varepsilon_F - v_o(\vec{r}))]^{1/2}$$

vanishes on a surface in the asymptotic region and is not defined outside. With the elimination of $k_F(\vec{r})$ in favour of $\rho(\vec{r})$ we have implicitly set up a continuation into the (semiclassically forbidden) outer region.

Next we consider the higher orders of the commutator expansion (56); we stick with the lowest order in the nonlocal operator $\hat{A}$. It is a straightforward matter to calculate the multiple commutators (57) yielding the density matrix in terms of $E_F(\vec{r})$, $\vec{\nabla} E_F(\vec{r})$, $\partial_{ij} E_F(\vec{r})$....
In the final step, the whole expansion is rearranged with respect to orders in the derivative of $E_F(\vec{r})$ where a first order derivative squared counts as a second order term. It is easy to see (though not manifestly visible in our unit system $\hbar = c = 1$ that the order of the

derivatives of E_F are equivalent to powers in $\hbar$. Thus, the commutator expansion (56) is regrouped into an $\hbar$ expansion. All the second order terms in $\hbar$ are contained in the first four terms of the commutator expansion. To this order, we obtain for the particle density and the kinetic and the exchange energy densities

$$\rho(\vec{r}) = \frac{k_F(\vec{r})^3}{3\pi^2} + \frac{1}{24\pi^2}\frac{\Delta k_F^2}{k_F} - \frac{1}{96\pi^2}\frac{(\nabla k_F^2)^2}{k_F^3} \tag{75}$$

$$\tau(\vec{r}) = \frac{k_F(\vec{r})^5}{10m\pi^2} - \frac{1}{48m\pi^2} k_F \nabla k_F^2 - \frac{1}{64m\pi^2}\frac{(\nabla k_F^2)^2}{k_F} \tag{76}$$

$$e_x(\vec{r}) = -\frac{e^2}{4\pi^3} k_F(\vec{r})^4 - \frac{e^2}{576\pi^3}\frac{(\nabla k_F^2)^2}{k_F^2} \quad . \tag{77}$$

In the lowest order, we obtain of course the same expressions as discussed before. Again we have a classical turning point problem at this stage. In the second order terms, this problem is more severe since k_F appears in the denominator, i.e. we have a divergence at the classical turning point. Again this problem can be overcome by elimination of the Fermi momentum in favour of the density: eq. (75) is inverted consistently to second order in h and inserted in (76) and (77). This yields

$$\tau(\vec{r}) = c_1\, \rho(\vec{r})^{5/3} + c_2 \frac{(\nabla\rho)^2}{\rho(\vec{r})} \tag{78}$$

$$e_x(\vec{r}) = -c_3\, \rho(\vec{r})^{4/3} - c_4 \frac{(\nabla\rho)^2}{\rho(\vec{r})^{4/3}} \tag{79}$$

with $c_1 = \frac{3}{10m}(3\pi^2)^{2/3} \qquad c_2 = \frac{1}{72m}$

$$c_3 = e^2\,\frac{3}{4}\left(\frac{3}{\pi}\right)^{1/3} \qquad c_4 = \frac{7e^2}{432\pi(3\pi^2)^{1/3}} \quad .$$

Of course, the results at this stage are not excitingly novel:the lowest order kinetic energy goes back to Thomas [58] and Fermi [59](TF), the lowest order exchange is due to Dirac [60] (D), the second order kinetic energy term has first been derived by von Weizsäcker [61](W) with a different coefficient, and the second order exchange contribution has first been used by Herman et al. [63] in an extended X_α calculation (a rigorous derivation has first been given by Sham [6]). However, the various details of the Kirzhnits method discussed above will be of importance a) in the consistent discussion of relativistic systems (section 5), and b) in the consistent inclusion of nonlocal effects which is addressed in the following.

So far, we have neglected the exchange part A of the HF potential which is represented by the matrix elements

$$< \vec{r}|\hat{A}|\vec{r}\,' > = - e^2 \frac{\rho(\vec{r},\vec{r}\,')}{|\vec{r}-\vec{r}\,'|} \tag{80}$$

In the following, we shall attempt to consistently include all first order contributions in $\hat{A}$. It should be noted that the exchange energy density

$$e_x(\vec{r}) = - \frac{e^2}{2} \int d^3r' \, \rho(\vec{r},\vec{r}\,') \frac{1}{|\vec{r}-\vec{r}\,'|} \rho(\vec{r}\,'\vec{r})$$

$$= \frac{1}{2} \int d^3r' < \vec{r}|\hat{A}|\vec{r}\,' > \rho(\vec{r}\,'\vec{r})$$

is inherently al least of first order in $\hat{A}$. Therefore, the density functional representation (79) of $e_x(\vec{r})$ remains unchanged to first order in $\hat{A}$. However, there will be additional contributions to the kinetic energy. In order to calculate these corrections we first consider the density operator which (to first order in $\hat{A}$) is given [62] by the following expansion

$$\hat{\rho} = \theta(E_F(\vec{r})-\hat{t}-\hat{A}) = \theta(E_F(\vec{r})-\hat{t})$$

$$+ \sum_{n=1}^{\infty} \theta^{(n)}(E_F(\vec{r})-\hat{t}) \frac{(-1)^n}{n!} \underbrace{[(E_F(\vec{r})-\hat{t}),[\ldots,[(E_F(\vec{r})-\hat{t}),\hat{A}]\ldots]]}_{(n-1)\,\text{times}} \tag{81}$$

This expansion will be evaluated consistently to second order in $\hbar$. Therefore, only a finite number of terms has to be calculated. To second order in $\hbar$, the result for the density and the kinetic energy density looks as follows

$$\rho(\vec{r}) = \frac{1}{3\pi^2}\,[k_F(\vec{r})^3 + \frac{3}{\pi}k_F(\vec{r})^2] \tag{82}$$

$$+\frac{1}{\pi^2}\,[\{\frac{1}{24}\frac{\Delta k_F^2}{k_F} - \frac{1}{96}\frac{(\nabla k_F^2)^2}{k_F^3}\} + \frac{1}{\pi}\{-\frac{1}{144}\frac{\Delta k_F^2}{k_F^2} + \frac{1}{288}\frac{(\nabla k_F^2)^2}{k_F^4}\}] \tag{83}$$

$$\tau(\vec{r}) = \frac{1}{10m\pi^2}\,[\,k_F(\vec{r})^5 + \frac{5}{\pi}k_F(\vec{r})^4\,] \tag{84}$$

$$+\frac{1}{m\pi^2}\,[\{-\frac{1}{48}k_F\Delta k_F^2 - \frac{1}{64}\frac{(\nabla k_F^2)^2}{k_F}\} + \frac{1}{\pi}\{-\frac{5}{288}\Delta k_F^2 + \frac{1}{576}\frac{(\nabla k_F^2)^2}{k_F^2}\}] \tag{85}$$

The second expression in each square bracket corresponds to first order in $\hat{A}$. In spite of these additional terms, the consistent elimination to first order in $\hat{A}$ yields the former result (78), i.e. there is no first order contribution in $\hat{A}$ to the kinetic energy density.

It should be added that the existence of a meaningful gradient expansion of the HF limit alone has be questioned [64,65] since a perturbative expansion of the second order exchange term leads to divergent expressions in the higher orders of e^2. This question is left open by our treatment since higher orders in e^2 are not considered. A consistent discussion of higher orders in e^2 on the basis of the Kirzhnits formalism will be attempted elsewhere.

4. Application to atoms and to two-center scattering systems

Once a density functional representation of the ground state energy has been derived within some approximation the next question is how it can be used to describe physical systems. There are essentially three options:

Option I: Derive a variational equation

$$\delta_\rho (E\,[\rho] + V_o \int \rho(\vec{r}) d^3r) = 0 \tag{86}$$

from which the density (and hence the ground state energy and other quantities derived from the density) can be determined. The disadvantage of this option lies in the fact that the usual approximations for the total energy fail to produce quantum mechanical shell structures. A classical example for this failure is the TF model which is obtained by neglecting the exchange-correlation energy and approximating the kinetic energy by the lowest order expression(73). In this case, the variational equation (86) can be transformed into the following well known differential equation for the total electrostatic potential V

$$\Delta V = \sigma \cdot (-V)^{3/2} \quad , \quad \sigma = - \frac{8\sqrt{2}}{3\pi} e^2 m^{3/2} \; . \tag{87}$$

(The Lagrangian multiplier V_o vanishes for neutral systems). The solution of this equation gives a rough picture of atomic and molecular properties which does not contain any shell effects.

Option II: If a more detailed analysis of the system is required we use only the density functional representation of the exchange-correlation energy to establish a set of KS-equations

$$[\hat{t} + \hat{v}_n + \frac{\delta}{\delta\rho} (E_{ee}[\rho] + E_{xc}[\rho])]\ \varphi_i = e_i \varphi_i \qquad (88)$$

$$\rho(\vec{r}) = \sum_i \varphi_i^*(\vec{r})\varphi_i(\vec{r}) \qquad (89)$$

which has to be solved selfconsistently.

One of our aims was to describe atomic scattering processes. In the case of heavy or even superheavy scattering systems the KS scheme would ultimately require a selfconsistent solution for approximately 200 electrons in a two-center set up.This problem is obviously very tedious if tractable at all. Therefore, we suggested a third possibility which is essentially a combination of the first two options:

Option III: By first solving the variational equation (86) we obtain an approximate density of the system. This density is then used to calculate the potential terms $\frac{\delta}{\delta\rho} (E_{ee}[\rho] + E_{xc}[\rho])$. It is hoped, that the subsequent solution of the orbital problem (88), without recourse to the selfconsistency aspect, will provide reasonably consistent results. This scheme was first proposed in 1955 by R. Latter [66] who obtained quite reasonable accuracy for the orbital energies of atoms. Our application to two-centre scattering systems consists of the following steps:

Step 1: Solve the variational equation (86) with two-center boundary conditions for fixed internuclear separation R. As a reasonable density functional representation of the total energy we use the TFDW model which contains the lowest and second order kinetic energy expressions (78) and the zero order exchange energy (74). The effects due to the second order exchange term will be presented elsewhere. [62] The resulting TFDW density $\rho(\vec{r},R)$ is used to contruct an effective single-particle potential by the prescription

$$u_{eff}(\vec{r},R) = e^2 \cdot \begin{cases} -\frac{Z_1}{r_1} - \frac{Z_2}{r_2} + \int \frac{\rho(\vec{r}'R)}{|\vec{r}-\vec{r}'|} d^3r' - [\frac{3}{\pi}\rho(\vec{r},R)]^{1/3} & \text{if } |u_{eff}| > e^2 \cdot \max(\frac{1}{r_1};\frac{1}{r_2}) \\ -\max(\frac{1}{r_1};\frac{1}{r_2}) & \text{if } |u_{eff}| < e^2 \cdot \max(\frac{1}{r_1};\frac{1}{r_2}) \end{cases}$$

(90)

where r_1 and r_2 denote the distance of point $\vec{r}$ from the two nuclei with charges Z_1 and Z_2. This definition includes an asymptotic self-energy correction which replaces the exponential fall off by a $1/r$ - tail.

Step 2: Solve the effective single-particle Schrödinger equation

$$\left(- \frac{\nabla^2}{2m} - u_{eff}(\vec{r},R)\right) \varphi_\nu(\vec{r},R) = \varepsilon_\nu(R)\varphi_\nu(\vec{r},R) \tag{91}$$

for fixed internuclear distance R. That way, one obtains a set of quasimolecular orbitals $\varphi_\nu(\vec{r},R)$and the corresponding single-particle energies $\varepsilon_\nu(R)$. If these energies are plotted in a so-called correlation diagram versus the internuclear separation R one can extract from avoided level crossings (their width and their position) the information which electronic transitions will be dominant during a particular collision process. In order to make contact with experimental data a third step is required:

Step 3: Solve the effective time-dependent Schrödinger equation

$$i \frac{d}{dt} \psi_\alpha(\vec{r},\vec{R}(t),t) = \left(- \frac{\nabla^2}{2m} + u_{eff}(\vec{r},\vec{R}(t))\right)\psi_\alpha(\vec{r},\vec{R}(t),t) \tag{92}$$

by expanding the wave functions ψ_α in the quasimolecular basis obtained in step 2

$$\psi_\alpha(\vec{r},\vec{R}(t),t) = \sum_\nu a_\nu^\alpha(t)\, e^{i\chi_\nu(\vec{r},t)} \varphi_\nu(\vec{r},\vec{R}(t)) \tag{93}$$

with $\chi_\nu(\vec{r},t)$ being a suitably chosen electron translation phase[67,68]. The time-dependent expansion coefficients $a_\nu^\alpha(t)$ then allow the calculation of transition probabilities and cross sections which can be compared with experiment. For slow collisions, the effective potential(90) (used as function of a given classical trajectory $\vec{R}(t)$) will be an acceptable choice. For fast collisions, an effective potential constructed from a time-dependent density functional theory would be more consistent. This was our motivation to study the time-dependent TF model[45-48] and possible extensions.[69]

In the following, we shall give a detailed discussion of numerical results obtained at the various steps : In order to check the accuracy of the TFDW model (step 1) we first take a look at atomic systems.[34,35] . Fig. 1 shows the TFDW density of the Kr atom in comparison to a HF calculation [70] . The TFDW density averages beautifully over the shell structure. For exactly this reason it will only be able to reproduce quantities that do not depend sensitively on shell effects.

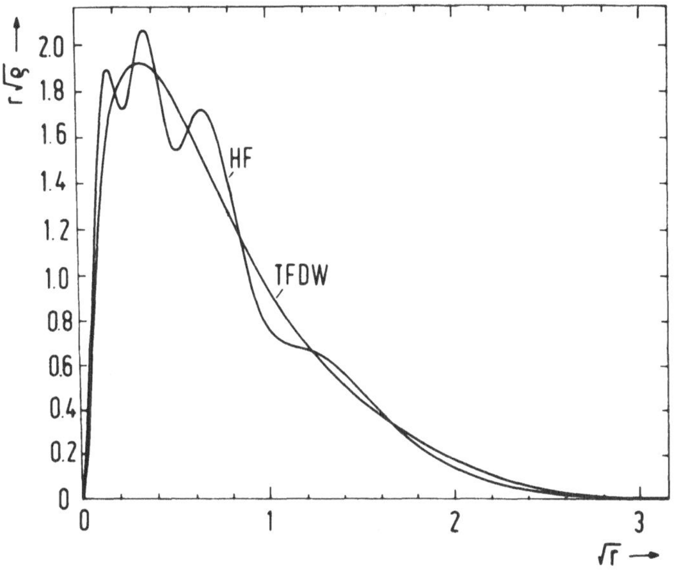

Fig. 1. Density of the Kr atom (all quantities are given in atomic units).

Table 1. Atomic ground state energies in atomic units: Comparison of HF-results (Ref. 74) with TF- and TFDW-results (1) using $c_2 = 1/72$ m, and (2) using $c_2 = 1/40$ m

Systems	$-E_{HF}$	$-E^{(1)}_{TFDW}$	$-E^{(2)}_{TFDW}$	$-E_{TF}$
Ne(10)	128.55	139.89	128.80	165.62
Ar(18)	526.82	561.82	524.75	652.76
Kr(36)	2752.1	2897.2	2744.3	3289.7
Xe(54)	7232.1	7558.7	7309.2	8472.9
Au(79)	17865	18575	17817	20586
Rn(86)	21867	22699	21799	25096
U(92)	25664	26618	25586	29373
Fm(100)	31283	32409	31187	35682
(120)	48203	49838	48069	54578

The total energy is a quantity that does not depend strongly on shell fluctuations. This is illustrated by the atomic binding energies listed in table 1. At this point, a few remarks are in order concerning the coefficient c_2 of the von Weizsäcker term (78). Setting $c_2 = \lambda/8m$, the value resulting from the gradient expansion discussed above is $\lambda = 1/9$ while the original number in von Weizsäcker's work [61] was $\lambda = 1$. If the coefficient is considered in the light of 1/Z- expansions, a different value enters the game: The TF energy of an atom is proportional to $Z^{7/3}$ while the leading correction should be proportional to Z^2. [71,72] According to Lieb [73], the coefficient of the Z^2- term is correctly reproduced within the TFW model if the value is chosen as $\lambda = 0.186$. A careful numerical analysis [35] indicates that atomic TFDW ground state energies agree best with HF values for the choice $\lambda = 0.2$. With this value the agreement is better than 0.4 % over the whole periodic table while the value $\lambda = 1/9$ (resulting from the gradient expansion) yields an agreement with HF [74] between 9 % and 3 % (see table 1). The TF energies, which are listed in the last column, are not satisfactory.

In the discussion of diatomic systems [36-39] we use the results obtained for atoms in the sense that we adopt the same coefficient $\lambda = 0.2$. In view of the universality of the exact HK kinetic energy functional this choice appears most reasonable. In Fig. 2 we show the corresponding TFDW energy as a function of the internuclear separation for the system N-N. The agreement with HF-groundstate energies [75] is better than 1.5 %. The corresponding electronic density is then used to carry through step 2 of our programme consisting of the solution of the single particle eigenvalue problem (91) for each internuclear separation. Fig. 3 shows the corresponding correlation diagram in comparison to HF [75] and to the variable screening model [76] which is based on an interpolation of the atomic HF potentials corresponding to the separated and the united atom limit. The agreement lies within a few percent.

It should be emphasized that within this approach we can with a very reasonable increase in computing time generate as accurate correlation diagrams for heavier systems. Fig. 4 shows the correlation diagram for the 26 electron system O-Ar and in Fig. 5 the correlation diagram for the 70 electron system I-Cl is presented. This system represents a borderline case for the nonrelativistic treatment. For heavier systems relativistic effects have to be taken into account.

Fig. 6 shows the final result of step 3: the total cross section for the 2p - 2s vacancy transition in a Ne^+ - Ne collision is plotted versus the collision energy. The result of the calculation [77,78] compares quite well with experimental data [79,80].

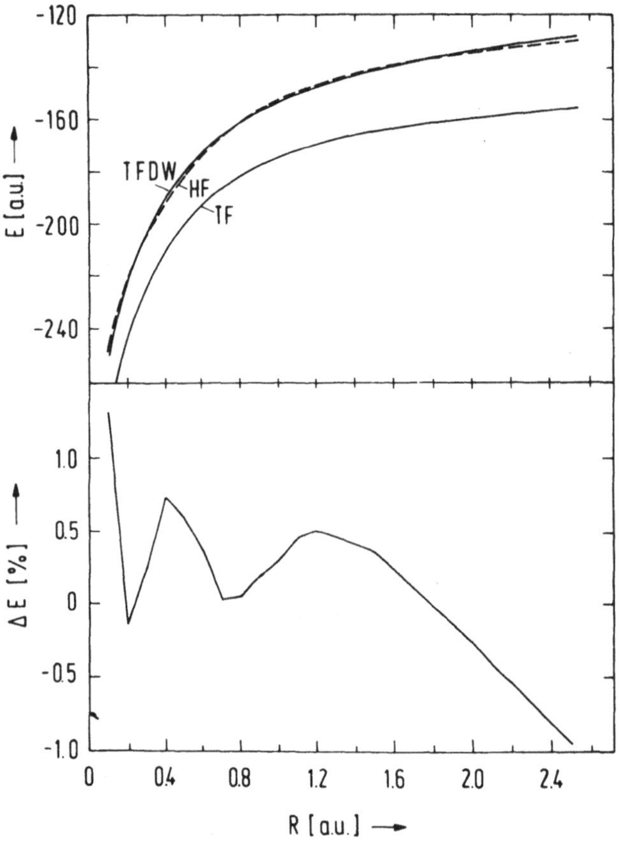

Fig. 2. Total electronic energy of the N-N system as function of the internuclear separation.

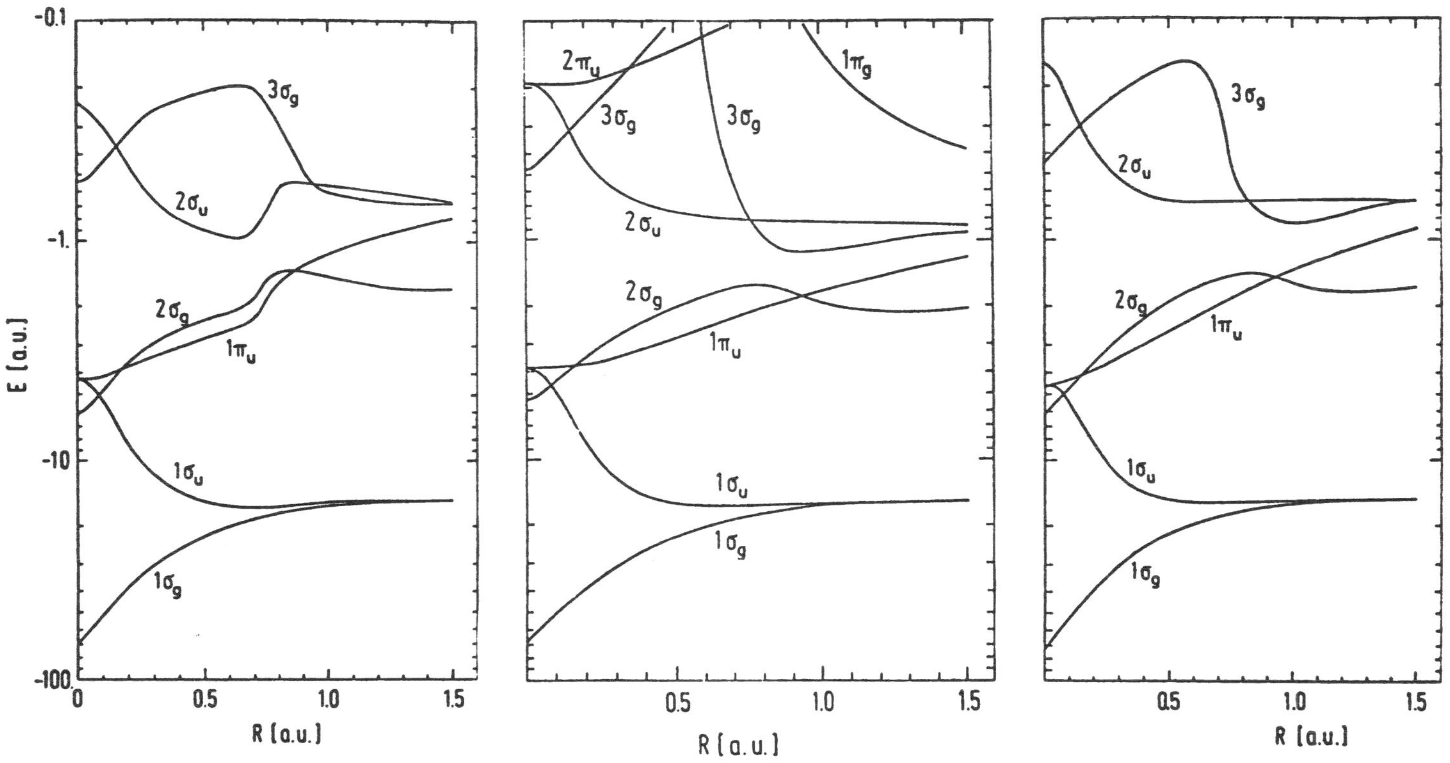

Fig. 3. Correlation diagrams for the N-N system based on different models. From left to right: HF calculation; result of the variable screening model; calculation on the basis of the TFDW approach.

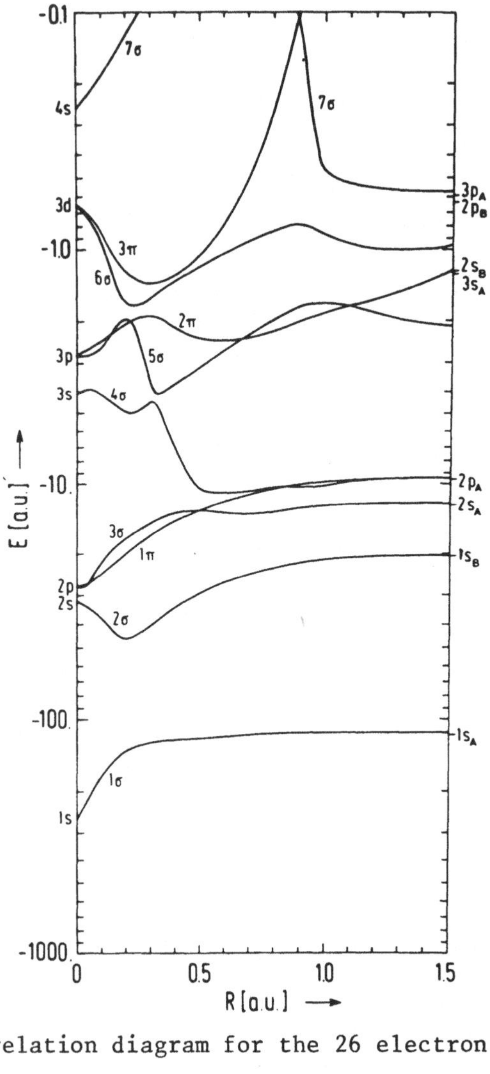

Fig. 4. Correlation diagram for the 26 electron system O-Ar.

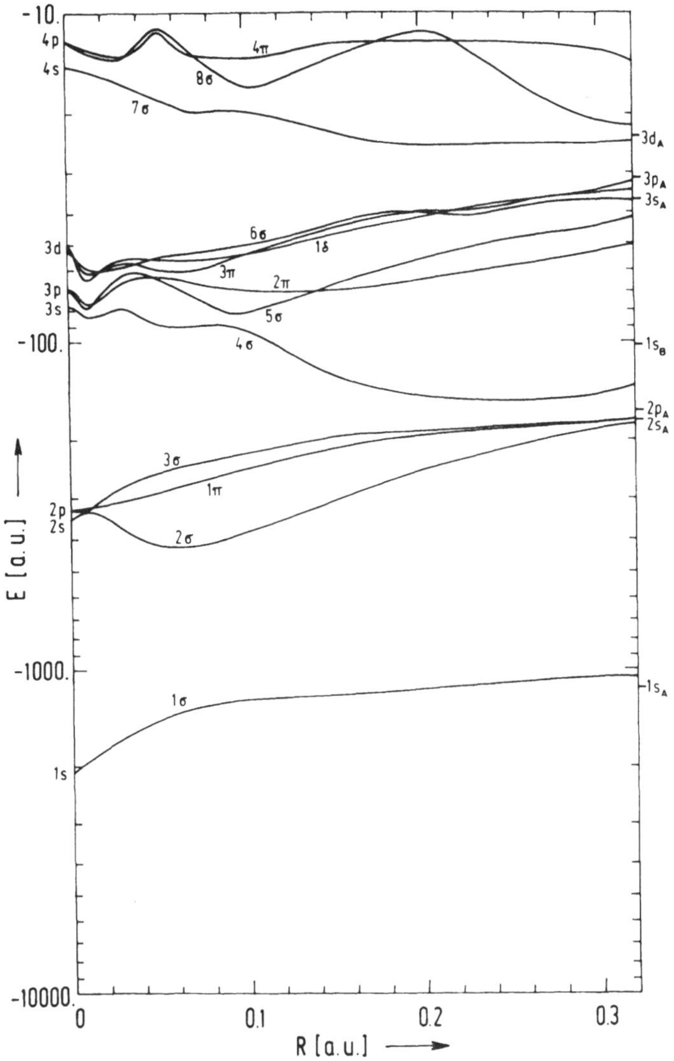

Fig. 5. Correlation diagram for the 70 electron system I-Cl.

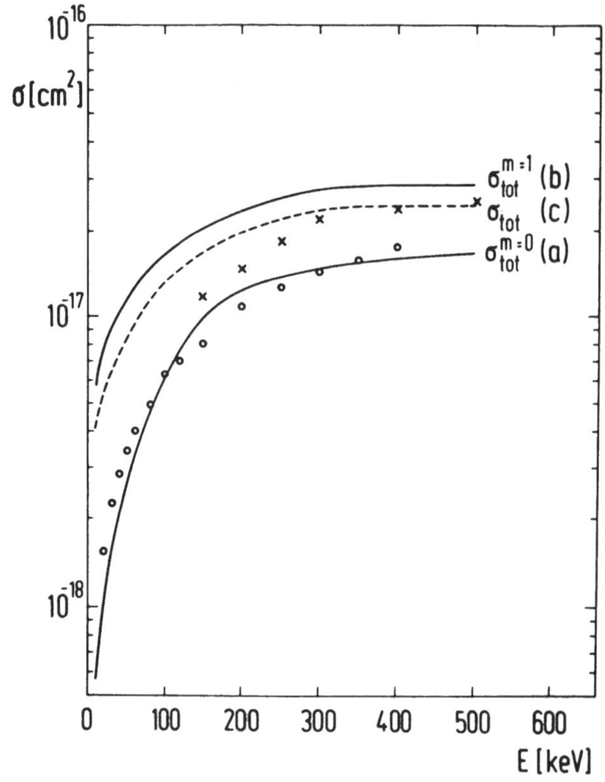

Fig. 6. Total cross section for the 2p-2s vacancy transition in a Ne^+ on Ne collision: (a) calculation with 2p hole initially in m=0 substate; (b) calculation with 2p hole initially in m=1 substate; (c) statistical superposition of (a) and (b): $\sigma_{tot} = \frac{1}{3}\sigma_{tot}^{m=o} + \frac{2}{3}\sigma_{tot}^{m=1}$; (o) experimental data from Ref. 79; (x) experimental data from Ref. 80.

5. Relativistic systems

In the discussion of relativistic systems, the main interest has been focussed on the investigation of higher order effects of the electron-electron interaction such as the Breit or the full transverse interaction. These terms yield additional contributions to the relativistic exchange-correlation potential. Their relative importance has been studied in great detail with relativistic KS calculations for atoms.[81,82] However, the approximations for the relativistic exchange-correlation energy are usually based on the homogeneous electron gas, i.e. the expressions for the homogeneous electron gas are used with a local density. In the following section, the main emphasis will be on inhomogeneity corrections. It will turn out that the use of homogeneous electron gas expressions with a local density is in general not sufficient for inhomogeneous systems: A fully relativistic treatment requires additional terms even in the lowest order, i.e. there will be nongradient corrections due to the inhomogeneity of the system.

In the following, we first give a brief review of the history of relativistic density functionals beginning in 1932 with the derivation of the relativistic TF (RTF) model.[83] After a short discussion of the various attempts to cure the diseases of the RTF model we shall derive a second order gradient correction of the kinetic energy density within the Foldy Wouthuysen (FW) approximation. This model turns out to give satisfactory results for weakly relativistic systems up to $Z \approx 90$. Finally a fully relativistic gradient expansion on the basis of the Dirac equation is discussed.

The RTF density is calculated in analogy to the nonrelativistic case from

$$\rho_o(\vec{r}) = \frac{2}{(2\pi)^3}\int d^3k\ \theta(E_F(\vec{r}) - E_k) \tag{94}$$

$$\text{with } E_k = [k^2 + m^2]^{1/2} \tag{95}$$

$$\text{and } E_F(\vec{r}) = \varepsilon_F - v_{eff}(\vec{r})\ . \tag{96}$$

This yields

$$\rho_o(\vec{r}) = \frac{1}{3\pi^2} \; k_F(\vec{r})^3 \tag{97}$$

$$\text{with } k_F(\vec{r}) = [\; E_F(\vec{r})^2 - m^2]^{1/2} \tag{98}$$

The kinetic energy density has to be calculated with the relativistic formula

$$\tau_o(\vec{r}) = \frac{2}{(2\pi)^3} \int d^3k \; \theta(E_F(\vec{r}) - E_k)E_k - m\rho(\vec{r})$$

$$= \frac{m^4}{\pi^2} \left(\frac{1}{8}[x(2x^2 + 1) \sqrt{x^2+1} - \text{Arsh } x] - \frac{1}{3} x^3 \right) \tag{99}$$

with $x = k_F/m$.

The elimination of k_F corresponds to the replacement

$$x = \frac{(3\pi^2)^{1/3}}{m} \quad \rho_o(\vec{r}).$$

The variational equation (86) corresponds to a functional relation between the total electrostatic potential V_{rel} and the density ρ. If this relation is inserted into Poisson's equation one obtains for neutral systems (i.e. $V_o = 0$)

$$\Delta V_{rel} = \sigma\cdot(-V_{rel})^{3/2} \; [1 - \frac{V_{rel}}{2m}]^{3/2} \;, \; \sigma = -\frac{8\sqrt{2}}{3\pi} e^2 m^{3/2}. \tag{100}$$

This nonlinear differential equation for the determination of V_{rel} (and hence ρ) contains, in comparison to the nonrelativistic TF equation (87), the additional factor $[\; 1 - \frac{V_{rel}}{2m}\;]^{3/2}$. This relativistic correction shows some reasonable features: It increases with increasing Z and it contributes particularly for small values of r, i.e. for the innermost electrons. On the other hand, however, it overestimates the relativistic effects: Since the right hand sides of equation (87) and (100) are proportional to the density we obtain in the vicinity of the atomic nucleus

$$\rho_{TF} \quad \underset{r \to 0}{} \quad r^{-3/2}$$

$$\rho_{RTF} \quad \underset{r \to 0}{} \quad r^{-3} \quad .$$

i.e. the RTF density is not integrable. It can in fact be shown that the RTF equation (100) does not possess solutions for pointlike nuclear boundary conditions.[36]

There is a very hesitating history of attempts to correct this deficiency: The simplest remedy already suggested by Jensen[84] in 1933 is the introduction of an extended nucleus instead of the point source. The problem is however: The nuclear dimensions are so small on the atomic scale that the incorrect $1/r^3$ behavior sets in before the cut-off procedure takes over.

Therefore, Ashby and Holzman[85] introduced a hybrid model by using simple relativistic K-shell orbitals to describe the inner part of the density distribution and matched the TF density at a distance r_0 from the point charge. As a condition to determine the point r_0 they required the kinetic energy functional to be continuous at this point. It turns out that r_0 moves inward with increasing Z so that no solution could be found for Z>87.

A look at the nonrelativistic case suggests that the deficiency can be remedied by considering the gradient terms. The first suggestion along these lines is due to Rudkjøbing[86] who proposed an old version of the once iterated Dirac equation as a starting point. In this equation the spin orbit term is made explicit and contributes to the density of states. The resulting correction for the variational equation is then

$$\Delta V_{rel} = - \frac{4e^2}{3\pi} \left[(m-V_{rel})^2 - m^2 - \left(r \frac{dV_{rel}}{dr} \right)^2 \right]^{3/2} \quad .$$

If the spin orbit term is neglected, this equation reduces to the RTF equation (100). In the limit $r \to 0$ the right hand side (i.e. the density) is proportional to $r^{-3/2}$. Unfortunately the solution does not only behave like the nonrelativistic solution at the origin, it agrees with the classical solution practically completely.

As long as the system in question may be regarded as weakly relativistic, it is possible to set up a gradient expansion [42] in perfect analogy to the nonrelativistic formalism. The basic idea is:

a) to represent the relativistic density operator as in the nonrelativistic case by

$$\hat{\rho} = \theta(\varepsilon_F - \hat{h}_{eff})$$

and

b) to use the Foldy-Wouthuysen representation of $\hat{h}_{eff}$ to second order in (p/m)

$$\hat{h}_{eff} = \beta[\hat{p}^2 + m^2]^{1/2} + \hat{v}_{eff} + \frac{1}{4m^2}\vec{\sigma}\cdot(\vec{\nabla}v_{eff})\times\hat{\vec{p}} + \frac{1}{8m^2}\Delta v_{eff}.$$

If the gradient expansion is carried through on the basis of this approximation the final result for the kinetic energy density looks as follows:

$$\tau_{FW}(\vec{r}) = \tau_o(\vec{r}) + \frac{1}{72m}\frac{(\nabla\rho)^2}{\rho(\vec{r})}B_{FW}(x) \qquad (101)$$

$$\text{with } B_{FW}(x) = \frac{1}{\sqrt{1+x^2}}\left[1 + \frac{x^4}{2} - \frac{x^2}{1+x^2} + \frac{2}{3}\frac{x^4}{(1+x^2)^2}\right] \qquad (102)$$

$$\text{and} \quad x = \frac{(3\pi^2)^{1/3}}{m}\rho(\vec{r})^{1/3} .$$

$\tau_o(\vec{r})$ is given by the lowest order expression (99). For small densities $x \ll 1$ one finds $B(x) \to 1$ which is the proper nonrelativistic limit. Some results obtained on the basis of this density functional are given in Table 2. The calculations are for neutral atomic systems. One notices (for the corrected coefficient $c_2 = 1/40m$) that the agreement with Dirac-Fock results is not as good as in the nonrelativistic case for light to medium atoms, but still acceptable as a first attempt. For heavier systems ($Z \gtrsim 90$) the deviation increases fairly rapidly, i.e. the weakly relativistic approximation inherent in the second order FW representation is not sufficient. Thus, we finally have to face a fully relativistic treatment.

Table 2. Atomic ground state energies in atomic units: Comparison of Hartree-Fock(HF) and Dirac-Fock(DF) values (both from Ref. 74) with results based on relativistic extensions of the TFDW-model; (1): FW-version, (2) : simplified fully relativistic version

Z	$-E_{HF}$	$-E_{DF}$	$-E^{(1)}_{RTFDW}$	$-E^{(2)}_{RTFDW}$
10	128.55	128.67	129.60	128.53
36	2752.1	2787.3	2862.0	2736.0
70	13392	14052	14202	13608
92	25664	28011	27363	26739
110	39225	44950	42012	42071
120	48203	57387	51773	52742

As in the nonrelativistic case, the starting point is the total groundstate energy within the single particle approximation (i.e. we neglect the correlation energy):

$$E^{(rel)} = \mathrm{tr}\ [\ \int d^3r \int d^3r'\ \delta(\vec{r}-\vec{r}')(-i\vec{\alpha}\cdot\vec{\nabla}_r+\beta m)\rho(\vec{r},\vec{r}')$$

$$+ \int d^3r\rho(\vec{r},\vec{r})\ v_n(\vec{r})$$

$$+ \frac{1}{2} \int d^3r \int d^3r'\rho(\vec{r},\vec{r})V(\vec{r},\vec{r}')\rho(\vec{r}',\vec{r}')$$

$$- \frac{1}{2} \int d^3r \int d^3r'\rho(\vec{r},\vec{r}')V(\vec{r},\vec{r}')\rho(\vec{r}',\vec{r})]\ . \qquad (103)$$

Again, $v_n(\vec{r})$ is the nuclear Coulomb potential. The particle-particle interaction is given by the series

$$V(\vec{r},\vec{r}') = \frac{e^2}{|\vec{r}-\vec{r}'|} - \frac{e^2}{2}\left[\frac{\vec{\alpha}\cdot\vec{\alpha}'}{|\vec{r}-\vec{r}'|} + \frac{\vec{\alpha}\cdot(\vec{r}-\vec{r}')\vec{\alpha}'\cdot(\vec{r}-\vec{r}')}{|\vec{r}-\vec{r}'|^3}\right] + \ldots$$

In the following we restrict ourselves to the lowest order, i.e. the nonrelativistic Coulomb interaction. The inclusion of higher orders, as e.g. the Breit interaction, does not pose principle difficulties.

The density matrix

$$\rho(\vec{r},\vec{r}') = \sum_{\substack{\text{occ.bound}\\ \text{states}}} \varphi_\nu(\vec{r})\ \varphi_\nu(\vec{r}')^+ \qquad (104)$$

consists of the above sum over the occupied bound states of the corresponding Dirac-Fock variational equation

$$\hat{h}_{eff}\ \varphi_\nu(\vec{r}) = \varepsilon_\nu\ \varphi_\nu(\vec{r})\ ,\ \hat{h}_{eff} = \hat{\vec{\alpha}}\cdot\hat{\vec{p}}+\hat{\beta}m+\hat{v}_{eff}\ . \qquad (105)$$

It should be noted that each element $\rho(\vec{r},\vec{r}')$of the density matrix (104), by definition, is a 4x4 matrix. Therefore, the groundstate energy (103) as well as the single particle density are obtained by taking the trace

$$\rho(\vec{r}) = \mathrm{tr}\, \rho(\vec{r},\vec{r}) \ . \qquad (106)$$

In order to approximate the kinetic energy and the exchange energy by functionals of the one-particle density (106) alone, we have to represent the density matrix in terms of an appropriately chosen density operator. The question is how this operator looks like in the relativistic case.

For nonrelativistic systems, the ground state occupation numbers of the HF single-particle states are given by $n_\nu = \theta(\varepsilon_F - \varepsilon_\nu)$ (see Fig. 7). This leads to the step function representation $\hat{\rho} = \theta(\varepsilon_F - \hat{h}_{eff})$ of the density operator. For relativistic systems, the single-particle spinors $\{\varphi_\nu(\vec{r})\}$ obtained from (105) consist of positive continuum states, bound states, and negative continuum states. If all levels with energy less than ε_F are considered to be occupied, i.e. if the nonrelativistic occupation numbers $n_\nu = \theta(\varepsilon_F - \varepsilon_\nu)$ are used (see Fig. 8), then the density matrix and all other quantities of interest will be divergent due to the infinite contribution of the negative continuum.

The correct HF density matrix as defined in (104) contains only the occupied <u>bound</u> states. This convention corresponds of course to the usual definition of the vacuum state for this situation. If the density matrix is to be reproduced correctly we have to extract a window from the relativistic spectrum which contains only the occupied bound states with energy less than ε_F (see Fig. 9). This is most easily achieved by a difference of two step functions

$$n_\nu = \theta(\varepsilon_F - \varepsilon_\nu) - \theta(-m - \varepsilon_\nu) \ .$$

From this, a representation of the relativistic density operator is readily obtained

$$\langle \vec{r}|\rho_{rel}|\vec{r}\,'\rangle = \sum_{\substack{\text{occ. bound}\\ \text{states}}} \varphi_\nu(\vec{r})\varphi_\nu(\vec{r}\,')^+$$

$$= \sum\!\!\!\!\!\!\int_\nu \, [\theta(\varepsilon_F - \varepsilon_\nu) - \theta(-m - \varepsilon_\nu)] \ \langle \vec{r}|\nu\rangle\langle\nu|\vec{r}\,'\rangle$$

$$= \langle \vec{r}| \ [\theta(\varepsilon_F - \hat{h}_{eff}) - \theta(-m - \hat{h}_{eff})] \ \sum\!\!\!\!\!\!\int_\nu |\nu\rangle\langle\nu|\vec{r}\,'\rangle \ .$$

Using the completeness of the Dirac-Fock single-particle basis $|\nu\rangle$ we obtain

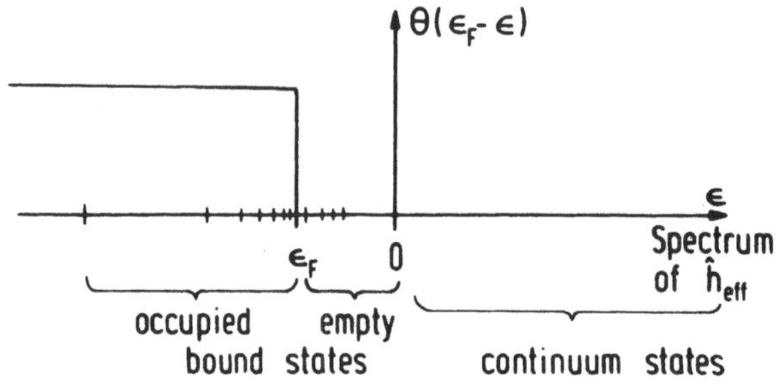

Fig. 7. Ground state occupation numbers for a nonrelativistic system.

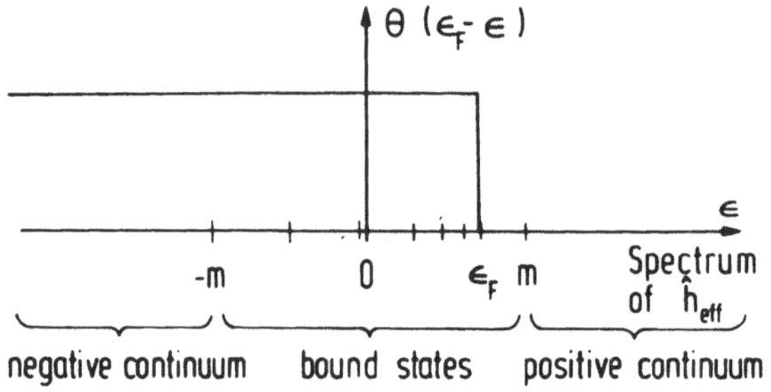

Fig. 8. Same occupation numbers as in Fig. 7 for a relativistic single-particle spectrum (leading to divergent contributions due to the negative continuum).

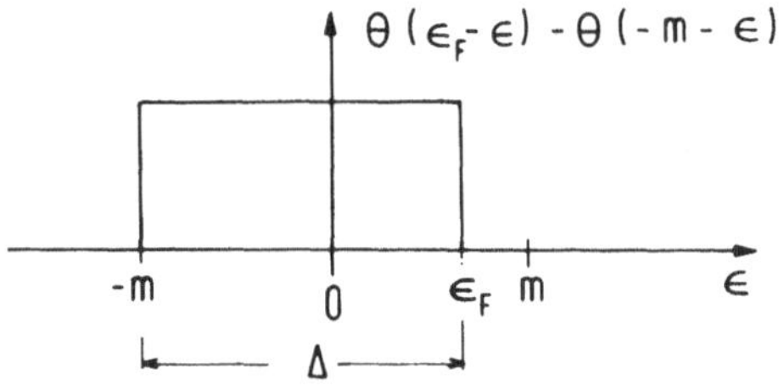

Fig. 9. Correct representation of the ground state occupation numbers for a relativistic system.

$$\hat{\rho}_{rel} = \theta(\varepsilon_F - \hat{h}_{eff}) - \theta(-m - \hat{h}_{eff}) \ . \tag{107}$$

For convencience we define

$$\hat{E}_F := \varepsilon_F - \hat{v}_{eff} \tag{108}$$

$$\hat{G}_F := -m - \hat{v}_{eff} \tag{109}$$

which leads to the final expression

$$\hat{\rho}_{rel} = \theta(\hat{E}_F - \hat{t}) - \theta(\hat{G}_F - \hat{t}) \tag{110}$$

with $\hat{t}$ being the Dirac operator

$$\hat{t} = \hat{\vec{\alpha}} \cdot \hat{\vec{p}} + \hat{\beta} m \ . \tag{111}$$

It should be pointed out that the difference between the operators $\hat{E}_F$ and $\hat{G}_F$ is a constant Δ(namely the width of the window) which amounts to approximately 2m.

$$\hat{E}_F - \hat{G}_F = \varepsilon_F + m =: \Delta(\approx 2m) \tag{112}$$

As a simple example let us now calculate the kinetic energy of a noninteracting homogeneous electron gas. In that case, the potential vanishes $v_{eff} \equiv 0$, and we have $\hat{E}_F = \varepsilon_F$ and $\hat{G}_F = -m$. In order to evaluate the density matrix

$$\rho(\vec{r},\vec{r}\,') = \langle \vec{r}|\theta(\varepsilon_F - \hat{t})|\vec{r}\,'\rangle - \langle \vec{r}|\theta(-m-\hat{t})|\vec{r}\,'\rangle \tag{113}$$

we insert the complete set of plane wave eigenfunctions of the Dirac operator

$$\hat{t}\ |\vec{k}\ s\ T\rangle = \varepsilon_{kT}|\vec{k}\ s\ T\rangle \ .$$

The standard representation in configuration space is given by

$$<\vec{r}|\vec{k}sT> = \frac{1}{(2\pi)^{3/2}} u_s^T(\vec{k}) \, e^{i\,\vec{k}\cdot\vec{r}} \tag{114}$$

$$\text{with } u_s^{(+)}(\vec{k}) = \left[\frac{E_k+m}{2E_k}\right]^{1/2} \begin{pmatrix} \chi_s \\ \frac{\vec{\sigma}\cdot\vec{k}}{E_k+m}\chi_s \end{pmatrix}$$

$$\text{and } \quad u_s^{(-)}(\vec{k}) = \left[\frac{E_k+m}{2E_k}\right]^{1/2} \begin{pmatrix} \frac{-\vec{\sigma}\cdot\vec{k}}{E_k+m}\chi_s \\ \chi_s \end{pmatrix}$$

The index $s=\pm 1/2$ denotes the spin, while $T=(+),(-)$ classifies positive and negative energy solutions:

$$\varepsilon_{kT} = \begin{cases} +E_k & \text{for } T = (+) \\ -E_k & \text{for } T = (-) \end{cases} \qquad \text{with } E_k = [k^2 + m^2]^{1/2} \quad . \tag{115}$$

This yields for the density matrix

$$\rho(\vec{r},\vec{r}\,') = \sum_{sT} \int \frac{d^3k}{(2\pi)^3} < \vec{r}|\theta(\varepsilon_F-\varepsilon_{kT})|\vec{k}sT> <\vec{k}sT|\vec{r}\,' >$$

$$- \sum_{sT} \int \frac{d^3k}{(2\pi)^3} < \vec{r}|\theta(-m-\varepsilon_{kT})|\vec{k}sT> <\vec{k}sT|\vec{r}\,' > \quad .$$

Writing out explicitly the positive and the negative energy contributions we obtain by summation over spin variables

$$\rho(\vec{r},\vec{r}\,') = \int \frac{d^3k}{(2\pi)^3}\, \theta(\varepsilon_F - E_k)\, e^{i\vec{k}\cdot(\vec{r}-\vec{r}\,')}\, M_o^{(+)}(\vec{k})$$

$$+ \int \frac{d^3k}{(2\pi)^3}\, \theta(\varepsilon_F + E_k)\, e^{i\vec{k}\cdot(\vec{r}-\vec{r}\,')}\, M_o^{(-)}(\vec{k})$$

$$- \int \frac{d^3k}{(2\pi)^3}\, \theta(-m + E_k)\, e^{i\vec{k}\cdot(\vec{r}-\vec{r}\,')}\, M_o^{(-)}(\vec{k})$$

$$- \int \frac{d^3k}{(2\pi)^3}\, \theta(-m - E_k)\, e^{i\vec{k}\cdot(\vec{r}-\vec{r}\,')}\, M_o^{(+)}(\vec{k}) \qquad (116)$$

$$\text{with } M_o^{(\pm)}(\vec{k}) = \frac{1}{2E_k} \begin{pmatrix} (E_k \pm m) & \pm\vec{\sigma}\cdot\vec{k} \\ \pm\vec{\sigma}\cdot\vec{k} & (E_k + m) \end{pmatrix}$$

In the first term, the step function leads to an upper boundary $k_F = \sqrt{\varepsilon_F^2 - m^2}$ for the momentum integration. The second term is clearly divergent since, for negative energy solutions, the step function retains the value 1 everywhere. However, this infinite contribution is exactly cancelled by the third term.The last term of (116) vanishes since $(-m-E_k) \leq 0$ for all $\vec{k}$. Thus only the first term remains

$$\rho(\vec{r},\vec{r}\,') = \frac{1}{(2\pi)^3} \int_0^{k_F} k^2 dk \int d\Omega_k\, e^{i\vec{k}\cdot(\vec{r}-\vec{r}\,')}\, M_o^{(+)}(\vec{k}). \qquad (117)$$

This leads to the common expression for the density

$$\rho(\vec{r}) = \mathrm{tr}\, \rho(\vec{r},\vec{r}) = \frac{1}{3\pi^2}\, k_F^3 \qquad (118)$$

and to the former result (99) for the kinetic energy density.

If we consider a homogeneous interacting electron gas within HF theory, then the plane wave spinors remain eigenfunctions of the single-particle Hamiltonian and we can therefore use the same density matrix (116) to calculate the exchange energy. The result, which was first given by Jancovici [87] and Salpeter [88], looks as follows

$$e_x^{(0)}(\vec{r}) = -\frac{e^2m^4}{(2\pi)^3} \cdot \left[x^4 + \frac{2}{3}x^2E_x^2 - \frac{2}{3}E_x^4 \ln(E_x^2) + \frac{4}{3}x^3E_x \mathrm{Arsh}\, x - (xE_x - \mathrm{Arsh}\, x)^2 \right] \qquad (119)$$

with $E_x = \sqrt{x^2+1}$.

Now let us consider an inhomogeneous system by means of the gradient expansion. Again, the effects due to the nonlocal part (80) of the HF potential will be neglected to lowest order, i.e. we assume a local representation of the operators $\hat{E}_F$ and $\hat{G}_F$:

$$E_F(\vec{r}) = \varepsilon_F - v_o(\vec{r}) \qquad (120)$$

$$G_F(\vec{r}) = -m - v_o(\vec{r}) \; . \qquad (121)$$

The gradient expansion of the relativistic density matrix is again obtained by insertion of the complete system of free Dirac spinors (114)

$$\rho(\vec{r},\vec{r}') = \sum_{sT} \int d^3k \, \langle\vec{r}|\theta(E_F(\vec{r})-\hat{t})|\vec{k}sT\rangle\langle\vec{k}sT|\vec{r}'\rangle - \sum_{sT} \int d^3k \, \langle\vec{r}|\theta(G_F(\vec{r})-\hat{t})|\vec{k}sT\rangle\langle\vec{k}sT|\vec{r}'\rangle \; . \qquad (122)$$

If the mathematical theorem (56) is applied to both expressions we obtain two formally identical series, one depending on $E_F(\vec{r})$, the other depending on $G_F(\vec{r})$. To lowest order in $\hbar$ this yields

$$\rho_0(\vec{r},\vec{r}') = \int \frac{d^3k}{(2\pi)^3}\, \theta(E_F(\vec{r})-E_k)\; e^{i\vec{k}\cdot(\vec{r}-\vec{r}')}\; M_0^{(+)}(\vec{k})$$

$$+ \int \frac{d^3k}{(2\pi)^3}\, \theta(E_F(\vec{r})+E_k)\; e^{i\vec{k}\cdot(\vec{r}-\vec{r}')}\; M_0^{(-)}(\vec{k})$$

$$- \int \frac{d^3k}{(2\pi)^3}\, \theta(G_F(\vec{r})+E_k)\; e^{i\vec{k}\cdot(\vec{r}-\vec{r}')}\; M_0^{(-)}(\vec{k})$$

$$- \int \frac{d^3k}{(2\pi)^3}\, \theta(G_F(\vec{r})-E_k)\; e^{i\vec{k}\cdot(\vec{r}-\vec{r}')}\; M_0^{(+)}(\vec{k}) \; . \qquad (123)$$

These 4 terms are formally identical with the homogeneous case(116). Again, the step function of the first term yields an upper boundary

$$k_F(\vec{r}) = \sqrt{E_F(\vec{r})^2 - m^2} \qquad (124)$$

for the momentum integration which can be considered as a local Fermi momentum. The next two contributions, both divergent, cancel each other. The step function in the last term again yields an upper boundary

$$q_F(\vec{r}) = \sqrt{G_F(\vec{r})^2 - m^2} \qquad (125)$$

for the momentum integration. That way, we obtain for the density matrix

$$\rho_0(\vec{r},\vec{r}') = \frac{1}{(2\pi)^3} \int_0^{k_F(\vec{r})} k^2 dk \int d\Omega_k\; e^{i\vec{k}\cdot(\vec{r}-\vec{r}')}\; M_0^{(+)}(\vec{k})$$

$$- \frac{1}{(2\pi)^3} \int_0^{q_F(\vec{r})} k^2 dk \int d\Omega_k\; e^{i\vec{k}\cdot(\vec{r}-\vec{r}')}\; M_0^{(+)}(\vec{k}) \qquad (126)$$

and for the density

$$\rho_0(\vec{r}) = \mathrm{tr}\,\rho_0(\vec{r},\vec{r}) = \frac{1}{3\pi^2} k_F(\vec{r})^3 - \frac{1}{3\pi^2} q_F(\vec{r})^3 \ . \tag{127}$$

In the case of the homogeneous electron gas (see eq.(118)) the second term did not occur since we had $G_F = -m$ and therefore $q_F = 0$. However, in the inhomogeneous case we obtain a nonvanishing contribution if $G_F{}^2 > m^2$. In terms of the potential (see eq.(121)) this means

$$v_0(\vec{r}) < -2m \ . \tag{128}$$

Thus, the additional contribution occurs in the highly relativistic region where the potential becomes more attractive than $-2m$. An estimate of the critical radius r_0 which characterizes the inset of the highly relativistic correction can be obtained if the effective atomic potential is approximated by the naked Coulomb potential $v_0(\vec{r}) \approx -Ze^2/r$. Table 3 shows r_0 in comparison to the position of the maximum of the $1s_{1/2}$ radial probability density.[74] For large values of Z the two radii become comparable.

It should be noted that $k_F(\vec{r})$ and $q_F(\vec{r})$ are not independent. Using the fundamental connection

$$G_F = E_F - \Delta = \sqrt{k_F^2 + m^2} - \Delta$$

we obtain for $v_0 < -2m$

$$q_F(k_F) = [\ (\sqrt{k_F^2 + m^2} - \Delta)^2 - m^2\ ]^{1/2} \ . \tag{129}$$

By means of (120) the condition $v_0 < -2m$ can be transformed into

$$k_F/m > \delta \tag{130}$$

with

$$\delta = [(1 + \Delta/m)^2 - 1]^{1/2} \simeq \sqrt{8} \ . \tag{131}$$

That way, the zero order density is expressed exclusively in terms of the local Fermi momentum $k_F(\vec{r})$:

Table 3. The critical radius r_o (pure Coulomb estimate calculated from $Ze^2/r_o = 2m$) in comparison to r_{max}, the radius of maximum radial density of the $1s_{1/2}$ orbital

Z	r_o[a.u.]	r_{max}[a.u.]
10	.0003	.1028
36	.0010	.0271
54	.0014	.0172
92	.0024	.0081
120	.0032	.0043

$$\rho_o(\vec{r}) = \rho_1(k_F(\vec{r})) - \rho_2(k_F(\vec{r})) \tag{132}$$

$$\text{with } \rho_1(k_F) = \frac{1}{3\pi^2} k_F^3 \tag{133}$$

$$\text{and } \rho_2(k_F) = \begin{cases} \frac{1}{3\pi^2} q_F(k_F)^3 & \text{if } k_F/m > \delta \\ 0 & \text{otherwise} \end{cases} \quad . \tag{134}$$

For the kinetic energy density we obtain twice the classical RTF-expression (99):

$$\tau_o(\vec{r}) = \tau_o(x) - \tau_o(y) \tag{135}$$

where the first term depends on ρ_1 via

$$x := k_F/m = (3\pi^2\rho_1)^{1/3}/m \tag{136}$$

while the second term depends on ρ_2 via

$$y := q_F/m = (3\pi^2\rho_2)^{1/3}/m \ . \tag{137}$$

The calculation of the exchange energy density is more complicated because a product of two density matrices and an additional integration have to be carried through (see eq. (51)). The final result in terms of (136) and (137) looks as follows:

$$\begin{aligned} e_x^{(0)}(\vec{r}) = -\frac{e^2m^4}{(2\pi)^3} \cdot \Bigg[& (x-y)(x^3-y^3) + \frac{2}{3}(xE_x-yE_y)^2 \\ & - \frac{2}{3} E_x^4 \ln(E_x^2) + \frac{2}{3}(E_x^4+E_y^4) \ln \left[\frac{E_xE_y+1+xy}{E_xE_y+1-xy} \right] - \frac{2}{3} E_y^4 \ln(E_y^2) \\ & - [(x^2-y^2)^2 + \frac{1}{6}(E_x-E_y)^4] \ln\left(\frac{x-y}{x+y}\right) \\ & + \frac{4}{3}(x^3E_x-y^3E_y)(\text{Arsh } x - \text{Arsh } y) \\ & - ([xE_x-yE_y] - [\text{Arsh } x - \text{Arsh } y])^2 \Bigg] \end{aligned} \tag{138}$$

with $E_x = \sqrt{x^2+1}$ and $E_y = \sqrt{y^2+1}$.

Now let us discuss some consequences of these results: First of all, it should be pointed out that the expressions for the density and for the kinetic and the exchange energy densities reduce to the homogeneous electron gas expressions (118), (99), (119) in the weakly relativistic domain characterized by $\rho_2 = 0$. Thus, for weakly relativistic systems (in the sense $v_o > -2m$) the use of the homogeneous electron gas expressions with a local density is perfectly justified.

If the potential becomes more attractive than $-2m$ the relativistic correction ρ_2 enters the game. With increasing k_F the correction becomes more and more important; and in the highly relativistic limit $k_F \to \infty$ ρ_2 even approaches ρ_1 since

$$q_F(k_F) = [(\sqrt{k_F^2 + m^2} - \Delta)^2 - m^2]^{1/2} \xrightarrow[k_F \to \infty]{} k_F . \tag{139}$$

For this reason, the leading terms cancel in the expressions for $\rho_o(\vec{r})$, $\tau_o(\vec{r})$, and $e_x^{(o)}(\vec{r})$. The density, e.g., becomes proportional to k_F^2 in the highly relativistic limit:

$$\rho_o = \frac{1}{3\pi^2} (k_F^3 - q_F(k_F)^3) \underset{k_F \to \infty}{\sim} k_F^2 . \tag{140}$$

An interesting consequence of this fact shows up in a corrected RTF equation which is derived from Poisson's equation

$$\Delta v = -4\pi \rho_o(v) \tag{141}$$

by insertion of the semiclassical expression

$$\rho_o(v) = \frac{1}{3\pi^2} (k_F(v)^3 - q_F(k_F(v))^3) . \tag{142}$$

Using

$$k_F(v) = [(\varepsilon_F - v)^2 - m^2]^{1/2} \xrightarrow[v \to \infty]{} |v|$$

one obtains from (140) in the vicinity of the atomic nucleus

$$\rho_o(v) \underset{v \to \infty}{\sim} v^2 \underset{r \to o}{\sim} r^{-2} \tag{143}$$

For a pointlike nucleus, the density is still divergent but, in contrast to the traditional RTF model (see discussion following eq. (100)), the density is at least normalizable. Thus, the correct definition of the relativistic density operator automatically cures the deficiency of the traditional RTF model. A schematic plot of the corrected RTF density is shown on Fig. 10.

So far, we have discussed only the semiclassical level, i.e. the dependence of ρ, τ, and e_x on the local Fermi momentum $k_F(\vec{r})$. In the final step, the transition to the corresponding density functionals is performed by eliminating the Fermi momentum in favor of the density. As long as ρ_2 vanishes, i.e. in the weakly relativistic domain, we simply make the usual replacement $k_F=(3\pi^2\rho)^{1/3}$. This is possible up to a critical density

$$\rho_{crit} = \frac{1}{3\pi^2}(m\delta)^3 \simeq 1.97 \times 10^6 \text{ a.u.} \tag{144}$$

For higher densities the highly relativistic correction $\rho_2(k_F)$ (see eq. (134)) has to be taken into account. From Fig. 10 it is obvious that the total density $\rho_1(k_F) - \rho_2(k_F)$ remains a monotonous, i.e. invertible function of k_F. However, the exact inversion of (132) requires the determination of the roots of a 5-th order polynomial. Probably this cannot be done analytically but numerical inversion is certainly possible. Thus, the zero order kinetic and exchange energy densities (135-138) can be considered as functions of the density, implicitly defined by the inverse of (132).

One particular point should be noted. At the semiclassical level we have <u>two</u> classical turning points: $k_F = 0$ and $q_F = 0$. By means of the elimination procedure only the first one at $k_F = 0$ disappears. The second one at $q_F = 0$ remains. It is precisely this point which characterizes the inset of the highly relativistic correction at the critical density (144).

Let us finally discuss the higher order terms of the gradient expansion. These are obtained by application of the mathematical theorem (56) to equation (122). This yields for the density

$$\rho(\vec{r}) = \sum_{sT} \int d^3k \sum_{n=o}^{\infty} \theta^{(n)}(E_F \pm E_k) \langle \vec{k}sT|\vec{r}\rangle \langle \vec{r}|\hat{O}_n|\vec{k}sT\rangle$$

$$- \sum_{sT} \int d^3k \sum_{n=o}^{\infty} \theta^{(n)}(G_F \pm E_k) \langle \vec{k}sT|\vec{r}\rangle \langle \vec{r}|\hat{O}_n|\vec{k}sT\rangle \; .$$

The operators $\hat{O}_n$ are calculated from (57) by insertion of the Dirac operator $\hat{a} = -\hat{t} = -\hat{\vec{\alpha}} \cdot \hat{\vec{p}} - \beta m$. As before, the operators $\hat{O}_n$ are evaluated up to second order gradients of $E_F(\vec{r})$ and $G_F(\vec{r})$.

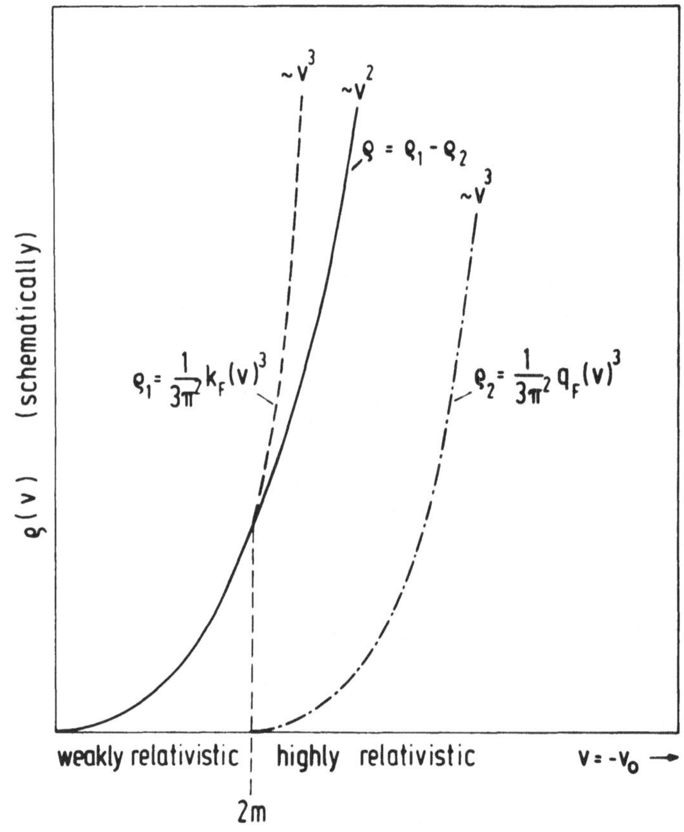

Fig. 10a. RTF density for an atomic system as function of the effective potential.

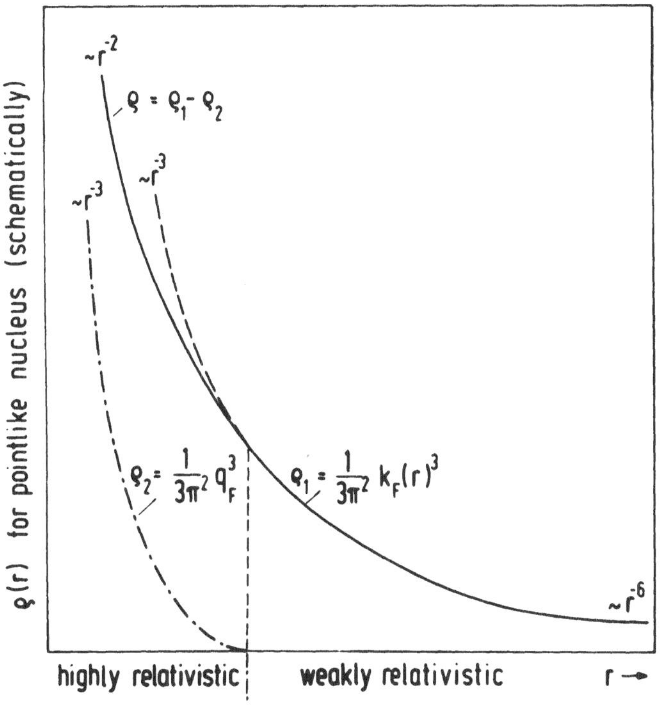

Fig. 10b. RTF density for an atomic system as function of the distance from the nucleus.

A detailed analysis shows that, in contrast to the nonrelativistic case, each operator $\hat{O}_n$ contains first and second order gradient terms. Therefore, the series over n have to be resummed. This problem is solved by use of the Taylor expansion of the step function

$$\theta(x+a) = \sum_{n=o}^{\infty} \frac{1}{n!} \theta^{(n)}(x-a)(2a)^n$$

which is a well defined object in the theory of distributions [89]. The result at the semiclassical stage looks as follows:

$$\rho(\vec{r}) = \rho_1 [E_F] - \rho_2 [G_F] \tag{145}$$

with $$\rho_1 [E_F] = \frac{1}{3\pi^2} (k_F^3+f(E_F)(\nabla E_F)^2+g(E_F)\Delta E_F)$$

$$\rho_2 [G_F] = \frac{1}{3\pi^2} (q_F^3+f(G_F)(\nabla G_F)^2+g(G_F)\Delta G_F)$$

and $$f(E_F) = -\frac{1}{8}\frac{E_F^2}{k_F^3} + \frac{3}{8}\frac{1}{k_F}$$

$$g(E_F) = \frac{1}{4}\frac{E_F}{k_F} + \frac{1}{2} \text{Arsh} \left(\frac{k_F}{m} \right) .$$

Again we obtain a main contribution $\rho_1[E_F]$ and a highly relativistic correction $\rho_2[G_F]$. The result for the kinetic energy density has a similar structure:

$$\tau(\vec{r}) = \tau_o(E_F) + \frac{1}{3\pi^2} (h(E_F)(\nabla E_F)^2 + k(E_F)\Delta E_F)$$

$$-\tau_o(G_F) - \frac{1}{3\pi^2} (h(G_F)(\nabla G_F)^2 + k(G_F)\Delta G_F)$$

$$\text{with } h(E_F) = \frac{1}{4}\frac{E_F}{k_F} - \frac{1}{8}\frac{E_F^3}{k_F^3} - \frac{1}{4} \text{Arsh} \left(\frac{k_F}{m} \right) \tag{146}$$

$$k(E_F) = \frac{1}{4} k_F + \frac{1}{4}\frac{E_F^2}{k_F} .$$

Again we have two classical turning points but, in contrast to the lowest order, the second order expressions are divergent for $k_F = 0$ and $q_F = 0$. By means of the relation $G_F(\vec{r}) = E_F(\vec{r}) - \Delta$, the right hand side of (145) depends only on $E_F(\vec{r})$ or, equivalently, on $k_F(\vec{r})$. Thus, we can invert (145) consistently to second order in $\hbar$ and subsequently eliminate $k_F(\vec{r})$ in favour of $\rho(\vec{r})$ in the kinetic energy expression (146). Similar to the lowest order, the first turning point disappears, while the second one at $q_F = 0$ remains. This fact causes the second order kinetic energy density to be divergent at the critical density $\rho = \rho_{crit}$ defined in eq.(144).

At first glance, another elimination procedure appears more attractive: Invert the functionals $\rho_1[E_F]$ and $\rho_2[G_F]$ separately and eliminate E_F in favour of ρ_1 and G_F in favour of ρ_2. This leads to

$$\tau(\vec{r}) = \tau_0(x) + \frac{1}{72m} B(x) \frac{(\nabla\rho_1)^2}{\rho_1} - \tau_0(y) - \frac{1}{72m} B(y) \frac{(\nabla\rho_2)^2}{\rho_2} \tag{147}$$

$$\text{with } B(z) = \frac{1}{\sqrt{1+z^2}} + \frac{2z}{1+z^2} \operatorname{Arsh} z \qquad \text{for } z = x,y.$$

The quantities x and y are given by (136) and (137).
Formally, the second turning point has disappeared. However, the quantities ρ_1 and ρ_2 may not be treated independently; and the turning point problem is recovered if the semiclassical connection

$$E_F[\rho_1] - G_F[\rho_2] = \Delta$$

is used. At present it is not clear how the divergence problem at the second turning point can be overcome.

So far, numerical results have been obtained only for a simplified model in which the ρ_2-dependent terms of (147) are completely neglected. The corresponding total atomic energies are given in the last column of Table 2. It turns out that for the heaviest systems the results are already slightly better than those obtained within the second order FW-approximation. Of course, this is not yet the final answer. However, the trend is favourable and, together with the qualitative improvement of the lowest order RTF model, this might be an indication that we are going into the right direction.

References

1. P. Hohenberg and W. Kohn, Phys. Rev. 136:B864 (1964).
2. W. Kohn and L.J. Sham, Phys. Rev. 140:A1133 (1965).
3. L.J. Sham and W. Kohn, Phys. Rev. 145:561 (1966).
4. G.A. Baraff and S. Borowitz, Phys. Rev. 121:1704 (1961).
5. S.-K. Ma and K.A. Brueckner, Phys. Rev. 165:18 (1968).
6. L.J. Sham, in "Computational Methods in Band Structure," p.458, P.M. Marcus, J.F. Janak, and A.R. Williams, eds., Plenum, New York (1971).
7. E.K.U. Gross and R.M. Dreizler, Z. Phys. A302:103 (1981).
8. D.J.W. Geldart and M. Rasolt, Phys. Rev. B13:1477 (1976).
9. D.C. Langreth and J.P. Perdew, Phys. Rev. B21:5469 (1980).
10. V. Sahni, J. Gruenebaum, and J.P. Perdew, Phys. Rev. B26:4371 (1982).
11. J.C. Stoddart, A.M. Beattie, and N.H. March, Int. J. Quant.Chem. 4:35 (1971).
12. A.M. Beattie, J.C. Stoddart, and N.H. March, Proc. Roy. Soc. A326:97 (1971).
13. O. Gunnarsson and B.I. Lundqvist, Phys. Rev. B13:4274 (1975).
14. O. Gunnarsson, M. Jonson and B.I. Lundqvist, Phys. Lett. 59A:177 (1976).
15. O. Gunnarsson, M. Jonson, and B.I. Lundqvist, Solid State Comm. 24:765 (1977).
16. D.A. Kirzhnits, "Field Theoretical Methods in Many-Body Systems," Pergamon, Oxford (1967).
17. C.H. Hodges, Can. J. Phys. 51:1428 (1973).
18. D.R. Murphy, Phys. Rev. 24:1682 (1981).
19. B. Grammaticos and A. Voros, Ann. Phys. 123:359 (1979).
20. S.F.J. Wilk, Y. Fujiwara, and T.A. Osborn, Phys. Rev. A24:2187 (1981).
21. Y. Fujiwara, T.A. Osborn, and S.F.J. Wilk, Phys. Rev. A25:14 (1982).
22. E. Tirapegui, F. Langouche, and D. Roekaerts, Phys. Rev. A27:2649 (1982).
23. V. Peuckert, J. Phys. C7:2221 (1974).
24. V. Peuckert, J. Phys. C9:809 (1976).
25. V. Peuckert, J. Phys. C9:4173 (1976).
26. C.Q. Ma and V. Sahni, Phys. Rev. B16:4249 (1977).
27. U. von Barth and L. Hedin, J. Phys. C5:1629 (1972).
28. G.L. Oliver and J.P. Perdew, Phys. Rev. A20:397 (1979).
29. L. Fritsche and H. Gollisch, in: "Local Density Approximations in Quantum Chemistry and Solid State Physics," J.P. Dahl and J. Avery, eds., Plenum, New York (1984).
30. G.E.W. Bauer, Phys. Rev. B27:5912 (1983).
31. K.F. Freed and M. Levy, J. Chem. Phys. 77:396 (1982).
32. G. Zumbach and K. Maschke, Phys. Rev. A28:544 (1983).
33. E. Runge and E.K.U. Gross, Phys. Rev. Lett. 52:997 (1984).
34. E.K.U. Gross and R.M. Dreizler, Phys. Rev. A20:1798 (1979).

35. W. Stich, E.K.U. Gross, P. Malzacher, and R.M. Dreizler, Z. Phys. A309:5 (1982).
36. E.K.U. Gross, A. Toepfer, B. Jacob, and R.M. Dreizler, in: "Proceedings of the XVII International Winter Meeting on Nuclear Physics in Bormio 1979," p. 68-94, I. Iori, ed., Instituto Nazionale di Fisica Nucleare, Milano (1979)
37. R.M. Dreizler, E.K.U. Gross, and A. Toepfer, Phys. Lett. 71A: 49 (1979).
38. A. Toepfer, E.K.U. Gross, and R.M. Dreizler, Phys. Rev. A20: 1808 (1979).
39. A. Toepfer, E.K.U. Gross, and R.M. Dreizler, Z. Phys. A298: 167 (1980).
40. E.K.U. Gross and J. Rafelski, Phys. Rev. A20:44 (1979).
41. R.M. Dreizler, E.K.U. Gross, M. Horbatsch, B. Jacob, H.J. Lüdde, and A. Toepfer, in:" Proceedings of the XVIII International Winter Meeting on Nuclear Physics in Bormio 1980," p.764-794, Ricerca Scientifica ed Educazione Permanente, supplemento n.13, Milano (1980).
42. E.K.U. Gross and R.M. Dreizler, Phys. Lett. 81A:447 (1981).
43. R.M. Dreizler and E.K.U. Gross, in:" Quantum Electrodynamics of Strong Fields," p. 383-412, W. Greiner, ed., Plenum, New York (1983).
44. E.K.U. Gross and R.M. Dreizler, in:" Local Density Approximations in Quantum Chemistry and Solid State Physics," p. 353-380, J.P. Dahl and J. Avery, eds., Plenum, New York (1984).
45. M. Horbatsch and R.M. Dreizler, Z. Phys.A300:119 (1981).
46. M. Horbatsch and R.M. Dreizler, Z. Phys.A308:329 (1982).
47. H.J. Lüdde, M. Horbatsch, E.K.U. Gross, and R.M. Dreizler, in: "Proceedings of the XVII International Winter Meeting on Nuclear Physics in Bormio 1979," p. 120-136, I. Iori, ed., Instituto Nazionale di Fisica Nucleare, Milano (1979).
48. M. Horbatsch, H.J. Lüdde, A. Henne, E.K.U. Gross, and R.M. Dreizler, in:" XII International Conference on the Physics of Electronic and Atomic Collisions (Gatlinburg, Tennessee, 1981), Abstracts of Contributed Papers I," p. 545-546, S. Datz, ed., Oak Ridge (1981).
49. G. Holzwarth, Phys. Lett. B66:29 (1977).
50. G. Holzwarth, Phys. Rev.C16:885 (1977).
51. For a similar treatment of nuclear giant resonances, see G. Holzwarth, this volume.
52. P. Malzacher and R.M. Dreizler, Z. Phys. A307:211 (1982).
53. S.C. Ying, Nuovo Cim. 23B:270 (1974).
54. G. Mukhopadhyay and S. Lundqvist, Nuovo Cim. 27B:1 (1975).
55. L.J. Bartolotti, Phys. Rev.A24:1661 (1981); ibid. A26:2243(1982).
56. M. Levy, Proc. Natl. Acad. Sci. USA 76:6062 (1979).
57. V. Peuckert, J. Phys. C11:4945 (1978).
58. L.H. Thomas, Proc. Camb. Phil. Soc. 23:542 (1926).
59. E. Fermi, Rend. Acad. Naz. Linzei 6:602 (1927).

60. P.A.M. Dirac, Proc. Camb. Phil. Soc. 26:376 (1930).
61. C.F.v.Weizsäcker,Z. Phys.96:431 (1935).
62. C. Schweitzer, R.M. Dreizler, and E.K.U. Gross, to be published.
63. F. Herman, J.P. Van Dyke, and I.B. Ortenburger, Phys. Rev. Lett. 22:807 (1969).
64. L. Kleinman, Phys. Rev.B10:2221 (1974).
65. D.J.W. Geldart, M. Rasolt, and C.O. Almbladh, Solid State Comm. 16:243 (1975).
66. R. Latter, Phys. Rev. 99:510 (1955).
67. S.B. Schneiderman and A. Russek, Phys. Rev. 181:311 (1969).
68. K. Taulbjerg, J. Vaaben, and B. Fastrup, Phys. Rev. A12:2325 (1975).
69. R.M. Dreizler, in:" Critical Phenomena in Heavy Ion Physics (Brasov International School 1980)," p.205-257, A.A. Raduta and G. Stratan,eds., Central Institute of Physics, Bucharest (1980).
70. E. Clementi and C. Roetti, At. Data Nucl. Data Tab. 14:177 (1974).
71. J. Scott, Phil. Mag. 43:859 (1952).
72. J. Schwinger, Phys. Rev. A22:1827 (1980).
73. E. Lieb, Rev. Mod. Phys. 53:603 (1981).
74. J.P. Desclaux, At. Data Nucl. Data Tab. 12:311 (1973).
75. W.C. Ermler, R.S. Mulliken, and A.C. Wahl, J. Chem. Phys. 66:3031 (1977).
76. J. Eichler and U. Wille, Phys. Rev. A11:1973 (1975).
77. A. Toepfer, B. Jacob, H.J. Lüdde, and R.M. Dreizler, Phys. Lett. 93A:18 (1982).
78. A. Toepfer, B. Jacob, H.J. Lüdde, and R.M. Dreizler, to be published in Z. Phys.
79. T. Andersen, E. Bøving, P. Hedegard, and O.J. Østgard, J. Phys. B11:1449 (1978).
80. R. Hippler and K.-H. Schartner, J. Phys. B8:2528 (1975).
81. For a broad review, see M.V. Ramana and A.K. Rajagopal, Adv. Chem. Phys. 54:231 (1983).
82. A.K. Rajagopal, this volume.
83. M.S. Vallarta and N. Rosen, Phys. Rev. 41:708 (1932).
84. H. Jensen, Z. Phys. 82:794 (1933).
85. N. Ashby and M.A. Holzman, Phys. Rev. A1:764 (1970).
86. M. Rudkjøbing , Kgl. Danske Videnskab Selskab, Mat.-Fys. Med. 27, No. 5 (1952).
87. B. Jancovici, Nuovo Cim. 25:428 (1962).
88. E.E. Salpeter, Astrophys. J. 134:669 (1961).
89. F. Constantinescu, " Distributionen und ihre Anwendungen in der Physik," p. 59 ff, Teubner, Stuttgart (1974).

DENSITY FUNCTIONAL THEORY IN CHEMISTRY

Robert G. Parr

Department of Chemistry
University of North Carolina
Chapel Hill, North Carolina 27514

In these lectures I will summarize some of the things that we have been doing in my laboratory with density functional theory.[1-63] Concepts are emphasized rather than computations, and exact theory rather than approximate methods. There are five previous reviews of our work.[64-68]

The job of quantum chemistry is to solve and intepret Schrödinger's two equations for chemical systems (N finite). In contrast with the impossibility of solving the equations quantitatively, as was understood in the 1930's, in the 1980's one takes for granted that quantitative solution is possible. As early as 1972 Clementi said, for example, "We can calculate everything.[69] This is only slightly an overstatement. Molecular properties are being computed to multidigit accuracy, and the rates of reaction are beginning to succomb as well.[70]

Accurate calculation is not synonomous with useful interpretation, however. To calculate a molecule is not to understand it. Indeed, a computer program which will perfectly compute any property of any molecule at will, while very valuable, is like a perfect experimental apparatus for determining properties, leaving the whys quite unfathomed. [To this one may argue that there remains no need to "fathom", but this attitude belies the science of chemistry itself, which, after all, is known to be highly systematic.]

The density functional theory is a powerful tool for elucidating the concepts of chemistry, as I hope to show in these lectures. It also is a powerful tool for calculations on chemical systems, which I will touch on only briefly but which is well illustrated by other contributions at this conference.

I will not describe the whole subject of density functional theory for finite systems, but rather will recount some of our own contributions to it. Our contributions have been disjoint, and these lectures also will be disjoint. Each section below deals with a different topic. I follow the lectures as orally presented at the conference except that I have suppressed some of the numerical and other results presented, which may be found in the references cited.

The research here described has been carried out in the main at the University of North Carolina at Chapel Hill. I have had many collaborators. Continuous support from the National Science Foundation and the National Institutes of Health is gratefully acknowledged.

1. The electron density in the ground state of an atom is monotonically decreasing.[71] [Try to prove it!] This comes as a surprise to most chemists, probably because they have been brought up on the importance of valence shells and pictures of radial distribution functions $r^2\rho$. Indeed, the simplest good characterization of atomic electron densities is that they are piecewise exponentially decaying, with one exponential per principal quantum shell.

One of our own first publications in density functional theory had to do with the fact just stated.[5] It turns out that if one constrains conventional Thomas-Fermi or Thomas-Fermi-Dirac or Thomas-Fermi-Dirac-Weizsacker energy functionals for atoms, by requiring that the density be piecewise exponential, the energies obtained are much better than the corresponding unconstrained results.[5,39] For the neon atom, for example, conventional Thomas-Fermi theory gives an energy 29% too low, but the exponentially constrained Thomas-Fermi method an energy only 2% low.

At about the same time, we found additional evidence that a good density put into statistical formulas can give good results.[4,13] Namely, if one computes the kinetic energy of an atom from the so-called gradient expansion,

$$T[\rho] = T_0[\rho] + T_2[\rho] + T_4[\rho] + \ldots, \tag{1}$$

with $T_0[\rho]$ the Thomas-Fermi term, $T_2[\rho]$ one-ninth of the Weizsacker $T_W[\rho] = (1/8)\int\nabla\rho(\nabla\rho/\rho)d\tau$, then good answers are obtained, the errors with one term being 5 to 10%, two terms about 1%, and with three terms for heavy atoms only a fraction of a percent. This does not mean, however, that the corresponding functional derivative $\delta T/\delta\rho$ is well represented by the corresponding sum.[11]

Very recently we have argued that a better representation of the kinetic energy of atoms than Eq. (1) is a formula of the form

$$T[\rho] = T_W[\rho] + \gamma T_0[\rho], \tag{2}$$

where γ is a predictable function of N.[25]

2. The chemical potential of density functional theory is an atomic or molecular property of considerable significance. It is, in fact, just the negative of the electronegativity of modern chemistry.[7]

Here, as always, when we talk the classical density functional language, we are speaking about ground states. The density functional description being essentially the "thermodynamic" description of an electronic species, one expects this restriction because classical thermodynamics is confined to equilibrium states.

Conventional wave mechanics for an electronic system with N electrons moving in an external potential field $V = \sum_\mu v(\mu)$ yields the energy of the ground state, a function of N and a functional of v, E[N,v]. The density functional theory yields the energy rather as a functional of the electron density ρ, $E[\rho]$, through solution of an equation

$$\delta\{E[\rho] - \mu(N[\rho] - N)\} = 0. \tag{3}$$

Here $N[\rho] = \int \rho d\tau$ and μ is a Lagrange multiplier ensuring the correct number of particles, the chemical potential of the system. It is elementary then to show that for a change of one ground state to another,

$$dE = \mu dN + \langle \rho dv \rangle. \tag{4}$$

Furthermore

$$\mu = (\partial E/\partial N)_v. \tag{5}$$

That this is the negative of the electronegativity, χ, follows from the fact that the latter quantity has for half a century been known by chemists to be approximately the negative slope of an E versus N plot at constant v. Very specifically, there is the famous formula of Mulliken,

$$\chi = \frac{I+A}{2} (= -\mu) \qquad (6)$$

where I and A are ionization potential and electron affinity, respectively. This is just the finite difference approximation to the negative slope.

With the equivalence established between chemical potential and electronegativity, properties of the latter follow from the density functional properties of the former.[7] For example, when two species unite to form a new species, there results a single chemical potential of the new species. Electronegativities have "equalized" or "neutralized", just as do chemical potentials in ordinary thermodynamics.

In the basic equation of density functional theory, Eq. (3), the Lagrange multiplier is the chemical potential, and this has the important physical implications just indicated. That is why density functional theory should be expected to be of great importance in chemistry. Further elucidation, beyond what I have sketched to this point, and beyond even what has so far been accomplished in the literature, is expected uniformly to feature physical language which is highly perspicuous.

In Eqs. (3) and (5), the functional $E[\rho]$ must be the correct functional for arbitrary N, not necessarily integral. More about this below. Suffice it here to say that in my opinion the resultant theory is not what Walter Kohn modestly at this conference called "formal and grandiose". Is it not rather profound and grand?

3. Now I want to say something about scaling and localization in energy functionals and their components.[66,72]

Any energy functional $E[\rho]$ is the sum of a kinetic energy component $T[\rho]$ and a potential energy component $V[\rho]$. Neither of these in general is known accurately, so that one must resort to approximations to them. Scaling properties delimit forms for the approximations.

Given a proper kinetic energy component $T[\rho]$, the corresponding scaled functional is $T[\rho_\zeta]$, where ρ_ζ is the scaled density, $\rho_\zeta(r) \equiv \zeta^3\rho(\zeta r)$. It follows from the fact that the kinetic operator is homogeneous of degree -2 in the coordinates, that[72]

$$T[\rho_\zeta] = \zeta^2 T[\rho]. \qquad (7)$$

This means that functionals $\int\rho^{5/3}d\tau$, $\int\nabla\rho(\nabla\rho/\rho)d\tau$, $C(N)\int\rho^{5/3}d\tau$, etc., are acceptable as kinetic energy contributions, but

$\int \nabla\rho(\nabla\rho/\rho^{4/3})d\tau$, for example, is not acceptable. Similarly, homogeneity of degree -1 for the potential energy for coulomb systems leads to the rule that potential energy contributions scale in accord with the formula

$$V[\rho_\zeta] = \zeta V[\rho], \tag{8}$$

and this validates terms $\int\rho^{4/3}d\tau$, $\int\nabla\rho(\nabla\rho/\rho^{4/3})d\tau$, etc. as potential energy contributions.

Consider in particular formulas of the type

$$T[\rho] = C(N)\int\rho^{5/3}d\tau + \ldots \tag{9}$$

or

$$K[\rho] = D(N)\int\rho^{4/3}d\tau + \ldots, \tag{10}$$

where K is the exchange energy. Is the condition that the leading terms give good approximations that the system in question behave "statistically"? No, the condition is that the functional derivatives $\delta T/\delta\rho$ and $\delta K/\delta\rho$ to a good approximation are local functions of ρ. If, for example, $\delta T/\delta\rho = t(\rho)$ + corrections, where $t(\rho)$ is a function, then $t(\rho) = C\rho^{2/3}$.

4. Since the first-order density matrix determines the density, there is an exact density-matrix functional theory in which the density matrix is the basic variable, and this theory has many interesting aspects.[8,10] A well-known approximation to this theory is the so-called $X\alpha$ method. We have used this method to compute the chemical potentials (electronegativities) of the elements.[17,60] The results are very good, exhibiting all trends expected from empirical data. One gets different formulas for two main cases. For an open-shell situation, the chemical potential is the highest occupied orbital energy for a neutral-atom transition state resembling though not always identical with the ground state. For a closed-shell situation, the chemical potential is the average of two orbital energies for a peculiar neutral-atom transition state having 1 1/2 electrons in one orbital (the ground-state HOMO) and 1/2 electrons in another (the ground-state LUMO).

5. An atom in a molecule can be defined using density functional theory.[7,69] If a molecule AB has density ρ_{AB} and chemical potential μ_{AB}, then atoms A and B <u>in the molecule</u> will be species A^* and B^* such that

$$\rho_{AB} = \rho_{A^*} + \rho_{B^*}, \tag{11}$$

$$\mu_{AB} = \mu_{A^*} = \mu_{B^*}. \tag{12}$$

The definitions are completed by imposing a minimum promotion energy criterion. Preliminary tests of this scheme have been carried out, but more work is needed in this problem area. Note that the species A^* and B^* in general will bear nonintegral charges, and that they are ground-state species--results of molecular field perturbations of the isolated atoms.

6. The foregoing picture of the atom in a molecule can be used to examine the possibility that when atoms A and B come together, having chemical potentials μ_A° and μ_B°, the final chemical potential of the molecule AB, μ_{AB}, is (or is not) approximately the geometric mean,

$$\mu_{AB} = (\mu_A^\circ \mu_B^\circ)^{1/2}. \tag{13}$$

It turns out, as we have shown in detail,[40] that Eq. (13) follows if each of the component atomic energies is exponentially decreasing with N, with the same decay parameter. Further, such exponential decay is reasonable for neutral species, and actual data support, very roughly, the idea of a universal decay parameter (value about 2.2).

7. There is a puzzle about the chemical potential of a pair AB. How can there be a common chemical potential as the pair dissociates, when the separate species clearly have different chemical potentials? The resolution has been given,[47] and is elaborated at length by John Perdew in his lectures at this conference. As long as there is equilibrium, one has a single μ value, and for large internuclear separation one finds

$$\mu_{AB} = -\frac{I_{HOMO} + A_{LUMO}}{2}, \tag{14}$$

where I_{HOMO} is the lower of I_A and I_B and A_{LUMO} is the higher of A_A and A_B. The proof is elementary.

I mention in passing another result concerning the chemical potential.[24] For an atom of charge Z and electron number N=Z, as N becomes very large, μ either becomes infinite or goes to zero like $Z^{-1/3}$. Actual data for the periodic table strongly imply the latter.

8. The Euler equation resulting from Eq. (3) is

$$\mu = \frac{\delta E}{\delta \rho}. \tag{15}$$

One must find the density which makes $\delta E/\delta\rho$ constant everywhere, and the value of the constant is the chemical potential. Here $\delta E/\delta\rho$ must allow change of the particle number N, even though in actual calculations on a system with N given, one deals with the functional derivative $(\delta E/\delta\rho)_N$ (as has been emphasized by Kohn at this conference). That these derivatives are not the same has caused some confusion in the literature. That they need not be the same can be seen from the following argument.[53]

Introduce the resolution of ρ into a number factor N and a shape factor σ:

$$\rho = N\sigma, \quad \int \sigma d\tau = 1. \tag{16}$$

Then for the energy $E[\rho] = E[N,\sigma]$ we have both

$$\Delta E = \left(\frac{\partial E}{\partial N}\right)_\sigma dN + \int \left(\frac{\delta E}{\delta \sigma}\right)_N \Delta\sigma d\tau \tag{17}$$

and

$$\Delta E = \int \frac{\delta E}{\delta \rho} \Delta\rho d\tau. \tag{18}$$

Introduce $\Delta\rho = \sigma\Delta N + N\Delta\sigma$ into Eq. (18) and subtract Eq. (17). There results

$$0 = \left[\left(\frac{\partial E}{\partial N}\right)_\sigma - \int \frac{\delta E}{\delta \rho}\sigma d\tau\right]\Delta N + \int \left[\left(\frac{\delta E}{\delta \sigma}\right)_N - N\frac{\delta E}{\delta \rho}\right]\Delta\sigma d\tau. \tag{19}$$

Hence we have

$$\left(\frac{\partial E}{\partial N}\right)_\sigma = \int \left(\frac{\delta E}{\delta \rho}\right)\sigma d\tau \tag{20}$$

and

$$\left(\frac{\delta E}{\delta \sigma}\right)_N = N\left(\frac{\delta E}{\delta \rho}\right)_N = N\frac{\delta E}{\delta \rho} + \text{constant}, \tag{21}$$

where the constant is arbitrary because of the normalization condition of Eq. (16). That is to say,

$$\frac{\delta E}{\delta \rho} = \left(\frac{\delta E}{\delta \rho}\right)_N + \text{constant}. \tag{22}$$

As a consequence, constant N calculations provide no access to numerical values of $\mu = \delta E/\delta\rho$.

9. In more detail, Eq. (15) reads $\mu = v + (\delta F/\delta\rho)$, where $F[\rho] = T[\rho] + V_{ee}[\rho]$ is the sum of kinetic energy and electron repulsion energy functionals. Or, separating out the classical coulomb energy from V_{ee}, leaving the exchange-correlation component $-K[\rho]$, we have

$$\mu = v^* + \left(\frac{\delta T}{\delta\rho}\right) - \left(\frac{\delta K}{\delta\rho}\right), \tag{23}$$

where v^* is the classical electrostatic potential as a function of position.

It is not helpful to try to use Eq. (23) to determine μ for a given electron distribution. The three terms on the right-hand side are of the same order of magnitude, generally each much bigger than μ. The functional derivatives are not at constant N and so are subject to the difficulties discussed in the last section.

One can employ Eq. (23) in another way, however, to obtain information about $(\delta T/\delta\rho)-(\delta K/\delta\rho)$ from an empirical μ [Eq. (6)] and calculated electrostatic potential:

$$\frac{\delta T}{\delta\rho} - \frac{\delta K}{\delta\rho} = \mu - v^*. \tag{24}$$

For an atom this reveals a fascinating fact: $\delta T/\delta\rho$ and $\delta K/\delta\rho$ are equal, or μ and v^* are equal, at a radial distance very close to the covalent radius of the atom.[56] Further study of this result is warranted; perhaps it provides a clue for a density functional model of chemical binding.

10. Just as in thermodynamics there are many useful reciprocal relations and stability conditions, so there are in density functional theory.[43,44,52] The fundamental differential expression for the energy is Eq. (4),

$$dE = \mu dN + \langle\rho dv\rangle. \tag{25}$$

From this follow the reciprocal relations

$$[\delta\mu/\delta v(1)]_N = [d\rho(1)/dN]_v \tag{26}$$

and

$$\delta\rho(1)/\delta v(2) = \delta\rho(2)/\delta v(1). \quad (27)$$

The importance of Eq. (26) we will return to later; Eq. (27) is a generalization of the well-known symmetry of the atom-atom polarizability tensor of Hückel theory. A stability condition going with Eq. (25) is

$$(\partial\mu/\partial N)_v \geq 0 \quad (28)$$

This result depends on convexity of the functional $E[\rho]$, not proved,[73] but it appears to be unexceptionable for coulomb systems.

Again as in thermodynamics, Legendre transformations are straightforward to define.[43,44,52] Let $Q[\mu,v] \equiv E - N\mu$, $F[N,\rho] \equiv E - \langle\rho v\rangle$, and $R[\mu,\rho] \equiv E - N\mu - \langle\rho v\rangle$. Then we have

$$dQ = -Nd\mu + \langle\rho dv\rangle. \quad (28)$$

$$dF = \mu dN - \langle v d\rho\rangle, \quad (29)$$

$$dR = -Nd\mu - \langle v d\rho\rangle. \quad (30)$$

These various quantities,and the reciprocal relations and stability conditions associated with them, have many applications.

11. Our "thermodynamic" description of the electrons in a system should admit of introduction of a local property behaving like a pressure.[15] Let $G[\rho] = T[\rho] - K[\rho]$ and take the gradient of Eq. (23). If we define a stress dyadic

$$\overleftrightarrow{\sigma} = \sum_{i=1}^{3} \sum_{j=1}^{3} \vec{\epsilon}_i \sigma_{ij} \vec{\epsilon}_j \quad (31)$$

such that

$$\nabla \cdot \overleftrightarrow{\sigma} = -\rho\nabla[\delta G/\delta\rho], \quad (32)$$

we find the equation for static equilibrium,

$$-\nabla \cdot \overleftrightarrow{\sigma} + \rho\nabla v^* = 0. \quad (33)$$

Then define the scalar pressure P by

$$P = -(1/3)\ \mathrm{trace}\ (\sigma_{ij}). \quad (34)$$

There result

$$\int \rho \vec{r} \cdot \nabla [\delta G/\delta \rho] d\tau = -2T[\rho] + K[\rho] \qquad (35)$$

and

$$3\int P d\tau = 2T[\rho] - K[\rho]. \qquad (36)$$

We also find, for the change from one equilibrium state to another,

$$N d\mu = \langle \rho dv^* \rangle + \langle dP \rangle + dX \qquad (37)$$

where

$$X[\rho] \equiv [\int \rho(\delta T/\delta \rho) d\tau - (5/3)T] - [\int \rho(\delta K/\delta \rho) d\tau - (4/3)K]. \qquad (38)$$

Equation (37) is purely classical except for the term dX, which vanishes identically for any purely local density functional model.

Other definitions of P are possible.[74]

12. The assumption that E(N) may be regarded as a continuous function of N, even for few-electron systems, is not trivial. Favorable indications that interpolation is reasonable include the following:[67] atomic and molecular properties generally are in fact smooth, bounded functions of N; non-integral N "atomic populations" have been used by chemists for decades; interpolating theories such as Thomas-Fermi theories already exist.

But the clearest argument that nonintegral N values should be accepted comes from the statistical mechanics of the grand canonical ensemble. This is implicit in Mermin's finite temperature extension of the Hohenberg-Kohn theory,[75] as well as in another early work by Gyftopoulos and Hatsopoulos.[76] The matter has recently been discussed in detail by Perdew et al.[47]

At 0°K, for atoms or molecules not in interaction the result is that fractional N means simply a statistical mixture. E(N) is a series of straight-line segments connecting integral N values. One finds from the statistical mechanics, in the limit of 0°K, for a species S with ionization potential I_s and electron affinity A_s,

$$\mu_s = \begin{Bmatrix} -I_s & \text{slightly position species} \\ -\frac{I_s + A_s}{2} & \text{netural species} \\ -A_s & \text{slightly negative species} \end{Bmatrix}. \qquad (39)$$

For species in interaction, as an atom in a molecule, one finds

$$A_s < -\mu_s < I_s. \tag{40}$$

Within these limits, the electronegativity of a species may be controlled by its environment, a fact which just might have implications for catalysis.

13. Recent studies of ours on the Hartree-Fock theory[31,50] strongly bear on density functional theory in that they show how a wave-function theory can be brought into close correspondence with a density theory by the introduction of complex orbitals.

Let the canonical Hartree-Fock orbitals for a 2N-electron closed-shell problem be denoted by λ_j:

$$\hat{F}\lambda_j = \epsilon_j \lambda_j. \tag{41}$$

Then the transformation to <u>circulant</u> <u>orbitals</u> ϕ_ℓ is given by[31]

$$\phi_\ell = N^{-1/2} \Sigma_j \lambda_j \omega^{(\ell-1)(j-1)} \tag{42}$$

where $\omega = \exp(2\pi i/N)$ is the complex Nth root of unity. The Hartree-Fock equations become

$$\hat{F}\phi_\ell = \Sigma_k \epsilon^\phi_{k\ell} \phi_k \tag{43}$$

where the matrix $\epsilon^\phi_{k\ell}$ is a <u>circulant</u> <u>matrix</u>: diagonal elements are equal and every row is a permutation of every other row. The electron density associated with each ϕ is very close to that associated with each other ϕ; circulant orbital densities oscillate around the average electron density $\bar{\rho} = \rho/N$. [To sense how close this correspondence is, note that in the atom Be, circulant orbital densities are identical with $\bar{\rho}$.] Among other properties, circulant orbitals minimize the sum of the average of the square distances of the various Fϕ from each other.

14. For the atom Be, in fact one can write the two circulant orbitals as

$$\phi_1 = (\rho/N)^{1/2} e^{i\theta},$$

$$\phi_2 = (\rho/N)^{1/2} e^{-i\theta}. \tag{44}$$

The Hartree-Fock energy functional can then be expressed in terms of integrals involving ρ and θ. Imposing orthogonality between ϕ_1 and ϕ_2 requires one Lagrange multiplier μ_c, normalization another, μ_N. The differential equations for ρ and θ then turn out to be[50]

$$\frac{1}{8}\frac{\nabla\rho\cdot\nabla\rho}{\rho^2} - \frac{1}{4}\frac{\nabla^2\rho}{\rho} + \frac{1}{2}\nabla\theta\cdot\nabla\theta - \frac{Z}{r} + \frac{3}{4}\phi - \frac{1}{4}\int\frac{\rho(2)\cos[2\theta(2)-2\theta(1)]}{r_{12}}d\tau_2 = \mu_N + \mu_c\cos 2\theta \qquad (45)$$

and

$$\frac{1}{\rho}\nabla(\rho\nabla\theta) - \frac{1}{2}\int\frac{\rho(2)\sin[2\theta(2)-2\theta(1)]}{r_{12}}\,d\tau_2 = \mu_c\sin 2\theta. \qquad (46)$$

Beginning with a guessed ρ, these equations can be solved by interation, thereby providing an explicit density functional formulation of the Hartree-Fock problem for Be.

More generally, one can imagine guessing ρ for a problem, writing down a complete set of orthonormal basis functions $\chi_p = (\rho/N)^{1/2}\exp(i\theta_p)$, where the phases would not have to be real functions, setting up a trial wavefunction in terms of the χ_p, and minimizing with respect to the θ_p. That would in principle constitute a totally density functional procedure!

15. The foregoing constitutes good reason to be interested in the following problem for a many-electron system: To find the best single-determinantal wavefunction for a system <u>such that all orbitals have the same electron density</u>.

Work of Harriman shows that this problem has an answer;[77] our work on circulants shows that there may be a surprisingly small energy loss associated with the equal density constraint.

One solution of this problem is as follows (unpublished). Define $Q_k = |\phi_k|^2[|\phi_k|^2 - (\rho/N)]$, $Q = \Sigma_k Q_k$, and

$$J = \iint\frac{Q(1)Q(2)}{r_{12}}\,d\tau_1 d\tau_2. \qquad (47)$$

Then since J=0 if and only if $Q\equiv 0$, we may introduce the constraint J=0 with a single global Lagrange multiplier λ. There result modified Hartree-Fock equations of the form

$$(\hat{F} + \lambda\hat{v}_Q)\phi_k = \sum_j \epsilon_{jk}\phi_j \qquad (48)$$

where $\hat{v}_Q$ is the potential due to Q. Solving these equations will give equidensity orbitals. This is not easy; the orbitals are complex and v_Q approaches zero at the solution point, so that λ approaches infinity at the solution point (except for the two-orbital case).

Substantial progress on this problem would amount to substantial progress on the problem of finding a differential equation for the electron density. As John Platt wrote me in 1968,[78] "We must find an equation for, or a way of computing directly, total electron density." One should try hard to solve this problem.

16. As we have already said, the first differential coefficient μ in Eq. (4) is of great physical importance: it is the chemical potential (or electronegativity) of the system of interest, itself a functional of N and v.

Very recently it has been shown[58] that the derivative of μ with respect to N is identifiable with what was called twenty years ago by Ralph Pearson the <u>hardness</u> of a species, η:[79]

$$2\eta = (\partial\mu/\partial N)_v = (\partial^2 E/\partial N^2)_v. \tag{49}$$

Just as we have the finite difference approximation for the first derivative or chemical potential,

$$-\mu \approx \frac{I+A}{2}, \tag{50}$$

we have the finite difference approximation for the second derivative or hardness,

$$\eta \approx \frac{I-A}{2}. \tag{51}$$

Note that for a metal, $\eta=0$; this is minimum hardness [Eq.(28)] or maximum softness. For an insulator or semiconductor, 2η is the band gap, a singularly important quantity.

That small η should represent great softness is clear from consideration of the disproportionation reaction

$$S + S \longrightarrow S^+ + S^-,$$

for which the energy change is just $I_S - A_S$. Also convincing is a glance at the thermodynamic identity

$$(\partial\mu/\partial N)_{TV} = (1/N\rho\beta) \tag{52}$$

where β is the compressibility.

Pearson's original hard and soft classification was for acids and bases. He enunciated the HSAB Principle: Hard acids prefer hard bases and soft acids prefer soft bases.[79] He and I have derived this principle from a theoretical model.[58] For an acid the hardness one should use is that of the neutral acid; for a base that of its positive ion. For numbers and details, one may refer to our paper. Suffice it that the evidence is overwhelming that $\partial\mu/\partial N$ is the hardness of chemistry.

17. Referring back to Eq. (25) once again, we see that one may expect importance not only of $(\partial\mu/\partial N)_v$, but also of the quantity $\delta\mu/\delta v$ of Eq. (26) and of the quantity $\delta\rho/\delta v$ of Eq. (27). The latter quantity has been studied quite a lot in density functional theory: it is sometimes called the linear response function. The former quantity we are now looking at in detail.

The story of any time-independent chemical happening is the story of Eq. (4): the quantities in it and their derivatives with respect to the system parameters. If we go to a corresponding time-dependent formulation,[32,45,63] we shall have the complete description of all ground-state chemistry. That is the goal toward which we are working.

REFERENCES

1. M. M. Morrell, R. G. Parr and M. Levy, Calculation of ionization potentials from density matrices and natural functions, and the long-range behavior of natural orbitals and electron density, J. Chem. Phys. 62, 549-554 (1975).
2. H. Nakatsuji and R. G. Parr, Variational principles which are functionals of electron density, J. Chem. Phys. 63, 1112-1117 (1975).
3. M. Levy and R. G. Parr, Long-range behavior of natural orbitals and electron density, J. Chem. Phys. 64, 2707-2708 (1976).
4. W.-P. Wang, R. G. Parr, D. R. Murphy and G. Henderson, Gradient expansion of the atomic kinetic energy functional, Chem. Phys. Lett. 43, 409-412 (1976).
5. W.-P. Wang and R. G. Parr, Statistical atomic models with piecewise exponentially decaying electron densities, Phys. Rev. A 16, 891-902 (1977).
6. P. W. Payne, Density response theory of nonbonded interactions, J. Chem. Phys. 68, 1242-1247 (1978).
7. R. G. Parr, R. A. Donnelly, M. Levy and W. E. Palke, Electronegativity: the density functional viewpoint, J. Chem. Phys. 68, 3801-3807 (1978).
8. R. A. Donnelly and R. G. Parr, Elementary properties of an energy functional of the first-order reduced density matrix, J. Chem. Phys. 69, 4431-4439 (1978).
9. N. K. Ray, L. Samuels and R. G. Parr, Studies of electronegativity equalization, J. Chem. Phys. 70, 3680-3684 (1979).

10. R. A. Donnelly, On a fundamental difference between energy functionals based on first- and on second-order density matrices, J. Chem. Phys. 71, 2874-2879 (1979).
11. D. R. Murphy and R. G. Parr, Gradient expansion of the kinetic energy density functional: local behavior of the kinetic energy density, Chem. Phys. Lett. 60, 377-379 (1979).
12. R. G. Parr, S. R. Gadre and L. J. Bartolotti, Local density functional theory of atoms and molecules, Proc. Natl. Acad. Sci. USA 76, 2522-2526 (1979).
13. D. R. Murphy and W.-P. Wang, Comparative study of the gradient expansion of the atomic kinetic energy functional-neutral atoms, J. Chem. Phys. 72, 429-433 (1980).
14. S. R. Gadre, L. J. Bartolotti and N. C. Handy, Bounds for Coulomb energies, J. Chem. Phys. 72, 1034-1038 (1980).
15. L. J. Bartolotti and R. G. Parr, The concept of pressure in density functional theory, J. Chem. Phys. 72, 1593-1596 (1980).
16. R. G. Parr and S. R. Gadre, On the basic homogeneity characteristic of atomic and molecular electronic energies, J. Chem. Phys. 72, 3669-3673 (1980).
17. L. J. Bartolotti, S. R. Gadre and R. G. Parr, Electronegativities of the elements from simple Xα theory, J. Am. Chem. Soc. 102, 2945-2948 (1980).
18. W.-P. Wang, Comparative study of the gradient expansion of the atomic kinetic energy functional-isoelectronic series, J. Chem. Phys. 73, 416-418 (1980).
19. N. K. Ray and R. G. Parr, Diamagnetic shieldings of atoms in molecules and their relation to electronegativity, J. Chem. Phys. 73, 1334-1339 (1980).
20. C. C. Shih, D. R. Murphy and W.-P. Wang, Gradient expansion of the exchange energy density functional: a complementary expansion of the atomic energy functional, J. Chem. Phys. 73, 1340-1343 (1980).
21. S. B. Sears, R. G. Parr and U. Dinur, On the quantum-mechanical kinetic energy as a measure of the information in a distribution, Israel J. Chem. 19, 165-173 (1980).
22. S. M. Valone, Consequences of extending 1 matrix energy functionals from pure-state representable to all ensemble representable 1 matrices, J. Chem. Phys. 73, 1334-1349 (1980).
23. S. M. Valone, A one-to-one mapping between one-particle densities and some n-particle ensembles, J. Chem. Phys. 73, 4653-4655 (1980).
24. N. H. March and R. G. Parr, Chemical potential, Teller's theorem, and the scaling of atomic and molecular energies, Proc. Natl. Acad. Sci. USA 77, 6285-6288 (1980).
25. P. K. Acharya, L. J. Bartolotti, S. B. Sears and R. G. Parr, An atomic kinetic energy functional with full Weizsacker correction, Proc. Natl. Acad. Sci. USA 77, 6978-6982 (1980).
26. G. A. Henderson, Variational theorems for the single-particle probability density and density matrix in momentum space, Phys. Rev. A 23, 19-20 (1981).

27. J. Katriel and M. R. Nyden, A comparison between hydrogenic and Thomas-Fermi expectation values, J. Chem. Phys. 74, 1221-1224 (1981).
28. R. G. Parr and A. Berk, The bare-nuclear potential as harbinger for the electron density in a molecule, in Molecular Electrostatic Potentials in Chemistry and Biochemistry, edited by P. Politzer and D. G. Truhlar (Plenum, 1981), pp. 51-62.
29. S. M. Valone and J. F. Capitani, Bound excited states in density functional theory, Phys. Rev. A 23, 2127-2133 (1981).
30. J. Katriel, R. G. Parr and M. R. Nyden, Concerning the chemical potential of few-electron systems, J. Chem. Phys. 74, 2397-2401 (1981).
31. R. G. Parr and M.-B. Chen, Circulant orbitals for atoms and molecules, Proc. Natl. Acad. Sci. USA 78, 1323-1326 (1981).
32. L. J. Bartolotti, Time-dependent extension of the Hohenberg-Kohn-Levy energy density functional, Phys. Rev. A 24, 1661-1667 (1981).
33. D. R. Murphy, The sixth-order term of the gradient expansion of the kinetic energy density functional, Phys. Rev. A 24, 1682-1688 (1981).
34. L. J. Bartolotti and R. G. Parr, Gradient expansion of the classical Coulomb energy of a charge distribution, J. Chem. Phys. 75, 4553-4555 (1981).
35. M. R. Nyden and P. K. Acharya, Coreless Thomas-Fermi models of atomic structure, J. Chem. Phys. 75, 4567-4571 (1981).
36. Y. Tal, L. J. Bartolotti and R. F. W. Bader, Universal scaling relations for free and bonded atoms, J. Chem. Phys. 76, 463-467 (1982).
37. J. F. Capitani, R. F. Nalewajski and R. G. Parr, Non-Born-Oppenhemier density functional theory of molecular systems, J. Chem. Phys. 76, 568-573 (1982).
38. Y. Tal and L. J. Bartolotti, The hydrogenic limit of many-electron atoms, J. Chem. Phys. 76, 2558-2564 (1982).
39. W.-P. Wang, Fixed-shell statistical atomic models with piecewise exponentially decaying electron densities, Phys. Rev. A 25, 2901-2912 (1982).
40. R. G. Parr and L. J. Bartolotti, On the geometric mean principle for electronegativity equalization, J. Am. Chem. Soc. 104, 3801-3803 (1982).
41. Y. Tal and L. J. Bartolotti, On the Z^{-1} and $N^{-1/3}$ expansions of Hartree-Fock atomic energies. J. Chem. Phys. 76, 4056-4062 (1982).
42. L. J. Bartolotti, A new gradient expansion of the exchange energy to be used in density functional calculation on atoms, J. Chem. Phys. 76, 6057-6059 (1982).
43. R. F. Nalewajski and R. G. Parr, Legendre tranforms and Maxwell relations in density functional theory, J. Chem. Phys. 77, 399-407 (1982).

44. R. F. Nalewajski and J. F. Capitani, Density functional theory: non Born-Oppenheimer Legendre transforms and Maxwell relations, equilibrium and stability conditions, J. Chem. Phys. 77, 2514-2526 (1982).
45. L. J. Bartolotti, Time-dependent Kohn-Sham density functional theory, Phys. Rev. A 26, 2243-2244 (1982); 2248 (1983).
46. L. J. Bartolotti and P. K. Acharya, On the functional derivative of the kinetic energy density functional, J. Chem. Phys. 77, 4576-4585 (1982).
47. J. P. Perdew, R. G. Parr, M. Levy and J. L. Balduz, Jr., Density functional theory for fractional particle number: Derivative discontinuities of the energy, Phys. Rev. Lett. 49, 1691-1694 (1982).
48. A. Berk, Complementary variational principle for density functional theories, Phys. Rev. A 27, 1-11 (1983).
49. P. K. Acharya, Comment on the derivation of atomic kinetic energy functionals with full Weizsacker correction, J. Chem. Phys. 78, 2101-2102 (1983).
50. M. R. Nyden and R. G. Parr, Restatement of conventional wavefunction determination as a density functional procedure, J. Chem. Phys. 78, 4044-4047 (1983).
51. M. R. Nyden, An orthogonality constrained generalization of the Weizsacker density functional model, J. Chem. Phys. 78, 4048-4051 (1983).
52. R. F. Nalewajski, Reduction of derivatives and simple applications of the Legendre transformed density functional theory, J. Chem. Phys. 78, 6112-6120 (1983).
53. R. G. Parr and L. J. Bartolotti, Some remarks on the density functional theory of few-electron systems, J. Phys. Chem. 87, 2810-2815 (1983).
54. R. K. Pathak and S. R. Gadre, Gradient-free representation of the Weizsacker term for atoms, Phys. Rev. A 28, 1808-1809 (1983).
55. A. Berk, Lower-bound energy functionals and their application to diatomic systems, Phys. Rev. A 28, 1908-1923 (1983).
56. P. Politzer, R. G. Parr and D. R. Murphy, Relationships between atomic chemical potentials, electrostatic potentials, and covalent radii, J. Chem. Phys. 79, 3859-3861 (1983).
57. R. K. Pathak, An upper bound to the exchange integral for Coulomb interactions, J. Chem. Phys., in press.
58. R. G. Parr and R. G. Pearson, Absolute hardness: companion parameter to absolute electronegativity, J. Am. Chem. Soc., in press.
59. R. K. Pathak, Bound excited states within the density functional formalism: the Levy functional, Phys. Rev. A, in press.
60. J. Robles and L. J. Bartolotti, Electronegativities, electron affinities, ionization potentials and hardnesses of the elements within spin polarized density functional theory, J. Am. Chem. Soc., submitted.

61. R. K. Pathak, Statistical electron angular correlation coefficients for atoms within the Hohenberg-Kohn-Sham theory, Phys. Rev. A, submitted.
62. R. K. Pathak, S. R. Gadre and S. P. Gejji, From molecular electron density to electron momentum density, Phys. Rev. A, submitted.
63. L. J. Bartolotti, Variation-perturbation theory within a time-dependent Kohn-Sham formalism: an application to the determination of multipole polarizabilities, spectral sums and dispersion coefficients, J. Chem. Phys., submitted.
64. R. G. Parr, Density functional theory of atoms and molecules, in Horizons of Quantum Chemistry, edited by K. Fukui and B. Pullman (Reidel, 1980), pp. 5-15.
65. R. G. Parr, Density functional theory, in Electron Distributions and the Chemical Bond, edited by M. B. Hall and P. Coppens (Plenum, 1982), pp. 95-100.
66. R. G. Parr, Density functional theory, Ann Rev. Phys. Chem. 34, 631-656 (1983).
67. R. G. Parr, Aspects of density functional theory, in Local Density Approximations in Quantum Chemistry and Solid State Theory, edited by J. Avery and J. P. Dahl (Plenum, 1983), in press.
68. R. G. Parr, Remarks on the concept of an atom in a molecule and on charge transfer between atoms on molecule formation, Int. J. Quantum Chem., submitted.
69. E. Clementi, Proceedings of the Welch Foundation Conferences on Chemical Research. XVI. Theoretical Chemistry (Welch Foundation, Houston 1973), Discussion remark on p. 117.
70. R. G. Parr, Proc. Natl. Acad. Sci. USA 72, 763 (1975).
71. H. Weinstein, P. Politzer and S. Srebrenik, Theoret. Chim. Acta 38, 159 (1975).
72. L. Szasz, I. Berrios-Pagan and G. McGinn, Z. Naturforsh. 30a, 1516 (1975).
73. E. Lieb, Int. J. Quantum Chem. 24, 243 (1983).
74. For example, O. H. Nielsen and R. M. Martin, Phys. Rev. Lett. 50, 697 (1983).
75. N. D. Mermin, Phys. Rev. 137, A 1441 (1965).
76. E. P. Gyftopoulos and G. N. Hatsopoulos, Proc. Nat. Acad. Sci. USA 60, 786 (1968).
77. J. E. Harriman, Phys. Rev. A 24, 680 (1981).
78. J. R. Platt, letter to R. G. Parr dated October 23, 1968.
79. R. G. Pearson, J. Am. Chem. Soc. 85, 3533 (1963).

A DENSITY FUNCTIONAL FORMALISM FOR CONDENSED MATTER SYSTEMS

A. K. Rajagopal

Department of Physics and Astronomy
Louisiana State University
Baton Rouge, Louisiana 70803

I. INTRODUCTION

A very large part of research in condensed matter physics may be considered as investigations of inhomogeneous electron systems. The inhomogeneities are due to nuclear charges located in certain geometrical forms (single nuclear charge for an atom, a small number of several nuclear charges distributed spatially to form small molecules, or a large number of nuclei arranged in a regular spatial three-dimensional array to form a solid etc.). We will not be interested here in amorphous systems even though a density functional scheme is being used in the liquid state research. Also we will not be interested in the detailed properties of the nuclei themselves and treat them as merely positively charged entities with no intrinsic character to them. One class of questions concerns the electronic properties including its spin effects and our main attention will be focussed on these. There have been some attempts to examine nuclear motions (molecular vibrations and phonons in solids) using density functional formalism; but this has not yet been explored fully as will be pointed out later. The central theme in the original density-functional formalism was that the ground state (equilibrium) properties depend only on the ground state (equilibrium) density of the electrons and the nuclei. Stated in this way, we may think of these problems as another facet of Relativistic Quantum Electrodynamics of many electrons and structureless charge compensating nuclei, which for all practical purposes can be treated as classical objects. We are <u>not</u> interested here in the very high density systems that occur in the astrophysical and nuclear matter contexts. These have been dealt with recently by relativistic field theoretical methods at finite temperatures.[1-8] We shall use these same methods for discussing our problems in a unified manner. We may however remark that the density functional schemes are general enough and could be used in studying matter at extremely high densities. I am sure some of this will be dealt with in the forthcoming lectures next week in this School. Let us state at the outset that the energies involved in our system ranges from a few thousand electron volts (corresponding to the total energy of a heavy atom) to fractions of an electron volt (as for the energy difference between different condensed systems). In the density functional view point, these different energies correspond to different portions of the electronic density distribution that partake in the corresponding physics. To set the stage of our discussions, let us define a few basic parameters. The Fermi momentum k_F of an electron gas is related to the electron density, n, by the relation $n = k_F^3/3\pi^2$ and the dimensionless relativistic parameter is often defined as $\beta = \hbar k_F/mc$ where $\hbar$ is the Planck constant divided by 2π, m is the rest mass of the electron and c is the velocity of light, because $\hbar k_F/m$ is a typical velocity in the system and in an atom this is replaced by the average velocity of the electron in its quantum state. For Krypton whose atomic number is 38, $\beta \simeq 1$ and in the solid state context, for Indium metal (Z=49) $\hbar k_F/mc \simeq 1$. The corresponding electron density is of order $6 \times 10^{29} cm^{-3}$. The other important parameter defining our temperature scale is the Fermi temperature, T_F, related to the electron density, n, by $T_F \cong 4 \times 10^{-11}\ n^{2/3}\ °K$ with n in cm^{-3}. The dimensionless temperature parameter is $t = T/T_F$, T being the equilibrium temperature of the system. One other important relationship worth knowing is the relative scale of an electron volt in terms of frequency and it is 2.4×10^5GHz ($= 2.4 \times 10^{14}$ Hz) and in terms of temperature, it is 1.16×10^4 °K. The fine structure constant $\alpha = e^2/\hbar c \simeq 1/137.07$ is a measure of the electromagnetic coupling constant.

In these lecture notes, we take the opportunity to update our own review articles on the density functional theory.[9,10,11] The original works of other authors are referred to in these articles. The present article is a supplement to these reviews and as such we shall repeat certain discussions of detail in these notes even though they will be discussed in the lectures. We follow our philosophy of presenting new material in our reviews not necessarily published before, so that it will serve the dual purpose of survey as well as exposition of new avenues and viewpoints which are sometimes the byproduct of our own extensions of results found in the literature. Companion review articles of Kohn and Vashishta,[12] Callaway and March,[13] and Williams and von Barth[14] should also be consulted. In these notes, we generalize the Levy-Lieb theorems[15,16] to temperature-dependent relativistic case for an electron-nuclei system. This leads us naturally to a discussion of the Born-Oppenheimer approximation which enables the original problem to be tackled as two problems, one for electrons and another for ions. Roughly speaking, the fourth root of the ratio of electron to ion mass, $(m/M)^{1/4}$, is a measure of the coupling between the electron and ion system. Such a scheme is relevant in discussing high density, high temperature plasmas[17,18] as well as low density, low temperature systems such as molecules and solids.[19] Typically, for a metal, $n \simeq 10^{24}$ cm^{-3}, at room temperature $T \sim 300°K$, $\beta \ll 1$ and $t \ll 1$ and so we can treat the system as a (zero temperature) degenerate, non-relativistic problem. However, for a metal like indium or mercury, the inner core electrons will have velocities such that $\beta \gtrsim 1$ and a (zero temperature) degenerate ($t \ll 1$) relativistic treatment suffices. This is also the case for highly stripped atoms such as lithium-like iron and tungsten. In Tokamaks, the electron densities are of order 10^{18} cm^{-3} and the temperature T is of order $10^6 °K$ so that $\beta \ll 1$, $t \gg 1$ so that the electron plasma can be treated as a classical non-relativistic, non-degenerate (high temperature) plasma. In a semiconductor, like silicon, $n \simeq 10^{18}$ cm^{-3} and at room temperature, $T \sim 300°K$, this corresponds to $t \simeq 1$ and $\beta \ll 1$ so that we need a non-relativistic, finite temperature analysis. In a laser-produced plasma, one has $n \simeq 10^{24}$ cm^{-3} or higher and $T \simeq 10^6 °K$ so that $t \simeq 1$ and $\beta \ll 1$ and it suffices to consider a non-relativistic but finite temperature (non-degenerate) theory. To need a finite temperature and relativistic theory, the electron density n must be of the order of 10^{29} cm^{-3} for which $\beta \simeq 1$ and the temperature T should be $10^8 °K$ so that $t \simeq 1$. This system is in the astrophysical domain and we will not consider this here. Thus, for condensed matter systems of our interest, the finite temperature, relativistic theory is not needed. We thus consider briefly the electronic properties of isolated light atoms and some simple molecules in the non-relativistic context and heavy atoms ($Z > 50$) as well as highly stripped atoms in the relativistic framework.[20] We will also discuss briefly the electronic properties of light atoms such as neon, placed in a high density, high temperature laser induced plasma where the finite temperature non-relativistic density functional theory is found to be promising. The spectral characteristics of such light atoms are expected to play a diagnostic role and we present the results of Perrot[21,22] who uses a modified form of our original scheme presented earlier by us.[23,24]

In order to discuss open-shell atoms and magnetic systems such as heavy rare-earths and actinides, one requires a spin-density functional scheme in a relativistic framework. We include this framework in our general presentation of the theory. The full use of this theory has not yet been made, but the non-relativistic spin-density functional scheme has proven quite successful in elucidating the open-shell atoms as well as magnetic properties of Ni, Fe, Co.[13,14]

Much of these discussions are based on a local density approximation. An analysis of the successes and failures of this approximation will only be brief here and the reader is referred to a critical analysis of this by von Barth.[25] Recently dipole susceptibilities of atoms from helium to mercury have been calculated using a time-dependent relativistic local-density scheme by Parpia and Johnson.[26] Mention should be made of the work of Mårtensson[27] on the calculation of the parity-violation in bismuth using a Dirac-Hartree-Fock scheme. We shall not discuss relativistic effects in the solid state context and the reader is referred to a review on this subject by Koelling and MacDonald.[28] This is a fascinating field and I am sure someone will talk about it in this School (Perdew or von Barth?). We will also only briefly discuss the question of improving the local density scheme and here again one may refer to a clear exposition of this by von Barth.[25,14] I will not discuss the electronic structure of an atom near surfaces of metals and semiconductors. This is an important area of physics with clear technological implications. One may refer to a recent review article by Lang[29] who is also lecturing in this School and I am sure he will discuss these aspects in great detail in his course. There are many other aspects of the density functional scheme, some of which will be mentioned here. For instance, improving the local density scheme by using a better representation of the homogeneous electron system[9,30]; the density functional theory for fractional particle number[31]; current density dependence of the functional to examine diamagnetic properties of closed shell systems[32] and the theory of relativistic spin-polarized systems.[33,34] These are some new works which I happened to hear in recent weeks. I am sure you will hear about many other advances and new problems concerning the density functional theory in this School.

In the next section, we will give a brief account of the formal framework of the relativistic density functional theory in its general setting. We will not give proofs of the basic theorems as they are patterned after the original works of Hohenberg, Kohn, and Sham, and Levy and Lieb, about which you have already heard from Professors Kohn and Lieb in this school. The notations will be the same as those used in our other papers.[9,10,11] In the last section, the various applications of the formalism, limitations of the scheme, and some of the new advances in the theory mentioned earlier in the Introduction will be given only briefly and for details the reader is referred to the original articles.

II. BASIC THEOREMS

The system under consideration is composed of N_e electrons and N nuclei which in general may itself be made up of several (K) types of nuclei with the number n_α of the α-type nuclei. Since the mass of the nuclei is three orders of magnitude of that of electrons, we need not consider them in the relativistic framework in our types of condensed matter contexts, nor do we need to consider them even in a quantum mechanical framework. They may be just thought of as structureless positively charged classical objects. However, there are circumstances of very high density matter as in the Astrophysical or Nuclear Matter contexts, where the nuclei will have to be treated appropriately quantum mechanically and even relativistically. We will not be interested in these systems here. At much lower density regimes, the quantum nature of the nuclei do become important in at least four cases, and at low temperatures. These cases are hydrogen, deuterium, helium 3, and helium 4 systems, where the intrinsic property (such as Bose or Fermi) of the nuclei become important. Also the electrons stay close to their parent nuclei and these atoms act as a whole. Of course, at higher temperatures, there are many atomic and molecular systems which occur as liquids and solids and the basic aggregates are ions and even groups of atoms. Depending on the level of questions raised, the gross properties of these systems may be handled by a corresponding density functional scheme. In the solid state, one has ions which are aggregates of a nuclei with a few electrons tightly bound to them. These "composite particles" are then the positively charged ions of interest and the rest of the electrons move about somewhat more freely. A complete theory which derives these ions and the electron disributions is perhaps not a useful framework to study these systems in order to answer the types of questions we are asking. However, what is important to observe here is that at this stage, certain remnant, basic properties of the atom, the ionic core, has to be first properly set up and this problem can be formulated fully field theoretically. We can even construct a field theory of the hydrogen plasma as a proton-electron system of Dirac fields plus the electromagnetic field with which they both interact. In the present context, we will present a field theory in which the electron system is described by a Dirac field, the nuclear system by a classical charged, spin-zero field, plus an electromagnetic field which provides the basic interaction between themselves and among them. The special case of the electron-proton system can be thought of as a two-Dirac field problem along with the electromagnetic interaction. With this we have set up the basic model field theory of interacting charges representing our condensed matter. Furthermore, we may decouple the electron and nuclear motions to a good approximation by using the so-called Born-Oppenheimer approximation. In a general setup, this approximation need not be made as was shown by Capitani et al.[19] Such a framework is important when discussing molecular vibrations, phonons in solids etc.

Before we outline our basic theorems, a few remarks are in order. We use the rest frame of the nuclei as our reference frame. Intrinsic to this field theoretical approach, there are two important features, one is the quantum electrodynamic renormalization effects which for our purposes can be incorporated by using physical mass and charge of the electron and small but significant shifts in the energy levels of the electrons (vacuum polarization effects) which also can be ignored in our context. Such procedures should be internally consistent and we follow them.[7] Second important feature is that the mutual interaction between two relativistically moving charged objects is not just the traditional Coulomb interaction but also new types of mutual interaction known as "Breit interaction". This will be dealt with here in a more general way and is due to the transverse photon-electron interaction and is usually of second order in the fine structure constant. In high density systems, this contribution can be very significant.[11] One further remark is to note that we will not consider situations where there is spontaneous electron-positron pair production. We will assume that the negative energy electron states are always filled so that there is an actual ground state of the condensed matter system, or if we are at a finite temperature, the system is in a statistical equilibrium ensemble.

The Hamiltonian operator of the system is, in the notation of Ref. 11,

$$H = T_e + T_n + H_{rad} + H_{e-rad} + H_{n-rad} + H_{e-ext} + H_{n-ext} \tag{1}$$

T_e is the kinetic energy of the Dirac field of the electrons, T_n is the kinetic energy of the nuclei in the classical form, $\sum_{\alpha=1}^{k} \int \phi_\alpha^\dagger \left(-\frac{\hbar^2}{2M_\alpha} \nabla^2 \phi_\alpha\right) d^3r$, where M_α is the mass of the α-type nucleus, ϕ_α is the field associated with it. H_{rad} is the Hamiltonian of the photon field representing the electromagnetic radiation, H_{e-rad} is the coupling the Dirac electron to the radiation, H_{n-rad} is the corresponding coupling of the nuclear field to the radiation, and H_{e-ext} and H_{n-ext} are the coupling of the electron and nuclear fields to the external fields of our choice. The external fields could be an external electric and magnetic fields etc. representing in general the interaction of the electron charge density, the electron current density and the electron spin density with the appropriate fields. For the nuclear-external field interaction, we take for simplicity just a coupling of the nuclear charge density with an external electrostatic potential. With the Hamiltonian given by Eq. (1), at a finite temprature, T, we define the statistical operator in the grand canonical ensemble,

$$P_o \equiv \frac{1}{Z} \exp -\beta(H-\mu N_e-\mu_n N), \tag{2}$$

where $Z = \mathrm{Tr}\exp-\beta(H-\mu N_e-\mu_n N)$, is the partition function for the system, μ is the electron chemical potential and μ_n is the corresponding nuclear chemical potential. The electron particle density, current density, and spin density are then respectively given by the following expressions:

$$n(\vec{r}) = -\, i\,\mathrm{tr}\{\gamma_o S_F^<(11^+)\} \tag{3}$$

$$\vec{J}(\vec{r}) = -\, i\,\mathrm{tr}\{\vec{\gamma} S_F^<(11^+)\} \tag{4}$$

$$\vec{m}(\vec{r}) = -\, i\,\mathrm{tr}\{\,\vec{\Sigma}\, S_F^<(11^+)\} \tag{5}$$

and the nuclear density of type α at a point $\vec{R}_a$ is

$$\rho_\alpha(\vec{R}_a) = -\, iG_\alpha(R_a t;R_a t^+) \tag{6}$$

Here

$$S_F(12) = (-i)\ \langle\tau(\psi(1)\overline{\psi}(2))\rangle \tag{7}$$

is the finite temperature Feynman propagator for the electron in the ensemble defined by (2) and

$$G_\alpha(R_a t;R_a' t') = (-i)\ \langle\tau(\phi_\alpha(R_a t)\phi_\alpha^\dagger(R_a, t'))\rangle \tag{8}$$

is the finite temperature Green's function for the nuclei of type α. τ is the time ordering symbol in the usual way. In Eqs. (3,4,5), tr stands for trace only the Dirac Spinor indices and $\langle\cdots\rangle$ stands for the average over the ensemble P_o.

We now state without proof the two theorems of Hohenberg and Kohn generalized here for our problem.

Theorem I: In the grand canonical ensemble at a given temperature the equilibrium state of the system is a functional of T and chemical potentials μ and μ_n, the components of the four current density of the electron $J_\mu(\vec{r}) \equiv (n(r), \vec{J}(r))$, the spin density of the electron, $\vec{m}(r)$, and the nuclear particle density $\rho(\vec{R})$ and they are determined by the static external four-vector potential $A^\mu_{ext}(r)$ acting on the electron and the static magnetic field $\vec{B}_{ext}(\vec{r})$, and the nuclear density $\rho_\alpha(\vec{R})$ is determined by the corrsponding external electrostatic field acting on the nuclei. The divergenceless condition $\partial^\mu J_\mu(r) = 0$ is obeyed by the electron four current.

Theorem II: The correct $J_\mu(\vec{r})$, $\vec{m}(\vec{r})$, and $\rho_\alpha(\vec{R})$ appropriate for the fixed external fields minimize the grand potential of the entire system. This grand potential may be written in the form

$$\Omega_{HK}[J_\mu,\vec{m},\rho_\alpha] = F_{HK}[J_\mu,\vec{m},\rho_\alpha] - e \int J_\mu(\vec{r})\, A^\mu_{ext}(r)d^3r$$

$$- \mu_B \int \vec{B}_{ext}(\vec{r}) \cdot \vec{m}(\vec{r})d^3r$$

$$- \int \phi_{ext}(R_a) \sum_{\alpha=1}^{K} \rho_\alpha(R_a)dR_a \tag{9}$$

where

$$F_{HK}[J_\mu,\vec{m},\rho_\alpha] = Tr\{P_0(\tilde{H} - \mu N_e - \mu_n N + k_B T\ln P_0)\} \tag{10}$$

is a universal functional where $\tilde{H} = H - H_{e-ext} - H_{n-ext}$ is the Hamiltonian without the external field. If Ω_0 is the equilibrium grand potential, then

$$\Omega_{HK} \geqslant \Omega_0. \tag{11}$$

The first theorem establishes the existence of the functional F_{HK} and the second theorem is a consequence of the Gibbs free energy variational principle, namely the minimal property of the grand potential with respect to statistical operator. It was independently pointed out by Levy and Lieb that these theorems are very very restrictive because one may define $J_\mu(\vec{r})$, $\vec{m}(\vec{r})$ and $\rho(\vec{R})$ for a wider class of ground canonical ensembles P_0 without reference to the existence of the external coupling fields to the system. i.e., we can in principle find the same $J_\mu(\vec{r})$, $\vec{m}(\vec{r})$, and $\rho_\alpha(\vec{R})$ for a wide clase of statistical operators, P, without reference to the special external coupling fields. Levy called those external potentials for which the Hohenberg-Kohn construction is possible as the v-representable. Levy then proposed a generalization of the above theorem in such a way that the Hohenberg-Kohn theorem is automatically obeyed when one has v-representability. The generalized Levy-Lieb theorems will now be stated. Define the Levy-Lieb functional

$$F_{LL}[J_\mu,\vec{m},\rho] \equiv \inf_{P\varepsilon m} F_P = \inf_{P\varepsilon m} \mathrm{Tr}\{P(\tilde{H}-\mu N_e-\mu_n N + k_B T \ln P)\} \tag{12}$$

where $P\varepsilon m$ implies all those ensembles which give the same $J_\mu(\vec{r})$, $\vec{m}(\vec{r})$, and $\rho(\vec{R})$ and inf implies the smallest such F_P. Such general density matrices are said to be N-representable. This terminology we retain here and it arises from the fact that in a many body system of N-particles, the particle density n(r) is by definition obtained by integrating over all but one coordinate of the modulus of the square of the N-particle wave function. In the early days of the HK theorem, the realization of this n(r) with the HK density constructed out of an external potential was not clearly understood. Out of all N-particle wave functions which give the same n(r), that N-particle wave function which gives the least expectation value of H is then the Levy-Lieb functional. In our context, the definitions of the expectation values were generalized to include grand canonical ensemble averages. The generalized Levy-Lieb theorems are now stated:

<u>Theorem III:</u>

$$\Omega_{LL}[J_\mu,\vec{m},\rho_\alpha] = F_{LL}[J_\mu,\vec{m},\rho_\alpha] - e\int J_\mu(\vec{r})\, A^\mu_{ext}(r)d^3r$$

$$- \mu_B \int \vec{B}_{ext}(\vec{r})\cdot\vec{m}(\vec{r})d^3r - \int \phi_{ext}(\vec{R}_a) \sum_{\alpha=1}^{K} \rho_\alpha(R_a)dR_a \tag{13}$$

<u>Theorem IV:</u>

$$\Omega_o \geqslant \Omega_{LL} \tag{14}$$

Theorem III is really a consequence of the definition of the grand potential, $\Omega[P] \equiv \mathrm{Tr}\{P(H-\mu N+k_B T\ln P)\}$ for a given H and N, and the definition of the Levy-Lieb potential, F_{LL}. Theorem IV is also a consequence of these definitions. If now for some N-representable cases, for which the Levy-Lieb construction is already made, it may be possible to achieve v-representability, then combining the two sets of theorems, we find $\Omega_{HK}[J_\mu,\vec{m},\rho] = \Omega_{LL}[P_0]$ and so the functional Ω is minimized by the equilibrium J_μ, $\vec{m}$, and ρ. The next step is to set up the one-particle equations from the minimum property of these functionals. Note that by its definition, Ω_{LL} need not obey the Euler equation, because of the definition (1). The Euler equations obtained here are $\frac{\delta\Omega}{\delta n(\vec{r})} = \mu$, $\frac{\delta\Omega}{\delta\vec{J}(\vec{r})} = 0$, $\frac{\delta\Omega}{\delta\vec{m}(r)} = 0$, and $\frac{\delta\Omega}{\delta\rho(R)} = \mu_n$.

We may point out here that the Born-Oppenheimer approximation is contained in the assumption of the existence of the external potentials $A_0(\vec{r})$ and $\phi_{ext}(R)$. In the case where the nuclei are non-responsive, fixed classical objects, $A_0(\vec{r})$ is really the attractive potential an electron feels due to the distribution of positive charges and $\phi_{ext}(\vec{R})$ is the corresponding attractive potential a nucleus feels due to the distribution of all electrons. Capitani et al.[19] assumed such A_0 and ϕ_{ext} in developing their non-Born-Oppenheimer density functional theory to obtain the HK-counterpart of the LL-functional. With this same caveat, we have here set up the Levy-Lieb functional quite generally in the non-Born-Oppenheimer context as Capitani et al.[19] but its HK equivalent can only be thought of as a version of the Born-Oppenheimer scheme. In a different context, Dharma-wardana et al.[17,18] constructed an H-K formalism for the hydrogen plasma and it should be evident from the discussion here that their results are subject to the Born-Oppenheimer approximation. These authors show that the hypernetted-chain scheme is a particular approximation to the density-functional equations for the classical proton system and they employ this scheme to examine a variety of questions concerning the hydrogen plasma. Following a time-dependent scheme of Peuckert as generalized by us in Ref. 9, we may, however, develop a time-dependent non-Born-Oppenheimer scheme. The equation for the nuclear motion will then be of the form of self-consistent-phonon theories.

We will from now on focus attention on the <u>electronic part only</u> and <u>state</u> without discussion the corresponding one-particle equation, which is a generalization of the Kohn-Sham equations of the non-relativistic density functional theory.[9]

<u>Theorem V</u>: The one-particle Dirac-like equations of the relativistic spin polarized inhomogeneous electron system are

$$[-i\vec{\alpha}\cdot\vec{\nabla} - m(1-\beta) + V_{eff}(\vec{r},J_\mu,\vec{m}) + \vec{\sum}\cdot\vec{W}_{eff}(\vec{r},J_\mu,\vec{m})]\phi_i(\vec{r})$$

$$= \varepsilon_i\phi_i(\vec{r}) \qquad (15)$$

where

$$V_{eff}(\vec{r},J_\mu,\vec{m}) = -\left(V_{ext}(r) + e^2 \int \frac{n(\vec{r}')}{|\vec{r}-\vec{r}'|} d\vec{r}' + \frac{\delta\Omega_{xc}}{\delta n(\vec{r})}\right)$$

$$- e\vec{\alpha}\cdot\left(\vec{A}_{ext}(r) + \int \frac{\vec{J}(\vec{r}')}{|\vec{r}-\vec{r}'|} d\vec{r}' + \frac{\delta\Omega_{xc}}{\delta\vec{J}(\vec{r})}\right) \quad (15a)$$

$$\vec{W}_{eff}(\vec{r},J_\mu,\vec{m}) = - e\left(\vec{B}_{ext}(r) + \frac{\delta\Omega_{xc}}{\delta\vec{m}(\vec{r})}\right) \quad (15b)$$

where $\vec{\alpha}$, β, $\vec{\Sigma}$ matrices are related to the usual Dirac matrices by $\vec{\gamma} = \beta\vec{\alpha}$, $\gamma_0 = \beta$, $\Sigma^k = \frac{1}{2}(\gamma^i\gamma^j - \gamma^j\gamma^i)$, i,j,k cyclic. The electron density

$$n(r) = \sum_i \text{tr} \langle\phi_i^*(r)\phi_i(r)\rangle , \quad (16a)$$

the current density

$$J_k(\vec{r}) = \sum_i \text{tr} \langle\phi_i^*(r)\alpha_k\phi_i(r)\rangle , \quad (16b)$$

and magnetization density

$$m_k(\vec{r}) = \sum_i \text{tr} \langle\phi_i^*(r)\Sigma_k\phi_i(r)\rangle , \quad (16c)$$

$\{\phi_i\}$ are a complete orthonormal set and the eigenvalues ε_i do not have a direct physical significance. Ω_{xc} is the exchange correlation contribution to the equilibrium free energy functional of the system.

We may also note that the free energy of the system can be expressed in terms of the eigenvalues $\{\varepsilon_i\}$ and additional terms involving Ω_{xc}, V_{eff} and W_{eff}. One may express the equilibrium state expectation values of operators of physical interest such as momentum density which is useful in computing Compton profiles, in terms of the $\{\phi_i\}$ and an appropriate derivative of Ω_{xc}.[35] For this formalism to be useful, it is clear that we need to obtain the universal functional Ω_{xc} in a suitable form ready for finding numerical solutions of our basic Eq. (15). Once this is accomplished, a large number of problems can be attacked as is clear from the review articles quoted earlier.[9,12,13,14,25,29] There is as yet no known clear procedure for obtaining Ω_{xc}. In summary, given the universal functional Ω_{xc}, and given the external potential which defines the problem we want to solve (for an atom, $V_{ext} = -Ze^2/r$, for molecules and solids $V_{ext} = -\sum_{\vec{\ell},k} e^2 Z_k/|r-R_{\ell k}|$ etc.), one solves self-consistently the equations (15) and obtains ε_i, ϕ_i and also the electron density $n(\vec{r})$ etc. as well as any other physical quantity of interest. One could calculate many other equilibrium state properties as well. In general the eigenvalues ε_i do not have any direct physical significance. A calculation of the binding energies based on the difference of two density-functional calculations, one for the system with N electrons and another for (N-1) electrons with the one electron taken out of the N-electron system by specifying its state, i, may be related under certain circumstances to, ε_i, the eigenvalues of the N-electron Kohn-Sham equation. For more discussion, see Ref. 25 of von Barth. One of the most popular and practical schemes for Ω_{xc} is to employ the local-density approximation. Here it is assumed that the system is composed of cells of uniform densities and from the known properties of electron gas, Ω_{xc} can be deduced. Using such a method, a large amount of work has been done. The scheme is not capable of systematic improvement as can be deduced from the efforts of all the attempts that have been made so far on this subject. But to a surprisingly large extent, the local density scheme gives reasonable answers even in cases where one expects it to fail. It is generally found that for all the ground-state properties, the local-density scheme gives "good" answers and for properties involving excitations such as binding energy, ionization energy etc. it gives "not so good" answers. One may suspect from this empirical observation that the latter depends on V_{xc} and W_{xc} which are derivatives of Ω_{xc} and hence more subtle and careful approximations are required than for Ω_{xc} itself. These aspects have become clearer recently and a discussion of these is a lecture in itself. One may refer to von Barth[25] for a clear account of this. The properties of the homogeneous electron gas are not known for all densities either and this is an active area of research by itself. What is required in the density functional scheme is a suitable representation of the most accurate determination of the free energy of an electron gas for a wide range of densities and temperatures. Only recently some of these have been evaluated for non-relativistic case in some approximation (see Ref. 10, for example) and for relativistic case at T = 0°K (see Ref. 11) using ring diagram sum approximation for the non-magnetic case and just the first order exchange-diagram approximation for the magnetic case.[34] The reader is referred to these works for details.

Recently Zumback and Maschke[36] have proposed a completely new approach to calculate in principle the Levy-Lieb functional, F_{LL}, in the T = 0°K case. They also suggest ways of obtaining upper bounds on this F_{LL}. They achieve this by explicitly constructing a complete set of orthonormal one-particle wave functions for a given particle density $n(\vec{r})$ (also see Lieb[16] for this construction) and construct a many-particle wave function by forming a determinantal wave function. Using this, they explicitly calculate F_{LL}. This construction at least demonstrates a method of obtaining F_{LL} and they also show that the wave function so constructed cannot be derived from a local one-particle potential as required in HK.[16] It appears that by taking linear combinations of these new determinantal wave functions one may be able to get a reliable exchange-correlation functional. These authors have not yet offered an alternative functional which can be used widely, but the procedure holds promise. It is not obvious just as yet how one could generalize this procedure to construct more general functionals for spin or relativistic or finite temperature situations. These authors, by their explicit construction of F_{LL}, have converted the existence theorem for F_{LL} into a feasible theory.

III. A FEW APPLICATIONS

We will focus attention on applications to major atomic problems in essentially two circumstances, one the relativistic, zero temperature case and another, the finite temperature calculation of an atom imbedded in a hot, dense plasma. Our discussion here will be brief and the original articles should be consulted for details. The basic ingredients in any such calculation are

(i) a model calculation of Ω_{xc} and V_{xc};

and (ii) self consistent solution of Eq. (14) with the given V_{ext} etc.

The relativistic Ω_{xc}, V_{xc} were calculated at zero temperature by Ramana and Rajagopal[11] using a local density scheme, and, for the spin-polarized case using only first order exchange processes by Xu, Rajagopal, and Ramana[34] (see also MacDonald[33]). The latter has not yet been applied to any problem yet. In the relativistic case one has besides the usual Coulomb interaction between electrons, an additional two-electron interaction originally discovered by Breit which is entirely relativistic in origin. This interaction arises from the transverse photon-electron interaction and is opposite in sign to that of the other term and at high densities corresponding to $\beta \gtrsim 2.52$, the new exchange energy overtakes the usual one and for $\beta > 1.9$, the corresponding local exchange potential changes sign. In the spin-dependent case, this region spreads out for different magnetizations as well.[34] The correlation contribution to these energies and potentials are not as dramatic but the final result is quite different from the non-relativistic answers for high densities. At one point, before our work,[11] it was thought that one may simply use the non-relativistic Ω_{xc} and V_{xc} in a relativistic scheme but this notion is now laid to rest. The results of calculations on lithium-like ions of carbon, iron, as well as Hg, U, Fm etc. have been published.[11,27] There is a new application of our relativistic potential to the calculation of dipole susceptibilities of atoms from helium to mercury by Parpia and Johnson.[26] The relativistic corrections are in general found to be about 5-10% to that obtained by merely using non-relativistic results in a Dirac-like theory. In the early works, the corrections were expected to be 2% or less.

The non-relativistic finite temperature local-density expressions for Ω_{xc} and V_{xc} were computed by Gupta and Rajagopal.[10] (Taylor and Dharma-wardana also reported a calculation of this independently. See Ref. 10 for a discussion.) These authors used the ring-diagram approximation to compute Ω_{xc} and V_{xc} as a function of n and t, and obtained some interesting features. These results were employed by Perrot[22] to study the shift in spectral lines of neon in an electron gas using a self-consistent scheme. See Perrot for details. In the computation of Ω, Gupta and Rajagopal used the ideal gas chemical potential μ_0 in place of the full chemical potential, μ. It can be shown[37] that the correction term due to this approximation is zero for zero temperature, and goes to zero like t^{-3} for large t, faster than Ω_{xc} which goes like $t^{-3/2}$. It is expected that the correction is not significant at any finite t. At finite temperatures, one computes the eigenvalues ε_i as well as the corresponding occupation probability $n_i = (\exp \beta(\varepsilon_i - \mu)+1)^{-1}$, where $\beta = 1/k_BT$ is proportional to the inverse temperature and μ is the corresponding chemical potential. The work of Perrot brings out these aspects in a self consistent scheme. The early work[10] employed a non-self-consistent scheme. It is found that self-consistency is important. In the lecture, we show slides exhibiting these various aspects concerning all the above calculations and in these notes we shall only make references to these original works in order to keep the length of the notes to a reasonable size.

We will also display the relative orders of magnitude of various contributions to the calculation of total energy or the state energy so as to bring out the importance or otherwise of Ω_{xc}, self-consistency, etc.

We would like to end this discussion with a few remarks. In the presence of a vector potential, the HK theorem would imply that the ground state is a functional of current density also.[9] This aspect is important in understanding diamagnetism in particular, as well as studying magnetic properties of closed shell systems. Harris and Cina[32] have calculated the energy of an inhomogeneous electron gas in the presence of a weak, slowly varying vector potential by an extension of the semi-classical approximation to this problem, and have expressed the ground state energy as a function of n(r) and $\vec{j}(r)$. This development will lead to some more new work in the basic as well as applied areas. Another area of interest is the spin-polarized relativistic systems which is expected to enlighten us concerning heavy open shell atoms and magnetism of actinides etc. Xu et al.[34] have calculated the exchange-energy functional for a model spin-polarized electron gas. This should be contrasted with an alternate theory of exchange in relativistic spin-polarized electron gas by MacDonald.[33] He puts in a spin-only magnetic field interaction and calculates the exchange energy of the system. In our scheme,[11,34] we construct the spin-dependent scheme by a covariant construction of spin in the relativistic theory and use it to set up the formalism. In our opinion, this is important because spin density is a subtle object in relativistic theory and as shown by us, does not follow the non-relativistic route of being generated by a magnetic field alone. The reason for this lies in the fact that only in a given frame, we have only the magnetic field present whereas in a proper relativistic scheme, one can define spin which properly transforms under a Lorentz transformation. One may use the Levy-Lieb theorem to compute the functional $\Omega_{LL}(n,\vec{m})$ without explicit use of an external potential as MacDonald seems to want. It should be pointed out that for vanishing magnetic fields, MacDonald expects to recover our results but as shown by us, this requires a lot of care.

We must end this discussion by stating that the "density functional" scheme requires much more investigation at a basic level. The meaning of ε_i, the extension of the theory to other than the ground state, the time-dependent formalism, how to construct better Ω_{xc} even for the ground state etc. are all avenues which will occupy us in the next few years. There exist partial schemes in each of the above areas[9] but some of these have become as "successful" as the "local-density scheme" for the ground state.[14]

ACKNOWLEDGMENTS

Thanks are due to my co-workers Drs. Ramana, Uday Gupta, and Bu-Xing Xu for close collaboration on the work described here. It is a pleasure to place on record our debt of gratitude to Dr. D. A. Liberman of Los Alamos Scientific Laboratory for all the generous help in setting up the numerical schemes both for the relativistic and for the finite temperature cases. These lectures were prepared during a visit to the Solid State Division of the Oak Ridge National Laboratory. The friendly hospitality and support of the Solid State Theory Group is acknowledged with many thanks. The expert typing of Ms. Velma Hendrix at Oak Ridge is appreciated.

REFERENCES

1. E. S. Fradkin, Proc. (Trudy) Lebedev Phys. Inst. 29, 1 (1967).
2. B. Bezzerides and D. F. Dubois, Ann. Phys. (N.Y.) 70, 10 (1972).
3. R. L. Bowers, J. A. Campbell, and R. L. Zimmerman, Phys. Rev. D 7, 2278 (1973).
4. C. W. Bernard, Phys. Rev. D 9, 3312 (1974).
5. L. Dolan and R. Jackiw, Phys. Rev. D 9, 3320 (1974).
6. S. Weinberg, Phys. Rev. D 9, 3357 (1974).
7. S. A. Chin, Ann. Phys. (N.Y.) 108, 301 (1977).
8. A. Bechler, Ann. Phys. (N.Y.) 135, 19 (1981). This paper gives references to many other studies on relativistic many-body theory of matter at high densities and temperatures relevant to astrophysical phenomena and nuclear matter.
9. A. K. Rajagopal, Adv. in Chem. Phys. 41, 59 (1980).
10. U. Gupta and A. K. Rajagopal, Phys. Rev. C 87, 259 (1982).
11. M. V. Ramana and A. K. Rajagopal, Adv. in Chem. Phys. 54, 231 (1983).
12. W. Kohn and P. Vashishta, review article in Physics of Solids and Liquids, (eds. S. Lundqvist and N. H. March) Plenum Press (to appear, 1983).
13. J. Callaway and N. H. March, review article in Advances in Solid State Physics, (eds. H. Ehrenreich, F. Seitz, and D. Turnbull) Academic Press (N.Y.) (to appear, 1983).
14. A. R. Williams and U. von Barth, review article in Physics of Solids and Liquids, (eds. S. Lundqvist and N. H. March) Plenum Press (to appear, 1983).
15. M. Levy, Proc. Natl. Acad. Sci. (USA) 76, 6062 (1979) and M. Levy, Phys. Rev. A 26, 1200 (1982).
16. E. H. Lieb, in Physics as Natural Philosophy, (ed. by A. Shimony and H. Feshbach), MIT Press, Cambridge, Mass., (1982).
17. M. W. C. Dharma-wardana and F. Perrot, Phys. Rev. A 26, 2096 (1982).
18. M. W. C. Dharma-wardana, F. Perrot, and G. C. Aers, Phys. Rev. A 28, 344 (1983).
19. J. Capitani, R. F. Nalewajski, and R. G. Parr, J. Chem. Phys. 76, 568 (1982).

20. M. P. Das, M. V. Ramana, and A. K. Rajagopal, Phys. Rev. A 22, 9 (1980).
21. F. Perrot, Phys. Rev. A 25, 489 (1982).
22. F. Perrot, Phys. Rev. A 26, 1035 (1982).
23. U. Gupta and A. K. Rajagopal, J. Phys. B: Atomic and Mol. Phys. 12, 2703 (1979); 14, 2309 (1981).
24. U. Gupta and A. K. Rajagopal, Phys. Rev. A 21, 2064 (1980) and A 22, 2792 (1980).
25. U. von Barth, review article in the Nato Advanced Study Institute held in Gent, July 1982.
26. F. A. Parpia and W. R. Johnson, J. Phys. B: At. and Mol. Phys. 16, L375 (1983), and private communication (1983).
27. Ann-Marie Martensson, Physica Scripta 21, 293 (1980).
28. D. D. Koelling and A. H. MacDonald, review article in the Nato Summer School Lectures, Burnaby B.C., Summer (1981).
29. N. D. Lang, review article in Physics of Solids and Liquids, (eds. S. Lundqvist and N. H. March) Plenum Press, (to appear, 1983).
30. R. Colle and O. Salvetti, J. Chem. Phys. 79, 1404 (1983).
31. J. P. Perdew, R. G. Parr, M. Levy, and J. L. Balduz, Jr., Phys. Rev. Lett. 49, 1691 (1982).
32. R. A. Harris and J. A. Cina, J. Chem. Phys. 79, 1381 (1983).
33. A. H. MacDonald, J. Phys. C: Solid State Phys. 16, 3869 (1983).
34. Bu-Xing Xu, A. K. Rajagopal, and M. V. Ramana, J. Phys. C: Solid State Phys. (to appear, 1983).
35. G. E. Bauer, Phys. Rev. B 27, 5912 (1983).
36. G. Zumbach and K. Maschke, Phys. Rev. A 28, 544 (1983).
37. A. K. Rajagopal, unpublished notes (1983).

DENSITY FUNCTIONALS FOR CORRELATION ENERGIES OF ATOMS AND MOLECULES

Hermann Stoll and Andreas Savin

Institut für Theoretische Chemie
Universität Stuttgart
D-7000 Stuttgart 80, West Germany

I. INTRODUCTION

The correlation energy, E_c, is usually defined as the difference of the exact (non-relativistic) energy, E, and the Hartree-Fock (HF) energy, E_{HF}.[1] E_c is a very small part of E only (1.4% for the He atom, 0.3% for Ne, 0.1% for Ar), but it is non-negligible in absolute value: for valence-shell removal, ΔE_c is 1.1 eV for He, 9.5 eV for Ne, and 9.3 eV for Ar. Inclusion of E_c is important in cases where the number of (strongly interacting) electron pairs is changed, for dissociation energies (D_e), ionization potentials and excitation energies, e.g.. Correlation is responsible for 23% of D_e in the case of H_2, and for 84% of D_e in the case of Li_2; Na_2 and K_2 are unbound at the HF level.

There is a number of methods for calculating E_c, among which are configuration interaction (CI), many-body perturbation theory (MBPT), and the density-functional (DF) method. Before concentrating on DF, a few remarks seem to be in order with regard to CI, the method which is most widely used in quantum chemistry nowadays.[2] In the CI wave-function, excited configurations are admixed to the HF wave-function, Φ_o, and the expansion coefficients are determined by energy minimization. The expansion is usually restricted to single and double substitutions (CI-SD). CI-SD is not size-consistent ($E_c \sim \sqrt{N}$ for a system of N non-interacting two-electron atoms), but unlinked-cluster effects can be introduced into CI-SD in a simple and efficient (although non-variational) way.[3] Already with few terms in the CI expansion, a substantial portion of the correlation contribution to dissociation energies can be obtained (70% for F_2, e.g., with two determinants, Φ_o and $\Phi(\sigma_g^2 \rightarrow \sigma_u^2)$, if orbitals are optimized[4]), but the convergence is extremely slow eventually;

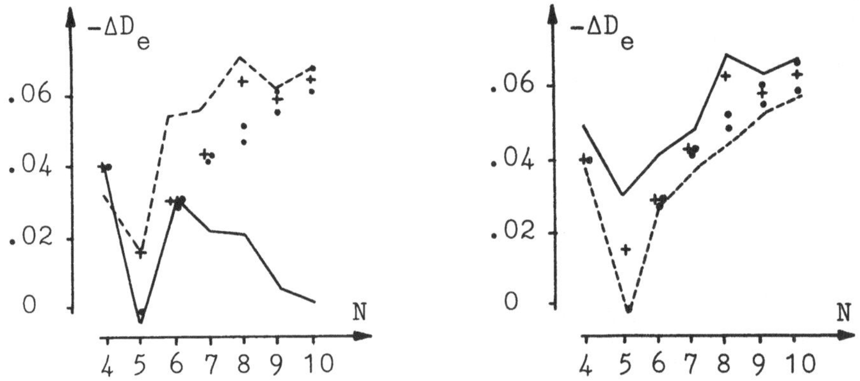

Fig. 1. Correlation contributions to dissociation energies, ΔD_e (au), for first-row monohydrides AH (N: number of electrons). + ,• : exp. values from Refs. 8,9.
a) $E_c(AH) \approx E_c(A^+)+E_c(H^-)$: ——— ; $E_c(AH) \approx E_c(A^-)$:----.
b) $E_c(A_{N-1}H) \approx E_c(A_N)$: ——— ; CI results from Ref.9:----.

up to $10^5..10^6$ determinants are included in current CI calculations. The convergence problem is intimately connected with the difficulty to describe the correlation cusp (cf. Sects.IV,VIIA) in a CI expansion. The accuracy of CI results for small molecules is impressive: for first- and second-row monohydrides, between 95% (LiH) and 85% (HCl) of the valence correlation energy is recovered;[1] deviations from experiment are ~0.003 Å for bond lengths r_e, ~14 cm^{-1} for vibrational frequencies ω_e, < 0.3 eV for D_e.[5]

The DF method provides an economical and physically appealing alternative to CI calculations. The exact density functional is not explicitly known (perhaps very complicated), but simple local approximations exist (cf. Sects.III,IV). Are they expected to work? If so, simple relations should exist between correlation energies and densities; E_c should be similar, in particular, for atoms and molecules with similar densities. There are such relations, indeed, and we just call attention to three of them: i) total correlation energies of (neutral) atoms and molecules have been found to increase, to a good approximation, linearly with the number of electrons, $E_c \approx -0.042\cdot(N-1)$ au;[6,7] actually, this is the most primitive form of a density functional for E_c. ii) For first-row monohydrides, a good estimate of the correlation contribution to D_e is obtained from $E_c(AH) \approx E_c(A^+)+E_c(H^-)$ at the beginning of the row, and from $E_c(AH) \approx E_c(A^-)$ at the end of the row[7,8] (cf.Fig.1a). This can easily be rationalized in terms of the charge transfer in AH molecules which is A→H at the beginning and H→A at the end of the row. iii) The charge densities of the monohydrides should not be too different, on the other hand, from those of the united atoms; thus we expect $E_c(A_{N-1}H) \approx E_c(A_N)$, where the index N refers to the electron number of the (neutral) atoms A. Again, this approximation proves to be rather satisfactory for the correlation contribution to D_e [8] (cf.Fig.1b).

II. HOHENBERG-KOHN THEOREM

The exact (non-relativistic) ground-state energy E of a N-electron system is obtained by minimizing the expectation value of the Hamiltonian H

$$E = \min_{\psi} \langle\psi|H|\psi\rangle = \min_{\varrho} \{ \min_{\psi_\varrho} \langle\psi_\varrho|H|\psi_\varrho\rangle\} = \min_{\varrho} E[\varrho] \quad , \qquad (1)$$
$$E[\varrho] = \int \varrho(r)u(r)dr + \min_{\psi_\varrho} \langle\psi_\varrho|T+V|\psi_\varrho\rangle \quad .$$

The minimization over all (normalized) N-electron wave-functions ψ is performed in two steps: the search is over all N-representable one-particle densities ϱ and, for given ϱ, over all wave-functions ψ_ϱ yielding this density.[10] The second term in the density functional $E[\varrho]$ contains the operators of kinetic energy, T, and electron-electron interaction, V; it is a universal functional of ϱ in the sense that it does not depend explicitly on the external potential u(r).[11] Restricting ψ and ψ_ϱ in (1) to the form of Slater determinants S, the HF density functional, $E_{HF}[\varrho]$, is defined

$$E_{HF}[\varrho] = \int \varrho(r)u(r)dr + \min_{\psi_\varrho \in S} \langle\psi_\varrho|T+V|\psi_\varrho\rangle \ . \qquad (2)$$

If the explicit form of $E[\varrho]$ (and $E_{HF}[\varrho]$) were known in terms of ϱ, HF as well as correlation energies could be obtained without any reference to wave-functions. The definitions (1) and (2) are not of immediate practical use for generating the explicit functionals; determination of $\min_{\psi_\varrho} \langle\psi_\varrho|T+V|\psi_\varrho\rangle$ for a variety of ϱ is by no means less difficult than solving the Schrödinger equation for a variety of external potentials. This is the reason why some quantum chemists seriously ask if the Hohenberg-Kohn (HK) theorem is of any use at all. The practical use of the HK theorem depends, of course, on the possibility to find simple but sufficiently accurate approximations to the functionals. The starting-point for the generation of approximations, both to $E[\varrho]$ and $E_{HF}[\varrho]$, is the decomposition[12]

$$E[\varrho] = \int \varrho(r)u(r)dr + \min_{\psi_\varrho \in P} \langle\psi_\varrho|T|\psi_\varrho\rangle + \frac{1}{2}\iint \frac{\varrho(r_1)\ \varrho(r_2)}{r_{12}} dr_1 dr_2$$
$$+ E_{xc}[\varrho] \ . \qquad (3)$$

P is the Hartree product form; the exchange-correlation functional $E_{xc}[\varrho]$ has to be replaced by an exchange functional, $E_x[\varrho]$, in the HF case. Alternatively, $E[\varrho]$ can be directly related to $E_{HF}[\varrho]$[12]

$$E[\varrho] = E_{HF}[\varrho] + E_c[\varrho] \ . \qquad (4)$$

Table 1. Correlation energies E_c (in 10^{-3} au) of atoms (ions), as obtained from different definitions:
a) $E_c = E[\rho] - E_{HF}[\rho_{HF}]$, b) $E_c = E[\rho] - E_{HF}[\rho]$.
The functional (15) is used for the evaluation of $E[\rho]$.
N: number of electrons, Z: nuclear charge.

Z \ N	2		3		4	
3	70.10	70.13	71.79	71.84	97.49	97.62
4	78.26	78.28	81.38	81.42	116.40	116.52
10	104.47	104.48	111.53	111.55	172.92	172.93
	a)	b)	a)	b)	a)	b)

Note that the correlation functional $E_c[\rho]$ does not yield, for the exact ρ, the correlation energy E_c as defined in Sect.I

$$E_c[\rho] = E[\rho] - E_{HF}[\rho] \neq E[\rho] - E_{HF}[\rho_{HF}] \quad . \tag{5}$$

The difference is very small numerically, however, at least if a local DF approximation is used for evaluating $E[\rho]$[7] (cf.Table 1). The discussion of $E_c[\rho]$ and various approximations to it will be the subject of the following sections. We conclude this section with two remarks, on the extension of the HK theorem to the spin-dependent case, and on the minimization of $E[\rho]$ in (4) with respect to ρ.

For spin-dependent external potentials u_α, u_β, (1) has to be replaced by[13]

$$\begin{aligned} E &= \min_{\rho_\alpha,\rho_\beta} E[\rho_\alpha,\rho_\beta] \\ E[\rho_\alpha,\rho_\beta] &= \int \rho_\alpha(r)u_\alpha(r)dr + \int \rho_\beta(r)u_\beta(r)dr \\ &\quad + \min_{\Psi_{\rho_\alpha,\rho_\beta}} \langle \Psi_{\rho_\alpha,\rho_\beta} | T+V | \Psi_{\rho_\alpha,\rho_\beta} \rangle \quad . \end{aligned} \tag{6}$$

ρ_α, ρ_β are partial charge densities for spin α and β (often called spin-densities), which add up to the correct electron number: $\int(\rho_\alpha(r) + \rho_\beta(r))dr = N$; the search is over wave-functions $\Psi_{\rho_\alpha,\rho_\beta}$ yielding these spin-densities. Eq.(6) is widely used for $\rho_\alpha \neq \rho_\beta$, even if $u_\alpha = u_\beta$, since the (exchange-)correlation functionals are more easily expressed in terms of ρ_α, ρ_β than in terms of ρ for non-vanishing spin-polarization. All the relations for $E_c[\rho]$ which will be given in the following sections can be easily generalized to corresponding ones for $E_c[\rho_\alpha,\rho_\beta]$.

For the minimization of $E[\rho]$ in (4) (or the corresponding $E[\rho_\alpha,\rho_\beta]$), the densities are written in terms of spin-orbitals $\rho_\sigma = \sum_i \varphi_i^{\sigma *}\varphi_i^\sigma$ ($\sigma = \alpha,\beta$). The HF energy is a well-known functional of the spin-orbitals. Thus

$$E = \min_{\rho_\alpha,\rho_\beta} (E_{HF}[\rho_\alpha,\rho_\beta] + E_c[\rho_\alpha,\rho_\beta])$$
$$= \min_{\{\varphi_i^\sigma\}} (E_{HF}[\{\varphi_i^\sigma\}] + E_c[\{\varphi_i^\sigma\}]) \quad . \tag{7}$$

and the minimization with respect to the φ_i^σ leads to the Kohn-Sham equations[12], which differ from the (unrestricted) HF equations only by local correlation potentials $\mu_{c\,\alpha,\beta} = \delta E_c[\rho_\alpha,\rho_\beta]/\delta\rho_{\alpha,\beta}$ which have to be added to the Fock operators $F_{\alpha,\beta}$.

III. LOCAL DENSITY FUNCTIONALS

We now turn to the determination of $E_c[\rho]$. Introducing a variable interelectronic interaction strength $V = \sum_{i<j} \lambda/r_{ij}$ in (1) leads to a λ-dependent density functional $E^\lambda[\rho]$, and

$$E_c[\rho] = E^{\lambda=1}[\rho] - E_{HF}^{\lambda=1}[\rho] = \int_0^1 d\lambda \left(\frac{\partial E^\lambda[\rho]}{\partial\lambda} - \frac{\partial E_{HF}^\lambda[\rho]}{\partial\lambda}\right) , \tag{8}$$

since the exact and the HF functional coincide for vanishing interelectronic interaction, $\lambda=0$. From the Hellmann-Feynman theorem

$$\frac{\partial E^\lambda[\rho]}{\partial\lambda} = \frac{1}{2}\iint \frac{\pi^\lambda(r_1, r_2)}{r_{12}} dr_1 dr_2$$
$$= \frac{1}{2}\iint \frac{\rho(r_1)\rho(r_2)\, g^\lambda(r_1,r_2)}{r_{12}} dr_1 dr_2 \tag{9}$$

where π is the two-particle density, and g is the pair-correlation function. From (8) and (9), a useful relation between $E_c[\rho]$ and the pair-correlation functions g^λ, g_{HF}^λ is obtained[14]

$$E_c[\rho] = \frac{1}{2}\iint \frac{\rho(r_1)\ \rho_c(r_1,r_2)}{r_{12}} dr_1 dr_2$$
$$\rho_c(r_1,r_2) = \rho(r_2)\int_0^1 d\lambda[g^\lambda(r_1,r_2) - g_{HF}^\lambda(r_1,r_2)] \tag{10}$$

An approximation to $E_c[\rho]$ can now be generated from (10) by using a suitable model system for the determination of ρ_c. If the homogeneous electron gas is chosen as model system,

$$E_c[\rho] = \int \rho(r_1)\ \epsilon_c(\rho(r_1))dr_1 \tag{11}$$

results, where $\epsilon_c(\rho)$ is the correlation energy per particle of the electron gas with density ρ. The electron-gas ϵ_c may be rather different from that of inhomogeneous systems such as atoms and molecules, but note that only the spherical average of $\rho_c(r_1,r_2)$ around r_1 contributes to $E_c[\rho]$; non-spherical parts, present in inhomogeneous systems but not in the electron gas, cancel out. Furthermore, the same sum-rule, $\int \rho_c(r_1,r_2)dr_2 = -1$, holds for the electron gas as well as for atoms and molecules.[14] In the spin-polarized case, the local-density (LD) approximation (11) has to be replaced by the local-spin-density (LSD) one

$$E_c[\rho_\alpha,\rho_\beta] = \int \rho(r_1)\ \epsilon_c(\rho_\alpha(r_1), \rho_\beta(r_1))dr_1 \tag{12}$$

where $\epsilon_c(\rho_\alpha, \rho_\beta)$ refers to the correlation energy (per particle) of the spin-polarized electron gas with (spin-)densities ρ_α, ρ_β. The correlation potentials corresponding to (11) and (12) are strictly local; they depend on the densities at a single point only

$$\mu_{c\alpha,\beta} = \epsilon_c(\rho_\alpha, \rho_\beta) + \rho\, \frac{\partial \epsilon_c(\rho_\alpha, \rho_\beta)}{\partial \rho_{\alpha,\beta}} \ . \tag{13}$$

Several parametrizations have been suggested for $\epsilon_c(\rho_\alpha, \rho_\beta)$ in the literature, the most accurate probably being that by Vosko et al.[15] Monte-Carlo calculations[16] for the para- and ferromagnetic electron gas are used here, the high- and low-density limits are carefully taken into account, RPA results are employed to model the dependence of ϵ_c on $\zeta = (\rho_\alpha - \rho_\beta)/\rho$. If not stated otherwise, the function by Vosko et al (VWN) underlies the numerical LSD results given in this paper. The deviation of the VWN parametrization from other ones, e.g. the widely used Gunnarsson-Lundqvist (GL)[14] parametrization, is significant for total correlation energies[17] (the differences are o.7 eV for He, 4.7 eV for Ne, and 8.7 eV for Ar; the VWN energies are smaller in magnitude). The deviations are mainly for high densities; valence properties are much less affected, therefore (differences in ionization potentials[17] are o.2 eV for He, o.4 eV for Ne, and o.3 eV for Ar; differences in D_e[18] are o.2 eV for H_2 and Li_2, and o.1 eV for F_2). Compared to "experimental" correlation energies[2,19], LSD correlation energies are too large[20] by a factor ~2 (with VWN[17], the factor is 2.7 for He, 1.9 for Ne, and 1.8 for Ar).

IV. SELF-INTERACTION CORRECTIONS

The overestimation of total correlation energies with LSD (eqs. 11,12) is connected to the fact that LSD ascribes a non-vanishing correlation energy even to a system of non-interacting one-electron atoms. Atomic and molecular correlation energies are contaminated by such spurious "self-correlation" terms in the LSD approximation.[21] The question is, how to eliminate them. An analysis of (10) in terms of different spin pairs $\sigma\sigma'$ is helpful here:

$$E_c[\rho_\alpha,\rho_\beta] = \frac{1}{2}\sum_{\sigma=\alpha,\beta}\sum_{\sigma'=\alpha,\beta}\iint \frac{\rho_\sigma(r_1)\ \cdot\ \rho_{c\,\sigma\sigma'}(r_1,r_2)}{r_{12}}\,dr_1 dr_2$$

$$\rho_{c\,\sigma\sigma'}(r_1,r_2) = \rho_{\sigma'}(r_2)\int_0^1 d\lambda\,[g^\lambda_{\sigma\sigma'}(r_1,r_2) - g^\lambda_{HF\,\sigma\sigma'}(r_1,r_2)] \tag{14}$$

The LSD approximation amounts to i) substituting $\varrho_{\sigma'}(r_2)$ by $\varrho_{\sigma'}(r_1)$, and ii) using the pair-correlation functions of the homogeneous electron gas with spin densities $\varrho_\alpha(r_1)$, $\varrho_\beta(r_1)$. Approximation i) is reasonable for $\sigma \neq \sigma'$, because the Coulomb hole $g_{\sigma\sigma'}^\lambda - g_{HF\,\sigma\sigma'}^\lambda = g_{\sigma\sigma'}^\lambda - 1$ is centered around r_1, with a cusp at that point. The difference between the exact and the HF Fermi hole, $g_{\sigma\sigma}^\lambda - g_{HF\,\sigma\sigma}^\lambda$, on the other hand, vanishes quadratically for $r_2 \to r_1$ (cf.Sect.VIIA); thus the substitution i) should be much less appropriate for $\sigma = \sigma'$.[22] We conclude that the LSD approximation should be better for antiparallel than for parallel spins. This is not too surprising. Two effects connected with the correlated motion of parallel-spin electrons, exchange and parallel-spin correlation beyond exchange, are treated on quite different levels of approximation in the LSD scheme (4),(11): exchange is treated at the HF level, while the remaining correlation effects are described by an electron-gas expression.

The alternatives are either to resort to the exchange-correlation variant of LSD (xc-LSD,eq.3), or to apply the LSD correlation functional (c-LSD) only to the antiparallel-spin case. In the latter case, the approximation for $E_c[\varrho_\alpha,\varrho_\beta]$ reads[22]

$$E_c[\varrho_\alpha,\varrho_\beta] = \int \varrho(r) \cdot \epsilon_c(\varrho_\alpha(r), \varrho_\beta(r)) dr \qquad (15)$$
$$- \int \varrho_\alpha(r) \cdot \epsilon_c(\varrho_\alpha(r),o) dr - \int \varrho_\beta(r) \cdot \epsilon_c(o,\varrho_\beta(r)) dr .$$

Self-interaction is effectively subtracted out in (15): $E_c=0$ for arbitrary one-electron systems. In atoms and molecules, (15) describes the leading correction to HF, since electrons of antiparallel spin are completely uncorrelated at the HF level. This point is illustrated by the following comparison of antiparallel-(parallel-)spin correlation energies (perturbational results[23]): Be o.o74 (o.oo2) au, N o.138 (o.o43) au, Ne o.298 (o.o86) au. The main contributions to parallel-spin correlation come from the degenerate p orbitals (o.o58 au for Ne).

Incorporation of parallel-spin correlation into (15) is possible, of course, by means of a CI expansion. There is another possibility yet which is due to Perdew and Zunger[24]

$$E_c[\varrho_\alpha,\varrho_\beta] = \int \varrho(r) \cdot \epsilon_c(\varrho_\alpha(r),\varrho_\beta(r)) dr$$
$$- \sum_{\sigma,i} \int \varrho_{\sigma i}(r) \epsilon_c(\varrho_{\sigma i}(r),o) dr \qquad (16)$$

Here $\varrho_{\sigma i}$ is the orbital density of spin-orbital φ_i^σ. By subtracting single-orbital contributions from the LSD correlation energy, self-correlation is removed, while, at the same time, some parallel-spin correlation is retained. E_c is no longer invariant, however, with respect to unitary transformations of the occupied spin-orbitals (which leave E_{HF} as well as (15) unchanged). For F_2, e.g., transformation from canonical to localized orbitals reduces the self-inter-

Table 2. Total correlation energies E_c (in 10^{-3}au), as obtained from various LSD versions. Exp. values from Refs. 19,9.

	LSD eq.(12)	LSD-SIC eq.(15)	eq.(16)	exp
C	360	176	187	158
N	432	205	223	188
O	539	269	292	258
BH	350	181	185	152
CH	425	216	225	197
NH	501	249	265	240

Table 3. Correlation contributions to ionization potentials (in 10^{-3} au), as obtained from various LSD versions. Exp. values from Refs. 19,25.

	LSD eq.(12)	LSD-SIC eq.(15)	eq.(16)	exp
N	41	14	20	21
O	77	51	55	64
BH	56	37	38	49
CH	41	17	22	22
NH	43	16	22	28
OH	71	44	49	55

action corrections by 0.003 au (10%) per orbital. The correlation potentials $\mu_{c\alpha,\beta}$ become orbital-dependent with (16).

Table 2 shows that total atomic and molecular correlation energies are largely improved with the self-interaction corrected LSD versions (15),(16) (LSD-SIC). The differences between (15) and (16), on the other hand, are only marginal. Table 3 gives results for ionization energies. Here, too, self-correlation corrections are important. If no change of electron number is involved (for dissociation or excitation energies, e.g.), SIC is less significant. Gunnarsson and Jones[26] give an example, where the original LSD approximation (12) is superior to LSD-SIC (eq.16); this is for the $2s^2$ pair-correlation energy of the 4-electron series Be, B^+, C^{2+},.. We shall demonstrate below (Sect.V), that in this case both LSD and LSD-SIC (and probably all DF approximations which are based on electron-gas data) are bound to fail.

V. TOTAL AND PAIR CORRELATION ENERGIES

Total correlation energies of first- and second-row atoms, calculated using the LSD-SIC approximations (15),(16), are compared to ex-

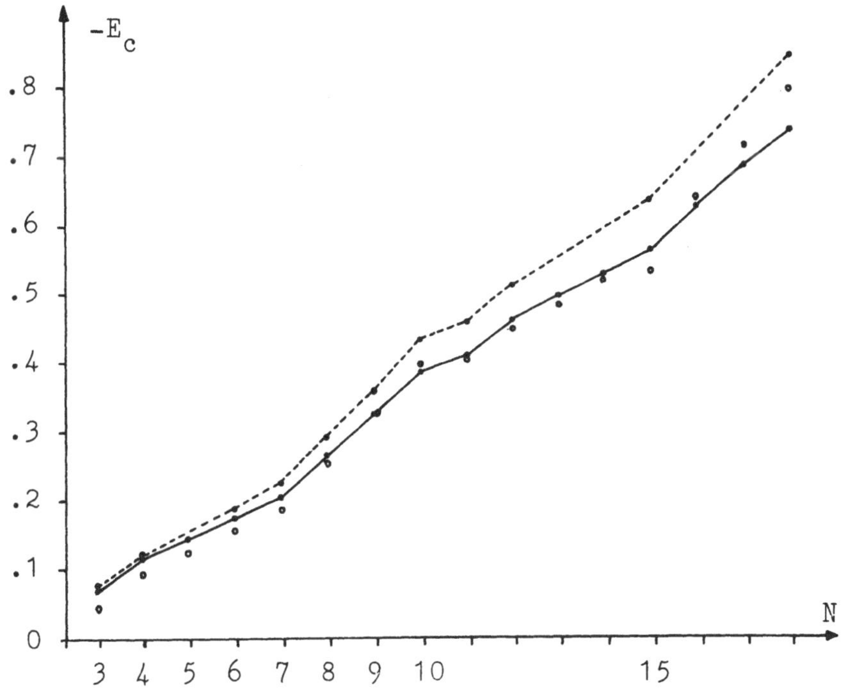

Fig. 2. Total correlation energies E_c (au) for the ground states of first- and second-row neutral atoms (N: electron number). LSD-SIC eq.15: ——, LSD-SIC eq. 16: ----, exp. values[19]: open circles.

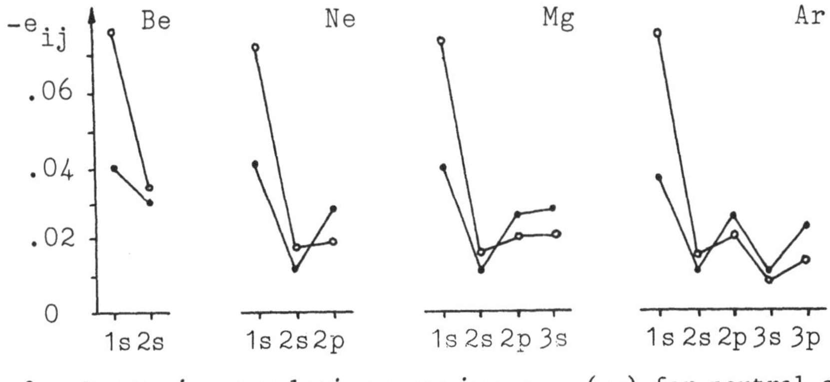

Fig. 3. Intrapair correlation energies e_{ij} (au) for neutral atoms. LSD eq. 18c: —o—o—, second-order perturbation theory[23,30]: —●—●—

perimental values in Fig.2. The mean deviation with eq.15 is 15% for the first, and 4% for the second row. With (16), E_c is consistently larger than experiment. This is (at least partly) due to the overestimation of core correlation energies in the LSD scheme. Consider the 2-electron series He, Li^+, Be^{2+}, ..[27,28] LSD performs well for small Z, but E_c^{LSD} increases as ln Z, while the true correlation energy approaches a constant value for $Z \to \infty$. This leads to an overestimation of E_c for Ne^{8+}, e.g., by a factor 2.3 with (15). For the 4-electron series Be, B^+, C^{2+}, .., on the other hand, E_c should be linear in Z (this is a near-degeneracy effect), but $E_c^{LSD} \sim \ln Z$ again.[28] This leads to an underestimation of the valence correlation energy for Ne^{6+} by a factor 2.o with (15). These examples nicely illustrate the fact that, for near-degeneracies and/or large density gradients, LSD behaves qualitatively incorrect.

In order to get further insight into the merits and shortcomings of LSD, we have performed a decomposition of total LSD correlation energies into contributions from electron pairs. Such a decomposition is widely used in CI calculations:

$$E_c = \sum_{i<j} e_{ij}^{CI}, \quad e_{ij}^{CI} = \sum_{a<b} \langle \Phi_0 | H | \Phi_{ij}^{ab} \rangle \, c_{ij}^{ab} \quad . \tag{17}$$

Here Φ_0 is the HF determinant, Φ_{ij}^{ab} a double substitution (i,j (a,b) denote spin-orbitals occupied (unoccupied) in Φ_0); c_{ij}^{ab} is the coefficient of Φ_{ij}^{ab} in a CI expansion which is normalized to $c_0=1$. A LSD partitioning of E_c[29] relies on the quantities $E_c[\rho]$, $E_c^o[\rho-\rho_i]$, $E_c[\rho-\rho_i-\rho_j]$, where ρ_i, ρ_j are spin-orbital densities; the LSD-SIC functional (16) is used. One assumes

$$\begin{aligned} E_c[\rho] &= \sum_{i<j} e_{ij} \\ E_c[\rho] - E_c[\rho-\rho_i] &= \sum_{k(\neq i)} e_{ik} \\ E_c[\rho] - E_c[\rho-\rho_i-\rho_j] &= \sum_{k(\neq i)} e_{ik} + \sum_{k(\neq j)} e_{jk} - e_{ij} \end{aligned} \tag{18}$$

Sums of pair correlation energies, involving a given spin-orbital φ_i^σ, can be obtained from (18b) (N equations), individual pair energies from (18c) (N(N-1)/2 equations).

Fig. 3 shows intra-pair correlation energies for Be, Ne, Mg and Ar[29], compared to values from second-order perturbation theory.[23,30] The trend is qualitatively correct with LSD, but i) $1s^2$ pairs have too large energies (as discussed above), and ii) the np^2 pair energies are too small. It is fair to say, though, that the convergence of the perturbational pair energies is very slow with respect to the angular quantum number l. For the Ar $3p^2$ energy, the LSD value (o.o14 au) is not so much smaller than the result from a perturbational calculation including s, p and d orbitals only (o.o17 au). Since ns^2 and np^2 densities are not too different, LSD predicts similar pair correlation energies for ns^2 and np^2 in a closed shell. In reality, however, the ns^2 energy is considerably smaller: in the closed L and M shells, angular correlation of the ns pair by np orbitals is suppressed. It is hard to see how such an exclusion effect could be properly taken into account by a local density functional. p inter-pair energies (for Ne, Mg, Ar) and d intra-pair energies (for Zn) show the same deficiencies as the p intra-pair energies.[7,29] Thus, valence correlation energies are underestimated in LSD-SIC, with the exception of those cases where an exclusion effect reduces the e_{ij}^{CI}.

VI. APPLICATIONS OF THE LSD METHOD

A. Energies *

Ionization energies (IE) of first-row atoms are shown in Fig. 4; c-LSD-SIC results (eq.15) are compared to xc-LSD (eq.3), HF and experimental results. The agreement with experiment is generally good for both c-LSD and xc-LSD. The Be IE is underestimated by both methods (cf.Sect.V). For B, C, N (singly occupied p orbitals), c-LSD is somewhat better; a self-interaction correction does not change the xc values significantly here[26], so interelectronic exchange is probably overestimated in xc-LSD. For O, F, Ne (doubly occupied p orbitals), the errors of c-LSD and xc-LSD are similar, but of opposite sign: c-LSD yields too small p intrapair energies, xc-LSD probably again overestimates exchange. Correlation contributions to IEs of one-valence-electron atoms (ions)[31] are compiled in Table 4. This is a severe test of LSD: the contributions are due to core-valence correlation only, and depend therefore, in LSD approximation, on the rather small overlap between core- and valence-orbitals. Without SIC, the deviations from experiment are of either sign. With SIC, the c-LSD values are consistently too small by a factor ~3; the trend is more satisfactorily reproduced now. The result is that LSD can qualitatively describe core-valence correlation, but the percentage recovered is rather small. Errors of this kind always come into play, when weakly overlapping densities are treated in LSD approximation; a possible improvement will be discussed in Sect.VIIB Correlation contributions to molecular ionization energies have already been given in Table 3. c-LSD without SIC leads to too large IEs, the SIC versions (eqs. 15, 16) yield similar results which are in good agreement with CI calculations.[5] The trend OH>BH>CH≈NH is correctly reproduced by all methods.

We now turn to electron affinities. With c-LSD-SIC (eq.15),

* Some of the following results for LSD correlation energies have been obtained from HF calculations with subsequent evaluation of $E_c[\rho=\rho_{HF}]$. The error of this non-self-consistent treatment is, for atomic energies, of the order of 10^{-4} au[7] (cf. also Table 1).

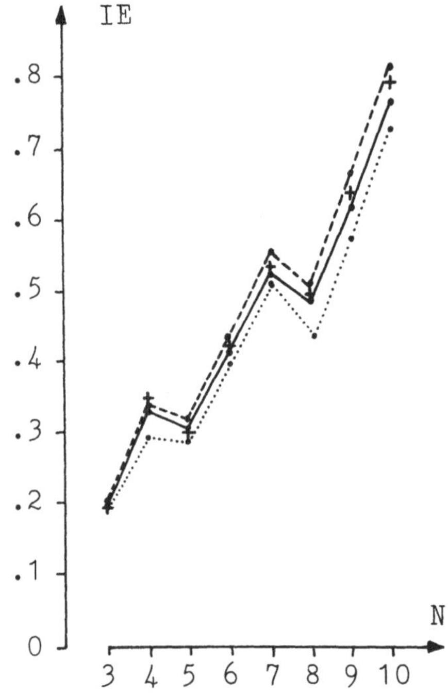

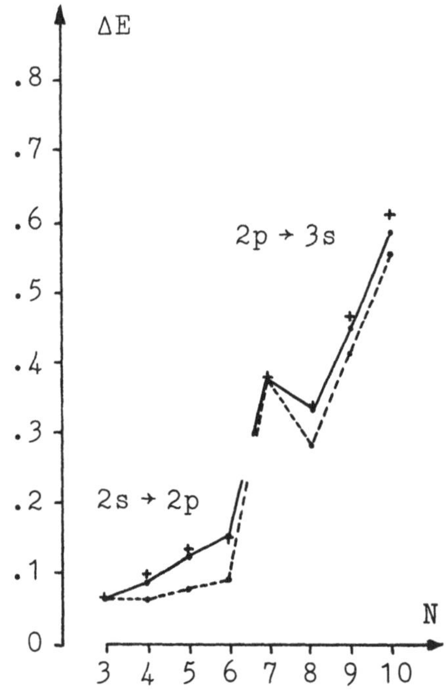

Fig. 4. Ionization energies IE (au) of first-row atoms (N: electron number). c-LSD-SIC eq.15: ——, xc-LSD[17]: – – –, HF: ······· ; exp. values are denoted by crosses.

Fig. 5. Excitation energies ΔE (au) of first-row atoms[27] (N: electron number). c-LSD-SIC eq. 15: ——, HF: - - - -; exp. values are denoted by crosses.

Table 4. Correlation contributions (in 10^{-3} au) to ionization energies of one-valence-electron atoms.[31] The "exp" values are differences of experimental and Dirac-Fock (DF) ionization energies.

	LSD, eq.12	LSD-SIC, eq.15	exp
K	17	4	12
Ca^+	23	7	20
Cu	28	11	43
Zn^+	33	13	41

reliable affinities are obtained for s electrons (H: o.71 eV (exp: o.75 eV), Na: o.57 eV[32] (exp:o.55 eV), Cu: 1.13 eV[33] (exp:1.23 eV)), but the p affinities are markedly too small (Cl: 3.3 eV (exp:3.7 eV)), as expected from the p intrapair defect discussed in Sect.V. In xc-LSD, negative ions are not stable due to exponential decrease of the electron-gas exchange potential for large r. With self-interaction correction[24], xc-LSD calculations give reasonable results[34], but p and, in particular, d affinities are too large.

Excitation energies for 2s→2p and 2p→3s transitions of first-row atoms[27] are depicted in Fig.5. Separate calculations for ground and excited (single-determinantal) states were performed; the GL parametrization for ϵ_c was used. Agreement with experiment is encouraging. The maximum deviation from experiment (~o.5 eV) occurs at the end of the row, where excitation is from p pairs. There are cases, however, where LSD performs less satisfactorily. For Ne^+, e.g., the two lowest states are $^2P(2s^2 2p^5)$ and $^2S(2s2p^6)$, both representable as single determinants. The HF excitation energy is 29.55 eV, c-LSD yields quite similar results (29.63 eV without, and 29.60 eV with SIC eq.15), but the experimental value is much smaller (26.88 eV). The error is due to the anomalously large correlation energy of the excited ^2S state, which strongly mixes with a Rydberg state slightly higher in energy[35]; the result is that the ^2S state of Ne^+ has even more correlation energy (o.426 au) than the ^1S ground state of the neutral Ne atom with one electron more (o.394 au). Such near-degeneracy effects cannot be described by a local DF. ns→(n+1)s, ns→np and ns→(n-1)d excitation energies of alkaline-earth atoms[36] are given in Fig. 6. The c-LSD-SIC values (eq.15) compare favourably with experiment (deviations ≲ o.1 eV), while considerably larger errors appear in the xc-LSD formalism, especially for the ^3D excitations.[37] In the latter method, electron-gas exchange artificially lowers the ^3D state with respect to the ^1S ground state by ~1 eV. A similar situation seems to arise for transition-metal atoms. For the splitting between the $d^{n-2}s^2$ and $d^{n-1}s^1$ states, HF results are quite acceptable as long as the d orbitals are singly occupied (Sc to Cr), while the xc-results favour the d rich state by ~1 eV.[37] Molecular excitation energies for the first-row monohydrides[38] are compiled in Table 5. We consider transitions of the type $n^2 \rightarrow n\pi$ for BH, CH^+, $n^2\pi \rightarrow n\pi\pi'$ for CH,NH^+ and $n^2\pi^2 \rightarrow n\,\pi\pi'$ for NH, OH^+. Correlation effects can be decisive here: HF as well as Xα[39] predict a $^4\Sigma^-(n\,\pi\pi')$ ground state for CH, while the experimental ground state is $^2\Pi(n^2\pi)$. Our results in Table 5 refer to single-determinantal states; these are pure states, with the exception of $n^2\pi^2$ which is composed of $^1\Sigma^+$ and $^1\Delta$: in this case we compare with the appropriate average of experimental energies. The correlation contributions are very similar with the various c-LSD variants (eqs.12,15, 16), because there is no change in the number of electrons. Compared to experiment, the c-LSD results deviate by few millihartrees. The $^4\Sigma$-$^2\Pi$ separation for CH is of the correct sign with c-LSD, and its magnitude (o.o27 .. o.o31 au) compares favourably with the CI calculation (4147 configurations) by Lie et al[40] (o.o23 au, estimate o.o23 .. o.o28 au). For NH^+, c-LSD predicts the $^2\Pi$ and $^4\Sigma^-$ states to be essentially degenerate (separation -o.oo3 .. +o.oo1 au); the best CI calculation[41] yields -o.oo3 au; experimentally, the $^2\Pi$ state is believed to be the ground state, i.e. the separation should be of positive sign.

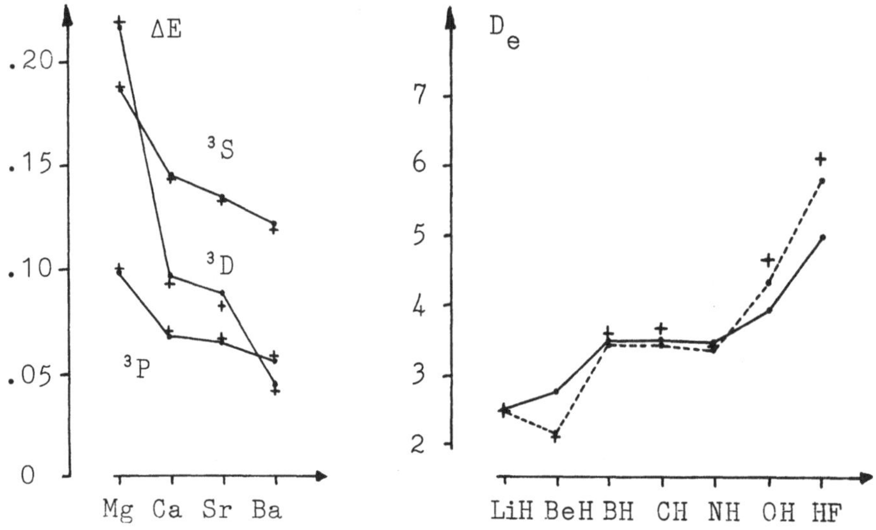

Fig. 6. Excitation energies ΔE (au) of group IIA atoms. c-LSD-SIC (eq.15, with GL parametrization; from valence-only calculations[36]):——. exp.values: crosses.

Fig. 7. Binding energies D_e(eV) of first-row monohydrides[42]. c-LSD-SIC (eq.15, with GL parametrization): ——— , CI[5]: – – – ; exp.values are denoted by crosses.

Table 5. Correlation contributions (in 10^{-3} au) to molecular excitation energies (cf. text) of first-row monohydrides.[38]

	LSD eq.12	LSD-SIC eq.15	LSD-SIC eq.16	exp
BH	27	29	26	
CH^+	28	30	26	(37)
CH	38	41	37	37
NH^+	34	37	33	38
NH	19	20	19	17..20
OH^+	20	19	20	22..26

Binding energies of first-row monohydrides, from valence-only calculations,[42] are shown in Fig.7. The results are compared to CI and experimental values. The LSD errors for OH and FH are in line with the underestimation of electron affinities by c-LSD discussed above. (With the VWN parametrization, the c-LSD value for FH in Fig.7 is improved by o.2 eV.) The deviation for BeH is due to an exclusion effect. The CI correlation energy for BeH is actually smaller than that of the Be atom. The latter has an exceptionally large correlation energy because of the near-degeneracy of the occupied s with

Table 6. Dissociation energies (eV) of alkali dimers, from valence-only calculations using the c-LSD (xc-LSD) functional with VWN (GL) parametrization.[43,44]

	c-LSD-SIC,eq.15	xc-LSD	exp
Li_2	1.00	0.83	1.06
Na_2	0.80	0.65	0.73
K_2	0.64	0.49	0.52

the unoccupied p orbitals. In BeH, the pσ orbital is (partially) occupied and can no longer be used as effectively for correlation. Dissociation energies of homonuclear alkali dimers (Li_2 to K_2) are given in Table 6. Here correlation contributions are exceptionally important: only Li_2 is bound at the HF level of approximation. It is seen that both c-LSD-SIC (eq.15) and xc-LSD are in good agreement with experiment. Be_2 is an interesting molecule, because large discrepancies exist here between different CI calculations. At the HF level, Be_2 is repulsive. With a multi-configuration-SCF (MC-SCF, 60 configurations)[45], a shallow minimum at r=10 au is found. Enlarging the MC-SCF to 816 configurations leads to two minima (at 5.1 and 8.2 au), with a depth of ~0.1 kcal/mole both. With a large-scale CI, including single and double substitutions with respect to the MC-SCF(60)-wave-function (10752 configurations), a single minimum at 4.9 au is obtained (depth 0.9 kcal/mole). The final CI estimate for D_e is 2.0 .. 2.3 kcal/mole.[45,46] c-LSD-SIC (eq.15) gives only the outer minimum (r_e=9 au, D_e=0.1 kcal/mole).[7] With xc-LSD, on the other hand, r_e=4.63 au and D_e=12 kcal/mole is found[18]. Thus xc-LSD is probably superior to c-LSD for Be_2, although the deviation from the CI estimate for D_e at 4.9 au is certainly smaller for the latter.

We conclude this section with a remark concerning bond-lengths. The deviations of HF bond-lengths from experiment can be of either sign: HF bond-lengths are longer than the experimental ones for alkali dimers, but smaller for the first-row monohydrides BeH to FH. c-LSD usually leads to a shortening of HF bond-lengths ($|E_c[\rho_1+\rho_2]| \geq |E_c[\rho_1]+E_c[\rho_2]|$ with eq.11). This means that c-LSD cannot be used for a consistent improvement of HF geometries. The same remark applies to xc-LSD. For a number of first-row molecules, Baerends and Ros[47] found Δr_e (HF)=0.07 au, and Δr_e(xc-LSD)=0.08 au.

B. Densities

In this section, we discuss radial densities $D(r)=\int\rho(r)r^2\sin\vartheta d\vartheta d\varphi$ for 2-, 3-, and 4-electron atoms. In particular, we consider differences $\Delta D(r)=D(r)-D_{HF}(r)$, which arise when including the LSD-SIC correlation potential (functional derivative of (15)) into the UHF equations. For 2-electron atoms, it is known from accurate CI calculations that density is shifted from the intermediate region to both short and

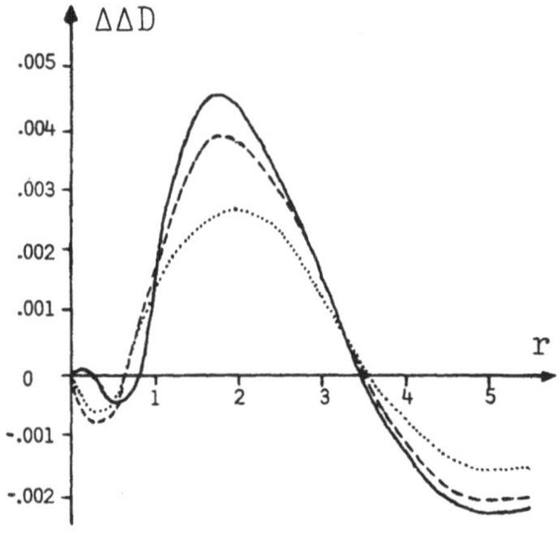

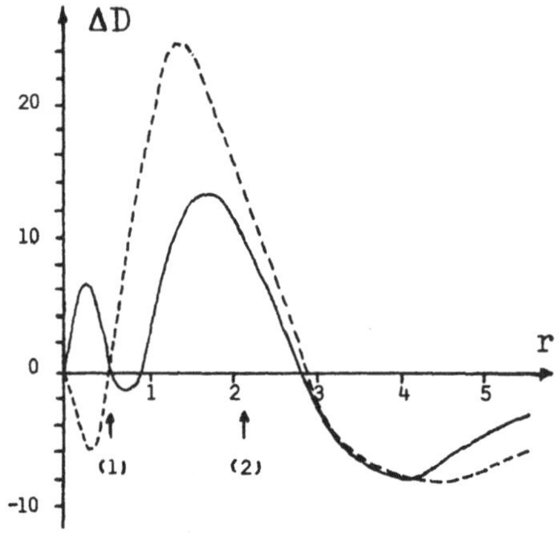

Fig. 8. Correlation contributions $\Delta\Delta D$ (in au) to the valence density of the Li atom, as a function of r (au).
c-LSD-SIC eq.15[50] : ——— ; CI, 11s4p1d GTO basis set: .... ; CI, 11s5p2d2f basis: — — — — .

Fig. 9. Correlation contributions ΔD (in 10^{-3} au) to the density of the Be atom, as a function of r (au).
c-LSD-SIC eq.15[7]: ——— ; CI[48] : — — — . The arrows indicate the $\langle r \rangle$ expectation values of the 1s and 2s orbitals.

long radial distances.[48] Enlarging the density near the nucleus (contraction) is caused by angular correlation, while the expansion effect is due to radial (in-out) correlation. c-LSD can describe only the former effect: the local correlation potential decreases monotonically with increasing density, and charge is shifted inwards only. This defect is certainly not limited to the specific LSD approximation (12) or (15). With an empirically adjusted local functional[27], it is possible to reproduce the correlation energies of the 2-electron series H^-, He, Li^+,... with remarkable accuracy, but the density shift with the corresponding μ_c is only towards the nucleus again.[7] Smith et al[49] determined the correlation potential μ_c for He by inverting the Kohn-Sham equation (cf.Sect.II; $\varphi=\sqrt{\rho/2}$ was taken from accurate CI calculations; under the assumption that μ_c approaches zero for $r\to\infty$, the eigenvalue ϵ of the modified HF equation is given by $-1/2 \lim_{r\to\infty} \nabla^2\varphi/\varphi$). The resulting μ_c is non-monotonic; it has a barrier at r=o.7 au and is negative only for $r \lesssim$ o.3 au. It is not clear how such a behaviour could be incorporated into a local electron-gas approximation.

In 3-electron atoms, valence-density differences are due to intershell correlation. Since intershell correlation is mainly of the angular type, c-LSD is expected to work better than in the 2-electron case. In Fig.8 $\Delta\Delta D = \Delta D_{Li} - \Delta D_{Li^+}$ is plotted.[50] The agreement with elaborate CI calculations is satisfactory, indeed.

An example of a 4-electron density difference $\Delta D(r)$ is given in Fig.9. The Be density shown in this figure exhibits quite different shifts in the core region with LSD and CI calculations, but in the outer valence region a similar (inward) shift is observed with both methods. This is surprising, in view of the qualitative differences for the He series. But note that the $2s^2 \to 2p^2$ (angular) correlation is exceptionally high in Be.

VII. EXTENSIONS OF THE LSD METHOD

A. Non-local Corrections

The LSD approximation for $E_c[\rho_\alpha, \rho_\beta]$, derived from data of the homogeneous electron gas, cannot be exact: atoms and molecules differ from the homogeneous electron gas in many respects: i) the number of electrons is finite; the excitation spectrum is discrete; ii) the density is not homogeneous; iii) the correlation hole does not depend on the local density at the center of the hole only. Points i) to iii) have been starting-points for various modifications of the local electron-gas expression. We shall discuss these modifications now.

Tong[51] pointed out that in the infinite electron gas a continuum of low-lying levels is available to adjust to the mutual Coulomb repulsion of the electrons; the system is "soft" and should have a higher correlation energy than a finite one with discrete levels, and this is indeed what was found with the (original) LD approximation (eq.11), as discussed in Sect.III. Tong therefore considers a model system with a finite number of electrons in a cubic box. The correlation energy is evaluated to second order perturbation theory, using a screened Coulomb potential. If the density in the box is chosen to match that of an average atom (Ne, Ar), a correction factor to the LD expression can be defined as $\epsilon_{c,model}(N)/\lim_{N\to\infty} \epsilon_{c,model}(N)$. This factor is about o.5 to o.6 for $N \approx 14..28$, bringing the LD estimate of total atomic correlation energies into reasonable agreement with experimental values for medium-sized atoms. An additional self-interaction correction is unnecessary here, of course; it is already included in the correction factor. Schneider[52] refined Tong's scheme and applied it to molecules (N_2, F_2). It is not clear, however, if this global model can be made sufficiently accurate to deal with small correlation-energy differences (contributions to dissociation energies, e.g.). The cubic box is certainly a more appropriate description for compact systems (atoms) than for more open ones (molecules).

Atoms and molecules have strongly inhomogeneous densities. For a comparison of correlation energies, the inhomogeneous electron gas would be, therefore, a more suitable starting-point than the homogeneous one. Assuming an expansion of E_c for the electron gas, perturbed by an external field, in powers of $\nabla\rho$, Ma and Brueckner[53] obtained the following expression for the correlation energy per particle

$$\epsilon_c(\rho) + B\ |\nabla\rho|^2\ \rho^{-7/3}$$

$$B = 4.23 \cdot 10^{-3}\ \text{au} \tag{19}$$

which is valid in the high-density limit. Applying (19) to atoms (O to K), Ma and Brueckner found an overcorrection of the LD error by a factor 5. Changing (19) empirically to

$$\epsilon_c(\rho)\cdot\left[1 - B\ |\nabla\rho|^2\ \rho^{-7/3}\ (\epsilon_c(\rho)\ y)^{-1}\right]^{-y} \tag{20}$$

agreement with experiment to ~8 % was obtained for O to K, with y=o.32.[53,52] While atomic correlation energies are too large in magnitude with (2o), the molecular $|E_c|$ was found to be too low for N_2 and F_2 (by ~1o %). This means that the correlation contributions to D_e are of wrong sign. In a recent paper, Langreth and Mehl[54] argued that the small-k part in a wave-vector decomposition of the prefactor of $|\nabla\rho|^2\ \rho^{-7/3}$ in (19) should be omitted, since it is effectively compensated by higher orders of the gradient expansion. With a single parameter f controlling the cut-off of k values in the interval $0 \leq k \leq f\ |\nabla\rho/\rho|$, and adjusting f empirically to atomic energies, Langreth and Mehl obtained the following formula for E_c/N

$$\epsilon_c(\rho) + \tilde{B}(\rho)\ |\nabla\rho|^2\ \rho^{-7/3}$$

$$\tilde{B}(\rho) = 4.28 \cdot 1o^{-3} \cdot \exp(-0.262\ |\nabla\rho|\ \rho^{-7/6})\ \text{au} \tag{21}$$

If (21) is applied to the calculation of correlation energies of closed-shell atoms, considerable improvement over the original LSD approximation (11) is achieved, but the agreement with experiment is still not very good. In their most recent paper[77], Langreth and Mehl suggest a repartitioning of the DF exchange-correlation energy into the exchange and correlation parts; according to their suggestion, $9f^2$ (with $f \approx 0.15$) has to be added to the exponential in (21b). Now experimental correlation energies of closed-shell atoms can be reproduced with an impressive accuracy. The method is very promising; note, however, that the gradient correction in (21) cannot be used to improve on the incorrect ln Z dependence of the LSD correlation energies for the 2-electron atoms.

The exact expressions (1o),(14) for the correlation energy E_c depend on the pair correlation functions $g^{\lambda}_{\sigma\sigma'}$. While $g^{\lambda}_{\alpha\alpha}$ and $g^{\lambda}_{\beta\beta}$ are essentially determined by the antisymmetry requirement already present in HF, $g^{\lambda}_{\alpha\beta}$ and $g^{\lambda}_{\beta\alpha}$ (which are constant (=1) in HF) are markedly changed by correlation beyond HF. The following properties apply to pair correlation functions in atoms and molecules, as well as to those of the homogeneous electron gas[55]

$$g^{\lambda}_{\sigma\sigma}(r_1,r_1)=0,\ g^{\lambda}_{\sigma\sigma'}(r_1,r_1)\geq 0 \qquad (\sigma\neq\sigma') \tag{22}$$

$$\int dr_2\ \rho_{\sigma'}(r_2)\ g^{\lambda}_{\sigma\sigma'}(r_1,r_2) = N_{\sigma}-\delta_{\sigma\sigma'} \tag{23}$$

$$\lim_{r_{12}\to 0}\frac{\partial}{\partial r_{12}}g^{\lambda}_{\sigma\sigma}=0,\ \lim_{r_{12}\to 0}\frac{\partial^2}{\partial r_{12}^2}g^{\lambda}_{\sigma\sigma}=\frac{2}{3\lambda}\lim_{r_{12}\to 0}\frac{\partial^3}{\partial r_{12}^3}g^{\lambda}_{\sigma\sigma}$$

$$\lim_{r_{12}\to 0}g^{\lambda}_{\sigma\sigma'}=\frac{1}{\lambda}\lim_{r_{12}\to 0}\frac{\partial}{\partial r_{12}}g^{\lambda}_{\sigma\sigma'} \qquad (\sigma\neq\sigma') \tag{24}$$

(Here N_{σ} is the number of electrons with spin σ.)
Ros[56] has shown that the Coulomb hole $g_{\alpha\beta}(r_1,r_2)-1$ in 2-electron ions can be understood in terms of simple concepts like shielding and polarization. If a perturbation of the form

$$V(r_1,r_2) = q\left[\frac{1}{r_{12}}-\frac{1}{2}\int\frac{\rho_{HF}(r_1)}{r_{12}}\,dr_1\right] \tag{25}$$

is added to the Fock operator $F(r_2)$, the resulting density change gives an accurate representation of $[g_{\alpha\beta}(r_1,r_2)-1]\cdot\rho_{\beta}(r_2)$ for a given r_1. (The parameter q is weakly dependent on r_1 ($q(0)=0.33$, $q(\infty)=1$).) In molecules, the following properties have been found for the $g_{\sigma\sigma'}$.[57] The depth of the Coulomb hole is rather small ($g_{\alpha\beta}(r_1,r_1)\approx 0.9\ ..\ 0.99$). At nuclear positions , the Coulomb-hole depth and range are inversely proportional to the nuclear charge Z. The range of the Coulomb hole is somewhat smaller in general than that of the Fermi hole.
The central quantity for calculating E_c is

$$\rho_{c\sigma\sigma'}(r_1,r_2) = \rho_{\sigma'}(r_2)\int_0^1 d\lambda\left[g^{\lambda}_{\sigma\sigma'}(r_1,r_2)-g^{\lambda}_{HF\sigma\sigma'}(r_1,r_2)\right] \tag{14'}$$

As discussed in Sects.III,IV, the LSD approximation is characterized by substituting $\rho_{\sigma'}(r_2)\to\rho_{\sigma'}(r_1)$, and using the pair correlation functions of the homogeneous spin-polarized electron gas with densities $\rho_{\alpha}(r_1)$, $\rho_{\beta}(r_1)$.
Keller and Gazquez[58] avoid the use of electron-gas data for evaluating (14'). They concentrate on antiparallel-spin correlation ($\sigma\neq\sigma'$). The replacement $\rho_{\sigma'}(r_2)\to\rho_{\sigma'}(r_1)$ is made, and for the integral in (14') the ansatz

$$-\exp(-cr_{12}/d)\cdot\cos(3\pi r_{12}/(2d)) \tag{26}$$

is used (note that $g^{\lambda}_{\alpha\beta}(r_1,r_1)=0$ with (26)). The constant d is identified with the Fermi-hole radius, and c is then determined from the sum rule (23). It is not clear why such a procedure (which constitutes a local DF approximation still) should be superior to the LSD-SIC approach (eqs. 15,16), and, in fact, the correlation energies calculated by Keller and Gazquez are not of very homogeneous quality. For He 12%, for Ne 80%, and for Ar 115% of the experimental E_c is obtained.

A more sophisticated approach has been suggested by Alonso and Balbás[59]. They start from the spin-integrated form for E_c (eq.10) and retain the exact density prefactor $\varrho(r_2)$ in

$$\varrho_c(r_1,r_2) = \varrho(r_2)\cdot\int_0^1 d\lambda\left[g^{\lambda}(r_1,r_2)-g^{\lambda}_{HF}(r_1,r_2)\right] \quad . \tag{10'}$$

For the integral, they make the ansatz

$$-\frac{1}{2}\exp(-\alpha r_{12})\cos(\beta r_{12}) \tag{27}$$

which is similar to (26) ($\varrho_\alpha = \varrho_\beta = \varrho/2$ is supposed here). The parameters α and β are adjusted to properties (ϵ_c, sum rule) of the homogeneous electron gas with density ϱ^*. Alonso and Balbás then suggest to choose, for each r_1, ϱ^* in such a way as to satisfy the sum rule (23) for $\varrho_c(r_1,r_2)$, using the (exact) density $\varrho(r_2)$ of the atom or molecule under consideration. Unfortunately, this prescription is non-unique, at least in the limit of constant density. The latter defect is absent in the closely related WD (weighted-density) method by Gunnarsson et al.[60] Here $E_x[\varrho]$ and $E_{xc}[\varrho]$ are calculated from the exchange- and exchange-correlation hole in a way similar to that described for $E_c[\varrho]$ above. Both WD approximations, for E_x as well as for E_{xc}, yield the exact solution for one-electron systems. But while exchange energies, in particular, are greatly improved with the WD method, this is not true for correlation energies $E_c=E_{xc}-E_x$. For Ne, e.g., 0.67 au is obtained (c-LSD, eq.12: 0.74 au, exp: 0.39 au).

In the approach suggested by Dobson and Rose[61], the non-local character of the functional is even more distinct. They do not only retain the density prefactor $\varrho_{\sigma'}(r_2)$ in (14'), but the exact HF two-particle density:

$$\begin{aligned}\varrho_{c\,\sigma\sigma'}(r_1,r_2) = \varrho_{\sigma'}(r_2)\cdot g^{\lambda=1}_{HF\,\sigma\sigma'}(r_1,r_2)\cdot \\ \cdot\int_0^1 d\lambda\left[g^{\lambda}_{\sigma\sigma'}(r_1,r_2)-g^{\lambda}_{HF\,\sigma\sigma'}(r_1,r_2)\right] / g^{\lambda=1}_{HF\,\sigma\sigma'}(r_1,r_2)\ .\end{aligned} \tag{28}$$

The prefactor of the integral is taken from the HF calculation of the atom or molecule under consideration, and the pair-correlation functions of the homogeneous electron gas (with densities $\bar{\varrho}_\sigma = (\varrho_\sigma(r_1) + \varrho_\sigma(r_2))/2$) are used in the integral. No numerical application has been performed so far. Self-interaction is corrected for in (28) through the use of g_{HF} as multiplicative factor, and the limiting case of the homogeneous electron gas is properly taken into account, but the sum rule (23) is not satisfied. Perhaps this could be amended by changing the prescription for $\bar{\varrho}_\sigma$ in a suitable way.

B. Adjustment to Atomic Data

In Sect.VIIA, non-local extensions of the LSD formalism have been discussed; in some cases atomic data were used complementary to the electron-gas data underlying the LSD expression. As an alternative, one can use atomic data exclusively, right from the beginning.
Lie and Clementi[62] employed the following ansatz for $\epsilon_c(\varrho)$:

$$\epsilon_c(\varrho) = a_1 \left[(a_2 + \varrho^{1/3})^{-1} \varrho^{1/3} + \ln(1+2.39\,\varrho^{1/3}) \right] . \qquad (29)$$

The parameters in (29) were adjusted to the experimental correlation energies of the He, Be, Ne atoms. It turns out that a LD calculation with (29) differs by roughly a factor 2 from a calculation with the LD electron-gas functional. Correlation contributions to dissociation energies, which are well described by the electron-gas expression, are too small now by a factor 2 with (29). Lie and Clementi found that the missing parts of molecular correlation energies can, to a good approximation, be calculated from a small valence MC-SCF (multi-configuration-SCF) which does not change the atomic energies. Such a valence MC-SCF has the additional advantage to ensure transition of the molecular state to the correct atomic dissociation products in the limit of large internuclear separations. The deviations from experiment of the dissociation energies computed by Lie and Clementi are $\lesssim 0.4$ eV for first-row monohydrides, and ≤ 1 eV for first-row dimers. The idea to separate correlation effects into internal ones (excitations into valence orbitals, which account for near-degeneracies and are not well described by a local DF) and a rest (which is connected to higher excitations and can be estimated from atomic correlation energies) is underlying also the method by Lievin et al.[63] The molecular density is decomposed here into atomic contributions $P(K_i)$ according to

$$P(K_i) = \prod_{q \in i} P_q^{\,n_q} \, (1-P_q)^{1-n_q} \qquad (30)$$

K_i is a configuration of atom i, P_q is the population of the atomic spin-orbital χ_q in the molecular density, and n_q is the occupation number of χ_q in K_i. Using the $P(K_i)$, the non-internal part of the molecular correlation energy is approximated by a weighted mean of atomic non-internal correlation energies; the internal part is calculated in the CI formalism. Results for dissociation energies of first-row dimers are (experimental values in parentheses): C_2 6.03 (6.33) eV, N_2 10.00 (9.91) eV, O_2 4.80 (5.21) eV, F_2 1.15 (1.65) eV.

An attempt to include internal as well as non-internal correlation effects in an atom-adjusted non-local DF scheme has been made by

Colle and Salvetti.[64] They start from a wave-function ansatz of the form

$$\psi(x_1,..,x_n)=\psi_{HF}(x_1,..,x_n)\cdot\prod_{i<j}(1-f(r_i,r_j))$$
$$f(r_i,r_j)=\exp(-\beta^2|r|^2)\cdot(1-g(R)\cdot[1+|r|/2]) \tag{31}$$

Here $r=r_i-r_j$, $R=(r_i+r_j)/2$. This ansatz leads to the following two-particle density-matrix

$$\pi(r_1,r_2;r_1',r_2')=\pi_{HF}(r_1,r_2;r_1',r_2')\cdot$$
$$\cdot(1-f(r_1,r_2)-f(r_1',r_2')+f(r_1,r_2)f(r_1',r_2')) \tag{32}$$

which satisfies the cusp conditions (24). It is assumed then, that π integrates to the HF one-particle density matrix:

$$\int\pi(r_1,r_2;r_1',r_2)\,dr_2 = (N-1)\,\rho_{HF}(r_1;r_1') \tag{33}$$

This amounts to neglecting the λ-dependence in (1o), and thus to omitting kinetic contributions to E_c. Such contributions play a distinct role in atoms and molecules (virial theorem!), so (33) is not wholly justified. From the requirement (33), a connection between g and β is established, and β (the inverse size of the correlation hole) is put equal to $q\rho^{1/3}$. The final formula for E_c depends on $\pi_{HF}(r_1,r_1)$ and $\nabla_r^2\pi_{HF}(r_1,r_2)|_{r=o}$; it contains a single parameter q which is adjusted using the experimental correlation energy of the He atom. Correlation energies of atoms and molecules can be calculated from this formula with an accuracy of $\sim$ 1o%. A self-interaction correction is included ($\pi_{HF}(r_1,r_1)=o$ for one-electron systems). Applying the formula to the paramagnetic electron gas[65] leads to a Wigner-like expression

$$\epsilon_c(\rho) = -(9.652 + 2.946\,\rho^{-1/3})^{-1}\ \text{au} \tag{34}$$

which is (for not too large densities) in remarkably good agreement with the VWN function (cf.Sect.III).
It is perhaps a consequence of approximation (33) that the wave-function (31), with the parameters determined by Colle and Salvetti, is no very good substitute to the true wave-function: Moskowitz et al[66] report that only 10% of the Be correlation energy is recovered with (31) (98% in the non-variational treatment of Colle and Salvetti).

In the methods discussed up to this point, E_c is calculated explicitly from atomic (molecular) densities. It is possible, however, to introduce, in addition, other atomic (molecular) properties into the functional. These properties should have, on the one hand, a more or less direct relation to E_c; they should be, on the other hand, accessible in the framework of the density-functional method, at least in principle. With such properties as intermediate quantities, it perhaps could become easier to find an accurate non-local DF for atoms and molecules.

Ang et al[67] have shown that there is a rather direct connection between correlation energies and $\langle r^2 \rangle$ expectation values. Atomic correlation energies can be economically fitted by

$$E_c = k\,(Z - \sigma)^{\gamma} \cdot \langle r^2 \rangle^{-1} \tag{35}$$

where the parameters k, σ, γ are constant within each isoelectronic family; the $\langle r^2 \rangle$ values can be taken from HF calculations. A relation such as (35) has also been used to predict protonation energies (within the series N^{3-}, NH^{2-}, .., NH_4^+, e.g.).

There is a very simple connection between core-valence correlation energies and core polarizabilities, which has been successfully exploited by a number of workers.[68 to 71]

$$E_{pol} = \langle \Psi_{val} \,|\, (-\frac{1}{2} \sum_{\lambda} \alpha_\lambda \vec{f}_\lambda^{\,2}) \,|\, \Psi_{val} \rangle$$

$$\vec{f}_\lambda = \sum_i \frac{\vec{r}_{\lambda i}}{r_{\lambda i}^3}\, g(r_{\lambda i}) - \sum_{\mu (\neq \lambda)} Q_\mu \frac{\vec{r}_{\lambda\mu}}{r_{\lambda\mu}^3} \tag{36}$$

E_{pol} is the polarization energy, which includes core-valence correlation (and core-polarization effects on the HF level); Ψ_{val} is the valence part of the wave-function; the α_λ are core dipole-polarizabilities, the Q_μ are core charges; $\vec{f}_\lambda$ is the field generated at the site of core λ by valence electrons and surrounding cores; $g(r_{\lambda i})$ is a cut-off-function, e.g.[70,69]

$$g(r) = (1 - e^{-\delta r^2})$$

or

$$g(r) = r^3\, (r^2 + r_o^2)^{-3/2} \tag{37}$$

which is necessary, since the polarization picture breaks down for valence-electron positions near and inside core λ. With the polarization potential

$$V_{pol} = -\frac{\alpha}{2}\, r^2\, (r^2 + r_o^2)^{-3} \tag{38}$$

derived from (36) and (37b), and reasonable r_o values (set equal to the $\langle r \rangle$ expectation value of the outermost core orbital) Migdalek and Baylis[69] obtained good results for correlation contributions to ionization energies of Cu, Ag, and Au, using (relativistic) HF ns valence-orbital densities only. The deviation from experiment of relativistic HF ionization energies is 1.16 eV for Cu, 1.24 eV for Ag, and 1.54 eV for Au. If, with (38), instantaneous polarization of the X^+ core by the valence electron is taken into account, the deviations are reduced to 0.25 eV for Cu, 0.04 eV for Ag, and 0.01 eV for Au.

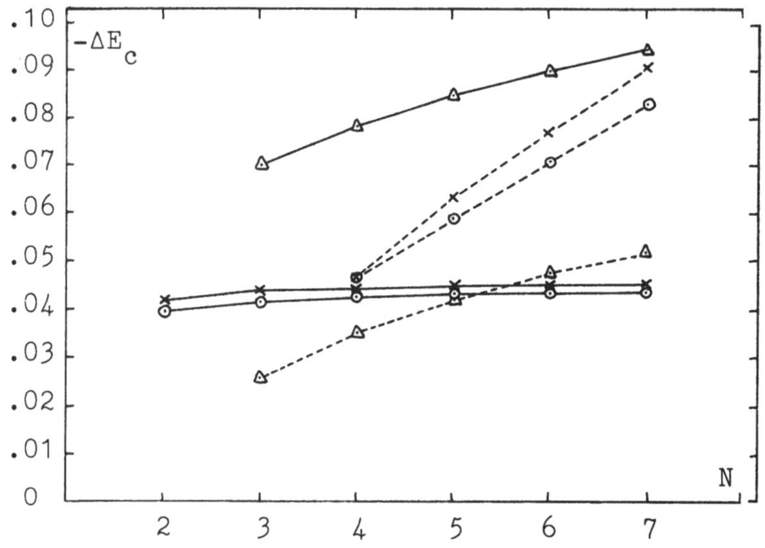

Fig. 10. Correlation energy differences $\Delta E_c = E_c(N) - E_c(N-1)$ (in au; N: electron number) for the 2-electron (solid lines) and the 4-electron series (broken lines). c-LSD-SIC eq.15: △ , polarization potential eq.38: ⊙ , exp: × .

Adjusting the single parameter of g(r) in (37a) to atomic ionization energies, HF deviations from experiment, of bond-lengths of alkali dimers could be reduced by about an order of magnitude (from 0.3 Å to ~0.04 Å for K_2^+, e.g.[70,71]). Thus when calculating valence properties of atoms and molecules, it is strongly advisable to apply the LSD formalism only to valence instead of total densities. Core-valence correlation can be accurately and economically determined using eq.36.

It is perhaps interesting to note that the polarization concept yields reasonable results not only in cases where the polarizing electron and the polarizable core are well separated in space. Migdalek and Baylis[72] calculated electron affinities of halogen atoms using correlation potentials of the type (38), where α is now the polarizability of the neutral atom, and r_o is the expectation value of the outermost p orbital. The extra electron in the negative ion experiences the attractive potential (38), which leads to a stabilization of the ion with respect to the neutral atom. The HF errors for the affinities, which are 2.07 eV for F, 1.09 eV for Cl, and 0.99 eV for Br, are reduced by this simple device to 0.02 eV for F, 0.13 eV for Cl, and 0.05 eV for Br.

Remember that LSD completely fails to reproduce the Z dependency of correlation energies for the 2- and 4-electron series. If E_c is determined from $\int \rho(r) V_{pol}(r) dr$, where ρ is the density of an electron in the outermost orbital, and the polarization potential (38) is that of the system consisting of the nucleus and the remaining electron(s), remarkable accuracy is obtained for both families[73] (Fig. 10). We feel that it could be worthwhile, in view of these results, to exploit the virtues of the polarization concept by coupling it to the LSD formalism, not only in the inter-shell case, but also for valence correlation energies. The long-range part, for example, of the LSD correlation potential (which is not too reliable, anyway[74] (cf. Sect. VIA)) could be substituted by a polarization potential.

C. LSD for Multi-determinantal States. Coupling with CI

LSD functionals are usually applied in the framework of unrestricted Hartree-Fock (UHF) (cf. Sect. II), i.e. α and β spin-orbitals are allowed to have different spatial parts, and the density-functional describes correlation beyond UHF. One could think of choosing a (spin-)restricted Hartree-Fock (RHF) reference state instead, or a PUHF state (where an eigenfunction of S^2 is projected out of the UHF wave-function), or even an extended Hartree-Fock (EHF) state, where the PUHF wave-function is energy-optimized. All these states are identical in the case of the homogeneous electron gas, from which the LSD correlation functionals (12), (15), (16) are derived. The following requirements should be met, however: i) the reference state should have densities and pair-correlation functions which are as close as possible to those of the true ground-state wave-function, and ii) the pair-correlation functions of the reference state should be similar to the electron-gas ones used in the LSD expression (cf. eq. 14). The first requirement generally rules out the use of RHF functions for large internuclear separations. For H_2, e.g., the RHF wave-function $|\varphi\bar{\varphi}|$ (with $\varphi = \chi_a + \chi_b$, where χ_a and χ_b are atomic orbitals (AOs)) is an inappropriate reference for c-LSD, not because the RHF density would be superior to that of the UHF function $|\chi_a \bar{\chi}_b|$, but because the pair density is qualitatively wrong with RHF; c-LSD can only describe local modifications of pair densities due to electron correlation, and the long-range behaviour of $g(r_1, r_2)$ (if electron 1 is at atom a, electron 2 must be at atom b) is correct with UHF but not with RHF. The second requirement means that PUHF or EHF wave-functions should not be used as reference, because for these multi-determinantal states electrons of different spin are correlated ($g_{\alpha\beta} \neq 1$), which is not true in the electron-gas case. LSD should only be applied to single determinants.[75]

If, for a given configuration, there is no single determinant with the angular and/or spin symmetry of the correlated state one is interested in, a transformation $\Phi_i = \sum_k \Psi_k c_{ki}$ should be performed[75,76], where the Φ_i are single determinants and the Ψ_k symmetry-adapted wave functions. If H is diagonal in the Ψ_k, the relation

$$\langle \Phi_i | H | \Phi_i \rangle = \sum_k E_k c_{ki}^2, \quad E_k = \langle \Psi_k | H | \Psi_k \rangle \tag{39}$$

Table 7. Correlation contributions (eV) to multiplet splittings of atoms.[76]

		c-LSD, eq.12	exp
C	$^3P \rightarrow {}^1D$	o.26	o.30
	$^3P \rightarrow {}^1S$	o.67	1.22
Si	$^3P \rightarrow {}^1D$	o.24	o.30
	$^3P \rightarrow {}^1S$	o.6o	o.78
N	$^4S \rightarrow {}^2D$	o.44	o.42
	$^4S \rightarrow {}^2P$	o.74	1.1o

Table 8. Energy lowering (in 10^{-3} au) of ground-state potential curves, relative to UHF.[7] c-LSD, eq.15 is applied a) to the UHF function directly, b) to the states on the rhs of (41). Exp. values from Ref.62

	r (au)	a)	b)	exp
H_2	1.4	49	49	41
	2.o	46	46	47
	4.o	5	19	14
	5.o	1	5	3
LiH	3.o	42	42	39
	4.o	39	39	39
	7.o	4	11	14
Li_2	5.o	28	28	37
	7.o	1o	19	23
	8.5	4	9	13
	1o.o	2	4	6

holds. If one assumes that an analogous formula also holds for the correlation energies, the LSD correlation energies of the Φ_i can be used to determine the unknown correlation energies of the symmetry-adapted states Ψ_k. Table 7 gives LSD correlation contributions to multiplet splittings in C, Si, and N.[76]
The above procedure can also be applied to the determination of ground-state potential curves. For H_2, e.g.,

$$|\chi_a \bar{\chi}_b| = \frac{1}{\sqrt{2}} \left[(1+s^2)^{1/2}\Psi_S + (1-s^2)^{1/2}\Psi_T\right], \quad |\chi_a \chi_b| = \Psi_T \quad . \tag{4o}$$

Here $s = \langle\chi_a|\chi_b\rangle$ is the overlap integral of the AOs, Ψ_S and Ψ_T are singlet and triplet wave-functions, respectively. From (39), (4o), the singlet ground-state energy E_S can be calculated

$$E_S = E(|\chi_a \bar{\chi}_b|) + \left[E(|\chi_a \bar{\chi}_b|) - E(|\chi_a \chi_b|)\right] \frac{1-s^2}{1+s^2} \quad . \tag{41}$$

In Table 8 correlation contributions (with respect to UHF) are given for ground-state potential curves of H_2, LiH, and Li_2.[7]

Near-degeneracy effects between states of the same symmetry cannot be properly accounted for with c-LSD (cf.Sect.V). A simple and economical coupling of LSD with the CI method would be highly desirable, therefore (small valence CI or MC-SCF accounting for near-degeneracies + LSD accounting for the bulk of higher excitations necessary to describe the correlation cusp (22)). Two possibilities to do such a coupling have already been discussed. In Sect.V, LSD pair-correlation energies were defined; the option exists to calculate E_c in the LSD formalism, and then to replace those pair energies, which are strongly affected by near-degeneracy effects, by CI pair energies. In Sect.VIIB it was shown how atom-adjusted LSD correlation energies could be used in connection with valence MC-SCF calculations. A third possibility for coupling LSD to CI is due to Colle and Salvetti[64]; β, the inverse size of the correlation hole (cf.eq.31), is multiplied by an (ad-hoc) factor, which is 1 for the HF reference, and approaches ∞, if Ψ_{HF} in (31) is replaced by reference functions of increasing quality. The last way for coupling which we want to mention here is the separation of the two-electron interaction in the Hamiltonian, $\sum_{i<j} r_{ij}^{-1}$, into a short-range and a long-range part

$$\sum_{i<j} \frac{1}{r_{ij}} = \sum_{i<j} \frac{1}{r_{ij}} \exp(-\lambda r_{ij}) + \sum_{i<j} \frac{1}{r_{ij}} [1-\exp(-\lambda r_{ij})] \quad (42)$$

Correlation effects due to the short-range part, involving the correlation cusp, could well be described by LSD (the screened interaction would modify, of course, the usual electron-gas expressions for $\epsilon_c(\rho)$); correlation connected with the long-range part could well be dealt with using the CI method.

VIII. ACKNOWLEDGMENTS

We are grateful to Prof. H. Preuss for continuous support and advice, to M. Dolg, J. Flad, P. Fuentealba and Dr. L.v. Szentpály for valuable discussions, and to H. Nestle for typing the manuscript.

REFERENCES

1. P.O.Löwdin, Quantum theory of many-particle systems I, Phys.Rev. 97:1474 (1955)
2. H.F.Schaefer III (ed.), "Methods of Electronic Structure Theory", Plenum Press, New York (1977)
3. W. Meyer, PNO-CI studies of electron correlation effects I, J.Chem.Phys. 58:1017 (1973)
4. M.A.Robb, Correlation energy as a stabilizing factor in molecular structure, in:"Molecular Structure and Conformation", I.G. Csizmadia, ed., Elsevier, Amsterdam (1982)

5. W.Meyer, P.Botschwina, P.Rosmus, and H.J.Werner, Computed physical properties of small molecules, in: "Computational Methods in Chemistry", J.Bargon, ed., Plenum Press, New York (198o)
6. D.Cremer, Thermochemical data from ab-initio calculations II, J.Comput.Chem. 3:165 (1982)
7. A.Savin, Die explizite Behandlung von Korrelationseffekten in Dichtefunktional- und Pseudopotentialmethoden, thesis, Stuttgart (1983)
8. A.C.Hurley, Thermochemistry in the Hartree-Fock approximation, Adv. Quantum Chem. 7:315 (1973)
9. W.Meyer and P.Rosmus, PNO-CI and CEPA studies of electron correlation effects III, J.Chem.Phys. 63:2356 (1975)
10. M.Levy, Universal variational functionals of electron densities, first-order density matrices, and natural spin-orbitals and solution of the v-representability problem, Proc.Natl.Acad.Sci. (USA) 76:6062 (1979)
11. P.Hohenberg and W.Kohn, Inhomogeneous electron gas, Phys.Rev. 136:B864 (1964)
12. W.Kohn and L.J.Sham, Self-consistent equations including exchange and correlation effects, Phys.Rev. 14o:A1133 (1965)
13. U.von Barth and L.Hedin, A local exchange-correlation potential for the spin-polarized case I, J.Phys. C 5:1629 (1972)
14. O.Gunnarsson and B.I.Lundqvist, Exchange and correlation in atoms, molecules and solids by the spin-density functional formalism, Phys.Rev. B 13:4274 (1976)
15. S.H.Vosko, L.Wilk, and M.Nusair, Accurate spin-dependent electron liquid correlation energies for local spin density calculations: a critical analysis, Can.J.Phys. 58:12oo (198o)
16. D.M.Ceperley and B.J.Alder, Ground state of the electron gas by a stochastic method, Phys.Rev.Lett. 45:566 (198o)
17. L.Wilk and S.H.Vosko, Estimates of non-local corrections to total ionisation and single-particle energies, J.Phys. C 15:2139 (1982)
18. G.S.Painter and F.W.Averill, Bonding in the first-row diatomic molecules within the local spin-density approximation, Phys.Rev. B 26:1781 (1982)
19. E.Clementi, Correlation energy for atomic systems, J.Chem.Phys. 38:2248 (1963); 39:175 (1963)
2o. B.Y.Tong and L.J.Sham, Application of a self-consistent scheme including exchange and correlation effects to atoms, Phys.Rev. 144:1 (1966)
21. R.D.Cowan, Atomic self-consistent-field calculations using statistical approximations for exchange and correlation, Phys. Rev. 163:54 (1967)
22. H.Stoll, C.M.E.Pavlidou, and H.Preuss, On the calculation of correlation energies in the spin-density functional formalism, Theoret.Chim.Acta 49:143 (1978)
23. E.Eggarter and T.P.Eggarter, Atomic correlation energies, J.Phys. B 11:1157,2o69,2969 (1978)
24. J.P.Perdew and A.Zunger, Self-interaction correction to density-functional approximations for many-electron systems, Phys.Rev. B 23:5048 (1981)

25. P.Rosmus and W.Meyer, PNO-CI and CEPA studies of electron correlation effects IV, J.Chem.Phys. 66:13 (1977)
26. O.Gunnarsson and R.O.Jones, Self-interaction corrections in the density-functional formalism, Solid State Commun. 37:249 (1981)
27. H.Stoll, E.Golka, and H.Preuss, Correlation energies in the spin-density functional formalism II, Theoret.Chim.Acta 55:29 (198o)
28. J.P.Perdew, E.R.McMullen and A.Zunger, Density functional theory of the correlation energy in atoms and ions: a simple analytic model and a challenge, Phys.Rev. A 23:2785 (1981)
29. A.Savin, H.Stoll and H.Preuss, Pair correlation energies and local spin-density functionals, in:"Local Density Approximations in Quantum Chemistry and Solid State Physics", J.Avery and J.P. Dahl, eds., Plenum Press, New York, in press
30. K.Jankowski, P.Malinowski and M.Polasik, Second-order correlation energies of Mg and Ar, J.Phys. B 12:3157 (1979)
31. A.Savin, P.Schwerdtfeger, H.Preuss, H.Silberbach, and H.Stoll, Relativistic effects on the contribution of the local-spin-density correlation energy to ionization potentials, Chem.Phys.Lett. 98:226 (1983)
32. P.Fuentealba, unpublished results
33. H.Stoll, P.Fuentealba, M.Dolg, J.Flad, L.v.Szentpály and H.Preuss, Cu and Ag as one-valence-electron atoms, J.Chem.Phys., submitted
34. L.A.Cole and J.P.Perdew, Calculated electron affinities of elements, Phys.Rev. A 25:1265 (1982)
35. G.Verhaegen and C.M.Moser, Hartree-Fock, correlation and term energies of valence excited states of atoms and ions of the second-row of the periodic table, Mol.Phys. 3:478 (197o)
36. P.Schwerdtfeger, H.Stoll and H.Preuss, A study of potential curve crossing in X-Ar complexes (X=Mg,Ca,Sr,Ba), J.Phys. B 15:1o61 (1982)
37. O.Gunnarsson and R.O.Jones, Extensions of the LSD approximation in density functional calculations, J.Chem.Phys. 72:5357 (198o)
38. A.Savin, in preparation
39. B.I.Dunlap, private communication
4o. G.C.Lie, J.Hinze, and B.Liu, Valence excited states of CH, J.Chem. Phys. 59:1872 (1973)
41. M.F.Guest and D.M.Hirst, Ab-initio potential curves for the valence states of the NH^+ ion, Mol.Phys. 34:1611 (1977)
42. H.Preuss, H.Stoll, U.Wedig, and Th.Krüger, A combination of pseudopotentials and density functionals, Int.J.Quantum Chem. 19:113(1981)
43. J.Flad, G.Igel, M.Dolg, H.Stoll and H.Preuss, A combination of pseudopotentials and density functionals: results for Li_n^{m+}, Na_n^{m+} and K_n^{m+} clusters ($n \leq 4$, m=o, 1), Chem.Phys. 75:331 (1983)
44. J.L.Martins, R.Car, and J.Buttet, Electronic properties of alkali trimers, J.Chem.Phys. 78:5646 (1983)
45. M.R.A.Blomberg, P.E.M.Siegbahn, and B.O.Roos, The ground-state potential curve of the Beryllium dimer, Int.J.Quantum Chem S 14:229 (198o)
46. B.Liu and A.D.McLean, Ab-initio potential curve for $Be_2(^1\Sigma_g^+)$ from the interacting correlated fragments method, J.Chem.Phys. 72:3418 (198o)

47. E.J.Baerends and P.Ros, Evaluation of the LCAO Hartree-Fock-Slater method, Int.J.Quantum Chem. S 12:169 (1978)
48. A.Gupta and R.J.Boyd, Density difference representation of electron correlation, J.Chem.Phys. 68:1951 (1978)
49. D.W.Smith, S.Jagannathan, and G.S.Handler, Density-functional theory of atomic structure I, Int.J.Quantum Chem. S 13:1o3 (1979)
5o. A.Savin, U.Wedig, H.Stoll, and H.Preuss, The correlated density of the Li atom: a test for density functionals and semi-empirical pseudopotentials, Chem.Phys.Lett. 92:5o3 (1982)
51. B.Y.Tong, Exchange- and correlation-energy calculations in finite systems, Phys.Rev. A 4:1375 (1971)
52. W.J.Schneider, Statistical correlation energies in atomic and molecular systems, J.Phys. B 11:2589 (1978)
53. S.K.Ma and K.A.Brueckner, Correlation energy of an electron gas with a slowly varying high density, Phys.Rev. 165:18 (1968)
54. D.C.Langreth and M.J.Mehl, Easily implementable non-local exchange-correlation energy functional, Phys.Rev.Lett. 47:446 (1981)
55. A.K.Rajagopal, Theory of inhomogeneous electron systems: spin-density-functional formalism, Adv.Chem.Phys. 41:59 (198o)
56. P.Ros, Exact and simulated Coulomb holes and Coulomb correlation potentials for the ground state of the He isoelectronic series, Chem.Phys. 42:9 (1979)
57. I.L.Cooper and C.N.M.Pounder, Effect of modified virtual orbitals on local correlation holes in FCN, Int.J.Quantum Chem. 23:257 (1983)
58. J.Keller and J.L.Gazquez, Self-consistent field electron-gas local-spin-density model including correlation for atoms, Phys.Rev. A 2o:1289 (1979)
59. J.A.Alonso and L.C.Balbás, A non-local approximation to the correlation energy of inhomogeneous electron systems, Phys.Lett. A 81A: 467 (1981)
6o. O.Gunnarsson, M.Jonson, and B.I.Lundqvist, Descriptions of exchange and correlation effects in inhomogeneous electron systems, Phys.Rev. B 2o:3136 (1979)
61. J.F.Dobson and J.H.Rose, Orbital self-interaction in Hartree-Fock and density functional theories, J.Phys. C 15:L1183 (1982)
62. G.C.Lie and E.Clementi, Study of the electronic structure of molecules XXI,XXII, J.Chem.Phys. 6o:1275,1288 (1974)
63. J.Lievin, J.Breulet, and G.Verhaegen, A method for molecular correlation energy calculations, Theoret.Chim.Acta 6o:339 (1981)
64. R.Colle and O.Salvetti, Approximate calculation of the correlation energy for the closed and open shells, Theoret.Chim.Acta 53:55 (1979)
65. R.McWeeny, Present status of the correlation problem, in: "The New World of Quantum Chemistry", B.Pullman and R.Parr, eds., Reidel, Dordrecht (1976)
66. J.W.Moskowitz, K.E.Schmidt, M.A.Lee, and M.H.Kalos, Monte-Carlo variational study of Be: a survey of correlated wave functions, J.Chem.Phys. 76:1o64 (1982)
67. M.H.Ang, K.Yates, I.G.Csizmadia, and R.Daudel, Relationship of correlation energy and size, Int.J.Quantum Chem. 2o:793 (198o)

68. A.Dalgarno, C.Boettcher, and G.A.Victor, Pseudopotential calculation of atomic interactions, Chem.Phys.Lett. 7:265 (197o)
69. J.Migdalek and W.E.Baylis, Core-polarization, relaxation, and relativistic effects in the first ionization potentials for some systems in Cu, Ag, and Au isoelectronic sequences, Can.J.Phys. 6o:1317 (1982)
7o. P.Fuentealba, H.Preuss, H.Stoll, and L.v.Szentpâly, A proper account of core-polarization with pseudopotentials, Chem.Phys. Lett. 89:418 (1982)
71. W.Müller, J.Flesch, and W.Meyer, Treatment of inter-shell correlation effects in ab-initio calculations by use of core polarization potentials, J.Chem.Phys., submitted
72. J.Migdalek and W.E.Baylis, Electron affinities for halogens calculated in the relativistic Hartree-Fock approach with atomic polarization, Phys.Rev. A 26:1839 (1982)
73. A.Savin and P.Fuentealba, in preparation
74. J.K.O'Connell and N.F.Lane, Nonadjustable exchange-correlation model for electron scattering from closed-shell atoms and molecules, Phys.Rev. A 27:1893 (1983)
75. T.Ziegler, A.Rauk, and E.J.Baerends, On the calculation of multiplet energies by the Hartree-Fock-Slater method, Theoret. Chim.Acta 43:261 (1977)
76. U.von Barth, Local-density theory of multiplet structure, Phys. Rev. A 2o:1693 (1979)
77. D.C.Langreth and M.J.Mehl, Beyond the local-density approximation in calculations of ground-state electronic properties, Phys.Rev. B 28:18o9 (1983)

DENSITY-FUNCTIONAL THEORY AND EXCITATION ENERGIES

Carl O. Almbladh and Ulf von Barth

Department of Theoretical Physics
University of Lund, Sölvegatan 14A
S - 223 62 Lund
Sweden

1. INTRODUCTION

During the past decade, the local-density (LD) approximation within density-functional (DF) theory has been the most important method for obtaining the electronic properties of realistic systems. The limitations of the method have to a large extent been computational in nature rather than theoretical. The conceptually simple one-particle equations arising in the method have often been too difficult to solve in systems with low symmetry such as amorphous systems or surfaces. However, due to our increased understanding of these systems the computational techniques are quickly developing, and we forsee an even greater importance of the method in the near future. The theoretical limitations of the method will then become more evident and result in an urgent need for improvements beyond the LD approximation. Such a need exists already today in many systems such as, e.g., atoms and molecules. The description of correlation effects in these systems is necessary and important and the answers provided by the LD approximations are often too poor to be of practical use. Unfortunately, for a long time, only minor theoretical advances were made since the modern version of DF theory was laid down by Hohenberg, Kohn, and Sham[1,2] in the mid sixties. There were some notable exceptions such as e.g. the concept of the exchange-correlation hole introduced into DF theory by Gunnarsson and Lundqvist.[3] It is therefore quite pleasing to see the new theoretical developments which have taken place during the last couple of years. The most important of these is, in our opinion, that due to Langreth, Perdew, and Mehl,[4,5] which focuses on a description of exchange and correlation in reciprocal space. The method has been tested in such diverse systems as atoms, surfaces, and solids, and found to be markedly better than the LD approximation. Using a modified version of the method it was possible to obtain atomic correlation energies with an accuracy of ~0.1eV, which is truly a remarkable achievement. Until this new method has been tested on many more systems we are not able to pass a final judgement on its utility, but it seems clear that it holds a lot of promise.

The second most important development was initiated by Williams and von Barth[6] and concerns the interpretation and physical significance of the one-electron eigenvalues which arise within DF theory. For a long time, these eigenvalues were, from a theoretical point of view, considered to be auxiliary parameters for obtaining the total energy of the system at hand. Their physical significance remained obscure. At the same time, these eigenvalues were used extensively and rather successfully to interpret e.g. photoemission spectra of solids. Investigations by others and by us, however, show that these eigenvalues are, in general, not to be identified with excitation energies. The exception is the highest occupied DF eigenvalue which, in all systems, equals the ionization potential. Furthermore, this recent line of work has led to an understanding of the kind of quantitative differences that can be expected between DF eigenvalues and excitation energies in various systems. The vitalized discussion of DF eigenvalues has also led to a need for improved exchange-correlation potentials as well as to clues how such improvements can be constructed.

As the third recent development we would like to mention the work by Levy[7] on the fundamental aspects of the theory. His work has clarified several unsettled questions concerning the formal justification of the theory but, more importantly, it has pointed to a much simplified way of how to think about the abstract functionals that are at the root of the theory. In this connection we consider the critical analysis offered by Lieb[8] as quite useful. This work distinguishes between that part of basic DF theory which hinges on physically plausible assumptions and that part which can be considered to be stringent in a more mathematical sense.

The renewed interest in DF theory that we have seen during the past couple of years has resulted in several review articles,[9-13] three of which have been written, at least in part, by one of us.[6,14,15] Therefore, these lecture notes are not intended as a comprehensive account of the subject. Instead we have chosen to comment on selected topics of current interest which have been hinted on in previous reviews or which have been brought up during some of the conferences on DF theory which have been organized during the last one or two years. Our comments will focus on the exchange-correlation potential and the DF eigenvalues.

2. SOME EXACT RESULTS

Recently it has become possible to make several exact statements concerning the properties of the exchange-correlation potential v_{xc} of DF theory and of the corresponding eigenvalues. We will here quote some of these results without proof but we will indicate the physical ideas behind the results.

As is well known, DF theory leads to a Schrödinger-like one-particle equation of the form[2]

$$[-\frac{1}{2}\nabla^2 + w(\vec{r}) + V_H(\vec{r}) + v_{xc}(\vec{r})]\phi_i(\vec{r}) = e_i\phi_i(\vec{r}) \ , \tag{1}$$

whose normalized solutions determine the particle density according to

$$n(\vec{r}) = \sum_1^N |\phi_i(\vec{r})|^2 \tag{2}$$

(we use atomic units such that $\hbar=e^2[/(4\pi\varepsilon_0)]=m=1$). The external potential $w(\vec{r})$ from e.g. the nucleus and the Hartree potential $V_H(\vec{r})= \int d^3r'v(\vec{r}-\vec{r}')n(\vec{r}')$ $(v(\vec{r})= 1/r)$ are explicitly known, but the exchange-correlation potential $v_{xc}(\vec{r})$; can only be obtained formally as the functional derivative with respect to density of the unknown functional E_{xc} which gives the exchange-correlation energy,

$$v_{xc}(\vec{r}) = \frac{\delta E_{xc}}{\delta n(\vec{r})} \ . \tag{3}$$

This suggests that a thorough knowledge of $E_{xc}[n]$ is necessary in order to deduce any exact result for v_{xc} and thus for the DF eigenvalues which are defined through Eq. (1). Fortunately, as we shall see, properties of v_{xc} can be obtained indirectly from a knowledge of the particle density $n(\vec{r})$. This follows from the uniqueness theorem by Hohenberg and Kohn[1] which states that there is a one-to-one correspondence between the densities n and the external potentials w. Thus, if we can find one potential v_{xc} which generates the known density n through Eqs. (1) and (2), we have found the exact v_{xc}. This idea was introduced[16] by Williams and von Barth[6] in order to obtain the exact v_{xc} and DF eigenvalues for the hydrogen atom and for the helium atom within the Hartree-Fock (HF) approximation. The same procedure was later used by von Barth and Car[17] for heavier atoms within HF theory and by Almbladh and Pedroza[18] for light atoms but including also full correlation.

In finite systems it is possible to obtain the asymptotically exact density profile far away from the system. This can actually be done without solving the full many-body problem[19] and, by again using the uniqueness theorem, we can deduce the asymptotically exact form for v_{xc} at large distances. We note that, in finite isolated systems, we can only do number-conserving variations of the density. Consequently v_{xc} from Eq. (3) is only determined to within an arbitrary constant. This is nothing but a reflection of the familiar fact that absolute values of energy have no meaning in particle-conserving physics. In order to get a unique definition of the eigenvalues of Eq. (1) one normally requires all three potentials in Eq. (1) to vanish at infinity. It is then easy to see that the individual orbitals ϕ_i decay as

$$\exp[-r(-2e_i)^{1/2}]$$

with distance r. Consequently, at far distances only the highest occupied orbital ϕ_N contributes to the density which thus decays as

$$\exp[-2r(-2e_N)^{1/2}].$$

On the other hand one can show quite independently from DF theory that the asymptotic density decays as

$$\exp[-2r(2I)^{1/2}],$$

where I is the ionization potential.[19,20] This allows us to make the identification[19]

$$e_N = -I \quad . \qquad (4)$$

This result is truly general and valid for systems of any size. It was e.g. shown already in the original work by Kohn and Sham[2] that the Fermi level in a metal is given correctly by the highest occupied DF eigenvalue. Eq. (4) also shows that the top of the valence band in an insulator is correctly reproduced by the corresponding DF eigenvalue.

The analysis of the asymptotic density profile can be carried further in order to deduce additional exact properties of v_{xc}. For instance, it can be shown directly from the many-electron Schrödinger equation that the density of a spherical atom can be expanded in powers of reciprocal distance according to[19]

$$n(r) = C\, r^{\beta} e^{-2\kappa r}\left(1 + \frac{a_1}{r} + \frac{a_2}{r^2} + \frac{a_3}{r^3} + \ldots\right). \tag{5}$$

Here, the parameters κ, β, a_1, and a_2 are simply related to the ionization potential of the atom and the electrostatic potential away from the corresponding singly ionized ion. For instance, $\kappa=(2I)^{1/2}$ and $\beta=(Z-N+1)/\kappa-1$. The coefficient a_3, on the other hand, also involves the polarizability α of the ion. Postulating that the exact DF theory must reproduce the asymptotic result in Eq. (5) we can infer the effective potential that produces a highest occupied orbital whose square gives this kind of asymptotic behavior. Assuming for simplicity that the ion is also spherical we find

$$v_{xc}(r) = -\frac{1}{r} - \frac{\alpha}{2r^4} \quad . \tag{6}$$

It is gratifying to see that the exact v_{xc} has the behavior one would expect from intuitive arguments. The first term, which has the same form for all systems, represents the attraction from the exchange hole. The second term represents the classical attraction between an electron and a polarizable ion. As we shall see below, however, v_{xc} can often not be interpreted in this physically appealing way. It is conceptually simple to work out the asymptotic form of v_{xc} also in more complicated systems without spherical symmetry and with more atoms involved. In such cases v_{xc} will contain additional powers of inverse distance as compared to Eq. (6). For instance, terms decaying as r^{-2} and r^{-3} can arise from a difference between the Hartree potentials of the system and the corresponding ion. The necessary labor to derive these results is, however, often substantial and for more details we refer the reader to Ref. 19.

One extension of the theory is of considerable interest. Since all systems with an odd number of electrons are spin polarized it is gratifying to know that the indirect procedure that we have discussed here for determining v_{xc} can be applied also to the spin-dependent version of the DF theory.[21] In finite systems, the two spin channels (the density matrix can easily be seen to be diagonal in spin indices) are completely decoupled in DF theory, and a knowledge of the exact spin-up and spin-down densities determines the exchange-correlation potential and the eigenvalues of the two channels. As in the unpolarized case asymptotic results for the two spin densities can be derived from the full many-electron Hamiltonian,[19] allowing us to determine the uppermost occupied eigenvalue and the asymptotic form of the exchange-correlation potential in each spin channel. In the sodium and lithium atoms, for instance, the ionization energy determines the uppermost eigenvalue of the majority channel, whereas that of the minority channel is determined by the energy of the singly ionized ion in its lowest state of triplet symmetry (of. Table 1).

Asymptotically exact results can also be derived for v_{xc} far outside solid surfaces. The theoretical analysis of the asymptotic density outside a macroscopic system is, however, much more involved[19] but the general ideas remain the same. The Fermi exchange term $-1/r$ and the terms decaying as r^{-2} and r^{-3} are in principle still present but are of order unity for a large system and vanish compared to the term containing the macroscopic polarizability, $-\alpha/(2r^4)$. At distances z from the surface which are small compared to the macroscopic dimensions of the system but still microscopically large, the polarizability term in v_{xc} goes over to the well-known image potential and we have the exact result[19,22,23]

$$v_{xc}(z) = -\frac{1}{4z}\,\frac{\varepsilon-1}{\varepsilon+1}\ . \tag{7}$$

Here, ε is the dielectric constant of the systems due to only the electrons.

3. THE INFINITE RANGE OF v_{xc}

We saw in the previous section that the exchange-correlation potential v_{xc} could be given a physical interpretation. In this section we will show that this is not always the case and that v_{xc}, sometimes, is merely a mathematical construction for obtaining the particle density. By studying two atoms, A and B, separated by a very large distance, we will see that v_{xc} has in fact an infinite range. It is not difficult to see that the ionization potential of the combined system is given by the smallest (I_A) of the two ionization potentials I_A and I_B. Thus, the exponential fall-off of the particle density is everywhere governed by I_A except in a region far but not too far away from atom B.[24] In this latter region the exponential tail of the density from atom B will be larger than the tail from atom A although it will have a faster fall-off governed by the ionization potential I_B ($I_B>I_A$) . We will further assume that each of the two atoms has one unpaired electron and that, therefore, the ground state of the combined system is a singlet with a vanishing spin density. In the DF description of this system the asymptotic density is determined by the highest occupied DF orbital which, for this unpolarized system, must be doubly occupied. As usual the eigenvalue of this orbital is $-I_A$, and we note that it must place half an electron on each atom. Consequently, the same orbital must describe both the exponential decay of the entire system (governed by I_A) and the exponential decay (governed by I_B) far but not too far away from atom B. It follows from the effective one-particle equations of DF theory that the decay in this later region is determined by the difference between the eigenvalue ($-I_A$) and the nearly constant v_{xc} in the region. Thus, $I_B=v_{xc}-(-I_A)$, or $v_{xc}=I_B-I_A$ in the region. In the absence of atom A, however, the highest occupied DF eigenvalue would be $-I_B$ and v_{xc} would nearly vanish in the region under discussion. We are thus led to the unphysical result that the exchange-correlation potential around an isolated atom can change by a finite amount upon introduction of a different[25] atom at an arbitrarily large distance.

This peculiar example shows that extreme care must be excercised in using physical intuition in order to deduce properties of the exchange-correlation potential v_{xc}. This convenient quantity is a logical construction which allows us to treat the full many-body problem with little more effort than that required for a one-particle problem. It is not inconceivable that a high price must be paid for such a simplification. There are actually more reasons to be surprised by the fact that peculiar situations like that descibed above are rarely encountered in practice.

It should be noted that the assumption of a singlet state of the "molecule" was an essential ingredient for obtaining the unphysical result of this section. If the ground state was instead a triplet state, the spin-polarized version of the DF theory would apply, and the corresponding spin-up and spin-down exchange-correlation potentials would not have any unphysical behavior. At a large separation between the atoms, an exponentially small magnetic field would suffice to make the triplet the ground state and remove the unphysical property of v_{xc}. This demonstrates another unphysical property of v_{xc}, namely that a very small external perturbation can cause large changes in v_{xc}. On the other hand, it shows that for a system with an unphysical v_{xc} there is often a "nearby" system for which physical intuition can be used as a guide for determining v_{xc}.

A related example which again illustrates the infinite range of v_{xc} is provided by an open-shell atom outside a non-magnetic metal surface. To be specific, we discuss the case of a hydrogen atom, and we consider the limit of large separation between the atom and the substrate. As in the example above, this system has no spin polarization. Since the atom has only one electron, the atomic 1s orbital is singly occupied. This implis that the atomic DF eigenvalue is pinned to the Fermi level μ of the substrate, and that v_{xc} in the neighborhood of the atom is shifted by an amount $\mu+I_H$ compared to its value for the isolated atom. The atomic excitation energy, on the other hand, approaches the value $-I_H= -1Ry$ and is thus unrelated to the corresponding DF eigenvalue.

In a more complete description, the charge on the atom is composed of continuum DF orbitals common to the atom and the substrate. At large separations, these orbitals(Ψ_k) can be expanded in unperturbed substrate orbitals (ϕ_k) and the atomic 1s orbital (ϕ_{1s}),

$$\Psi_k(\vec{r}) = \sum_{k'} U_{k'k}\phi_{k'}(\vec{r}) + v_k\phi_{1s}(\vec{r}) .$$

The problem of finding the correct mixture of substrate and atomic states is identical to that encountered in the Fano-Anderson problem[26] and has a well-known analytical solution. For small overlaps, $|v_k|^2$ is a Lorentzian function of energy centered at the discrete state energy e_{1s}. The width of this Lorentzian tends to zero with the overlap between the unperturbed substrate states and the atomic state. The resonance contains exactly one state, i.e.,

$$\sum_{\text{all } k} |v_k|^2 = 1.$$

Since the system is unpolarized, each occupied orbital Ψ_k of the combined system contributes $2|v_k|^2$ electrons to the atom, which contains only one electron. Thus, the resonance is pinned to the Fermi level and we must have $e_{1s} = \mu$ as stated above.

As in the case of the two separated atoms, an exponentially small magnetic field is enough to remove the pinning and to shift v_{xc} by a finite amount in the neighborhood of the atom.

4. EXCITATION ENERGIES AND DF EIGENVALUES OF LOCALIZED SYSTEMS

As shown above, the highest occupied DF eigenvalue is always equal to the ionization potential. This fact immediately leads us to wonder about a possible connection between other DF eigenvalues and the corresponding excitation energies. In section 2 we discussed a fitting procedure by which, in principle, all the exact DF eigenvalues of a finite system could be determined from a knowledge of the exact particle density. In ref. 14 we pointed out some numerical problems encountered in this procedure. They were related to the difficulty in obtaining very accurate densities far away from the system which, in turn, produced large uncertainties in $v_{xc}(r) - e_{DF}$ as $r \to \infty$. By using our exact theorem concerning the highest occupied DF eigenvalue these difficulties were, however, easily overcome, and in Table 1 we present some of our results for the atoms Li, Be, and Ne. The results for Ne[17] are based on Hartree-Fock (HF) densities, but for an atom the inclusion of correlation gives no major numerical changes. This is seen in the Be and the Li atoms where both HF results and fully correlated results [18] are given. The differences are actually still smaller in the Ne atom as illustrated by quoting values obtained from a density produced by the exact exchange potential plus a correlation correction due to Langreth and Mehl.[5]

We see that the DF eigenvalues are always above the corresponding excitation energies and that this effect becomes more pronounced with the localization of the corresponding orbital. The HF eigenvalues, on the other hand, are always below the excitation energies. Thus, thinking of HF as a self-interaction-free approximation, we can use common self-interaction jargon and say that there should <u>not</u> be a complete cancelation of the self-interaction in exact <u>DF</u> theory. For comparison, we have in Table 1 also given the results obtained from the local-density (LD) approximation. Evidently, this approximation places the eigenvalues even higher than DF theory, and also this effect increases with the localization of the orbital. It should be noted, however, that a large part of the error introduced into the DF eigenvalues by the use of the LD approximation is a constant shift. If we, for instance, shift all the LD eigenvalues by an amount required to make the highest occupied LD eigenvalue agree with the corresponding exact DF eigenvalue, the errors in the 2s and 1s eigenvalues in Ne become only 1.0 and 2.5eV compared to originally 9.0 and 10.5V. This point was first discussed by Perdew and Norman.[29]

TABLE 1. Orbital eigenvalues in different theoretical schemes, obtained from refs. 17 and 18, compared with measured excitation energies for the Li, Be, and Ne atoms. Here DF refers to the exact density-functional theory, LD to its local-density approximation, HF to the Hartree-Fock approximation, DFX to the exact density-functional treatment of exchange effects only, and LDX refers to the corresponding local-density approximation. e_k is the eigenvalue for orbital k, and n_k its occupancy. Energies in eV.

Atom, orbital	n_k	$-e_k$					Excitation energy
		DF	LD	HF	DFX	LDX	
Li, 1s+	1	57.6	51.0	67.7	56.0		66.2
Li, 2s+	1	5.4	3.2	5.3	5.3		5.4[a]
Li, 1s-	1	64.5	50.8	67.2	67.2		64.5[a]
Li, 2s-	0	5	1.9	---	8		0.6[b]
Be, 1s	2	115.1	104.9	128.8	112.2	103.2	123.6[a]
Be, 2s	2	9.3	5.6	8.4	8.4	4.6	9.3[a]
Be, 2p	0	5.7	2.1	---	4.9		---
Ne, 1s	2	835.2[c]	824.7	891.7	838.8	822.7	870.2[a]
Ne, 2s	2	45.1[c]	36.1	52.5	46.7	34.5	48.5[a]
Ne, 2p	6	21.6[c]	13.6	23.1	23.1	12.1	21.6[a]
Ne, 3s	0	4[c]	0.2	---	5.2		---

[a]Siegbahn and Karlsson, ref. 27. [b]Patterson *et al.*, ref. 28.
[c]The effects of correlation were estimated using an approximate correlation potential as described in the text.

In connection with the discussion of band gaps in insulators we need to know the energy position of the lowest unoccupied DF eigenvalue. We remind the reader that the exact v_{xc} always has a $-1/r$ tail which assures that, even in the case of a neutral atom or molecule, the effective one-particle equation of DF theory has infinitely many bound states below the continuum.[14] Thus, even atoms with no electron affinity such as e.g. Be and Ne have bound unoccupied orbitals (cf. Table 1). As a consequence the lowest unoccupied DF eigenvalue for these atoms and presumebly for all atoms (cf. Li in Table 1) is below the negative of the electron affinity.[14,17,18]

5. BANDGAPS AND DF EIGENVALUES OF EXTENDED SYSTEMS

In the case of finite systems we håve seen that only the uppermost occupied DF eigenvalue gives an excitation energy, whereas the lower occupied and the unoccupied levels lie respectively above and below the corresponding excitation energies. The fact that the deviations between DF eigenvalues and excitation energies decrease with increasing extent of the corresponding orbitals might induce a hope that the two sets of energies could be the same in an extended system. Unfortunately, this is not the case. This point was addressed by e.g. Williams and von Barth[6], who considered the case of a homogeneous but interacting electron gas as an illustrative example. The translational invariance of this system immediately allows us to write down the dispersion of the DF eigenvalues e_k. The correct excitation energies ε_k are also relatively easy to obtain from the Dyson equation involving the energy and momentum dependent self-energy. Near the Fermi level, the dispersion of the excitation energies ε_k is determined by the many-body effective mass m^*, whereas the dispersion of the DF eigenvalues is determined by the bare electronic mass m. Thus, the DF eigenvalues deviate from the corresponding excitation energies immediately away from the Fermi level. Not even the density of states at the Fermi level (which is proportional to m^*) is given correctly by DF theory. Due to a fortunate cancellation between the momentum and energy dependence of the self-energy of free-electron-like systems there is, however, often less than a 10 % difference between the m^* and m. This is the reason behind the successful interpretation of photoemission and other spectra of simple metals in terms of the LD approximation. For these systems, with relatively slowly varying densities , the latter should give results not too far from those of the exact DF theory. In systems with more localized electrons such as e.g. transition metals we can, at this stage, not really tell whether the discrepancies which are sometimes observed between measured and calculated spectra are due to the use of the LD approximation or to the more fundamental difference between the DF eigenvalues and excitation energies.

In semiconductors and insulators the situation is somewhat different due to the existence of a finite energy gap just above the highest occupied one-particle excitation energy-normally the top of the valence band. As discussed in Sec. 2 this level is correctly reproduced by the corresponding DF eigenvalue, but what about the bottom of the conduction band ? This case has similarities with that encountered in an atom, with the top of the valence band and the bottom of the conduction band corresponding respectively to the ionization potential and the affinity level of the atom. In the atom we saw[14,17,18] that the affinity level is substantially above the corresponding DF eigenvalue. On the basis of this observation Perdew and Levy[30] conjectured that the same discrepancy does prevail in the solid and consequently the fundamental bandgap is not given by the spectrum of DF eigenvalues. A similar claim has been made by Sham and Schlüter,[31] but on different grounds. It is, of course, necessary to show that the discrepancy does not vanish as the size of the system is made very large. We will here try to convince the reader that the discrepancy does indeed persist, but we will first make some historical remarks on the topic of DF band gaps.

For different reasons, it has long been customary to introduce fractional occupation numbers as a tool in the practical applications of DF theory. The prescription has simply been to renormalize each DF orbital ϕ_i from 1 to $(n_i)^{1/2}$ and to consider the total energy E and the DF eigenvalues e_i as functions of the "occupation" numbers n_i. It is quite easy to show[32] that the relation

$$e_i = \frac{\delta E}{\delta n_i} \tag{8}$$

is valid in all approximations of DF theory which are presently in use. This relation can, however, also be proven correct in the exact DF theory, provided the functional E_{xc} can be generalized to a continuous and differentiable function of the occupation numbers. The validy of Eq. (8) leads, however, to the following paradox.[15] It follows readily from this equation[6] that the fundamental gap is given by the corresponding DF eigenvalues. On the other hand, consider an infinite solid made up of a regular lattice of identical atoms. By choosing a very large lattice constant we can obviously arrange to have the top of the valence band ε_v of the solid arbitrarily close to the negative of the atomic ionization potential $-I$,

$$\varepsilon_v = -I + \delta_v .$$

When the solid is formed, the highest occupied DF eigenvalue e_{ao} of the atom will shift and broaden into a band. The broadening effect ($W_v/2$) can again be made arbitrarily small by choosing a large interatomic separation, but due to the long-range nature of v_{xc} (cf. Sec. 3) we can not preclude a constant shift Δv_{xc} remaining at infinite separation. Note that v_{xc} is the only term in the effective one-particle DF equation which can produce a remaining shift. Note also that such a shift must affect all eigenvalues equally because in the limit of large separation the total density must tend to a superposition of atomic densities in all regions with any appreciable density. Thus, the top of the valence band of DF eigenvalues is given by

$$e_v = e_{ao} + \Delta v_{xc} + \frac{1}{2}W_v ,$$

where W_v can be made as small as desired. The same chain of arguments can now be applied to the case of the atomic affinity level A, the bottom of the two conduction bands (ε_c and e_c) and the lowest unoccupied DF level e_{au} of the atom. We obtain

$$\varepsilon_c = -A + \delta_c$$

and

$$e_c = e_{ao} + \Delta v_{xc} + \frac{1}{2}W_c$$

where δ_c and W_c can be made arbitrarily small. We have here used the same shift Δv_{xc} as above because, as noted before, v_{xc} can not change its shape in the vicinity of the atom. In the limit of large separation we obtain the fundamental gap

$$\Delta\varepsilon_{cv} = \varepsilon_c - \varepsilon_v = I - A \tag{9}$$

whereas DF theory gives

$$\Delta e_{cv} = e_c - e_v = e_{au} - e_{ao} \quad . \tag{10}$$

From the discussion in Sec. 4 we know that e_{ao}=−I, whereas e_{au}≠−A, and we must conclude that DF theory does not give the fundamental gap. This result is in clear contradiction with Eq.(8),[6] and the paradox has been resolved by Perdew and Levy. Mermin[33] has presented a generalization of DF theory valid at finite temperatures and for an arbitrary non-integral number of particles. The zero-temperature limit of the Mermin theory is the natural generalization of DF theory which can incorporate fractional occupation numbers. This limit was obtained by Perdew et al.,[34] who found that Eq. (8) is valid when the occupancy n of the highest DF level lies in the interval $0<n\leq 1$. When n=0, however, the derivative is ill-defined, and the eigenvalue has a discontinuous jump, which precludes the possibility of obtaining the affinity level from this equation. For further details we refer the reader to Sec. 7 and to the article by Perdew in this volume.

Knowing now which part of the paradox is in error, we can take the second part of the paradox as a proof that the gap is not given by the DF eigenvalues,

$$\Delta \varepsilon_{cv} \neq \Delta e_{cv} \tag{11}$$

In a monoatomic solid with a large lattice parameter we know that the discrepancy is given by A+e_{au}. The case of solid Ne is actually not too far from this limit. The ionization potential of the atom is 21.6 eV and, since there is no electron affinity, the bottom of the conduction band of the sparse solid is at the bottom of the atomic continuum. Thus, the band gap at infinite separation is 21.6 eV whereas that of solid Ne is 21.3 eV.[35] From Table 1 the LD bandgap (13.4 eV) at infinite separation is, however, 4.2 eV more narrow than the DF gap (17.6 eV), which suggests that the exact DF theory would correct almost half the bandgap error found in calculations based on the LD approximation. We have then assumed the error to scale with the bandgap, an assumption which is not unreasonable in view of the fact that the error must vanish with the gap. Thus, in Si the exact DF gap would be ~0.8 eV compared to an exact gap of 1.1 eV and a LD gap of 0.5 eV. Perdew[36] has, however, put forth an argument against such expectations. The new exchange-correlation potential by Langreth, Perdew, and Mehl[4,5] gives eigenvalue differences in atoms which are superior to those of the LD approximation. In a system like bulk silicon with a much more slowly varying particle density the new method[5] should be even more superior to the LD approximation, yet the two methods produce nearly the same band-gap in silicon (~ 0.5-0.6 eV).[37] Thus, it could be that the entire discrepancy (0.6 eV) between the measured and the calculated gap is due to the more fundamental incapability of DF theory to provide a correct gap and not due to use of the LD approximation. Another argument in support of this view can be obtained by considering a sparse solid of Be atoms. Be has no electron affinity, and by the argument we used above in the case of Ne we conclude that the fundamental gap of the Be solid is 9.3 eV at a large lattice constant. The DF gap and the LD gap are 3.6 eV and 3.5 eV, respectively (cf. Table 1), demonstrating that, in this case, the exact DF theory represents almost no improvement over the LD approximation. It is tempting to guess that the qualitatively different results obtained for the Ne and Be atoms are due to the fact that in Be the valence band and the conduction band originate from the same shell (2s,2p) whereas in Ne they do not (2p,3s). From these considerations it is, however, difficult to draw any conclusions regarding silicon. Here the valence and the conduction bands originate from the same atomic shell (3s,3p), but the gap is a consequence of the energy separation between bonding and anti-bonding orbitals rather than between atomic 3s and 3p orbitals. Much more research is needed in order to resolve this interesting issue. In this context it is worth-while to try to use our newly acquired knowledge of the exact v_{xc} for several atoms in an attempt to construct better exchange-correlation potentials for solids.

6. THE FERMI SURFACE IN A METAL

The fact that the uppermost occupied DF eigenvalues of a metal equal the Fermi energy has been known since the original work by Kohn and Sham.[2] It is, however, still an unresolved question whether or not the "DF Fermi surface" is identical to the correct Fermi surface for quasi-particles as obtained from e.g. a de Haas-van Alphen measurement. On this latter surface in k-space the quasi-particle energies ε_k are equal to the Fermi energy μ

$$\varepsilon_k = \mu \ , \tag{12}$$

whereas the "DF Fermi surface" is defined by replacing the quasi-particle energies with the corresponding DF eigenvalues,

$$e_k = \mu \ . \tag{13}$$

We notice that the two Fermi surfaces have the same symmetry and enclose the same volume in k space. Furthermore, from the following argument[38] it is tempting to guess that the two surfaces are in fact the same. As in our previous discussion, the basic idea is to study the density profile of a suitably chosen system. In this case, we consider a point impurity in the metal. The long-range behaviour of the density induced by the impurity can be obtained directly from the many-electron Hamiltonian without resorting to DF theory. As is well known, the induced density has Friedel oscillations which can be shown to give a direct measure of the Fermi-surface dimensions. Within DF theory, which by definition reproduces the correct density, the induced oscillations must be a consequence of the DF Fermi surface. This fact induces hope that the two surfaces are identical.

7. REMARKS ON FRACTIONAL OCCUPATION NUMBERS IN DF THEORY

Macroscopic systems can not even in principle be prepared to be in a fixed quantum state with a fixed number of particles although, in many cases, correct results can be obtained using such an idealization. Macroscopic systems are most naturally described in terms of ensembles in which the number of particles are allowed to fluctuate. Such an ensemble theory can, in principle, also be applied to finite systems. Sometimes a treatment of this kind can give useful analogies between finite and macroscopic systems. In particular, the notion of fractional occupation numbers can be given a rigorous meaning. It is, however, important to realize that, for finite systems, the usual quantum mechanics for a fixed number of particles gives a complete description. Thus, extensions to the case of a fractional number of particles does not add any new physics.

The generalization of DF theory to the case of ensembles has been given by Mermin.[33] This theory gives, in principle, the exact grand canonical potential Ω and the ensemble-averaged particle density $n(\vec{r})$ by a procedure analogous to that used in the ground-state theory.[1,2] In order to have a well-defined generalization to the domain of fractional occupation, only the zero-temperature limit of the Mermin theory is needed. This limit has been obtained by Perdew, Levy, Parr, and Balduz.[34]

As in the usual DF theory, Mermin introduced a basic functional $F[n]$ which gives the sum of the kinetic energy and the interaction energy, and a functional $T_0[n]$ which gives the kinetic energy of a non-interacting system. At finite temperatures F and T_0

also include entropy terms. The exchange-correlation energy is defined in the usual way as the difference

$$E_{xc}[n] = F[n] - T_o[n] - \frac{1}{2}\int n(\vec{r})V_H(\vec{r})d^3r \ .$$

The functionals F and T_o are most conveniently defined through Levy's procedure of a restricted search.[7] Given the functional F, the grand potential for any electronic system can be obtained as the minimum of

$$\Omega_w[n] = F[n] + \int[w(\vec{r})-\mu]n(\vec{r})d^3r$$

for fixed external potential w, chemical potential μ, and temperature. As in the usual DF theory, we express F in T_o to obtain an equivalent one-electron problem,

$$\frac{\delta F[n]}{\delta n(\vec{r})} + w(\vec{r}) - \mu = \frac{\delta T_o[n]}{\delta n(\vec{r})} + V_H(\vec{r}) + v_{xc}(\vec{r}) + w(\vec{r}) - \mu = 0. \tag{14}$$

This leads to a simple one-electron equation (Eq. 1), whose solutions give the particle density according to

$$n(\vec{r}) = \sum_i n_i|\phi_i(\vec{r})|^2 \ . \tag{15}$$

Here, the occupation number n_i is given by a Fermi function of the corresponding DF eigenvalue e_i. The exchange-correlation potential,

$$v_{xc}(\vec{r}) = \delta E_{xc}/\delta n(\vec{r}),$$

is now defined on an absolute energy scale since arbitrary density variations are allowed. Making the conventional choice of effective potential for the one-electron problem, i.e. $V_H+v_{xc}+w$, we see that the chemical potential of this problem is the same as that of the interacting problem, μ.

In the case of finite systems, it was shown by Perdew et al.[34] that the grand ensemble at zero temperature reduces to a simple mixture of N- and (N+1)-electron ground states. More specifically, one finds that only N-particle ground states are involved when the chemical potential μ is chosen between the N- and (N+1)-electron ionization energies, $-I_N<\mu<-I_{N+1}$, and that both N- and (N+1)-particles states may be involved when $\mu=-I_{N+1}$. In the latter case, the ensemble can have a fractional number of electrons $N+\rho$ $(0<\rho<1)$, and the particle density and total energy are superpositions of N- and (N+1)-particle ground-level averages,

$$E = (1-\rho)E(N) + \rho E(N+1), \tag{16}$$

and

$$n(\vec{r}) = (1-\rho)\langle N|\hat{n}(\vec{r})|N\rangle + \rho\langle N+1|\hat{n}(\vec{r})|N+1\rangle . \tag{17}$$

In the case of fractional occupation the chemical potential must also coincide with the ionization energy of the equivalent non-interacting system, i.e. with e_{N+1}. Otherwise this system can not reproduce the correct number $(N+\rho)$ of electrons. Thus,

$$e_{N+1} = -I_{N+1} . \tag{18}$$

Note that this result[34] is a simple consequence of choosing the effective one-particle potential at infinity in such a way as to have the same chemical potential for both the interacting and the non-interacting problem. In this way the eigenvalue e_{N+1} is defined by Eq. (18), and the physical problem lies in finding the value of the effective potential far away from the system, i.e. the value of $v_{xc}(\infty)$. (Note that $V_H(\infty)=w(\infty)=0$.) We will now show that $v_{xc}(\infty)=0$, although suggestions to the contrary have appeared in the literature.[34,39] It has so far not been possible to obtain $v_{xc}(\infty)$ directly from its definition as a variational derivative in order to give an alternative proof of our eigenvalue theorem discussed in Sec. 2. Instead, we use this theorem to infer the correct value of $v_{xc}(\infty)$. In the interval $0<\rho<1$ we find from Eq. (17) and the known long-range behaviour of the ground-level densities that the exponential fall-off of the density is given by the smallest (I_{N+1}) of the two ionization potentials involved. In the DF description of this density, the exponential fall-off is given by the highest occupied DF eigenvalue relative to the continuum edge, $v_{xc}(\infty)$. Thus, $I_{N+1}=v_{xc}(\infty)-e_{N+1}$. Comparing with Eq. (18) we find

$$v_{xc}(\infty) = 0 \tag{19}$$

for all densities that correspond to a fractional occupancy. When the mean number of particles is integral, the variational derivatives are not unambiguously defined. In this case we define v_{xc} as the limit when the occupancy tends to an integral value from below. With this choice we adopt the same convention for the fractional theory as for the theory with a fixed number of particles. Similar considerations for a spin-polarized ensemble shows that v_{xc} of both spins vanishes far away from a finite system.

We conclude this section by discussing the discontinuous behaviour of v_{xc} and the DF eigenvalues when the number $(N+\rho)$ of particles tends to an integral value from above. When ρ is small, the ensemble density is close to the N-electron ground-level density out to some large distance $R(\rho)$(cf. Eq. 17). Sufficiently far away, however, the (N+1)-electron component of the density will dominate because it has a slower exponential fall-off. In the DF description the uppermost orbital has a very small occupation number (ρ) and gives a negligible contribution to the density as long as $r<R(\rho)$, but it gives the dominating contribution when $r>>R(\rho)$. In order to correctly reproduce the density in the region $r<R(\rho)$, the exchange-correlation potential $v_{xc}(\vec{r},N+\rho)$ for the ensemble must have the same shape as that of the N-electron system. There is, however, a possibility that it is shifted by a constant Δ. Thus,

$$v_{xc}(\vec{r},N+\rho) = v_{xc}(\vec{r},N) + \Delta, \; r < R(\rho). \tag{20}$$

For any $\vec{r}$ we can always choose ρ such that $\vec{r}$ is in the region $(r<R(\rho))$ where Eq. (20) is valid. Thus, in the limit $\rho\to 0^+$, this equation is valid everywhere. Note that the order of limits here is important. If r becomes large enough at a finite ρ both $v_{xc}(N+\rho)$ and $v_{xc}(N)$ tend to zero as discussed above. The highest occupied eigenvalue e_{N+1} corresponding to $v_{xc}(N+\rho)$ is independent of ρ and equal to $-I_{N+1}$ (of. Eq. (18)). Thus, it follows from Eq. (20) that

$$-I_{N+1} = e^{(N+\rho)}_{N+1} = e^{(N)}_{N+1} + \Delta, \tag{21}$$

where $e_{N+1}^{(N)}$ is the lowest unoccupied eigenvalue of the N-electron system. As discussed in Sec. 2, this eigenvalue is normally quite different from the corresponding excitation energy $-I_{N+1} = -A_N$ (cf. Table 1), demonstrating that the discontinuity Δ is indeed finite. (A_N is the elctron affinity of the system). Intuitively one would expect this discontinuity to vanish in the macroscopic limit. In insulators, however, the difference between the fundamental gap and the corresponding difference of DF eigenvalues shows that the discontinuity remains. Again, this emphasizes that v_{xc} is primarily a convenient construction for reproducing the exact density and that extreme care must be exercized in order to deduce its properties from simple physical arguments.

ACKNOWLEDGMENTS

We wish to thank R. Car and A.C. Pedroza for fruitful collaboration, and the organizers of the NATO Advanced Study Institute, Professor R.M. Dreizler and Professor J. da Providencia, for arranging a most stimulating summer school. We also want to thank John Perdew and Mel Levy for exciting discussions during the summer of 1983. The present work was supported by the Swedish Natural Science Research Council.

REFERENCES

1. P. Hohenberg and W. Kohn, Phys. Rev. 136:B864 (1964).
2. W. Kohn and L.J. Sham, Phys. Rev. 140:A1133 (1965), for extensions
3. O. Gunnarsson and B.I. Lundqvist, Phys. Rev. B13:4274 (1976).
4. D.C. Langreth and J.P. Perdew, Phys. Rev. B15:2884 (1977); Solid State Commun. 31:567 (1979).
5. D.C. Langreth and M.J. Mehl, Phys. Rev. B28:1809 (1983).
6. A.R. Williams and U. von Barth, in " Theory of the Inhomogeneous Electron Gas," S. Lundqvist and N.H. March, ed., Plenum Press, New York(1983).
7. M. Levy, Proc. Natl. Acad. Sci. USA 76:6062 (1979).
8. E.H. Lieb in " Physics as a Natural Philosophy," A. Shimony and H. Feshbach, ed., MIT press, Cambridge Mass (1982).
9. A.K. Rajagopal, in " Advances in Chemical Physics," I. Prigogine and S.A. Rice, ed., vol.41, p.59 Wiley, New York (1980).
10. "Theory of the Inhomogenous Electron Gas," S. Lundqvist and N.H. March, ed., Plenum New York (1983).
11. J. Callaway and N.H. March, in " Solid State Physics," H. Ehrenreich, F. Seitz, and D. Turnbull, ed., Academic, New York, to be published.
12. "NATO Advanced Study Institute on the Electronic Structure of Complex Systems," W. Temmerman and P. Phasiseau, ed., Plenum, New York, to be published.
13. Proceedings of the workshop on "Many-Body Phenomena at Surfaces," D.C. Langreth and H. Suhl, ed., Academic, New York, to be published.
14. U. von Barth, in " NATO Advanced Study Institute on the Electronic Structure of Complex Systems," W. Temmerman and P. Phasiseau, ed., Plenum, New York, to be published.
15. U. von Barth, in the proceedings of the workshop on " Many-Body Phenomena at Surfaces," D.C. Langreth and H. Suhl, ed., Academic, New York, to be published.

16. A similar procedure has been used independently by D.W. Smith, S. Jagannathan, and G.S. Handler, Int. J. Quant. Chem. S13:103 (1979) for the case of the He atom.

17. U. von Barth and R. Car, to be published.

18. C.-O.Almbladh and A.C. Pedroza, Phys. Rev. A, to be published.

19. C.-O. Almbladh and U. von Barth, to be published.

20. The long-range behaviour of the density to leading order, i.e. the exponential fall-off, has been found earlier by M. Levy, unpublished report 1975, and by J. Katriel and E.R. Davidson, Proc. Nat. Acad. Sci. USA 77:4403 (1980).

21. U. von Barth and L. Hedin, J. Phys. C5:1629 (1972); A.K. Rajagopal and J. Callaway, Phys. Rev. B7:1912 (1973).

22. D.C. Langreth, to be published.

23. L.J. Sham, to be published.

24. It is not difficult to show, using e.g. perturbation theory, that the exponential fall-off of atom B in this region is the same as for the free atom.

25. Notice that this unphysical result would not arise if the two atoms were identical.

26. U. Fano, Phys. Rev. 124:18866 (1961); P.W. Anderson, Phys. Rev. 124:41 (1961).

27. H. Siegbahn and L. Karlsson, in " Handbuch der Physik ", vol.31, W. Mehlhorn, ed., Springer, Berlin, p. 215 (1982).

28. T.A. Patterson, H. Hotop, A. Kasdan, D.W. Norcross, and W.C. Lineberger, Phys. Rev. Lett. 32:189 (1974).

29. J.P. Perdew and M.R. Norman, Phys. Rev. B26:5445 (1982).

30. J.P. Perdew and M. Levy, Phys. Rev. Lett. 14:1884 (1983).

31. L.J. Sham and M. Schlüter, Phys. Rev. Lett. 14:1888 (1983).

32. J.C. Slater, Adv. Quantum Chem. 6:1 (1972); C.-O. Almbladh and U. von Barth, Phys. Rev. B13:3307 (1976); J.F. Janak, Phys. Rev. B18:7165 (1978).

33. N.D. Mermin, Phys. Rev. 137:A1441 (1965).

34. J.P. Perdew, R.G. Parr, M. Levy, and J.L. Balduz, Phys. Rev. Lett. 49:1691 (1982).

35. U. Rössler, in " Rare-gas solids ", M.L. Klein and J.A. Venables, ed., Academic, N.Y. (1975).

36. J.P. Perdew, (Private communication).

37. U. von Barth and R. Car, (unpublished).

38. C.-O. Almbladh (unpublished).

39. J.P. Perdew and M.R. Norman, Phys. Rev. B26:5445 (1982).

DENSITY FUNCTIONALS AND THE DESCRIPTION OF METAL SURFACES

Norton D. Lang

IBM Thomas J. Watson Research Center
Yorktown Heights, New York 10598

INTRODUCTION

We consider here the use of the density-functional formalism to study the electronic structure of metal surfaces, and atoms chemisorbed on these surfaces. It is convenient to begin by collecting several of the density-functional results which we will need.[1-3]

The total energy $\tilde{E}_v$ of a system of N electrons in a static external potential $v(\vec{r})$ (produced e.g. by an array of nuclei) is a minimum for the correct electron number density $n(\vec{r})$, considering densities all of which correspond to N electrons. Thus

$$\delta \tilde{E}_v[n] = 0 \tag{1a}$$

subject to

$$\int n(\vec{r})d\vec{r} = N \ . \tag{1b}$$

Hence for the correct $n(\vec{r})$,

$$\frac{\delta \tilde{E}_v[n]}{\delta n(\vec{r})} = \mu \ , \tag{2}$$

with μ the chemical potential (in a large system). With $G[n]$ defined as the non-electrostatic part of the total energy, i.e.*

* We use atomic units, with $|e| = m = \hbar = 1$.

$$\tilde{E}_{\mathrm{v}}[n] = \int \mathrm{v}(\vec{r})n(\vec{r})d\vec{r} + \frac{1}{2}\int \frac{n(\vec{r})n(\vec{r}\,')}{|\vec{r}-\vec{r}\,'|}d\vec{r}d\vec{r}\,' + G[n] \,, \tag{3}$$

we can write

$$\phi(\vec{r}) + \frac{\delta G[n]}{\delta n(\vec{r})} = \mu \,, \tag{4}$$

where $\phi(\vec{r})$ is the total electrostatic potential (as seen by a negative test charge). The functional $G[n]$ can be expanded in a series in gradients of the density, as discussed in Refs. 1-3, and substituted into Eq. (4). In the simplest case in which only the first (non-gradient) term in the series is retained, and in which only the kinetic-energy part of this term is kept, this equation (together with Poisson's equation) yields the Thomas-Fermi equation. Including more of the expression for $G[n]$ leads to extensions of the Thomas-Fermi method.

Defining an exchange-correlation energy functional $E_{xc}[n] \equiv G[n]-T_s[n]$, where T_s is the kinetic energy of a non-interacting system that has the same density $n(\vec{r})$ as the interacting system [the $\mathrm{v}(\vec{r})$ in the non-interacting system would in general have to be different], permits the wave-mechanical formulation of the density-functional approach.[4] A local potential v_{eff} is defined as

$$\mathrm{v}_{eff}[n;\vec{r}] \equiv \phi(\vec{r}) + \frac{\delta E_{xc}[n]}{\delta n(\vec{r})} \;; \tag{5}$$

then the density distribution $n(\vec{r})$ can be determined by solving the set of one-particle equations

$$\left\{-\frac{1}{2}\nabla^2 + \mathrm{v}_{eff}[n;\vec{r}]\right\}\Psi_i(\vec{r}) = \varepsilon_i\Psi_i(\vec{r}) \tag{6a}$$

together with

$$n(\vec{r}) = \sum_i |\Psi_i(\vec{r})|^2 n_i \;, \tag{6b}$$

where the Ψ_i are orthonormal solutions of Eq. (6a) and n_i is 1 for the N lowest-lying solutions and zero otherwise. The total energy is given by

$$\tilde{E}[n] = \sum_i \varepsilon_i n_i - \int v_{eff}[n;\vec{r}]n(\vec{r})d\vec{r}$$

$$+ \int v(\vec{r})n(\vec{r})d\vec{r} + \frac{1}{2}\int \frac{n(\vec{r})n(\vec{r}\,')}{|\vec{r}-\vec{r}\,'|}d\vec{r}d\vec{r}\,' + E_{xc}[n] \ . \tag{7}$$

(This does not include the self-energy of the static charge distribution which gives rise to $v(\vec{r})$, e.g. the lattice self-energy; when this self-energy is included, we will write E instead of $\tilde{E}$.) Approximations to v_{eff} can be obtained by expanding $E_{xc}[n]$ in a series in gradients of the density, and dropping certain of the terms. The simplest ("local-density") approximation consists in keeping only the first (non-gradient) term:

$$E_{xc}[n] \doteq \int g_{xc}(n(\vec{r}))d\vec{r} \ , \tag{8}$$

where $g_{xc}(n)$ is the exchange-correlation part of the energy density of a homogeneous electron gas of density n. The wave-mechanical approach treats the kinetic-energy part $T_s[n]$ exactly, in contradistinction to the Thomas-Fermi type methods.

Sometimes it is convenient for purposes of discussion to study the density of eigenstates in Eq. (6a). In this connection, we define a total density of eigenstates as

$$n(\varepsilon) = \sum_i \delta(\varepsilon - \varepsilon_i) \ . \tag{9}$$

Work Function

Consider a crystal with several different macroscopic faces labelled by the index i. The work function Φ_i of face i is the minimum energy required to remove an electron from the crystal to any point $\vec{r}_i$ which is a distance outside of face i small compared with the face dimensions but large compared with the lattice spacing (and is at a macroscopic distance from the face edges). There are electric fields extending well outside the crystal such that moving an electron from $\vec{r}_i$ to $\vec{r}_j$ requires an amount of work $\Phi_j - \Phi_i$. The work function of a metal is therefore given by

$$\Phi_i = [\phi(\vec{r}_i) + E_{N-1}] - E_N \ , \tag{10}$$

with E_N the ground-state energy of the neutral N-electron crystal, E_{N-1} the ground-state energy of the singly ionized crystal, and $\phi(\vec{r}_i)$ the electrostatic potential at $\vec{r}_i$. Since the chemical potential μ is $E_N - E_{N-1}$,

$$\Phi_i = \phi(\vec{r}_i) - \mu \ . \tag{11}$$

The two terms in Eq. (11) are each referred to the arbitrary zero of potential; it is convenient for discussion purposes to choose this reference to be the mean electrostatic potential in the metal,

$$\bar{\phi} \equiv \Omega^{-1} \int_{\text{metal}} \phi(\vec{r}) d\vec{r} \ , \tag{12}$$

where Ω is the volume of the metal. Thus we write

$$\bar{D}_i \equiv \phi(\vec{r}_i) - \bar{\phi} \ , \tag{13a}$$

$$\bar{\mu} \equiv \mu - \bar{\phi} \ , \tag{13b}$$

so that

$$\Phi_i = \bar{D}_i - \bar{\mu} \ . \tag{14}$$

Now if we average both sides of Eq. (4) over the volume of the metal, we find that

$$\bar{\mu} = \mu - \bar{\phi} = \Omega^{-1} \int_{\text{metal}} \frac{\delta G[n]}{\delta n(\vec{r})} d\vec{r} \ . \tag{15}$$

We see therefore that $\bar{\mu}$ is a bulk property of the metal. It is the bulk chemical potential, relative to the mean electrostatic potential in the metal; its independence of this potential follows from the definition of $G[n]$ as the total non-electrostatic energy. Equation (14) thus divides the work function into bulk ($-\bar{\mu}$) and surface ($\bar{D}_i$) components.[5,6]

All many-body effects are contained in the exchange and correlation contributions to $\bar{\mu}$ and in their effect on the electrostatic surface barrier $\bar{D}_i$. In particular, the image force effect on Φ_i may be regarded as contained in the disappearance of part of the correlation energy when one electron is moved away from the surface.

Often it is convenient to use a reference other than the mean potential $\bar{\phi}$ for the two terms in Eq. (14). In a Wigner-Seitz calculation, for example, a natural reference is the electrostatic potential at the Wigner-Seitz sphere, ϕ^{WS}. Thus we would write

$$D_i^{WS} \equiv \phi(\vec{r}_i) - \phi^{WS} , \tag{16a}$$

$$\mu^{WS} \equiv \mu - \phi^{WS} , \tag{16b}$$

and hence

$$\Phi_i = D_i^{WS} - \mu^{WS} . \tag{17}$$

Surface Energy

The surface energy σ_i of a crystal face i is the work required, per unit area of new surface formed, to split the crystal in two along the ith crystal plane. We consider a macroscopic crystal, and take the two fragments into which the crystal is split to be identical. Let A_i be the area of the newly exposed face on each fragment, E_i the total energy of each fragment, and E the total energy of the unsplit crystal. Then

$$\sigma_i = (2A_i)^{-1}(2E_i - E). \tag{18}$$

Note that the energies E differ from the energies $\tilde{E}$ discussed above in that they include the electrostatic self-energy of the lattice of nuclei.*

UNIFORM-BACKGROUND MODEL OF A METAL SURFACE

We turn now to consider the uniform-background (jellium) model of a metal surface. This is the surface analogue of a model that is widely used in the study of the bulk properties of simple (s-p bonded) metals. We discuss the case of a semi-infinite crystal, imagining that the charge on the ionic lattice has been smeared out into a homogeneous positive background of density $\bar{n}$ that terminates abruptly at a plane. This charge density [which gives rise to the external potential $v(\vec{r})$ discussed earlier] is thus taken to have the form

$$n_+(z) = \begin{cases} \bar{n} & z \leq 0 \\ 0 & z > 0 . \end{cases} \tag{19}$$

The density $\bar{n}$ is the mean density of positive charge in the ionic lattice; a convenient measure of this density is the radius r_s defined by $(4/3)\pi r_s^3 \equiv \bar{n}^{-1}$ (r_s ranges from about 2 to 6 bohrs for simple metals). This is probably the most elementary model of a metal surface that yields quantitatively

---

* We omit the contribution to the surface energy from changes of the zero-point lattice vibrations.

accurate information on such basic properties as the work function.

The electron density distribution n(z) that is obtained by solving Eqs. (6) self-consistently for the uniform-background model[7,8] has the form shown in Fig. 1. Note the Friedel oscillations in the density. These are characteristic of any disturbance in an electron gas (in this case the surface acts as the disturbance), and arise as a consequence of the sharpness of the Fermi surface (not of the positive background). The electric double layer formed by the spreading out of the electron distribution past the edge of the positive background means that in general the electrostatic potential in the vacuum $[\phi(\infty)]$ will be higher than it is in the metal interior $[\phi(-\infty)]$. Thus, an electron trying to leave the metal encounters an electrostatic barrier of height

$$D = \phi(\infty)-\phi(-\infty) = 4\pi\int_{-\infty}^{\infty} z[n(z)-n_+(z)]dz. \tag{20}$$

Since $\bar{\phi} = \phi(-\infty)$ for this case, this is just the quantity defined in Eq. (13a). By Eqs. (14) and (15), the work function is given by

$$\Phi = D-\bar{\mu} \tag{21}$$

with

$$\bar{\mu} = \frac{d\bar{n}\varepsilon(\bar{n})}{d\bar{n}} \quad ; \tag{22}$$

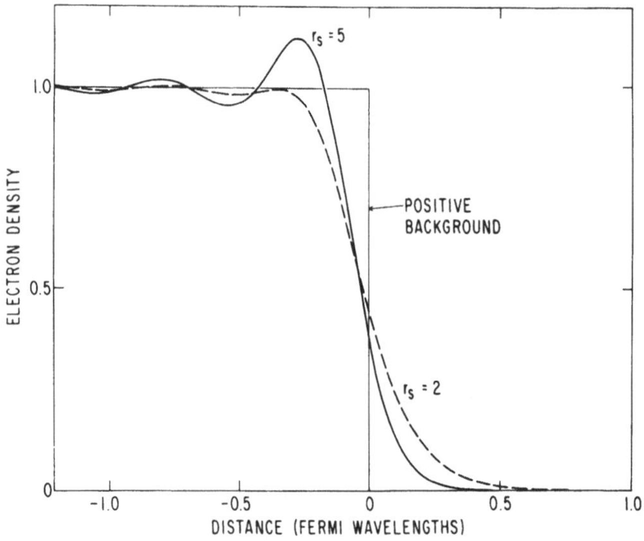

Fig. 1. Electron density in surface region of uniform-background model. One Fermi wavelength is equal to $2\pi/k_F$. (From Ref. 7.)

here $\varepsilon(\bar{n})$ is the average energy per particle of a uniform electron gas of density $\bar{n}$.

We mention here two sum rules derived by Budd and Vannimenus[9,10] for the uniform-background model of a surface: one for the surface potential, the other for the rate of change of surface energy with bulk density. They are

$$\phi(0)-\phi(-\infty) = \bar{n}\frac{d\varepsilon(\bar{n})}{d\bar{n}} \tag{23a}$$

$$\frac{d\sigma(\bar{n})}{d\bar{n}} = \int_{-\infty}^{0} [\phi(-\infty)-\phi(z)]dz \ . \tag{23b}$$

The derivation of these sum rules is given in Ref. 10. Note that Eq. (23a) provides an exact relation between surface and bulk properties.

Thomas-Fermi Solution

The Thomas-Fermi approximation consists of taking G[n] in Eq. (4) to be given by

$$G[n] \doteq \int g_s(n(\vec{r}))d\vec{r} \ , \tag{24}$$

where $g_s(n)$ is the kinetic energy density of a non-interacting electron gas of density n; the resulting equation is then solved simultaneously with Poisson's equation. By evaluating Eq. (4) directly at $z = \pm\infty$, we see immediately using Eqs. (21) and (22) [note $g(n) = n\varepsilon(n)$] that the Thomas-Fermi work function vanishes for all r_s:

$$\Phi_{TF} = 0 \ , \ \forall r_s \ . \tag{25}$$

Solving the Thomas-Fermi equation for the uniform-background model[11] shows that

$$n(z) \ \propto \ z^{-6} \ \ (z\rightarrow\infty) \tag{26a}$$

$$1 - n(z)/\bar{n} \ \propto \ e^{z/\lambda_{TF}} \ \ (z\rightarrow-\infty) \ , \tag{26b}$$

with λ_{TF} the Thomas-Fermi length $[0.640\sqrt{r_s}]$. The Thomas-Fermi approximation, as expected, yields a density distribution inside the metal without Friedel oscillations. In the vacuum, the density shows a power-law decrease, just as in the Thomas-Fermi atom,[12] while the true density at a surface is expected to decay exponentially.

The inclusion of exchange and correlation in Eq. (24) (still omitting terms in gradients of the density however) yields a density distribution that has the property of dropping discontinuously to zero at a certain distance from the surface[13-15] (just as in the corresponding treatment of atoms[16]). The work function for this case again has a value independent of r_s [13,14] (in the range for which the density discontinuity occurs outside the positive-background region[14]).*

The surface energy for the uniform-background model in the Thomas-Fermi approximation is found to be[2,11]

$$\sigma = -0.0763 r_s^{-9/2} \ . \tag{27}$$

The fact that σ is negative (i.e. that the crystal cleaves spontaneously) reflects a deficiency not only of the Thomas-Fermi approximation, but of the use of a model without a discrete ionic lattice, as will be seen below.

Extended Thomas-Fermi Solution

We saw above that omitting all gradient terms from the density-gradient series for $G[n]$ precluded obtaining a good account of surface electronic structure. Smith[17] considered the question of whether this defect would be remedied when at least some gradient terms were included. He retained the first gradient term in the series and used the simplest approximation for its coefficient, in which only kinetic-energy effects are included. Thus he took

$$G[n] \doteq \int \left[g(n(\vec{r})) + \frac{1}{72 n(\vec{r})} |\vec{\nabla} n(\vec{r})|^2 \right] d\vec{r} \ , \tag{28}$$

where $g(n)$ is the energy density for the homogeneous electron gas of density n, including exchange and correlation. He parameterized the electron density as

$$n(z) = \begin{cases} \bar{n} - \frac{1}{2}\bar{n}e^{\beta z} & z<0 \\ \frac{1}{2}\bar{n}e^{-\beta z} & z>0 \end{cases} \tag{29}$$

and minimized the expression for the total energy $E_v[n]$ with respect to β. [The resulting surface energy is negative for $r_s<2.6$, a shortcoming which again is only remedied when discrete lattice effects are included.]

The work function Φ can be computed from Eq. (21), with $D = 4\pi\bar{n}\beta^{-2}$. The results showed approximate agreement with experimental work functions for

---

* For example, if only exchange is included (Thomas-Fermi-Dirac), this value is 1.3 eV.

a large number of metals. While exchange-correlation effects were found to be important in all cases, ordinary electrostatic effects were seen to be quite strong at higher densities. The use of a parametric form for the density is significant in obtaining good results; it is not obvious that the direct solution of the Euler equation will yield such results.*

Wave-Mechanical Solutions

A wave-mechanical treatment is most directly effected by self-consistently solving the set of equations (6). For the uniform-background model of a surface, the potential v_{eff} in Eq. (6a) is a function only of z, and is constant everywhere in the metal except the surface region. The eigenfunctions of this equation can therefore be written

$$\Psi_{\vec{k}}(\vec{r}) = \Psi_{k_z}(z)e^{i(k_x x+k_y y)} , \tag{30}$$

where for $z \to -\infty$,

$$\Psi_{k_z}(z) = \sin\,[k_z z - \gamma(k_z)] , \tag{31}$$

with γ determined uniquely by the requirements that $\gamma(0) = 0$ and $\gamma(k_z)$ be continuous. (The origin for z is taken at the positive background edge, as before, in defining γ .)

Lang and Kohn[7,8] have given details of the self-consistent solution of Eqs. (6) for the uniform-background surface model,† using the local-density approximation for exchange and correlation. The local-density approximation has given good results in the study of atoms, molecules, and solids.[26-28] Its adequacy in the treatment of metal surfaces is considered further below.

The charge density obtained by Lang and Kohn is shown in Fig. 1. Deep in the metal, the density has the Friedel-oscillation form

$$n(z) = \bar{n} + \frac{A\cos\,(2k_F z + \alpha)}{z^2} + O\left(\frac{1}{z^3}\right) , \tag{32}$$

---

* The equation appears in fact not to have a solution for certain regions of r_s (W. Kohn, private communication).

† We note a number of other calculations of this general class, including the very early calculation of Bardeen,[18] for surfaces, both spin-unpolarized[18-24] and spin-polarized.[25] The calculations of Sahni and coworkers[23] for surfaces have been done largely analytically by using simple parametric forms (such as a step or a truncated linear potential) for the surface barrier in conjunction with theoretical constraints such as Eq. (23a).

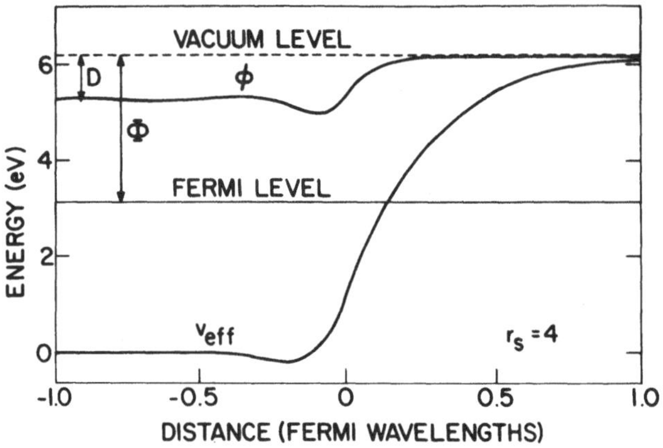

Fig. 2. Effective one-electron potential v_{eff}, with electrostatic part ϕ, in surface region of uniform-background model ($r_s = 4$). One Fermi wavelength is equal to $2\pi/k_F$. (Data from Ref. 8.)

a result which can be obtained by inserting the form for the wave function given in Eqs. (30) and (31) into Eq. (6b). For low bulk densities (r_s~5), the Friedel oscillations are large, but for high bulk densities (r_s~2), they are greatly reduced, and the density begins to resemble the monotonically decreasing Thomas-Fermi form, except in the tail region. There, the wave-mechanical density decreases exponentially, in contrast to the z^{-6} decrease of the Thomas-Fermi solution.

Figure 2 gives an example of the computed potentials. Note that in this case (r_s=4), the electrostatic potential barrier is only a small part of the total effective barrier (as given by v_{eff}). This is generally true for low and intermediate bulk densities: the exchange-correlation part of the surface barrier is of major importance.*[18]

When one of the metal electrons passes out through the surface region, its exchange-correlation hole stays behind, flattening out on the surface and assuming an image-charge form.[30-33] This clearly non-local effect is not reproduced by the local-density approximation, according to which the exchange-correlation potential has an exponential, rather than an image, form. This circumstance, however, mainly affects the details of the electron density well out

---

* This does not however mean that if exchange and correlation are omitted, the electrons will no longer be confined at the surface. In this case, the surface electron distribution can be expected to spread out further, but this raises the electrostatic potential barrier in a self-consistent way (cf. Ref. 18). Peuckert[29] indicates that the Hartree work function is small but positive (high-density limit).

in the vacuum tail of the distribution and is of little importance[8,33] to the calculation of quantities such as the work function.

The work function associated with the density distributions discussed above is obtained by adding, according to Eq. (21), the electrostatic barrier height (the surface part of Φ) and the negative of the bulk chemical potential (the bulk part of Φ). The computed work function[6] and its two components are shown as a function of r_s in Fig. 3. As r_s decreases, Φ reaches a maximum and begins to decrease. The components of Φ diverge as $r_s \to 0$, but Φ itself, according to Peuckert,[29] tends to a finite limit. This latter type of behavior was seen in the Thomas-Fermi case (with $\Phi \equiv 0$ for all r_s). We remind the reader of the arbitrary character of the split-up of Φ into components.

Figure 4 compares the $\Phi(r_s)$ computed by Lang and Kohn[6] with experimentally measured work functions for polycrystalline simple metals.[2] (The uniform-background model is not intended to describe transition or noble metals.) The agreement between theory and experiment is quite reasonable, and is improved when discrete lattice effects are taken into account using perturbation theory, which furthermore permits the discussion of the crystal-face anisotropy of the work function.

It will be recalled that the total energy was divided into electrostatic and non-electrostatic parts, with the latter, denoted $G[n]$, further divided into $T_s[n]$ (non-interacting kinetic) and $E_{xc}[n]$ (exchange-correlation). This division leads to a corresponding division in the surface energy. For the uniform-background model, it yields

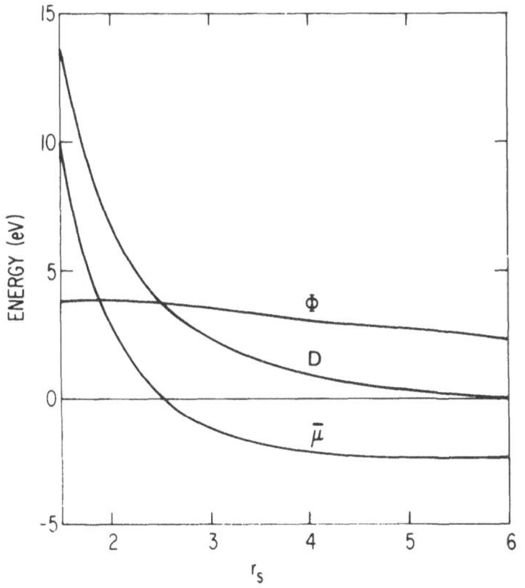

Fig. 3. Components of the work function in the uniform-background model. $\Phi = D - \bar{\mu}$. (From Ref. 34.)

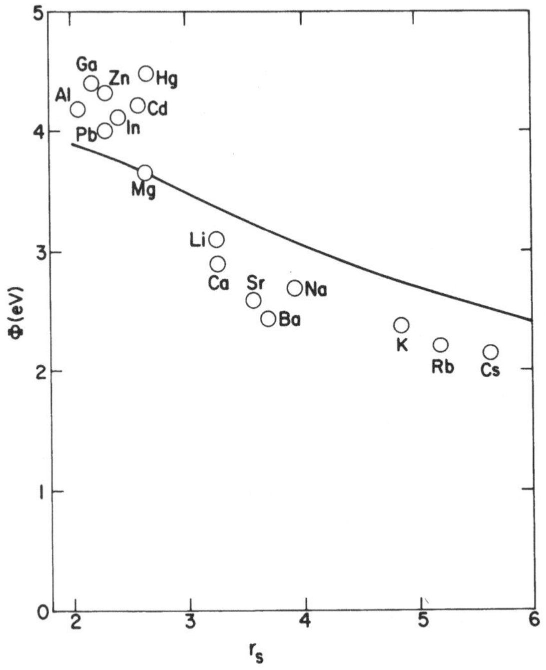

Fig. 4. Solid line: work function for the uniform-background model. Circles: measured work function for polycrystalline samples. (From Ref. 2.)

$$\sigma_{es} = \frac{1}{2}\int_{-\infty}^{\infty} \phi(z)[n(z)-n_{+}(z)]dz \ , \tag{33a}$$

$$\sigma_s = \int_0^{\varepsilon_F} \varepsilon\eta^s(\varepsilon)d\varepsilon - \int_{-\infty}^{\infty} \mathrm{v}_{eff}[n;z]n(z)dz \ , \tag{33b}$$

$$\sigma_{xc} = \int_{-\infty}^{\infty} [g_{xc}(n(z)) - g_{xc}(\bar{n}\theta(-z))]dz \ , \tag{33c}$$

where we use the local-density form for the exchange-correlation term, and where (in σ_s) we continue to choose our zero of energy so that $\mathrm{v}_{eff}[n;-\infty] = 0$. The surface density of eigenstates in (33b) is [2,35,36]

$$\eta^s(\varepsilon) = \frac{1}{\pi^2}[\gamma(\sqrt{2\varepsilon}) - \frac{\pi}{4}] \ , \tag{34}$$

with γ the phase shift of Eq. (31). The values of these three terms,[8] and their total, are shown as a function of r_s in Fig. 5. As $r_s \to 0$, kinetic-energy effects dominate, and σ becomes negative. This behavior was seen earlier in the Thomas-Fermi case, where σ was negative for all r_s.

Now the surface energy of a high-density metal is positive (otherwise the metal would cleave spontaneously), in contrast to the result shown in Fig. 5. It has been demonstrated in a variety of ways that it is the absence of the discrete lattice which leads to a negative surface energy, and that only a much smaller part of the underestimate of σ is ascribable to the use of the local-density approximation. A review of this question has been given by Langreth.[37]

In the paper of Lang and Kohn,[8] the discrete lattice was re-introduced into the calculation of σ using first-order pseudopotential perturbation theory, with the Madelung energy taken into account exactly. This yielded a positive σ in reasonable agreement with experiment for a number of simple metals, except Pb (which has a particularly strong pseudopotential).* The results of this are shown in Fig. 6. A similar calculation in which part of the z-axis variation of the ionic pseudopotential is taken into account non-perturbatively has been presented by Monnier and Perdew.[39] The results for σ are generally fairly close to those of Lang and Kohn except for the case of Pb, where the calculated value is brought into agreement with experiment. This work suggests therefore that the non-local exchange-correlation effects, which were omitted, are not extremely large.

A complementary way of exploring this question is to obtain and evaluate forms for σ_{xc} that go beyond the local-density approximation, avoiding arguments based on agreement with experiment.

Schmit and Lucas,[40] Craig,[41] and Peuckert[42] discussed a particular non-local contribution to σ_{xc} that can be expected not to be well represented in the local-density approximation. When a crystal is cleaved, the bulk plasmon modes of longest wavelength (along the cleaved surface normal) become unallowed, but new surface plasmon modes are introduced. The contribution to σ_{xc} discussed by these authors is the difference in zero-point energies of these modes. This work represented a very interesting conceptual point in the discussion of surface energies; but the actual magnitude of the contribution does not appear to be very large. This has been demonstrated by Langreth and Perdew.[43]

Langreth and Perdew[43] take the Coulomb part of the Hamiltonian to be a function of a dimensionless coupling constant λ varying between 0 and 1, with the electron-electron interaction simply proportional to λ, and the external

---

* As the Miller indices of the surface plane considered are raised, the surface energy computed in this way increases quickly, however. The surface-energy anisotropy of metals is found experimentally to be rather small however. The origin of this discrepancy, as pointed out by Manninen and Nieminen,[38] is the fact that as the surface becomes sparser, the use of an electron density that is constrained to be uniform in the surface plane (first-order perturbation theory applied to the uniform-background model) becomes less and less appropriate.

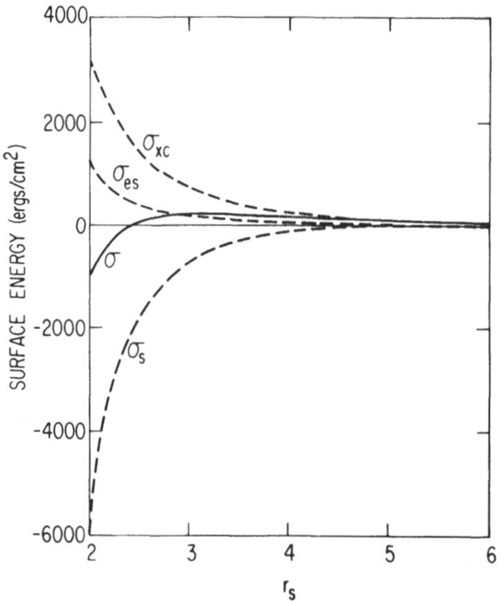

Fig. 5. Components of the surface energy in the uniform-background model. $\sigma = \sigma_s + \sigma_{es} + \sigma_{xc}$. (From Ref. 34.)

potential taken to depend on λ in such a way that the electron density distribution does not change as λ is varied. If the equal-wave-vector Fourier transform of the density-density correlation function is defined:

$$S_\lambda(\vec{k}) = \frac{1}{N}\int e^{i\vec{k}\cdot(\vec{r}\,'-\vec{r})}<[n(\vec{r})-<n(\vec{r})>][n(\vec{r}\,')-<n(\vec{r}\,')>]>_\lambda d\vec{r}d\vec{r}\,' \quad (35)$$

(N is the number of electrons), then it is seen that σ_{xc} can be written

$$\sigma_{xc} = \int \sigma_{xc}(\vec{k})\frac{d\vec{k}}{(2\pi)^3} \quad , \quad (36)$$

with

$$\sigma_{xc}(\vec{k}) = \frac{2\pi N}{Ak^2}\int_0^1 [S_\lambda(\vec{k})-S_\lambda^B(\vec{k})]d\lambda \quad . \quad (37)$$

Here S_λ^B is the density-density correlation function appropriate to a bulk uniform electron gas, and A is the surface area. It is most convenient to discuss the spherical average of $\sigma_{xc}(\vec{k})$, denoted $\sigma_{xc}(k)$; only this average enters σ_{xc}. Langreth and Perdew are able to show that the long-wavelength limit of $\sigma_{xc}(k)$ is given by†

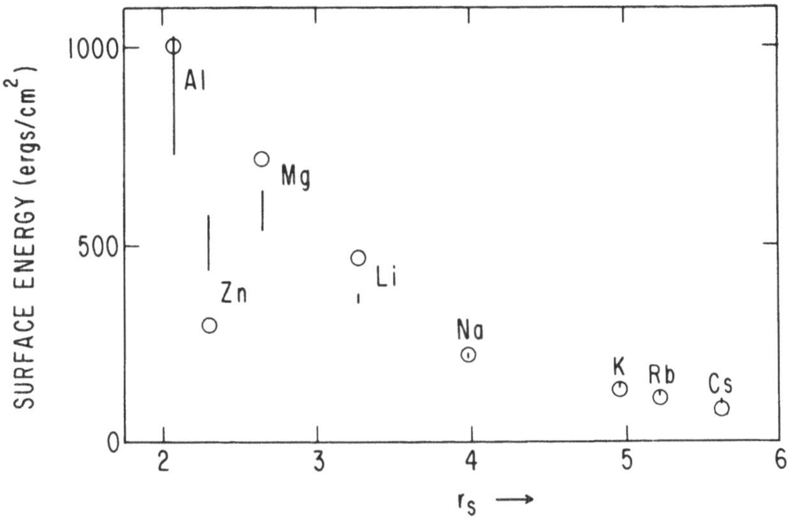

Fig. 6. Comparison between computed values of the surface energy and extrapolations to zero temperature of measured liquid-metal surface tensions. Vertical lines give computed values: the lower end-point represents the value appropriate to an fcc lattice, the upper end-point that appropriate to a bcc lattice. The surface plane in both cases is taken to be the most closely packed plane of the lattice. Note that the lines are contracted almost to points for the lower-density alkali metals. (After Ref. 8.)

$$\lim_{k\to 0} \sigma_{xc}(k) \to \frac{\pi}{4k}[\omega_s(\bar{n})-\tfrac{1}{2}\omega_p(\bar{n})] \quad ; \tag{38}$$

it is dominated by the change in zero-point energies of the plasmons,[40] and is independent of the details of the surface.

Now σ_{xc} in the local-density approximation can also be decomposed according to wave-vector. In the long-wavelength limit,

$$\lim_{k\to 0} \sigma_{xc}^{LDA}(k) \to \frac{1}{2}\int_{-\infty}^{\infty} [\omega_p(n(z))-\omega_p(\bar{n})\theta(-z)]dz \; . \tag{39}$$

While this is dominated by shifts in plasmon zero-point energies, it sharply underestimates the importance of these effects (it does not diverge as $k\to 0$, as the exact limit does), and, again in contrast to the exact limit, depends on the form of the surface density profile. On the other hand, Langreth and Perdew

† $\omega_p(n) = \sqrt{4\pi n}$ (bulk plasmon frequency) and $\omega_s = \omega_p/\sqrt{2}$ (surface plasmon frequency).

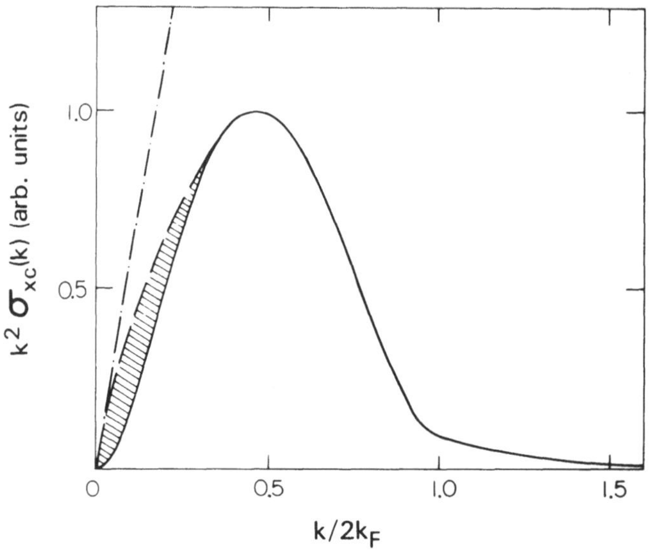

Fig. 7. Wave-vector analysis of the exchange-correlation part of the surface energy in the uniform-background model ($r_s = 4$). Solid curve: local-density approximation; dash-dot line: exact $k \rightarrow 0$ asymptote [Eq. (38)]; dashed arc: interpolation. Value of σ_{xc} is proportional to the area under the curve; shaded area represents the correction for effects beyond the local-density approximation. (After Ref. 43.)

show that $\sigma_{xc}^{LDA}(k)$ is exact in the large-k limit, which can be described by saying that $\sigma_{xc}(k)$ represents the energy of a packet of excitations whose spatial extent is $O(k^{-1})$ and thus for large k the excitations have a small extent and do not "know" the density is varying.[37]

Given the electron density distribution, $\sigma_{xc}^{LDA}(k)$ is straightforward to evaluate (an example is shown in Fig. 7). Now let it be assumed that $\sigma_{xc}^{LDA}(k)$ (which is exact for $k \rightarrow \infty$) represents a good approximation to $\sigma_{xc}(k)$ even down to $k \gtrsim k_F$ (the region of the maximum in the curve shown). Langreth and Perdew[43] employ a simple interpolation between the exact $k \rightarrow 0$ limit (dot-dash straight line) and the maximum in the $k^2\sigma_{xc}^{LDA}(k)$ curve (Fig. 7). Using this interpolation, together with $\sigma_{xc}^{LDA}(k)$ for larger k values, should therefore provide accurate values of $\sigma_{xc}(k)$ for all k. This in turn will yield an accurate value for σ_{xc}, which is proportional to the area under the curve of $k^2\sigma_{xc}(k)$. We see in particular that the error in the local-density approximation is not large; this is a direct result of the presence of the phase-space factor k^2, which has the effect of minimizing the actual numerical importance of collective effects [that dominate $\sigma_{xc}(k)$ at small k].

We consider now density-functional analyses of the response of the metal surface to a static perturbing charge distribution.[6,44-52] We describe the simple

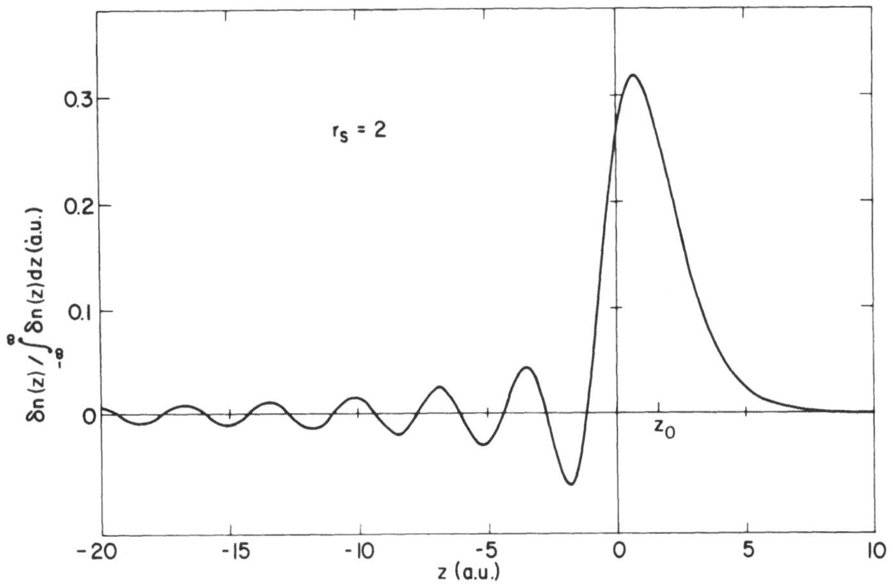

Fig. 8. Screening charge in the surface region of the uniform-background model induced by a weak uniform normal electric field. The position of the center of gravity of the screening charge is given by z_0. (The background occupies the $z < 0$ half-space.) (After Ref. 47.)

case of the charge density $\delta n(z)$ induced by a weak uniform electric field $\mathcal{E}$ normal to the surface. An example of such a distribution is shown in Fig. 8.[47] In our discussion below, we take $\delta n(z)$ to be normalized to unity: $\int_{-\infty}^{\infty} \delta n(z)dz = 1$.

One of the most important features of the induced density distribution is the position of its center of gravity

$$z_0 = \int_{-\infty}^{\infty} z\delta n(z)dz \ . \tag{40}$$

This determines the effective location of the metal surface in the sense that if we take the total electrostatic potential inside the metal to be fixed, then well outside the surface (with the field pointing out of the metal),

$$\delta\phi(z) = \mathcal{E}(z - z_0) \ . \tag{41}$$

It can also be shown that z_0 gives the location of the image plane for the surface.[47] Using the local-density approximation for exchange-correlation, it is found[47] that z_0 (relative to the positive background edge) ranges from 1.9 bohr for $r_s = 1.5$ to 1.2 bohr for $r_s = 6$.*

A relation between the rate of change of the work function with bulk density and the field-induced distributions $\delta n(z)$ has been obtained by Budd and Vannimenus[52]:

$$\frac{d\Phi(\bar{n})}{d\bar{n}} = 4\pi \int_{-\infty}^{0} dz\, z \int_{-\infty}^{z} dz' \delta n(z') \ , \qquad (42)$$

where $\delta n(z)$ is the distribution appropriate to $\bar{n}$. This is an analogue of Eq. (23b) for the surface energy.

LATTICE MODELS OF A METAL SURFACE

We now discuss briefly non-perturbative studies of the electronic structure of metal surfaces which at the outset take the external potential to be that due to a lattice of nuclei (or a lattice of ionic pseudopotentials). The presence of the discrete lattice, as noted earlier, permits the consideration of the anisotropy of quantities such as the work function and surface energy (and leads to a positive surface energy for high interior electron densities, in contrast to the uniform-background model). Analysis of such features as surface states, and the quantitative study of the surfaces of transition metals (and semiconductors and insulators) are possible using non-perturbative treatments of the lattice. We also note a number of density-functional calculations of semiconductor surfaces.[53-60]

These analyses[61-73] have generally proceeded by solving self-consistently the equations of Kohn and Sham [Eqs. (6)] in the local-density approximation, with the external potential taken to be either the full nuclear potential or a superposition of ionic pseudopotentials (in which case the core states of the metal are not obtained). We do not discuss here studies in which a small cluster of metal atoms is used to simulate a semi-infinite metal (or is studied in its own right), but consider only those in which the metal is taken to be infinite parallel to the surface.

Figure 9 shows the charge densities for Cu(100) calculated by Gay, Smith and Arlinghaus.[62] The charge density contours below the surface plane are very similar to those of the bulk. Note the way in which the charge smooths parallel to the surface as it spreads into the vacuum, as originally suggested by Smoluchowski.[74]

---

* Note that the semi-infinite uniform-background model represents an ionic lattice whose outermost lattice plane lies half an interplanar spacing behind the background edge. Thus, if we denote the interplanar spacing by d and if we express z_0 relative to the background edge, then the effective surface in a metal will be a distance $z_0 + \frac{1}{2}d$ in front of the outermost lattice plane. (We neglect here any small distortion of the metal lattice at the surface.)

Cu(100) CHARGE DENSITY

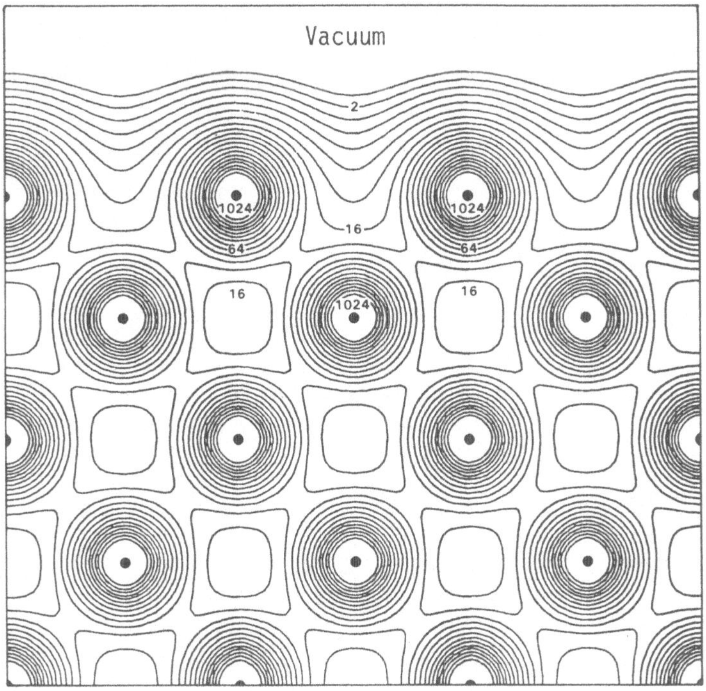

Fig. 9. Conduction-band charge density at a Cu(100) surface plotted on a plane normal to the surface (passing through a line connecting a surface atom with one of its near neighbors in the second plane of atoms). The units of charge density are 2.44×10^{-3} electrons/bohr^3. Charge densities on successive contours differ by a factor $\sqrt{2}$. (After Ref. 62.)

Now surface states (as opposed to surface resonances) can occur only in gaps in the projection of the bulk band structure onto the surface of interest (see Ref. 69 e.g. for details). In such a gap, a state localized in the surface region will have no bulk states at the same energy into which it can decay. The bands of surface states (and strong surface resonances) calculated by Louie et al.[68] for Nb(100) are shown in Fig. 10, together with the projected bulk band structure.

Surface states below the Fermi level can be observed experimentally using angle-resolved photoemission. Calculations of the type described in this section, using the local-density approximation for exchange-correlation, are able to yield results in quite reasonable agreement with these experiments. (See the articles by Almbladh and von Barth and by Perdew et al. in these Proceedings

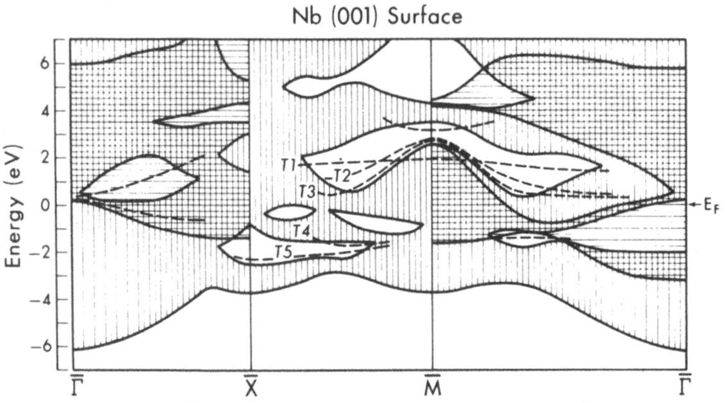

Fig. 10. Surface bands (dashed curves) and projected bulk band structure (crosshatched areas) for Nb(001). (Type of crosshatching is related to symmetry of the bulk states.) (From Ref. 68.)

for a discussion of the relation between the eigenstates of the Kohn-Sham equations, using the local-density approximation for exchange and correlation, and actual excitation energies.) For example, a surface state is observed at the center of the 2-dimensional Brillouin zone on Cu(111) with an energy of -0.4 eV relative to the Fermi level; the calculation of Appelbaum and Hamann places this state at -0.5 eV.[61] The full self-consistency of the calculation is found to be important to the proper positioning of these states.

The value of the work function is a particularly sensitive test of the self-consistency of calculations of the type described here. The most recent of these give values in reasonable agreement ($\lesssim 0.2$ eV) with measured work functions.

CHEMISORPTION ON METAL SURFACES

The density-functional formalism has proven to be a very useful tool in the study of the chemical bond between an atom or molecule (the adsorbate) and a metal surface (the substrate). Some of the analyses treat the case of a single atom (or molecule) on the surface; the adsorbate in this instance destroys the periodicity of the substrate. Other studies consider the case of a layer of atoms on the surface, usually with the same lattice periodicity as the substrate, in which case the problem retains the symmetry of the bare substrate.

It is often of particular interest to calculate the difference between the electron density for the metal-adsorbate system (MA) and that for the bare metal (M):

$$\delta n(\vec{r}) = n^{MA}(\vec{r}) - n^{M}(\vec{r}) \ , \tag{43}$$

as well as the corresponding difference of eigenstate densities [cf. Eq. (9)]:

$$\delta n(\varepsilon) = n^{MA}(\varepsilon) - n^{M}(\varepsilon) \ . \tag{44}$$

The relation between these state densities and actual densities of one-electron excitations is discussed elsewhere in these Proceedings, as noted above.

Uniform-Background Model

We describe first treatments of the classic chemisorption problem of a single atom bonded to a semi-infinite metal surface, using a uniform-background model for the substrate. Use of this model for the metal leads to a problem with cylindrical symmetry, which represents an important simplification.*

We denote by d the distance between the adatom nucleus and the positive background edge. This distance is fixed by minimizing the total computed energy. The model is completely specified by only two numbers: the density parameter r_s of the uniform-background substrate, and the adatom nuclear charge Z.

The problem of an atom chemisorbed on a uniform-background substrate has been treated wave-mechanically [i.e. by solving Eqs. (6)] by Lang and Williams,[76-78] and by Gunnarsson, Hjelmberg, Johansson and Lundqvist.[79-84†]

Figure 11 shows $\delta n(\varepsilon)$ for Li, Si and *Cl* chemisorbed on a high-electron density substrate ($r_s = 2$), as calculated by Lang and Williams.[76] The states constituting the *Cl* 3p resonance are below the Fermi level, and are therefore occupied; those constituting the Li 2s resonance are essentially empty. This implies that charge transfer has taken place, toward the *Cl* and away from the Li, as would be expected from the electronegativities of these atoms. The prohibitively large energy required either to fill or empty the Si 3p level forces this resonance to straddle the Fermi level, resulting in the formation of a covalent, rather than ionic bond.

The electron densities associated with the three fundamental bond types are exhibited for comparison in Fig. 12. In the top row of the figure are con-

* See also Ref. 75, in which both the substrate and an adsorbed layer are represented using the uniform-background model.

† Kahn and Ying[85] have done an extended Thomas-Fermi calculation, using the approximation of linear response of the substrate to the adsorbate potential, for alkali adsorption within this model. The calculated ionic desorption energies and dipole moments are in reasonable agreement with values found for adsorption on transition metals.

tours of constant total electron density. Note first the way in which these contours rapidly regain their bare-metal form away from the immediate region of the atom. This is a graphic illustration of the short range of metallic screening. Note also the way in which the metal contours bend toward the positively charged Li and away from the negatively charged C*l*. The very elongated outermost closed contour in the Si plot is characteristic of such plots for covalently bonded diatomic molecules.

The detailed charge rearrangements associated with chemical-bond formation are displayed most clearly by the contour maps of the difference between the electron density in the chemisorption system and the superposition of bare-metal and free-atom densities shown in the second row of Fig. 12. The solid contours indicate regions of charge accumulation; broken contours indicate regions of charge depletion.

The density-difference contours in the case of Li reveal the complexity underlying the notion of charge transfer. Taken alone, this difference plot indicates only that electrons have been displaced from the vacuum side to the metal side of the atom. The plot does not distinguish between the two possible interpretations of the displacement: polarization of the atom, or ionization (followed by metallic screening). The state-density graph in Fig. 11, however, resolves this ambiguity: the fact that virtually none of the states in the Li 2s resonance are occupied indicates that the ionization/screening interpretation is the correct one.

The kidney-shaped depletion contour on the vacuum side of the Li adatom, and even the reverse-dipole contours in the core region, are very similar to those found by Bader and co-workers[86] in difference plots for the LiH and LiF molecules. Note also that for this case, as well as for Si and C*l*, the sequence of contours continues into the metal in the form of Friedel oscillations induced by the perturbing atom.

The difference contours for C*l* show clearly that charge has been transferred from the metal to form a polarized negative ion. In the case of Si, there is a central region of charge depletion, and accumulations on both the bond and vacuum sides. The same general configuration of contours seen here for Si is found in difference plots for covalently bonded diatomic molecules in which p orbitals play a significant role.[86]

While the calculations of Lang and Williams and of Gunnarsson, Hjelmberg and Lundqvist predict equilibrium metal-adatom separations, the results of calculations performed for other separations yield information important to discussing the dynamics of atoms at surfaces, and provide theoretical insight into the distance dependence of the metal-adatom interaction. Of particular interest are the variations of the state-density difference $\delta n(\varepsilon)$.[87]

In Fig. 13, $\delta n(\varepsilon)$ is shown for a chemisorbed hydrogen atom at three different distances. At the largest distance, the metal-adatom interaction is not strong, and the resonance that is present below the Fermi level is relatively narrow. When the atom is moved closer, this interaction increases, and the resonance widens considerably. It also moves further below the Fermi level,

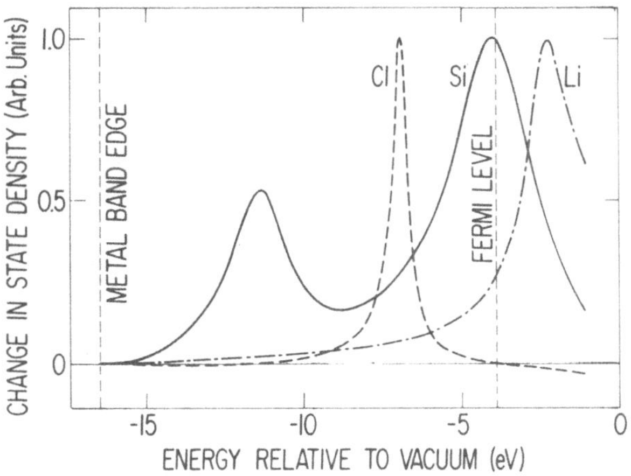

Fig. 11. Change in state density $\delta n(\varepsilon)$ due to chemisorption of one atom on a uniform-background substrate ($r_s = 2$). Curves correspond to metal-adatom distance which minimizes the total energy. Note that the lower Si resonance corresponds to the 3*s* level of the atom; for C*l* this is a discrete state below the bottom of the band (band edge). (From Ref. 76.)

showing a tendency to follow the bare-metal surface potential. When the atom is moved still closer, the broadening due to increasing metal-adatom interaction is overtaken by narrowing due to the decreasing density of metal states seen by the resonance as it moves down toward the bottom of the metal band, and the resonance narrows again.*

In connection with Fig. 13, we show in Figure 14 the effective potential v_{eff} for chemisorbed hydrogen (calculated by Hjelmberg[82] for an $r_s = 2.07$ substrate). It is particularly noteworthy that there is only a very small potential barrier between the metal and the adatom. From Fig. 13 we see that the resonance position is above the barriers shown (along the normal direction), implying a strong mixing between metal and adatom states.

Nørskov, Houmøller, Johansson and Lundqvist[89] have studied the adsorption and dissociation of H_2 on a uniform-background substrate (including also the lattice in first-order perturbation theory), using an extension[90] of the calculational scheme of Gunnarsson, Hjelmberg, and Lundqvist.[79-84] Lang[91] has studied the adsorption of rare-gas atoms on a uniform-background surface, using the chemisorption formalism of Lang and Williams, and has obtained results for the heats of adsorption, dipole moments and core-level binding energies in good agreement with experimental data. It is argued that the bonding of the atom *at its equilibrium distance from the surface* is more correctly understood using a local-density description of exchange-correlation effects than the asymptotic van der Waals description.

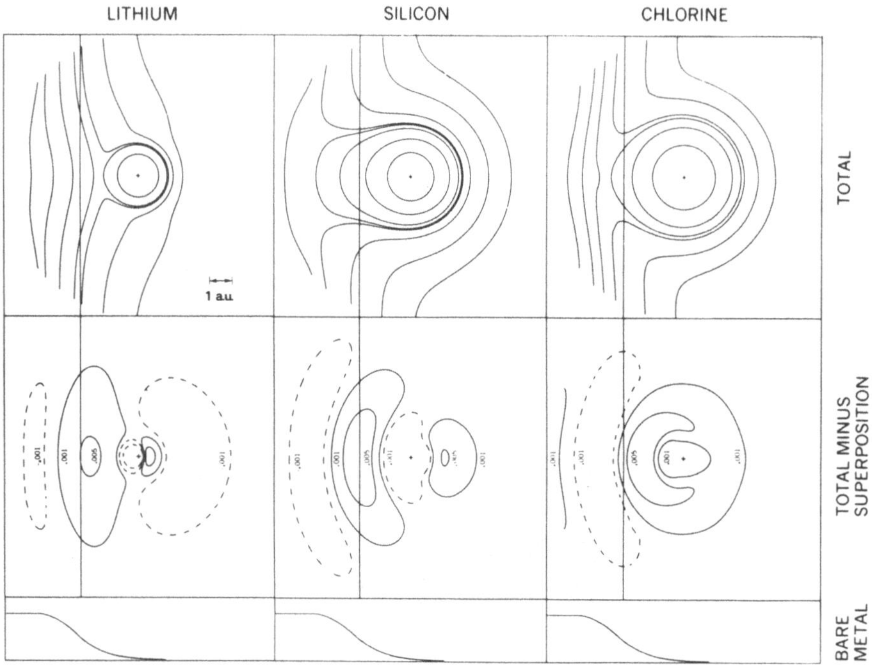

Fig. 12. Electron-density contours for chemisorption of one atom on a uniform-background substrate ($r_s = 2$). Metal-adatom distances shown minimize the total energy. Upper row: Contours of constant density in (any) plane normal to the metal surface containing the adatom nucleus (indicated by +). Metal is to the left-hand side; positive-background edge indicated by vertical line. Contours are not shown outside the inscribed circle of each square; contour values were selected to be visually informative. Center row: total electron density minus the superposition of atomic and bare-metal electron densities (electrons/bohr^3). The polarization of the core region, shown for Li, was deleted for Si and *Cl* because of its complexity. Bottom row: bare-metal electron-density profile (shown to establish physical distance scale). (For reference, the bulk metal density is 0.03 electrons/bohr^3.) (From Ref. 76.)

Lattice Model

* The fact that the resonance position tends to follow the surface potential can be easily demonstrated using first-order perturbation theory (see Ref. 78). A more general analysis of changes in resonances with distance has been given by Gunnarsson, Hjelmberg, and Nørskov.[88]

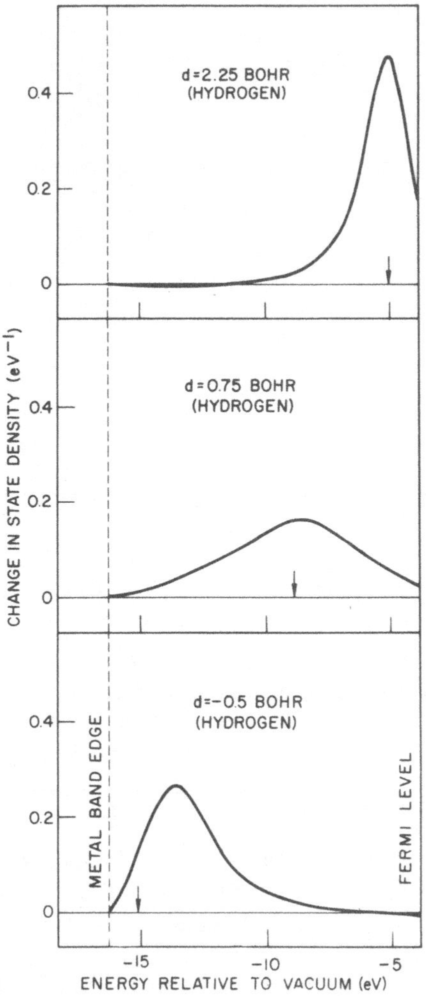

Fig. 13. State density change $\delta n(\varepsilon)$ due to chemisorption of a hydrogen atom on a uniform-background substrate ($r_s = 2$). Curves are shown for three different metal-adatom distances d. Arrow gives value of bare-metal potential $v_{eff}[n^M; z]$ at adatom nucleus; zero of potential is set so that arrow falls under peak of resonance for largest distance. This shows the way in which the resonance position roughly follows the surface potential. (If the arrows had been drawn for the bare-metal electrostatic potential v_{es} instead of v_{eff}, the arrow in the bottom panel would have been to the right of the peak, reflecting the fact that the actual behavior of the resonance is intermediate between the two potentials.) (From Ref. 78.)

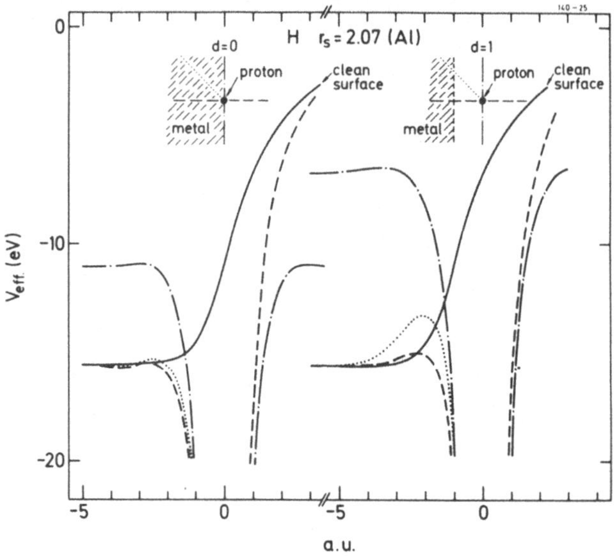

Fig. 14. The effective potential $v_{eff}[n^{MA}; \vec{r}]$ for a hydrogen atom chemisorbed on a uniform-background substrate (r_s = 2.07, corresponding to A*l*) at two different metal-adatom separations (d = 0 and d = 1 bohr). The curves show v_{eff} as a function of distance from the proton along lines normal (– – –), parallel (– • – • –) and at an angle of 45° (• • •) to the surface. The solid line shows the effective potential for the bare metal surface. (A.u. ≡ bohr.) (From Ref. 82.)

A number of the effects of the discrete substrate lattice in chemisorption can be studied *via* a perturbative reintroduction of the lattice into the uniform-background substrate. As in the bare-surface case, this perturbative treatment is suitable only for simple-metal surfaces. Most such treatments have used first-order perturbation theory, and, accordingly, the result they yield is the value of the total system energy as a function of binding site and metal-adsorbate separation. The minimum value of this total-energy function indicates the most favorable binding site and equilibrium adsorbate-substrate bond length, and the adsorbate binding energy (heat of adsorption). The details of such analyses are given by Gunnarsson, Hjelmberg and Lundqvist,[79-84] and by Lang and Williams.[78]

There has been a number of self-consistent density-functional analyses of chemisorption on metal surfaces in which the full discrete-lattice potential (or pseudopotential) of the metal is taken into account.[70,92-97] We note a number of important density-functional studies of chemisorption on semiconductors.[98-104] We mention also the so-called effective-medium theory, in which the energy to embed an atom in an arbitrary inhomogeneous electron

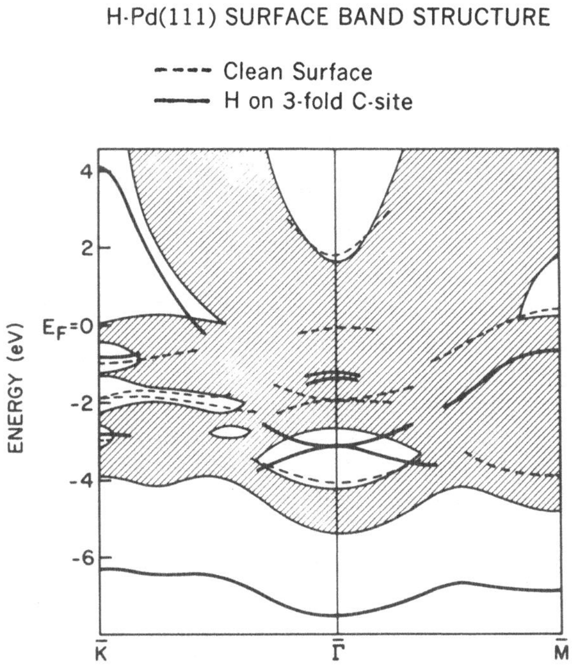

Fig. 15. Localized states (solid curves) and the projected bulk band structure (shaded areas) in the $\overline{K} = (2\pi/a)(0,-\frac{2}{3},\frac{2}{3})$ and $\overline{M} = (2\pi/a)(-\frac{1}{3}, -\frac{1}{3}, \frac{2}{3})$ directions for hydrogen chemisorbed at the centered three-fold sites on Pd(111) (a is the bulk-crystal lattice constant). (The C site is the centered site over an atom in the second Pd layer.) Also indicated are the surface states for the clean surface (dashed curves). (After Ref. 93.)

distribution (such as a surface) is obtained in terms of the energy of embedding in a bulk *homogeneous* electron gas.[105-108]

An example of results of such calculations is given in Figure 15, which shows the states localized in the surface region for H/Pd(111), as well as for bare Pd(111), as calculated by Louie[93]. The prominent adsorbate-induced features are an H-Pd bonding adsorbate band below the Pd bulk d bands and a dispersive band of antibonding H-Pd states just above the Fermi level in a gap in the projected band structure (near $\overline{K}$). [Chemisorption also affects the intrinsic surface states, as is evident in the figure.] These results do in fact give a rather reasonable account of the photoemission spectrum for this system.*

* See also e.g. the paper of Himpsel et al.[109]

1. W. Kohn and P. Vashishta, in: Theory of the Inhomogeneous Electron Gas (S. Lundqvist and N. H. March, eds.), Plenum, New York (1983).
2. N. D. Lang, in: Solid State Physics (F. Seitz, D. Turnbull, and H. Ehrenreich, eds.), Vol. 28, p. 225, Academic Press, New York (1973).
3. P. Hohenberg and W. Kohn, Phys. Rev. 136, B864 (1964).
4. W. Kohn and L. J. Sham, Phys. Rev. 140, A1133 (1965).
5. E. Wigner and J. Bardeen, Phys. Rev. 48, 84 (1935).
6. N. D. Lang and W. Kohn, Phys. Rev. B 3, 1215 (1971).
7. N. D. Lang, Solid State Commun. 7, 1047 (1969).
8. N. D. Lang and W. Kohn, Phys. Rev. B 1, 4555 (1970).
9. H. F. Budd and J. Vannimenus, Phys. Rev. Lett. 31, 1218 and 1430 (erratum) (1973).
10. J. Vannimenus and H. F. Budd, Solid State Commun. 15, 1739 (1974).
11. J. Frenkel, Z. Phys. 51, 232 (1928); A. Samoilovich, Acta Physicochim. URSS 20, 97 (1945) (notes certain errors in the Frenkel paper); B. Mrowka and A. Recknagel, Phys. Z. 38, 758 (1937).
12. P. Gombas, Z. Phys. 121, 523 (1943).
13. J. R. Smith, Ph. D. Thesis, Ohio State University, Columbus, 1968 (unpublished).
14. J. Vannimenus, These de Doctorat d'État, Université Pierre et Marie Curie, Paris, 1976 (unpublished).
15. J. Goodisman, J. Chem. Phys. 63, 4437 (1975).
16. P. Gombas, in: Handbuch der Physik (S. Flügge, ed.), Vol. 36, p. 109, Springer-Verlag, Berlin (1956).
17. J. R. Smith, Phys. Rev. 181, 522 (1969).
18. J. Bardeen, Phys. Rev. 49, 653 (1936).
19. A. J. Bennett and C. B. Duke, in: The Structure and Chemistry of Solid Surfaces (G. A. Somorjai, ed.), p. 25-1, Wiley, New York (1969).
20. F. K. Schulte, Surf. Sci. 55, 427 (1976).
21. J. H. Rose, Jr. and H. B. Shore, Solid State Commun. 17, 327 (1975).
22. G. D. Mahan, Phys. Rev. B 12, 5585 (1975).
23. V. Sahni and J. Gruenebaum, Phys. Rev. B 15, 1929 (1977); V. Sahni, J. B. Krieger, and J. Gruenebaum, Phys. Rev. B 15, 1941 (1977); V. Sahni and J. Gruenebaum, Solid State Commun. 21, 463 (1977); V. Sahni, C. Q. Ma, and J. S. Flamholz, Phys. Rev. B 18, 3931 (1978).
24. J. E. van Himbergen and R. Silbey, Phys. Rev. B 18, 2674 (1978).
25. R. L. Kautz and B. B. Schwartz, Phys. Rev. B 14, 2017 (1976).
26. A. R. Williams and U. von Barth, in: Theory of the Inhomogeneous Electron Gas (S. Lundqvist and N. H. March, eds.), Plenum, New York (1983).
27. O. Gunnarsson, J. Harris, and R. O. Jones, J. Chem. Phys. 67, 3970 (1977); J. Harris and R. O. Jones, J. Chem. Phys. 68, 1190 (1978); 70, 830 (1979).
28. V. L. Moruzzi, J. F. Janak, and A. R. Williams, Calculated Electronic Properties of Metals, Pergamon Press, New York (1978).

29. V. Peuckert, J. Phys. C 7, 2221 (1974); 9, 809 (1976).
30. T. L. Loucks and P. H. Cutler, J. Phys. Chem. Solids 25, 105 (1964).
31. J. Bardeen, Surf. Sci. 2, 381 (1964).
32. J. C. Inkson, J. Phys. C 10, 567 (1977).
33. J. E. Inglesfield and I. D. Moore, Solid State Commun. 26, 867 (1978).
34. N. D. Lang, in: Electronic Structure and Reactivity of Metal Surfaces (E. G. Derouane and A. A. Lucas, eds.), Plenum, New York (1976).
35. G. Paasch and H. Wonn, Phys. Stat. Sol. (b) 70, 555 (1975).
36. V. E. Kenner and R. E. Allen, Phys. Rev. B 11, 2858 (1975).
37. D. C. Langreth, Comments Solid State Phys. 8, 129 (1978).
38. M. Manninen and R. M. Nieminen, J. Phys. F 8, 2243 (1978).
39. R. Monnier and J. P. Perdew, Phys. Rev. B 17, 2595 (1978).
40. J. Schmit and A. A. Lucas, Solid State Commun. 11, 415 (1972).
41. R. A. Craig, Phys. Rev. B 6, 1134 (1972).
42. V. Peuckert, Z. Phys. 241, 191 (1971).
43. D. C. Langreth and J. P. Perdew, Solid State Commun. 17, 1425 (1975); Phys. Rev. B 15, 2884 (1977).
44. L. I. Schiff, Phys. Rev. B 1, 4649 (1970).
45. A. K. Theophilou and A. Modinos, Phys. Rev. B 6, 801 (1972); A. K. Theophilou, J. Phys. F 2, 1124 (1972).
46. J. A. Appelbaum and D. R. Hamann, Phys. Rev. B 6, 1122 (1972).
47. N. D. Lang and W. Kohn, Phys. Rev. B 7, 3541 (1973).
48. V. E. Kenner, R. E. Allen, and W. M. Saslow, Phys. Rev. B 8, 576 (1973).
49. J. R. Smith, S. C. Ying, and W. Kohn, Phys. Rev. Lett. 30, 610 (1973).
50. S. C. Ying, J. R. Smith, and W. Kohn, Phys. Rev. B 11, 1483 (1975).
51. P. W. Lert and J. H. Weare, J. Chem. Phys. 68, 5010 (1978).
52. H. F. Budd and J. Vannimenus, Phys. Rev. B 12, 509 (1975).
53. K. C. Pandey, Phys. Rev. Lett. 47, 1913 (1981); 49, 223 (1982).
54. J. A. Appelbaum and D. R. Hamann, Phys. Rev. Lett. 32, 225 (1974); Phys. Rev. B 12, 1410 (1975).
55. J. A. Appelbaum, G. A. Baraff, and D. R. Hamann, Phys. Rev. B 11, 3822 (1975); 12, 1410 and 5749 (1975); 14, 588 and 1623 (1976); 15, 2408 (1977).
56. M. Schlüter, J. R. Chelikowsky, S. G. Louie, and M. L. Cohen, Phys. Rev. Lett. 34, 1385 (1975); Phys. Rev. B 12, 4200 (1975).
57. J. R. Chelikowsky and M. L. Cohen, Phys. Rev. B 13, 826 (1976).
58. J. R. Chelikowsky, Phys. Rev. B 15, 3236 (1977).
59. G. P. Kerker, S. G. Louie, and M. L. Cohen, Phys. Rev. B 17, 706 (1978).
60. S. Ciraci and I. P. Batra, Phys. Rev. B 15, 3254 (1977).
61. J. A. Appelbaum and D. R. Hamann, Solid State Commun. 27, 881 (1978).
62. J. G. Gay, J. R. Smith, and F. J. Arlinghaus, Phys. Rev. Lett. 38, 561 (1977); 42, 332 (1979).

63. F. J. Arlinghaus, J. G. Gay and J. R. Smith, Phys. Rev. B 21, 2055 (1980).
64. J. A. Appelbaum and D. R. Hamann, Phys. Rev. B 6, 2166 (1972).
65. K.-P. Bohnen and S. C. Ying, Phys. Rev. B 22, 1806 (1980).
66. G. P. Alldredge and L. Kleinman, Phys. Rev. B 10, 559 (1974).
67. J. R. Chelikowsky, M. Schlüter, S. G. Louie, and M. L. Cohen, Solid State Commun. 17, 1103 (1975).
68. S. G. Louie, K.-M. Ho, J. R. Chelikowsky, and M. L. Cohen, Phys. Rev. B 15, 5627 (1977).
69. S. G. Louie, Phys. Rev. Lett. 40, 1525 (1978).
70. P. J. Feibelman, Phys. Rev. B 26, 5347 (1982).
71. O. Jepsen, J. Madsen and O. K. Andersen, Phys. Rev. B 18, 605 (1978).
72. K. Mednick and L. Kleinman, Phys. Rev. B 22, 5768 (1980).
73. C. S. Wang and A. J. Freeman, Phys. Rev. B 24, 4364 (1981).
74. R. Smoluchowski, Phys. Rev. 60, 661 (1941).
75. N. D. Lang, Phys. Rev. B 4, 4234 (1971).
76. N. D. Lang and A. R. Williams, Phys. Rev. Lett. 34, 531 (1975); 37, 212 (1976).
77. N. D. Lang and A. R. Williams, Phys. Rev. B 16, 2408 (1977).
78. N. D. Lang and A. R. Williams, Phys. Rev. B 18, 616 (1978)
79. O. Gunnarsson and H. Hjelmberg, Physica Scripta 11, 97 (1975); O. Gunnarsson, H. Hjelmberg, and B. I. Lundqvist, Phys. Rev. Lett. 37, 292 (1976).
80. O. Gunnarsson, H. Hjelmberg, and B. I. Lundqvist, Surf. Sci. 63, 348 (1977).
81. H. Hjelmberg, O. Gunnarsson, and B. I. Lundqvist, Surf. Sci. 68, 158 (1977).
82. H. Hjelmberg, Physica Scripta 18, 481 (1978).
83. P. Johansson and H. Hjelmberg, Surf. Sci. 80, 171 (1979).
84. H. Hjelmberg, Surf. Sci. 81, 539 (1979).
85. L. M. Kahn and S. C. Ying, Surf. Sci. 59, 333 (1976).
86. R. F. W. Bader, W. H. Henneker, and P. E. Cade, J. Chem. Phys. 46, 3341 (1967); R. F. W. Bader, I. Keaveny, and P. E. Cade, J. Chem. Phys. 47, 3381 (1967).
87. M. L. Yu and N. D. Lang, Phys. Rev. Lett. 50, 127 (1983).
88. O. Gunnarsson, H. Hjelmberg, and J. K. Nørskov, Physica Scripta 22, 165 (1980).
89. J. K. Nørskov, A. Houmøller, P. K. Johansson, and B. I. Lundqvist, Phys. Rev. Lett. 46, 257 (1981).
90. J. K. Nørskov, Solid State Commun. 25, 995 (1978).
91. N. D. Lang, Phys. Rev. Lett. 46, 842 (1981).
92. J. R. Smith, F. J. Arlinghaus, and J. G. Gay, Solid State Commun. 24, 279 (1977).
93. S. G. Louie, Phys. Rev. Lett. 42, 476 (1979).

94. D. M. Bylander, L. Kleinman and K. Mednick, Phys. Rev. Lett. 48, 1544 (1982).
95. D.-S. Wang, A. J. Freeman and H. Krakauer, Phys. Rev. B 26, 1340 (1982).
96. R. Richter and J. W. Wilkins, J. Vac. Sci. Technol. A 1, 1089 (1983).
97. E. Wimmer, A. J. Freeman, J. R. Hiskes and A. M. Karo, Phys. Rev. B 28, 3074 (1983).
98. J. A. Appelbaum and D. R. Hamann, Phys. Rev. Lett. 34, 806 (1975); Phys. Rev. B 15, 2006 (1977).
99. J. A. Appelbaum, H. D. Hagstrum, D. R. Hamann, and T. Sakurai, Surf. Sci. 58, 479 (1976).
100. J. A. Appelbaum, D. R. Hamann, and K. H. Tasso, Phys. Rev. Lett. 39, 1487 (1977).
101. J. A. Appelbaum, G. A. Baraff, D. R. Hamann, H. D. Hagstrum, and T. Sakurai, Surf. Sci. 70, 654 (1978).
102. M. Schlüter, J. E. Rowe, G. Margaritondo, K. M. Ho, and M. L. Cohen, Phys. Rev. Lett. 37, 1632 (1976).
103. K. M. Ho, M. L. Cohen, and M. Schlüter, Phys. Rev. B 15, 3888 (1977).
104. M. Schlüter and M. L. Cohen, Phys. Rev. B 17, 716 (1978).
105. M. J. Stott and E. Zaremba, Phys. Rev. B 22, 1564 (1980).
106. J. K. Nørskov and N. D. Lang, Phys. Rev. B 21, 2131 (1980).
107. J. K. Nørskov, Phys. Rev. B 26, 2875 (1982).
108. N. D. Lang and J. K. Nørskov, Phys. Rev. B 27, 4612 (1983).
109. F. J. Himpsel, K. Christmann, P. Heimann, D. E. Eastman and P. J. Feibelman, Surf. Sci. 115, L159 (1982).

WHAT DO THE KOHN-SHAM ORBITAL ENERGIES MEAN?

HOW DO ATOMS DISSOCIATE?

John P. Perdew

Department of Physics and Quantum Theory Group
Tulane University
New Orleans, Louisiana 70118, U.S.A.

1. INTRODUCTION

Hohenberg and Kohn[1] have demonstrated the existence of a functional $E_v[n]$ which, when minimized over trial densities $n(\underset{\sim}{r})$ integrating to N electrons, yields the exact ground-state energy E and density for N electrons subject to an external potential $v(\underset{\sim}{r})$. The Euler equation for the variational principle is

$$\delta \{E_v[n] - \mu \int d^3r\, n(\underset{\sim}{r})\} = 0, \tag{1}$$

or

$$\delta E_v / \delta n(\underset{\sim}{r}) = \mu. \tag{2}$$

The Lagrange multiplier μ is evidently the chemical potential:

$$\mu = \partial E / \partial N. \tag{3}$$

Kohn and Sham[2] have shown how to implement this variational principle in practical calculations: Divide $E_v[n]$ into pieces,

$$E_v[n] = K_0[n] + \int d^3r\, v(\underset{\sim}{r})\, n(\underset{\sim}{r}) + U[n] + E_{xc}[n], \tag{4}$$

where $K_0[n]$ is the ground-state kinetic energy of hypothetical non-interacting electrons of density $n(\underset{\sim}{r})$, U[n] is the classical repulsion

$$U[n] = \frac{1}{2} \int d^3r \int d^3r' \; n(\underset{\sim}{r}) \; n(\underset{\sim}{r}')/|\underset{\sim}{r}-\underset{\sim}{r}'|, \tag{5}$$

and $E_{xc}[n]$ is the exchange-correlation energy. The true n(r) is found from the self-consistent solution of the one-electron equations

$$[-\frac{1}{2}\nabla^2 + v_{eff}([n];\underset{\sim}{r})] \; \psi_i(\underset{\sim}{r}) = \varepsilon_i \; \psi_i(\underset{\sim}{r}) \tag{6}$$

$$n(\underset{\sim}{r}) = \sum_i f_i \; |\psi_i(\underset{\sim}{r})|^2, \tag{7}$$

where f_i is an occupation number in the range $0 \leq f_i \leq 1$. The effective potential of the Kohn-Sham equation (6) is

$$v_{eff}([n];\underset{\sim}{r}) = v(\underset{\sim}{r}) + u([n];\underset{\sim}{r}) + \delta E_{xc}/\delta n(\underset{\sim}{r}), \tag{8}$$

where

$$u([n];\underset{\sim}{r}) = \delta U/\delta n(\underset{\sim}{r}) = \int d^3r' \; n(\underset{\sim}{r}')/|\underset{\sim}{r}-\underset{\sim}{r}'|. \tag{9}$$

Thus the many-electron problem has in principle been reduced exactly to self-consistent-field form.

A. What Do the Kohn-Sham Orbital Energies Mean?

The Kohn-Sham orbital energies ε_i of Eq. (6) appear as formal Lagrange multipliers. Do they have any exact or approximate physical meaning? This question is in fact a crucial one for solid state physics, where the Kohn-Sham formalism has become the *de facto* foundation of most band-structure calculations for crystals. Is there anything like an analog of Koopmans' theorem, i.e., can any or all of the ε_i's be interpreted as physical electron removal energies? (Although the question has been formulated here for an infinite system, the answer will be valid for a system of any size.)

These questions cannot be answered definitively by a comparison between calculated band structures and experiment, because practical calculations must rely on some approximation (see section 4), such as the local density approximation[2,3] (LDA), to the unknown exact functional $E_{xc}[n]$. However, the results of such a comparison within LDA are at least easily summarized: The work functions Φ of metals, which typically fall between 2 and 6eV, are approximated accurately[4,5] (within about 0.2eV) by $-\varepsilon_{max}^{LDA}$, where ε_{max}^{LDA} is the greatest occupied LDA orbital energy measured from the vacuum level. The LDA valence bands of metals also display small (but systematic) deviations from experiment.[6] However, the LDA band-structures underestimate the fundamental band gaps by about a factor of two in many insulators[7] and semiconductors[8], and place the deeper orbitals too high in energy[7]. Would similar discrepancies arise from the unknown exact Kohn-Sham orbital energies?

It has been known[9,10] for some time that the exact work function of a metal is $-\varepsilon_{max}$, where ε_{max} is the greatest occupied exact Kohn-Sham orbital energy. More recently, Janak[10] (following arguments given by Slater[3]) proved that

$$\varepsilon_i = \partial \tilde{E}/\partial f_i, \tag{10}$$

where

$$\tilde{E} \equiv \sum_i f_i \langle\psi_i| - \frac{1}{2}\nabla^2|\psi_i\rangle + \int d^3r\, v(\underset{\sim}{r})\, n(\underset{\sim}{r})$$

$$+ U[n] + E_{xc}[n]. \tag{11}$$

The proof requires only self-consistency, and so is valid for the exact $E_{xc}[n]$ or for any approximation to it. From (10), Janak also derived an "aufbau" principle: $\tilde{E}$ is minimized when $f_i = 1$ for all $\varepsilon_i < \mu$, and 0 for all $\varepsilon_i > \mu$, where μ is the chemical potential. For fixed electron number N, $\tilde{E}$ at its minimum is just the ground-state energy E(N). Let the orbital energies of the N-electron system be ordered as

$$\varepsilon_1\ (N) \leq \varepsilon_2\ (N) \leq \cdots \tag{12}$$

Then, for the J-electron system (J = any positive integer), the first ionization potential and electron affinity are respectively

$$I(J) = E(J-1) - E(J) = -\int_0^1 df\ \varepsilon_J(J-1+f) \tag{13}$$

$$A(J) = I(J+1) = E(J) - E(J+1) = -\int_0^1 df\ \varepsilon_{J+1}(J+f). \tag{14}$$

Janak's theorem, in the form of Eqs. (13) and (14), makes an important connection between physical energy differences and Kohn-Sham orbital energies. The price of this connection is the introduction of non-integer electron number N.

The original Hohenberg-Kohn theorem was proved[1,11-14] for a closed system with a fixed, integer number of electrons. Mermin[15] extended the theorem to the grand potential and density of an open system in equilibrium at temperature T and chemical potential μ, while Kohn and Sham[2] derived self-consistent equations of the form of Eqs. (6) and (7) for this extension. In an open system, the electron number can fluctuate around a non-integer average value N, as electrons pass between the system and its "reservoir." The zero-temperature limit of Mermin's theory, which was reformulated and explored by Perdew, Parr, Levy and Balduz,[16-18] will be discussed here in full detail.

B. How Do Atoms Dissociate?

The physical extension of the ground-level energy E(N) to non-integer electron number N should also explain the following simple facts about the dissociation limit: (i) When any diatomic molecule $\alpha\beta$ dissociates adiabatically, the dissociation products are separated neutral atoms α and β. There is no doubt that this is so, since the greatest electron affinity of all the neutral atoms [A(Cℓ) = 3.62eV] is smaller than the least ionization potential [I(Cs) = 3.89eV]. (ii) When a chemisorbed atom (with atomic ionization potential I and electron affinity A) desorbs adiabatically from a metal surface (with work function Φ), the charge q on the desorbed atom is

$$q = \begin{cases} +1 & (I < \Phi) \\ 0 & (A < \Phi < I) \\ -1 & (\Phi < A). \end{cases} \tag{15}$$

The ground-state density varies discontinuously with the parameter Φ. Here is a real discontinuity of nature which must be reflected somehow in the exact density functional theory. But how?

Consider, for definiteness, case (ii). If the desorbed atom and the metal are regarded together as a closed system, then the proper desorption limit is "hidden" in the unknown Hohenberg-Kohn functional $E_v[n]$: This functional samples various number-conserving trial densities, including those which put fractional charge on the desorbed atom, but minimizes somehow at the correct integer charge. It will be more instructive to regard the desorbed atom as an open system, dynamically isolated from but free to exchange electrons with the metal or "reservoir" of chemical potential $\mu = -\Phi$. In fact the open-system density functional theory of Mermin, even in its zero-temperature limit, will prove to be a useful complement to the closed-system theory of Hohenberg and Kohn.

2. SHORT ANSWERS

A. The Energy as a Function of Non-integer Electron Number

Consider an open system very far from, but free to exchange electrons with, a reservoir r. Write a trial wavefunction for the combined K-electron system:

$$\Psi^c(1\ldots K) = \sum_J \Sigma_{\mu\nu}\, C_{\mu\nu}(J)\, \hat{A}\, \Psi_\mu(1\ldots J)\, \Psi^r_\nu(J+1\ldots K). \qquad (16)$$

The $\{C_{\mu\nu}(J)\}$ are complex numbers, and $\hat{A}$ is an operator which antisymmetrizes and normalizes. $\Psi_\mu(1\ldots J)$ and $\Psi^r_\nu(J+1\ldots K)$ are bound stationary states of the open system and reservoir, respectively. As the distance separating the open system from the reservoir tends to infinity, no matrix element can link one to the other [i.e., $\int d^3r_i\, \Psi^*_\mu(\ldots i\ldots)\, \hat{O}\, \Psi^r_\nu(\ldots i\ldots) = 0$]. Thus

$$\langle\Psi^c|\Psi^c\rangle = 1 = \sum_J \Sigma_{\mu\nu}\, p_{\mu\nu}(J), \qquad (17)$$

where $p_{\mu\nu}(J) = |C_{\mu\nu}(J)|^2$ lies between 0 and 1 inclusive. The energy and density of the combined system are expressible in terms of those of the eigenstates:

$$\langle\Psi^c|\hat{H}^c|\Psi^c\rangle = \sum_J \Sigma_{\mu\nu}\, p_{\mu\nu}(J)\, [E_\mu(J) + E^r_\nu(K-J)] \qquad (18)$$

$$\langle\Psi^c|\hat{n}^c(\underset{\sim}{r})|\Psi^c\rangle = \sum_J \Sigma_{\mu\nu}\, p_{\mu\nu}(J)\, [n_\mu(J;\underset{\sim}{r}) + n^r_\nu(K-J;\underset{\sim}{r})], \qquad (19)$$

and the electron number in the open system is clearly

$$N = \sum_J \Sigma_{\mu\nu}\, p_{\mu\nu}(J)\, J. \qquad (20)$$

Thus, in the infinite-separation limit, the important wavefunction expectation values reduce to ensemble expectation values: $p_{\mu\nu}(J)$ can be interpreted as the probability that the open system contains J electrons in state μ, while the reservoir contains the remaining (K-J) electrons in state ν.

To find the ground-state energy and density of the open system as a function of its average electron number N, it is only necessary to find the $p_{\mu\nu}(J)$ which minimize (18), subject to the constraints (17) and (20). Evidently the excited states will receive zero probability, leaving the problem of minimizing the weighted-sum of ground-state energies

$$\sum_J p(J)\, [E(J) + E^r(K-J)], \qquad (21)$$

subject to the constraints

$$\sum_J p(J) = 1, \qquad \sum_J p(J)J = N. \qquad (22)$$

Further progress requires some information about E(J) for the open system and its reservoir. Since each is an electronic system, its first ionization energy presumably increases with decreasing electron number:

$$E(J-1) - E(J) > E(J) - E(J+1). \qquad (23)$$

This inequality can be rearranged to read

$$\frac{1}{2}\,[E(J+1) + E(J-1)] > E(J), \qquad (24)$$

i.e., E(J) is a convex or bowl-shaped function. The inequality is strict, except that it may tend to an equality in the limit of infinite volume. (If the latter limit is needed, it will be taken last in any sequence of limits.) All electronic systems appear to be convex, although there is no proof[13] that they must be.

Now it is easy to see that, for N between the integers J-1 and J, the energy and density of the open system are[16,17]

$$E(N) = (J-N)\,E(J-1) + (N-J+1)\,E(J) \qquad (25)$$

$$n(N;\underset{\sim}{r}) = (J-N)\,n(J-1;\underset{\sim}{r}) + (N-J+1)\,n(J;\underset{\sim}{r}). \qquad (26)$$

Eqs. (25) and (26) are the principal results of the ground-level density functional theory of open systems,[16-18] and the basis for an answer to the two questions of section 1. As a function of the continuous variable N (time-averaged electron number), the ground-level energy E(N) of the open system is a linkage of straight-line segments, with possible derivative discontinuities at integer values of N. Due to the convexity of E(J), the same result is obtained for E(N) whether the reservoir can bind infinitely-many electrons (grand canonical ensemble) or just one (mini-canonical ensemble).[16]

If the constraint on N of Eq. (22) is now relaxed, expression (21) will normally minimize with integer N. Electron number N between the integers J-1 and J will be found only when there is a degeneracy between the situations in which the open system has J-1 and J electrons, i.e., only when

$$E(J-1) + E^{r}\,(K-J+1) = E(J) + E^{r}\,(K-J). \qquad (27)$$

Indeed the results (25) and (26) are most easily obtained by mentally adjusting the external potential in the reservoir until (27) is satisfied. Then, since $E(J) + E^{r}(K-J)$ is convex in the integer variable J, expression (21) is clearly minimized by putting all the weight onto the two lowest, degenerate energies of Eq. (27).

B. The Kohn-Sham Orbital Energies

Kohn-Sham theory[2] hypothesizes a system of non-interacting electrons having the same ground-state density and chemical potential μ as the real system. When Eqs. (10) and (12) - (14) are combined with Eqs. (3) and (25), the result is

$$\mu = \begin{cases} \varepsilon_J(N) = -I(J) & (J-1 < N < J) \\ \varepsilon_{J+1}(N) = -A(J) & (J < N < J+1). \end{cases} \tag{28}$$

The energy eigenvalue of the highest partly-occupied Kohn-Sham orbital is the chemical potential. The electron number N may be brought arbitrarily close to the integer J through the introduction of a positive infinitesimal δ:

$$\mu = \begin{cases} \varepsilon_J(J-\delta) = -I(J) & (J-1 < N < J) \\ \varepsilon_{J+1}(J+\delta) = -A(J) & (J < N < J+1). \end{cases} \tag{29}$$

As N increases through the integer J, the chemical potential μ and the highest partly-occupied Kohn-Sham orbital energy jump discontinuously from one physical value (minus the ionization potential) to another (minus the electron affinity). The other Kohn-Sham orbital energies have no precise physical meaning, as discussed in section 5A.

What happens when N increases through the integer J? Eq. (26) shows that the ground-level density $n(N;\underset{\sim}{r})$ changes only infinitesimally, and so the effective potential of Eq. (8) can change only by an infinitesimal plus a possible finite constant C:[16,17]

$$\lim_{\delta\to 0}\left[\left.\frac{\delta E_{xc}}{\delta n(\underset{\sim}{r})}\right|_{J+\delta} - \left.\frac{\delta E_{xc}}{\delta n(\underset{\sim}{r})}\right|_{J-\delta}\right] = C. \tag{30}$$

The subscript $N = J\pm\delta$ means that the functional derivative is to be evaluated for the N-electron ground-level density. Thus

$$\begin{aligned} I(J) - A(J) &= \varepsilon_{J+1}(J+\delta) - \varepsilon_J(J+\delta) + C \\ &= \varepsilon_{J+1}(J-\delta) - \varepsilon_J(J-\delta) + C \\ &= \varepsilon_{J+1}(J) - \varepsilon_J(J) + C, \end{aligned} \tag{31}$$

where $\varepsilon_J(J)$ and $\varepsilon_{J+1}(J)$ are, respectively, the highest occupied and lowest unoccupied Kohn-Sham levels of the J-electron system.

The derivative discontinuity C of Eqs. (30) and (31) is positive in an atom, and probably also in an insulating or semiconducting crystal[19,20] where it equals the difference between the physical band gap and the gap in the exact Kohn-Sham band structure (section 5B). In an open-shell $[\varepsilon_{J+1}(J) = \varepsilon_J(J)]$ atom, all of the positive quantity I(J) - A(J) comes from the derivative discontinuity C. In a closed-shell atom, von Barth[21] has given a convincing argument that $I(J) - A(J) > \varepsilon_{J+1}(J) - \varepsilon_J(J)$, so that again $C > 0$; this conclusion will be confirmed numerically for the neon atom in section 5A.

The long-range behavior of the Kohn-Sham potential is[17]

$$\lim_{r\to\infty} \left. \frac{\delta E_{xc}}{\delta n(\underset{\sim}{r})} \right|_N = 0 \qquad \text{(all N)} \tag{32}$$

$$\lim_{\delta\to 0}\ \lim_{r\to\infty} \left. \frac{\delta E_{xc}}{\delta n(\underset{\sim}{r})} \right|_{J\pm\delta} = 0 = \lim_{r\to\infty}\ \lim_{\delta\to 0} \left. \frac{\delta E_{xc}}{\delta n(\underset{\sim}{r})} \right|_{J-\delta}, \tag{33}$$

but

$$\lim_{r\to\infty}\ \lim_{\delta\to 0} \left. \frac{\delta E_{xc}}{\delta n(\underset{\sim}{r})} \right|_{J+\delta} = C. \tag{34}$$

These results will be derived and explained in section 4B.

The Kohn-Sham orbital energies for integer electron number are somewhat ill-defined. For positive C, they must satisfy the inequality $- I(J) \le \varepsilon_J(J) \le \varepsilon_{J+1}(J) \le - A(J)$. Although this inequality can be satisfied in many ways, there are two preferred choices: (1) Eqs. (32) - (34) suggest that it is most natural to use the limit as N approaches J from below, i.e.,

$$\mu = \varepsilon_J(J) = \varepsilon_J(J-\delta) = - I(J). \tag{35}$$

(2) Alternatively, one can set N = J before taking the zero-temperature limit in the Mermin formalism; the result (derived in section 3) is then[17,18]

$$\mu = \frac{1}{2}\,[\varepsilon_J(J) + \varepsilon_{J+1}(J)] = - \frac{1}{2}\,[I(J) + A(J)]. \tag{36}$$

Eq. (36), which averages over the discontinuity in $\delta E_{xc}/\delta n(r)$, is roughly what should be expected from approximations in which $\delta E_{xc}/\delta n(\underset{\sim}{r})$ is continuous, e.g., the local density approximation.

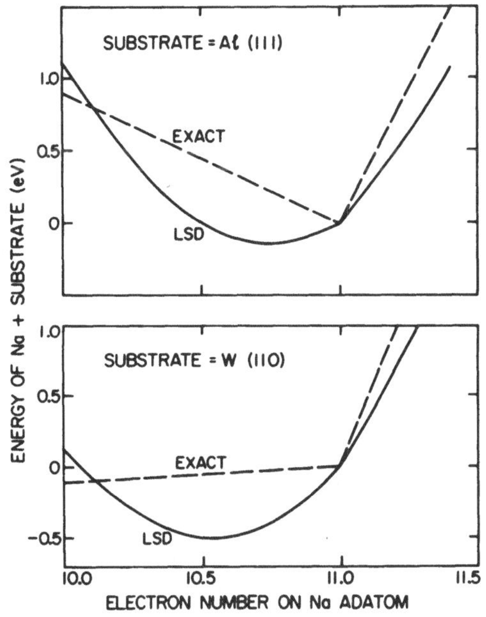

Fig. 1. Total energy of a metallic substrate plus a desorbed sodium atom, as a function of the electron number on the atom. (From Ref. 22).

With this choice, the right-hand side of Eq. (32) must be replaced by the positive number C/2 when N = J.

C. The Dissociation Limit

Equations (25) and (28) provide a simple explanation for the dissociation limit.

Figure 1 (dashed curve), from the work of Perdew and Smith,[22] shows the total energy of the combined system composed of a semi-infinite metal (work function Φ) and a distant or desorbed sodium atom [$I(11) = 5.1$eV, $A(11) = 0.6$eV], as a function of the number N of electrons on the atom. The energy of the atom is a linkage of straight line segments, with slopes $-I(11)$ ($10 < N < 11$) and $-A(11)$ ($11 < N < 12$). The energy of the metal is a straight line of slope Φ. The total energy has been zeroed at $N = 11$. In the upper frame, the metallic substrate is aluminum ($\Phi = 4.25$eV); since $A(11) < \Phi < I(11)$, the total energy minimizes correctly at $N = 11$, and the desorbed atom is neutral. In the lower frame, the

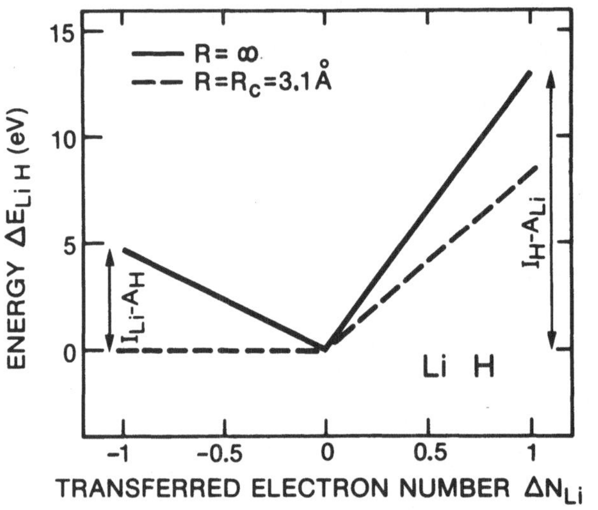

Fig. 2. Total energy of the diatomic system Li H, at infinite and finite separations R, as a function of the number of electrons transferred to the Li atom. (After Ref. 16).

metallic substrate is tungsten (Φ = 5.25eV); since $\Phi > I(11)$, the total energy minimizes correctly at N = 10, and the desorbed atom is Na^{+1}. (The solid curve shows how the local spin density approximation[22] to the exchange-correlation energy minimizes improperly, with fractional charge on the desorbed atom, as discussed in section 5C.)

Figure 2 (solid curve) shows the total energy of the diatomic system Li H at infinite separation, as a function of the number of electrons transferred from the neutral Li to the neutral H. The energy properly minimizes at the neutral-atoms configuration. (The dashed curve shows how the ground-state can switch over[16] to the ionic configuration $Li^{+1}\ H^{-1}$ when the interatomic separation is small enough.) More generally, any pair of atoms $\alpha\beta$ at infinite separation will assume the neutral configuration,[16] since $I_\alpha - A_\beta$ and $I_\beta - A_\alpha$ are both positive for any choice of neutral atoms α and β from the periodic table.

According to Eqs. (25) and (28), the energy E(N) of an open system displays cusps or derivative discontinuities at integer values of N. The total energy of open system plus distant reservoir tends to minimize nonanalytically at one of these cusps. The cusps, of course, will be rounded either by finite temperature (section 3B), or by finite interaction with the reservoir, as discussed below:

Consider a diatomic system with nuclear charges Z_α and Z_β separated by a large distance R, and K electrons. Choose the z-axis along the line from α to β, and define

$$\hat{\zeta} = Z_\beta - Z_\alpha - \sum_{i=1}^{K} \frac{\hat{z}_i}{(R/2)} . \tag{37}$$

Eichler[23] has observed that $\langle\hat{\zeta}\rangle$ tends as $R \to \infty$ to the charge difference between atoms α and β. But $\langle\hat{\zeta}\rangle$ is also the dipole moment p (about the midpoint) divided by R/2, so the polarizability of the $\alpha\beta$ system along its axis is just[23]

$$\alpha_{||} = (\partial^2 E^{\alpha\beta}/\partial p^2)^{-1} = (\partial^2 E^{\alpha\beta}/\partial\langle\hat{\zeta}\rangle^2)^{-1} (R/2)^2 . \tag{38}$$

In the limit $R \to \infty$, the ground-state has $\langle\hat{\zeta}\rangle = 0$ and finite positive polarizability. Thus $\partial^2 E^{\alpha\beta}/\partial\langle\hat{\zeta}\rangle^2$ at $\langle\hat{\zeta}\rangle = 0$ (and hence $\partial^2 E/\partial N^2$) must tend to $+\infty$ like R^2, a clear signal that $\partial E/\partial N = \mu$ is developing a discontinuity as $R \to \infty$. Since the discontinuity is rounded at finite separation, it may be legitimate to speak about a differentiable E(N) for, say, an atom in a molecule.[24]

3. THE DENSITY FUNCTIONAL THEORY OF EQUILIBRIUM STATES

A. Quantum Statistical Mechanics

The most general description of the state of a system is via an ensemble (density matrix, statistical mixture) - a set of pure states $\{\chi_n\}$ and their respective probabilities $\{p_n\}$. Mathematically, an ensemble $\hat{\Gamma}$ is a self-adjoint operator with non-negative eigenvalues and unit trace. In spectral representation,

$$\hat{\Gamma} = \sum_n p_n \,|\, \chi_n \rangle \langle \chi_n \,| , \tag{39}$$

where

$$p_n \geq 0, \qquad \sum_n p_n = 1, \qquad \langle\chi_n|\chi_m\rangle = \delta_{nm}. \tag{40}$$

The ensemble expectation value of an operator $\hat{O}$ is

$$\mathrm{Tr}\, \hat{\Gamma}\, \hat{O} = \sum_n p_n \langle\chi_n|\hat{O}|\chi_n\rangle. \tag{41}$$

In particular, for the entropy $\hat{S} = -k_B \ln \hat{\Gamma}$,

$$S = -k_B \text{ Tr } \hat{\Gamma} \ln \hat{\Gamma} = -k_B \sum_n p_n \ln p_n \ . \tag{42}$$

The equilibrium ensemble $\hat{\Gamma}_{eq}$ maximizes S, subject to the constraints Tr $\hat{\Gamma}$ $\hat{H}$ = E (a given average energy) and Tr $\hat{\Gamma}$ $\hat{N}$ = N (a given average electron number). It can be found from the unconstrained minimization of the grand potential

$$\Omega\ [\hat{\Gamma}] = \text{Tr } \hat{\Gamma}\ (\hat{H} - \mu\hat{N} + k_B T \ln \hat{\Gamma}), \tag{43}$$

where the Lagrange multipliers are the temperature T and chemical potential μ:

$$\Omega\ (\mu,T) = \Omega\ [\hat{\Gamma}_{eq}] = \min\ \{\Omega\ [\hat{\Gamma}]\}. \tag{44}$$

It is easy to find the unique stationary point $\hat{\Gamma}_{eq}$ of $\Omega\ [\hat{\Gamma}]$:

$$\delta\Omega\ [\hat{\Gamma}] = \text{Tr }(\delta\hat{\Gamma})\ (\hat{H} - \mu\hat{N} + k_B T \ln \hat{\Gamma}_{eq}) + k_B T\, \text{Tr}(\delta\hat{\Gamma}) \tag{45}$$

must be zero for infinitesimal variations $\delta\hat{\Gamma}$ about $\hat{\Gamma}_{eq}$ such that $\delta(\text{Tr}\hat{\Gamma}) = \text{Tr}(\delta\hat{\Gamma}) = 0$. Thus $\hat{H} - \mu\hat{N} + k_B T \ln \hat{\Gamma}_{eq}$ must be a constant, from which

$$\hat{\Gamma}_{eq} = e^{-(\hat{H}-\mu\hat{N})/k_B T} / \Xi\ , \tag{46}$$

when

$$\Xi = \text{Tr } e^{-(\hat{H}-\mu\hat{N})/k_B T} \tag{47}$$

is the grand partition function. Mermin[15] has shown that the grand canonical ensemble (46) is the absolute minimum of $\Omega\ [\hat{\Gamma}]$, not merely the unique stationary point.

The eigenvectors of $\hat{\Gamma}_{eq}$ are the simultaneous eigenvectors of $\hat{H}$ and $\hat{N}$, so

$$\Xi = \sum_n e^{-(E_n - \mu J_n)/k_B T} \ . \tag{48}$$

All thermodynamic functions may be obtained from $\Xi(\mu,T)$ and its derivatives, e.g.,

$$\Omega\ (\mu,T) = -k_B T \ln \Xi \tag{49}$$

$$N\ (\mu,T) = k_B T \frac{\partial}{\partial \mu} \ln \Xi\ . \tag{50}$$

How does the chemical potential μ depend upon N at a low finite temperature T? As $T \to 0$, the contribution to the partition function (48) from each electron number J will be dominated by the ground-state eigen-value E(J) with degeneracy g_J. Since E(J) is convex [Eq. (24)], so is $E(J) - \mu J$. Let M be the integer closest to N, i.e., $M - \frac{1}{2} < N < M + \frac{1}{2}$. Then the partition function can be found from a three-state model, involving the (M-1), M- and (M+1)-electron ground states. Eq. (50) becomes a quadratic equation for $e^{-\mu/k_BT}$, with the solution

$$\mu = -k_BT \, \ell n \left[\frac{-g_M\Delta + \left\{ g_M^2 \Delta^2 + 4 g_{M-1} g_{M+1} (1-\Delta^2) e^{-[I(M)-A(M)]/k_BT} \right\}^{\frac{1}{2}}}{2 g_{M-1} (1+\Delta) e^{-I(M)/k_BT}} \right] \quad (51)$$

where $\Delta = N - M$. Eq. (51) may be expanded to order T as

$$\mu = \begin{cases} -I(M) + k_BT \, \ell n \left[\dfrac{g_{M-1}}{g_M} \left(\dfrac{1}{|\Delta|} - 1 \right) \right] & (\Delta < 0) \\ -\dfrac{1}{2} [I(M) + A(M)] - \dfrac{k_BT}{2} \ell n \left(\dfrac{g_{M+1}}{g_{M-1}} \right) & (\Delta = 0) \\ -A(M) - k_BT \, \ell n \left[\dfrac{g_{M+1}}{g_M} \left(\dfrac{1}{|\Delta|} - 1 \right) \right] . & (\Delta > 0) \end{cases} \quad (52)$$

Results similar to Eq. (52) have been presented by Gyftopoulos and Hatsopoulos.[25] At T = 0, Eq. (52) reduces to Eqs. (29) and (36) of section 2.

Figure 3 shows μ vs. N for a lithium atom [$I(3) = 5.4$eV, $A(3) = 0.6$eV, $g_3 = 2$, $g_2 = g_4 = 1$], at $T = 0^0$K and $T = 2320^0$K ($k_BT = 0.2$eV). Also shown is the free energy

$$F(N,T) = \Omega(\mu,T) + \mu N = E - TS, \quad (53)$$

which is related to μ through the identity

$$\mu = \left(\frac{\partial F}{\partial N} \right)_T . \quad (54)$$

(The free energy has been zeroed at E(3)). At $T = 2320^0$K, the deviation of F from its $T = 0^0$K behavior arises almost entirely from the -TS term in (53).

For a neutral system in the zero-temperature limit, the chemical potential $\mu = -\frac{1}{2}(I+A)$ of Eq. (52) equals minus the Mulliken electronegativity.[24,25] Parr and Pearson[26] have identified a com-

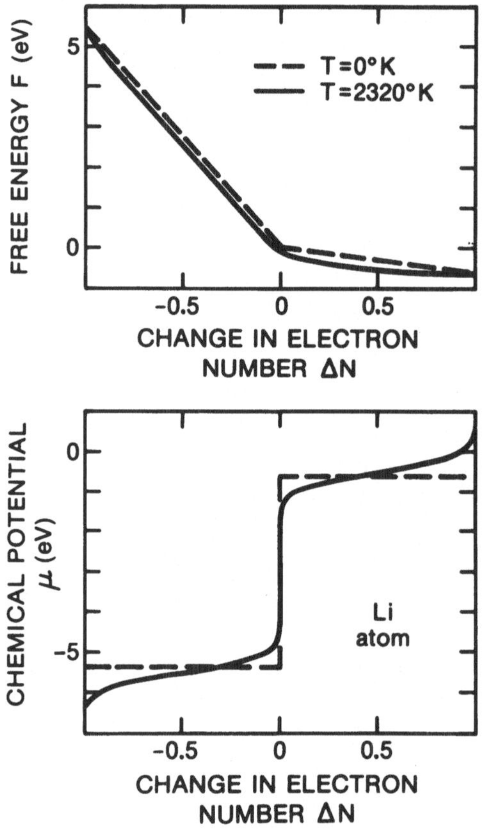

Fig. 3. Free energy and chemical potential of a lithium atom, as a function of the change in electron number from its neutral -atom value 3.

panion parameter, ½ (I-A), as the "chemical hardness". In a crystal, this "chemical hardness" is just half of the fundamental band gap, which will be discussed in section 5B.

B. Hohenberg-Kohn-Mermin Theory

Here the quantum statistical mechanics of the preceding section will be reduced to a density functional theory, following the transparent "constrained search" approach of Levy.[11] This approach has been applied to pure states,[11] ensembles of pure states with fixed electron number,[13,27] and ensembles with fluctuating electron number.[16,21]

The idea is simply to minimize $\Omega[\hat{\Gamma}]$ of Eq. (43) in two steps: First, choose a density $n(\underset{\sim}{r})$ and minimize over all ensembles $\hat{\Gamma}$ yielding that density. The result is the density functional for the grand potential,

$$\Omega_v[n] = \min \{\Omega[\hat{\Gamma}] \mid \mathrm{Tr}\, \hat{\Gamma}\, \hat{n}(\underset{\sim}{r}) = n(\underset{\sim}{r})\}. \tag{55}$$

$\Omega_v[n]$ is defined for every $n(\underset{\sim}{r})$ which can be realized by an ensemble. Second, minimize $\Omega_v[n]$ over all trial densities $n(\underset{\sim}{r})$:

$$\Omega\,(\mu,T) = \Omega_v[n_{eq}] = \min \{\Omega_v[n]\}. \tag{56}$$

The minimizing density is the equilibrium density $n_{eq}(\underset{\sim}{r})$. The variational principle (56) can also be expressed as

$$\delta\Omega_v/\delta n(\underset{\sim}{r}) = 0. \tag{57}$$

It also follows that

$$\delta\Omega_v/\delta v(\underset{\sim}{r}) = n(\underset{\sim}{r}). \tag{58}$$

From the definitions (43) and (55),

$$\Omega_v[n] = G[n] + \int d^3r\, [v(\underset{\sim}{r}) - \mu]\, n(\underset{\sim}{r}) \tag{59}$$

where $G[n]$ is a universal functional of the density (depending also on T):

$$G[n] = \min \{\mathrm{Tr}\, \hat{\Gamma}\, [\hat{K}+\hat{V}_{ee}+k_B T\, \ell n\, \hat{\Gamma}] \mid \mathrm{Tr}\, \hat{\Gamma}\, \hat{n}(\underset{\sim}{r}) = n(\underset{\sim}{r})\}. \tag{60}$$

$\hat{K}$ is the kinetic energy, and $\hat{V}_{ee}$ is the electron-electron interaction. The ensemble search of Eq. (60) is truly immense, and the minimizing ensemble is the equilibrium ensemble for an electronic system with equilibrium density $n(\underset{\sim}{r})$ (assuming that such a system exists). In the zero-temperature limit, the search can be restricted[16] to ensembles of J-electron pure states (if N = J) or to ensembles of (J-1)- and J-electron pure states (if J-1 < N < J).

By extension of (53), a density functional can also be found for the free energy:

$$F_v[n] = G[n] + \int d^3r\, v(\underset{\sim}{r})\, n(\underset{\sim}{r}), \tag{61}$$

$$\delta F_v/\delta n(\underset{\sim}{r}) = \mu\,. \tag{62}$$

By Eq. (46), the equilibrium density $n_{eq}(\tilde{r})$ has the same symmetry as the external potential $v(\tilde{r})$: No symmetry-breaking can occur in the infinite-time or equilibrium-ensemble averages. In particular, if there is no external magnetic field, the equilibrium spin densities[28,29] $n_\uparrow(\tilde{r})$ and $n_\downarrow(\tilde{r})$ must be equal. For N = J and $T \to 0$, the free energy $\tilde{F}(N,T)$ tends to the ground-state energy E(J), and the equilibrium density tends to the equal-weight average of all the degenerate ground-state energies.

C. Kohn-Sham-Mermin Theory

In order to reduce the many-electron problem to self-consistent-field form, break the universal functional G[n] of Eq. (60) into pieces:

$$G[n] = G_0[n] + U[n] + F_{xc}[n], \tag{63}$$

where

$$G_0[n] = \min \{\mathrm{Tr}\ \hat{\Gamma}\ (\hat{K} + k_B T\ \ell n\ \hat{\Gamma}) \Big|\ \mathrm{Tr}\ \hat{\Gamma}\ \hat{n}(\tilde{r}) = n(\tilde{r})\}. \tag{64}$$

U[n] is defined by Eq. (5). Eq. (63) defines $F_{xc}[n]$, the exchange-correlation free energy. In Eq. (64), the minimizing ensemble is the equilibrium ensemble for a system of <u>non-interacting</u> electrons with equilibrium density n(r). In the zero-temperature limit, $G_0[n]$ of Eq. (64) reduces to a definition of the "non-interacting kinetic energy functional" proposed[7] some time ago to justify the use of fractional occupation numbers for orbitals with $\varepsilon_i = \mu$.

The Euler equation (62) becomes

$$\delta G_0/\delta n(\tilde{r}) + v_{eff}([n];\tilde{r}) = \mu, \tag{65}$$

where

$$v_{eff}([n];\tilde{r}) = v(\tilde{r}) + u([n];\tilde{r}) + \delta F_{xc}/\delta n(\tilde{r}) \ . \tag{66}$$

u([n];r) is defined by Eq. (9). Eq. (65) is just the Euler equation for a fictional system of <u>non-interacting</u> electrons moving in an external potential $v_{eff}([n];\tilde{r})$ at temperature T and chemical potential μ.

The solution of Eq. (65) for the equilibrium density is well-known: Just solve the Kohn-Sham equations (6) and (7) self-consistently, using the fermion occupation numbers

$$f_i^{eq} = [e^{(\varepsilon_i - \mu)/k_B T} + 1]^{-1} . \tag{67}$$

In the $T \to 0$ limit, orbitals with $\varepsilon_i = \mu$ will be fractionally occupied. In particular, the boron atom $[(1s)^2(2s)^2(2p)^1]$ will have $f^{eq} = 1/6$ for each of its six 2p spin orbitals, hence a spherical density. The "central field approximation" for atoms is <u>exact</u> in this theory.

The "non-interacting" part of the free energy is readily evaluated for the equilibrium density:

$$G_0[n_{eq}] = K_0 - TS_0 \tag{68}$$

$$K_0 = \sum_i f_i^{eq} \langle\psi_i|-\tfrac{1}{2}\nabla^2|\psi_i\rangle$$

$$= \sum_i f_i^{eq} \varepsilon_i - \int d^3r \; v_{eff}([n_{eq}];\underset{\sim}{r}) \; n_{eq}(\underset{\sim}{r}) \tag{69}$$

$$S_0 = - k_B \sum_i \{f_i^{eq} \ln f_i^{eq} + (1-f_i^{eq}) \ln (1-f_i^{eq})\}. \tag{70}$$

Finally, there is a finite-temperature generalization of Janak's theorem, Eq. (10). Define

$$\tilde{\Omega} = \sum_i f_i \langle\psi_i|-\tfrac{1}{2}\nabla^2|\psi_i\rangle + k_B T \sum_i \{f_i \ln f_i$$

$$+ (1-f_i) \ln (1-f_i)\} + U[n] + F_{xc}[n]$$

$$+ \int d^3r \; [v(\underset{\sim}{r}) - \mu] \; n(\underset{\sim}{r}), \tag{71}$$

where the $\{\psi_i(\underset{\sim}{r})\}$ and $n(\underset{\sim}{r})$ are obtained from the self-consistent solution of Eqs. (6) and (7). A little algebra reveals that

$$\partial\tilde{\Omega}/\partial f_i = (\varepsilon_i - \mu) + k_B T \ln [f_i/(1-f_i)], \tag{72}$$

where the partial derivative is taken with the other ($j \neq i$) occupation numbers (but not the orbitals) held fixed. Definition (71) and equation (72) are valid for <u>any</u> choice of $\{f_i\}$. However, the equilibrium f_i's of Eq. (67) make $\tilde{\Omega} = \Omega(\mu,T)$ and $\partial\tilde{\Omega}/\partial f_i = 0$.

4. EXCHANGE AND CORRELATION

The Kohn-Sham-Mermin equations of section 3C can be closed

only by an expression for $F_{xc}[n]$, the exchange-correlation free energy. While some exact statements about F_{xc} are possible (sections 4A and B), approximations are ultimately unavoidable (sections 4C and 4D).

A. Coupling Constant Integration and Exchange-Correlation Hole

Here is a generalization to open systems and finite temperatures of an exact expression[30-32] for the ground-state exchange-correlation energy:

For coupling constants λ between 0 and 1, define

$$\hat{H}_\lambda = \hat{K} + \int d^3r\, v_\lambda(\underset{\sim}{r})\, n(\underset{\sim}{r}) + \frac{1}{2}\int d^3r \int d^3r' \frac{\lambda \hat{n}(\underset{\sim}{r})}{|\underset{\sim}{r}-\underset{\sim}{r}'|}[\hat{n}(\underset{\sim}{r}')-\delta(\underset{\sim}{r}-\underset{\sim}{r}')] \tag{73}$$

$$\hat{\Gamma}_\lambda = e^{-(\hat{H}_\lambda-\mu\hat{N})/k_BT} \Big/ \mathrm{Tr}\, e^{-(\hat{H}_\lambda-\mu\hat{N})/k_BT} \tag{74}$$

$$\Omega_\lambda = \mathrm{Tr}\, \hat{\Gamma}_\lambda\, (\hat{H}_\lambda - \mu\hat{N} + k_BT\, \ell n\, \hat{\Gamma}_\lambda). \tag{75}$$

The potential $v_\lambda(\underset{\sim}{r})$ is chosen so that the equilibrium density, for each λ, equals the true one:

$$\mathrm{Tr}\, \hat{\Gamma}_\lambda\, \hat{n}(\underset{\sim}{r}) = n_{eq}(\underset{\sim}{r}) \ . \tag{76}$$

Thus Ω_1 is the grand potential of the real system, while Ω_0 is the grand potential of the Kohn-Sham system of non-interacting electrons.

Now consider a change $\delta\Omega_\lambda$ resulting from an infinitesimal change $\delta\lambda$ of coupling constant. The piece of $\delta\Omega_\lambda$ associated with $\delta\hat{\Gamma}_\lambda$ is zero by (45), so

$$\delta\Omega_\lambda = \mathrm{Tr}\, \hat{\Gamma}_\lambda\, (\delta\hat{H}_\lambda), \tag{77}$$

i.e.,

$$\Omega_1 = \Omega_0 + \int_0^1 d\lambda\, \mathrm{Tr}\, \hat{\Gamma}_\lambda \frac{d\hat{H}_\lambda}{d\lambda} \ . \tag{78}$$

The integration may be performed explicitly, with the result

$$\Omega_1 = \Omega_0 - \int d^3r\, n_{eq}(\underset{\sim}{r})\, v_0(\underset{\sim}{r})$$

$$+ \int_0^1 d\lambda\ \mathrm{Tr}\, \hat{\Gamma}_\lambda\ \tfrac{1}{2}\int d^3r \int d^3r'\ \frac{\hat{n}(\underset{\sim}{r})}{|\underset{\sim}{r}-\underset{\sim}{r}'|}\ [\hat{n}(\underset{\sim}{r}') - \delta(\underset{\sim}{r}-\underset{\sim}{r}')]$$

$$+ \int d^3r\, n_{eq}(\underset{\sim}{r})\, v_1(\underset{\sim}{r})\ . \tag{79}$$

But the expressions of section 3C imply that

$$\Omega_1 = \Omega_0 - \int d^3r\, n_{eq}(\underset{\sim}{r})\, v_0(\underset{\sim}{r}) + \tfrac{1}{2}\int d^3r \int d^3r'\ \frac{n_{eq}(\underset{\sim}{r}) n_{eq}(\underset{\sim}{r}')}{|\underset{\sim}{r}-\underset{\sim}{r}'|}$$

$$+ F_{xc}[n_{eq}] + \int d^3r\, n_{eq}(\underset{\sim}{r})\, v_1(\underset{\sim}{r}). \tag{80}$$

Therefore,

$$F_{xc}[n_{eq}] = \tfrac{1}{2}\int d^3r\, d^3r'\ \frac{1}{|\underset{\sim}{r}-\underset{\sim}{r}'|}\ \{\int_0^1 d\lambda\ \mathrm{Tr}\, \hat{\Gamma}_\lambda$$

$$[\hat{n}(\underset{\sim}{r}) - n_{eq}(\underset{\sim}{r})]\ [\hat{n}(\underset{\sim}{r}') - n_{eq}(\underset{\sim}{r}')] - n_{eq}(\underset{\sim}{r})\ \delta(\underset{\sim}{r}-\underset{\sim}{r}')\}. \tag{81}$$

Equation (81) can be interpreted as the electrostatic interaction between the electron density at each point r and the density $\rho_{xc}(\underset{\sim}{r},\underset{\sim}{r}')$ of the exchange-correlation hole around $\underset{\sim}{r}$:

$$F_{xc}[n_{eq}] = \tfrac{1}{2}\int d^3r \int d^3r'\ \frac{n_{eq}(\underset{\sim}{r})\ \rho_{xc}(\underset{\sim}{r},\underset{\sim}{r}')}{|\underset{\sim}{r}-\underset{\sim}{r}'|}\ . \tag{82}$$

From equations (81) and (82), the hole obeys the sum rule

$$\int d^3r\ \rho_{xc}(\underset{\sim}{r},\underset{\sim}{r}') = -\,1 + \int_0^1 d\lambda\ \mathrm{Tr}\, \hat{\Gamma}_\lambda\ \frac{\hat{n}(\underset{\sim}{r})}{n_{eq}(\underset{\sim}{r})}\ (\hat{N}-N)\ . \tag{83}$$

If the electron number does not fluctuate, the right hand side of (83) reduces to -1, the familiar closed-system result. Thus, in the combined system (open system plus reservoir), the hole around each point r represents a deficit of one electron. However, part of the hole around a point $\underset{\sim}{r}$ in the open system may be located in the far-away reservoir.

Evaluation of (83) is straight-forward in the T → 0 limit.

For N between the integers J-1 and J, the equilibrium ensemble for each λ becomes a mixture of the (J-1)- and J-electron ground states, and (83) becomes

$$\int d^3r' \, \rho_{xc}(\underset{\sim}{r},\underset{\sim}{r}') = -1 + f(1-f)\int_0^1 d\lambda \, \frac{[n_\lambda(J;\underset{\sim}{r})-n_\lambda(J-1;\underset{\sim}{r})]}{n(N;\underset{\sim}{r})} \tag{84}$$

Here $f = N - (J-1)$, and $n_\lambda(M;r)$ is the ground-state density for M electrons, each with charge $-\lambda e$, moving in the potential $v_\lambda(\underset{\sim}{r})$. The second term of (84) vanishes for an infinite system, and also for a finite one with <u>integer</u> N.

B. Long-Range Behavior of the Exchange-Correlation Potential

This section derives and explains the exact results of equations (32) - (34) for the long-range behavior of the zero-temperature exchange-correlation potential $\delta E_{xc}/\delta n(\underset{\sim}{r})$.

In the asymptotic ($r \to \infty$) limit,[21,33-39] the ground-level density of a many-electron system decays exponentially:

$$n^{\frac{1}{2}}(\underset{\sim}{r}) \to r^{\beta} e^{-\alpha r} . \tag{85}$$

For N = J [and hence, by Eq. (26), for $J-1<N\leq J$], it is rigorous[33,38] that

$$\alpha \geq \sqrt{2I(J)} , \tag{86}$$

where I(J) is the ionization potential of Eq. (13). Convincing arguments[21,34-38] can also be made that, for many or perhaps all real systems, (86) is actually an equality:

$$\alpha = \sqrt{2I(J)} \tag{87}$$

$$\beta = (Z-J+1)/\alpha - 1 , \tag{88}$$

where Z is the total nuclear charge of the system.

The fictional Kohn-Sham system of non-interacting electrons has the same ground-level density as the real system, and hence the same asymptotic decay:[7,16,35,37-39]

$$\alpha = \{2\,[v_{eff}([n];\infty) - \varepsilon_J(N)]\}^{\frac{1}{2}} \quad (J-1<N\leq J) . \tag{89}$$

By Eqs. (28) and (35), $-\varepsilon_J(N) = I(J)$. Thus[17,39] $v_{eff}([n];\infty) = 0$ for all N, from which Eqs. (32) and (33) follow. More precisely, as $r \to \infty$,

$$\delta E_{xc}/\delta n(\underset{\sim}{r}) \rightarrow - f/r, \qquad f = N - (J-1) \ . \tag{90}$$

How can Eqs. (32) and (33) be reconciled with Eq. (30)? The answer is Eq. (34), which shows what happens in the special order of limits in which the electron number tends to the integer J from above, before r tends to ∞. In order to understand Eq. (34), let $N = J+\delta$ where δ is a very small positive number. By Eq. (26), the density will be dominated by that of the J-electron system, except at very large r where the density of the (J+1)-electron system will eventually prevail due to its slower spatial decay. Let r_0 be the radius of a sphere outside which the electrostatic potential is essentially constant, and let r_c be the larger radius at which $(1-\delta)n(J;\underset{\sim}{r}) = \delta\, n(J+1;\underset{\sim}{r})$. Then $\delta E_{xc}/\delta n(\underset{\sim}{r})$ will tend to C in the region $r_0 < r < r_c$, and to zero in the region $r > r_c$. If $\delta \rightarrow 0$ (and hence $r_c \rightarrow \infty$) before $r \rightarrow \infty$, then r can never exceed r_c and Eq. (34) results.

In order to visualize the Kohn-Sham potential as a function of non-integer N, consider the following model of a "one-dimensional hydrogen atom": The external potential is the delta function $v(x) = -\delta(x)$, so for N = 1 the density is $n(1;x) = e^{-2|x|}$ and the ionization potential is $I(1) = 0.5$ hartree. The electron affinity A(1) is assumed to be 0.083 hartree, and for N = 2 the density is taken to be $n(2;x) = 2be^{-2b|x|}$ where $b^2/2 = A(1)$. For $1 < N < 2$, the Kohn-Sham potential $v_0(x)$ is obtained from the Sternheimer construction

$$v_0(x) = \frac{d^2\psi}{dx^2} \,/\, (2\psi) + \varepsilon \ , \tag{91}$$

where $\psi = \sqrt{n(N;x)/N}$. The density n(N;x) and orbital energy ε are constructed according to Eqs. (26) and (28), respectively.

Model results for the density and Kohn-Sham potential are displayed in Figure 4 for a several values of N between 1 and 2. Note in particular how the Kohn-Sham potential for $x \neq 0$ passes over continuously from $C = I(1) - A(1)$ (for $N = 1+\delta$) to 0 (for $N = 2-\delta$). This model illustrates Eq. (34). In particular, for N equal to 1 plus a small positive δ, the critical radius r_c, outside which the Kohn-Sham potential tends to zero, becomes

$$|x_c| = \frac{1}{2(1-b)} \ln \left[\frac{(1-\delta)}{2b\delta}\right] \ . \tag{92}$$

When $\delta \rightarrow 0$, $|x_c|$ diverges logarithmically.

C. Approximations for Exchange and Correlation

The popular local density approximation[2] (LDA) for the

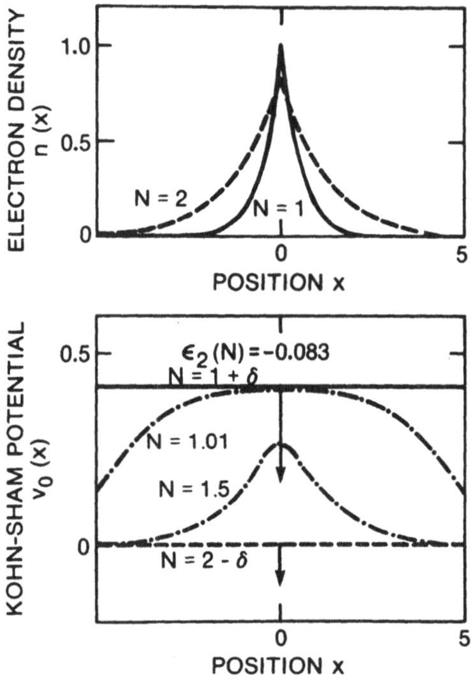

Fig. 4. Electron density and Kohn-Sham potential for a model "one-dimensional hydrogen atom", with electron number N between 1 and 2. Downward arrows at x = 0 indicate delta functions.

exchange-correlation free energy is exact in the limit of slowly-varying $n(\underset{\sim}{r})$:

$$F_{xc}^{LDA}[n] = \int d^3r\; n(\underset{\sim}{r})\; f_{xc}(n(\underset{\sim}{r})). \tag{93}$$

The input $f_{xc}(n)$, the exchange-correlation free energy per particle in an electron gas of uniform density n at temperature T, has been calculated recently in the random phase approximation.[40] The corresponding potential is

$$\delta F_{xc}^{LDA}/\delta n(\underset{\sim}{r}) = \mu_{xc}(n(\underset{\sim}{r})) \;, \tag{94}$$

where $\mu_{xc}(n) = \partial[nf_{xc}(n)]/\partial n$. The quantitative successes of LDA calculations (usually carried out in the zero-temperature limit) have been impressive.[41]

The LDA exchange-correlation hole satisfies[32]

$$\int d^3r' \, \rho_{xc}^{LDA} (\underset{\sim}{r},\underset{\sim}{r}') = -1 \tag{95}$$

for all N and T. Consider a finite open system in the zero-temperature limit: LDA obeys the exact sum rule (84) only for integer electron number N. Consequently LDA gives a good account of the energy of a finite system with integer N, but fails to imitate the exact straight-line behavior of Eq. (25) for N between two integers, as shown in Figure 1. In the asymptotic ($r \to \infty$) limit, the zero-temperature potential $\delta E_{xc}^{LDA}/\delta n(\underset{\sim}{r})$ decays to zero exponentially (like $n^{1/3}$), instead of tending to the proper limit $C/2 - f/r$ [as discussed at the end of section 2B and in Eq. (90)].

For a system with integer electron number, the errors of the LDA total energy and density are sometimes acceptably small. These small errors may be significantly reduced via the generalized gradient expansion of Langreth and Mehl:[42,43]

$$E_{xc}^{LM} [n] = E_{xc}^{LDA} [n] + a \int d^3r \frac{|\nabla n|^2}{n^{4/3}} \left\{ 2e^{-b|\nabla n|/n^{7/6}} - \frac{7}{9} \right\}, \tag{96}$$

where $a = 4.287 \times 10^{-3}$, $b = 0.2618$.

In the local spin density (LSD) approximation,[2,28,29]

$$E_{xc}^{LSD} [n_\uparrow,n_\downarrow] = \int d^3r \, n(\underset{\sim}{r}) \, \varepsilon_{xc}(n_\uparrow(\underset{\sim}{r}),n_\downarrow(\underset{\sim}{r})), \tag{97}$$

$$\delta E_{xc}^{LSD}/\delta n_\sigma(\underset{\sim}{r}) = \mu_{xc}^\sigma \, (n_\uparrow(\underset{\sim}{r}),n_\downarrow(\underset{\sim}{r})), \tag{98}$$

where $\varepsilon_{xc}(n_\uparrow,n_\downarrow)$ is the exchange-correlation energy per particle of an electron gas with uniform spin densities $n_\uparrow$ and $n_\downarrow$, and $\mu_{xc}^\sigma = \partial[n\varepsilon_{xc}(n_\uparrow,n_\downarrow)]/\partial n_\sigma$. Like the LDA (to which it reduces when $n_\uparrow = n_\downarrow = n/2$), LSD has trouble with non-integer electron number and with the long-range behavior of the potential. Moreover, LSD often breaks the symmetry of the exact equilibrium density by spontaneously forming a net spin moment in the absence of an external magnetic field - a sacrifice of logical rigor in favor of physical realism.

The principal error of LSD in localized systems (like atoms) is the self-interaction error, i.e., the spurious interaction of an electron with itself. This error is removed by the self-interaction correction[7] (SIC):

$$E_{xc}^{SIC} = E_{xc}^{LSD} [n_\uparrow,n_\downarrow] - \sum_i \{U[n_i] + E_{xc}^{LSD}[n_i,0]\}, \tag{99}$$

$$v_{xc}^{SIC,i}(\underset{\sim}{r}) = \mu_{xc}^{\sigma_i}(n_\uparrow(\underset{\sim}{r}),n_\downarrow(\underset{\sim}{r})) - u([n_i];\underset{\sim}{r}) - \mu_{xc}^{\uparrow}(n_i(\underset{\sim}{r}),0). \tag{100}$$

In Eqs. (99) and (100), $n_i(r) = f_i|\psi_i(r)|^2$ is the i-th orbital density, while U[n] and $u([\tilde{n}];r)$ are defined by Eqs. (5) and (9). The numerical successes of SIC, especially for atoms, have been reviewed elsewhere.[44] Alternative forms of self-interaction correction are possible.[45] Note that the SIC potential (100) depends on the quantum numbers <u>and</u> occupation number of the orbital on which it acts. The SIC total energy tends to minimize with each orbital fully occupied or fully empty,[7] permitting at most one orbital to carry a fractional occupation f. This behavior of course breaks the symmetry of the external potential.

Although SIC is not properly a density functional approximation, it mimics the exact properties of sections 4A and 4B rather well. The SIC exchange-correlation hole obeys a sum rule[7] which can be expressed as

$$\int d^3r' \, \rho_{xc}^{SIC}(\underset{\sim}{r},\underset{\sim}{r}') = -1 + f(1-f)\,\frac{\left[n_0(J;\underset{\sim}{r})-n_0(J-1;\underset{\sim}{r})\right]}{n(N;\underset{\sim}{r})}, \tag{101}$$

the non-interacting or $\lambda \to 0$ approximation to the exact sum rule (84). Consequently, SIC is much more successful than LDA or LSD in its description of fractional electron number, as shown for the hydrogen atom in Figure 5. This figure was constructed from self-consistent numerical calculations with fractional electron numbers. Note that LDA displays no derivative discontinuity at N = 1, while LSD displays a small discontinuity due to the distinction it makes between the 1s↑ and 1s↓ orbitals. Only SIC mimics the exact behavior of Eqs. (25) and (28).

Janak's theorem, Eq. (10), is also satisfied by SIC. Moreover, the SIC potential (100) tends at large r to $-f_i/r$, much as the exact Kohn-Sham potential does according to Eq. (90).

D. <u>Optimum Local Potential for Hartree-Fock and SIC Approximations</u>

Talman and Shadwick[46] have solved the following problem: Given an expression $E[\{\psi_i\}]$ for the functional dependence of the total energy on the orbitals, find the hamiltonian $-\frac{1}{2}\nabla^2 + V(r)$ whose eigenstates $\psi_i(r)$ minimize $E[\{\psi_i\}]$. They thereby constructed an optimum local potential for the Hartree-Fock atom, which can be interpreted[42,47,48] as the exact Kohn-Sham potential for exchange alone. Recently Norman and Koelling[49] have used the Talman-Shadwick solution to construct an optimum local potential for the

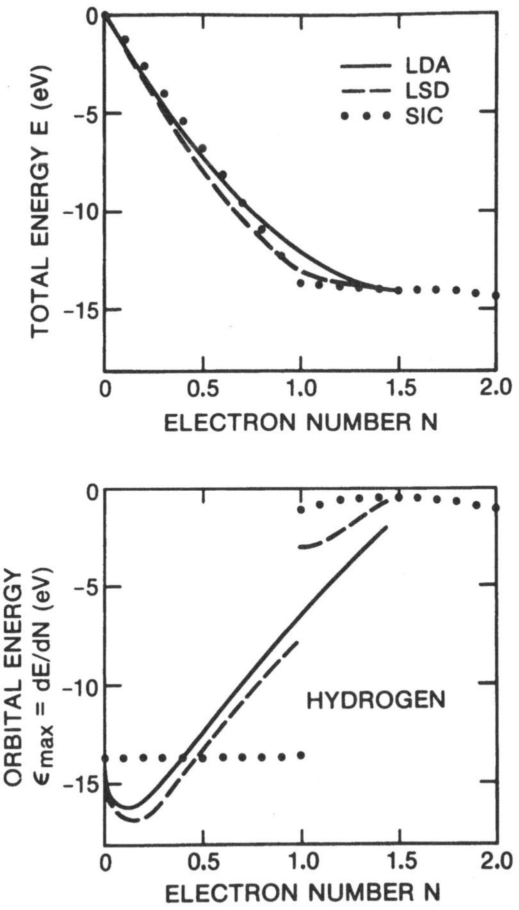

Fig. 5. Total energy, and eigen-energy of the highest partly-occupied orbital, for a real hydrogen atom with electron number N between 0 and 2.

SIC atom, using Eq. (99) for $\tilde{E}[\{\psi_i\}]$.

The basic variational equation is

$$\frac{\delta \tilde{E}}{\delta V(\underset{\sim}{r})} = \sum_i \int d^3r' \, \frac{\delta \tilde{E}}{\delta \psi_i(\underset{\sim}{r}')} \, \frac{\delta \psi_i(\underset{\sim}{r}')}{\delta V(\underset{\sim}{r})} = 0 \,. \tag{102}$$

Let $V_i(r,r')$ be either the nonlocal Hartree-Fock potential $V(r,r')$ or the orbital-dependent SIC potential $V_i(\underset{\sim}{r})\delta(\underset{\sim}{r}'-\underset{\sim}{r})$, constructed from the orbitals $\{\psi_i\}$. In either case,

$$\frac{\delta \tilde{E}}{\delta \psi_i(\underset{\sim}{r}')} = \{-\tfrac{1}{2}\nabla^2\psi_i^*(\underset{\sim}{r}') + \int d^3r'' V_i(\underset{\sim}{r}',\underset{\sim}{r}'')\psi_i^*(\underset{\sim}{r}'')\}f_i$$

$$= \{[\varepsilon_i - V(\underset{\sim}{r}')]\psi_i^*(\underset{\sim}{r}') + \int d^3r'' V_i(\underset{\sim}{r}',\underset{\sim}{r}'')\psi_i^*(\underset{\sim}{r}'')\}f_i . \tag{103}$$

By first-order perturbation theory,

$$\delta\psi_i(\underset{\sim}{r}) = -\int d^3r' \; G_i(\underset{\sim}{r},\underset{\sim}{r}')\psi_i(\underset{\sim}{r}')\delta V(\underset{\sim}{r}'), \tag{104}$$

where $G_i(\underset{\sim}{r},\underset{\sim}{r}')$ is the Green's function for non-interacting electrons in the potential $V(r)$:

$$G_i(r,r') = \sum_{j\neq i} \frac{\psi_j(\underset{\sim}{r})\psi_j^*(\underset{\sim}{r}')}{\varepsilon_j - \varepsilon_i} . \tag{105}$$

Thus

$$\frac{\delta\psi_i(\underset{\sim}{r}')}{\delta V(\underset{\sim}{r})} = -\, G_i(\underset{\sim}{r}',\underset{\sim}{r})\psi_i(\underset{\sim}{r}) . \tag{106}$$

Substitution of (103) and (106) into (102) yields the desired connection between $V(\underset{\sim}{r})$ and $V_i(\underset{\sim}{r},\underset{\sim}{r}')$:

$$\sum_i f_i \int d^3r' \; V(\underset{\sim}{r}') \; G_i(\underset{\sim}{r}',\underset{\sim}{r}) \; \psi_i^*(\underset{\sim}{r}')\psi_i(\underset{\sim}{r})$$

$$= \sum_i f_i \int d^3r' \int d^3r'' \; V_i(\underset{\sim}{r}',\underset{\sim}{r}'') \; G_i(\underset{\sim}{r}',\underset{\sim}{r}) \; \psi_i^*(\underset{\sim}{r}'')\psi_i(\underset{\sim}{r}). \tag{107}$$

Eq. (107) may be inverted to find the optimum local potential $V(\underset{\sim}{r})$.

Sham and Schlüter[20] have obtained a relationship, between the exact Kohn-Sham potential and the Dyson self-energy, which can be regarded as a generalization of Eq. (107) to include correlation.

E. Exact Description of Kinetic and Exchange Energies in the Zero-Temperature Limit

This section will develop an exact expression for the density-functional exchange energy of an open system, and for its derivative discontinuity. The approach will be illustrated first for the simpler kinetic energy.

The zero-temperature kinetic energy functional $K_0[n]$ is defined by Eqs. (64) or (69). Here is a prescription[7] for the functional $K_0[n]$: Given a density $n(\underset{\sim}{r})$, find the potential $v_0(\underset{\sim}{r})$ of a system of non-interacting electrons with ground-level density $n(\underset{\sim}{r})$, and hence the orbitals $\{\psi_i(\underset{\sim}{r})\}$, orbital energies $\{\varepsilon_i\}$, and occupation numbers $\{f_i\}$ of this fictional system. Then construct the kinetic energy via

$$K_0[n] = \sum_i f_i \langle\psi_i|-\tfrac{1}{2}\nabla^2|\psi_i\rangle \ . \qquad (108)$$

Now consider the energy $E_0 = K_0 + \int v_0 n$ of this fictional system, as a function of its time-averaged electron number N. For fixed $v_0(\underset{\sim}{r})$, $E_0(N)$ is a semi-convex linkage of straight line segments with possible derivative discontinuities at integer values of N, by Eq. (25). But $\partial E_0/\partial N = \mu_0 = \delta E_0/\delta n(\underset{\sim}{r}) = \delta K_0/\delta n(\underset{\sim}{r}) + v_0(\underset{\sim}{r})$, and $v_0(\underset{\sim}{r})$ is independent of N by hypothesis. Thus $\delta K_0/\delta n(\underset{\sim}{r})$ jumps by a non-negative constant C_0 as N increases through the integer J:

$$\lim_{\delta\to 0}\left[\left.\frac{\delta K_0}{\delta n(\underset{\sim}{r})}\right|_{J+\delta} - \left.\frac{\delta K_0}{\delta n(\underset{\sim}{r})}\right|_{J-\delta}\right] = C_0 \ . \qquad (109)$$

Consider an infinitesimal variation δN of the electron number near the integer number J, and the resulting variation δK_0 of the kinetic energy. For small positive δ,

$$\left.\frac{\delta K_0}{\delta N}\right|_{J-\delta} = \langle\psi_J|\left\{\left.\frac{\delta K_0}{\delta n}\right|_{J-\delta}\right\}|\psi_J\rangle = \langle\psi_J|-\tfrac{1}{2}\nabla^2|\psi_J\rangle \qquad (110)$$

$$\left.\frac{\delta K_0}{\delta N}\right|_{J+\delta} = \langle\psi_{J+1}|\left\{\left.\frac{\delta K_0}{\delta n}\right|_{J+\delta}\right\}|\psi_{J+1}\rangle = \langle\psi_{J+1}|-\tfrac{1}{2}\nabla^2|\psi_{J+1}\rangle \ . \qquad (111)$$

Since $\delta K_0/\delta n(\underset{\sim}{r})|_{J-\delta} = \mu_0(J-\delta) - v_0(\underset{\sim}{r}) = \varepsilon_J - v_0(\underset{\sim}{r})$, Eq. (110) simply says that

$$\langle\psi_J|-\tfrac{1}{2}\nabla^2 + v_0(\underset{\sim}{r}) - \varepsilon_J|\psi_J\rangle = 0 \ . \qquad (112)$$

Eq. (111), combined with Eq. (109), states that

$$C_0 = \langle\psi_{J+1}|-\tfrac{1}{2}\nabla^2 + v_0(\underset{\sim}{r}) - \varepsilon_J|\psi_{J+1}\rangle = \varepsilon_{J+1} - \varepsilon_J \ . \qquad (113)$$

The derivative discontinuity of the kinetic energy is responsible for all of the "gap" $\varepsilon_{J+1} - \varepsilon_J$ in this fictional system.[19,20] It is also clear from Eqs. (112) and (113) that a non-zero "gap" for non-interacting electrons can arise only when the J-electron system is closed-shell, i.e., only when $\psi_{J+1}(\underset{\sim}{r})$ differs nontrivially from $\psi_J(\underset{\sim}{r})$.

The zero-temperature exchange functional will be defined as

$$E_x[n] = \lim_{\lambda\to 0} \lim_{T\to 0} \left\{ \mathrm{Tr}\, \hat{\Gamma}_\lambda \hat{V}_{ee} \right\} - U[n] \ . \qquad (114)$$

$\hat{V}_{ee}$ is the electron-electron repulsion, and $\hat{\Gamma}_\lambda$ is defined by Eqs. (73) and (74) so that $\mathrm{Tr}\, \hat{\Gamma}_\lambda \hat{n}(\underset{\sim}{r}) = n(\underset{\sim}{r})$. The order of limits in Eq. (114) is deliberate.

Eq. (73) now defines $v_0(\underset{\sim}{r})$ and hence the orbitals $\{\psi_i(r)\}$, orbital energies $\{\varepsilon_i\}$ and occupation numbers $\{f_i\}$ as functionals of n(r); the definitions are the same as those used in the construction of the kinetic energy functional $K_0[n]$. Consider the exchange energy for fixed $v_0(r)$ and electron number N between the integers J-1 and J. Eq. (114) becomes

$$E_x(N) = (J-N)\, E_x(J-1) + (N-J+1)\, E_x(J), \qquad (115)$$

where

$$E_x(M) = -\tfrac{1}{2} \sum_{i=1} \sum_{j=1} \langle \hat{n}_i \hat{n}_j \rangle_M \, \delta_{s_i, s_j} \int d^3r\, d^3r' \frac{\psi_i^*(\underset{\sim}{r})\psi_j(\underset{\sim}{r})\psi_i(\underset{\sim}{r}')\psi_j^*(\underset{\sim}{r}')}{|\underset{\sim}{r}-\underset{\sim}{r}'|} \qquad (116)$$

The orbital occupations for the M-electron ground-level are computed by elementary probability theory[3]: $\langle \hat{n}_i \rangle_M = f_i(M)$ and

$$\langle \hat{n}_i \hat{n}_j \rangle_M = \begin{cases} f_i(M) f_j(M) & \text{unless } \varepsilon_i = \varepsilon_j = \mu_0 \\[2ex] \dfrac{n_M}{g} & \text{if } \varepsilon_i = \varepsilon_j = \mu_0 \text{ and } i = j \\[2ex] \dfrac{n_M}{g}\left(\dfrac{n_M - 1}{g-1}\right) & \text{if } \varepsilon_i = \varepsilon_j = \mu_0 \text{ and } i \neq j, \end{cases} \qquad (117)$$

where g is the number of orbitals with energy $\varepsilon_i = \mu_0$, and n_M is the number of electrons available for distribution over those degenerate orbitals.

It is now straightforward to evaluate the discontinuity

$$\lim_{\delta\to 0} \left[\left.\frac{\delta E_x}{\delta n(\underset{\sim}{r})}\right|_{J+\delta} - \left.\frac{\delta E_x}{\delta n(\underset{\sim}{r})}\right|_{J-\delta} \right] = C_x \ . \qquad (118)$$

To simplify the algebra and results, consider a system with very large (infinite) volume, in which the manifold of orbitals degenerate at the Fermi level becomes a set of measure zero. In this case, Eq. (116) may be used for all N, without invoking Eq. (118) and with $\langle \hat{n}_i \hat{n}_j \rangle = f_i f_j$. Define the non-local exchange potential

$$\Sigma_x(\underset{\sim}{r},\underset{\sim}{r}') = -\tfrac{1}{2} \sum_j f_j \; \psi_j(\underset{\sim}{r})\psi_j^*(\underset{\sim}{r}')/|\underset{\sim}{r}-\underset{\sim}{r}'| \; . \tag{119}$$

Consider a variation δE_x associated with a change δN of electron number, holding $v_0(\underset{\sim}{r})$ fixed. For variations near the integer electron number J,

$$\left.\frac{\delta E_x}{\delta N}\right|_{J-\delta} = \langle\psi_J| \left\{ \left.\frac{\delta E_x}{\delta n}\right|_{J-\delta} \right\} |\psi_J\rangle = \langle\psi_J|\Sigma_x|\psi_J\rangle \tag{120}$$

$$\left.\frac{\delta E_x}{\delta N}\right|_{J+\delta} = \langle\psi_{J+1}| \left\{ \left.\frac{\delta E_x}{\delta n}\right|_{J+\delta} \right\} |\psi_{J+1}\rangle = \langle\psi_{J+1}|\Sigma_x|\psi_{J+1}\rangle \; . \tag{121}$$

Eq. (120) implies that

$$\langle\psi_J| \left\{ \Sigma_x - \left.\frac{\delta E_x}{\delta n}\right|_{J-\delta} \right\} |\psi_J\rangle = 0 \; , \tag{122}$$

and Eq. (121) combined with Eq. (118) yields

$$C_x = \langle\psi_{J+1}| \left\{ \Sigma_x - \left.\frac{\delta E_x}{\delta n}\right|_{J-\delta} \right\} |\psi_{J+1}\rangle \; . \tag{123}$$

Clearly, C_x will be zero in a metal, but non-zero in an insulator or semiconductor where the lowest unoccupied orbital $\psi_{J+1}(\underset{\sim}{r})$ differs nontrivially from the highest occupied one $\psi_J(\underset{\sim}{r})$. Eq. (123) is the same as Sham and Schlüter's[20] expression (14), evaluated to order e^2 where e is the electronic charge.

Finally, it should be observed that, if correlation is neglected, then even this exact description of exchange is somewhat anomalous. For example, Eqs. (25), (28) and (87) do not hold, except for $N \leq 1$ and in the $N \to \infty$ (infinite-volume) limit. Even worse, the exchange-only scheme is not size-consistent: When applied to a single hydrogen atom, it is exact; but when applied to two hydrogen atoms at infinite separation, it gives the familiar total-energy error of the spin-restricted Hartree-Fock approximation.[50] Moreover, as in spin-restricted Hartree-Fock theory, there is an improper dissociation of a heteronuclear diatomic molecule to fractionally-charged fragments.

Table 1. Highest occupied orbital energy $\varepsilon_{max} = \varepsilon_J(J)$, vs. measured ionization potential I and electron affinity A in open-shell atoms. (eV)

Atom	I^a	$-\varepsilon_{max}^{SIC}$	$\frac{1}{2}(I+A)^{a,b}$	$-\varepsilon_{max}^{LDA}$
H	13.6	13.6	7.2	6.4
Li	5.4	5.3	3.0	2.9
B	8.3	8.3	4.3	3.7
C	11.3	11.6	6.3	5.4
O	13.6	14.5	7.5	9.2
F	17.4	18.7	10.4	11.3
K	4.3	4.3	2.4	2.4
Cr	6.8	6.8	3.7	3.1
Cu	7.7	7.7	4.5	4.9
Ag	7.6	7.6	4.4	4.7
Au	9.2	9.2	5.8	6.0

[a]Ref. 52 [b]Ref. 53

5. DETAILED ANSWERS

A. Numerical Study of the Orbital Energies in Atoms

The only Kohn-Sham orbital energy with a precise physical meaning is that of the highest occupied (or partly-occupied) orbital,[16,21,37] as derived in section 2B. Table 1, for open-shell $[\varepsilon_{J+1}(J) = \varepsilon_J(J)]$ neutral atoms, shows that the local density approximation (LDA) of section 4C roughly obeys Eq. (36), as expected of an approximation which neglects derivative discontinuities; the same can be expected of the LM approximation of Eq. (96), which yields orbital energies close[42] to those of LDA. Table 1 also reveals that the SIC approximation of section 4C obeys Eq. (35) rather well. Although the SIC potential is orbital-dependent, its highest occupied orbital energy is essentially the same[49] as that of the optimum local potential for SIC (section 4D), which can be regarded as an estimate of the exact Kohn-Sham potential in an atom. The central-field LDA and SIC calculations were performed self-consistently and semi-relativistically,[51] to facilitate comparison with measured ionization potentials[52] I and electron affinities[53] A.

What can be said of the deeper occupied Kohn-Sham orbital

Table 2. All occupied orbital energies $\varepsilon_{n\ell}$ in the argon atom, vs. measured electron removal energies $\Delta E_{n\ell}$. Two different estimates for the exact Kohn-Sham (KS) orbital energies are shown: those of Langreth and Mehl (LM) and those of Norman and Koelling (NK). (hartrees)

$n\ell$	$-\varepsilon_{n\ell}^{SIC}$	$\Delta E_{n\ell}$ [a]	$-\varepsilon_{n\ell}^{KS}$ [b]	$-\varepsilon_{n\ell}^{KS}$ [c]	$-\varepsilon_{n\ell}^{LDA}+0.20$
1s	118.29	117.31	114.44	115.17	114.00
2s	11.59	11.91	11.22	11.25	10.99
2p	9.41	9.16	8.81	8.87	8.64
3s	1.11	1.07	1.17	1.08	1.08
3p	0.58	0.58	0.66	0.58	0.58

[a]Ref. 54 [b]LM, Ref. 42 [c]NK, Ref. 49

energies ε_i? How well does $-\varepsilon_i$ represent the measured electron removal energy ΔE_i? A first answer was provided by von Barth[21,37] and by Perdew and Norman,[48] who examined the orbital energies of the exact Kohn-Sham potential for exchange alone (sections 4D and E). They found that the deeper Kohn-Sham orbital energies satisfy

$$-\varepsilon_i < \Delta E_i \,, \tag{124}$$

while of course the deeper Hartree-Fock orbital energies (associated with a nonlocal potential) satisfy

$$-\varepsilon_i^{HF} > \Delta E_i \,. \tag{125}$$

The difference between the two sides of the inequality (124) is too large to be made up by correlation effects. From exchange-only studies, Perdew and Norman[48] also concluded that the deeper Kohn-Sham orbital energies are fairly well approximated by the LDA orbital energies, once the latter are all shifted down by the same additive constant.

Table 2 illustrates these statements for the argon atom, with correlation effects now included. This table shows two different estimates for the exact non-relativistic Kohn-Sham orbital energies: The first, due to Langreth and Mehl[42] (LM), is based on the exact or Talman-Shadwick[46] exchange potential (section 4D), plus an approximation for the correlation potential obtained from Eq.

Table 3. **Lowest unoccupied minus highest occupied orbital energy in the neon atom, vs. measured ionization potential I minus electron affinity A. (eV)**

LDA	LM	KS-NK[a]	SIC	I-A[b,c]
13.5	15.7	17.4	22.9	$\geq$ 21.6

[a]Ref. 49 [b]Ref. 52 [c]Ref. 53

(96). The second, due to Norman and Koelling[49] (NK), is based on the optimum local potential for SIC (section 4D). These estimates tend to agree with one another, and with the shifted non-relativistic LDA orbital energies. They do not agree so well with the electron removal energies measured by photoemission.[54] [Small relativistic contributions Δ have been subtracted from the measured 1s (Δ = 0.40 hartree) and 2s (Δ = 0.07 hartree) removal energies.] A third way to construct an accurate Kohn-Sham potential for an atom, using the electron density from a configuration-interaction calculation, has been employed by Almbladh and von Barth[37,39] for the lighter atoms.

Finally, what can be said about the unoccupied Kohn-Sham orbital energies? Special interest attends the difference between the lowest unoccupied and highest occupied orbital energies; this difference in an insulating crystal is just the fundamental gap in the Kohn-Sham band structure. Table 3 displays this difference for the neon atom. The neon atom "gap" in the Kohn-Sham orbital energies (estimated by Norman and Koelling[49] from the optimum local potential for SIC) is 17.4eV, considerably less than the physical quantity I-A $\geq$ 21.6eV = I. The SIC orbital energy difference, by contrast, gives a good estimate of I-A. Note that the Kohn-Sham "gap" of 17.4eV is itself underestimated by the LDA, and to a lesser extent by the Langreth-Mehl[42] (LM) approximation of Eq. (96).

In summary, the exact Kohn-Sham orbital energies differ significantly from physical electron removal energies, with the exception of the one for the least-bound electron. (A consequence is that detailed calculations of the atomic photoabsorption cross section misplace the thresholds.[55]) The correct potential for the calculation of electron removal energies is of course the nonlocal, complex and energy-dependent self energy.[9,56-60] The orbital-dependent SIC potential may be regarded[48,61] as a rough approximation to the real part of the self-energy. Alternatively, accurate electron removal energies from small systems may be obtained, in practice if not always in principle, from ΔSCF calculations[62-68] of changes in relaxed total energies, using density-functional approximations for the energy.

B. Band Gaps in Crystals

The fundamental gap of an insulator or semiconductor is the difference[69] between the ionization potential I and the electron affinity A of the neutral J-electron crystal, i.e., the discontinuity of the chemical potential as the electron number increases through the integer J:

$$I(J) - A(J) = \lim_{\delta\to 0} \left[\left.\frac{\delta E_v}{\delta n(\underset{\sim}{r})}\right|_{J+\delta} - \left.\frac{\delta E_v}{\delta n(\underset{\sim}{r})}\right|_{J-\delta} \right] . \qquad (126)$$

With the energy functional $E_v[n]$ decomposed as in Eq. (4), it is clear that the discontinuity of Eq. (126) can arise only from the kinetic energy $K_0[n]$ [see Eqs. (109) and (113)] and the exchange-correlation energy $E_{xc}[n]$ [see Eqs. (118) and (123)]. Consequently, as asserted in Eqs. (30) and (31), the physical gap of Eq. (126) differs[16,19,20] from the gap $\varepsilon_{J+1}(J) - \varepsilon_J(J)$ of the exact Kohn-Sham band structure by a finite exchange-correlation correction C.

This difference may come as no surprise to many-body theorists, since $-I(J)$ and $-A(J)$ are eigenvalues of a one-electron Schrödinger equation like Eqs. (6) and (8), but with a nonlocal, energy-dependent self-energy[9,56] Σ_{xc} in place of $\delta E_{xc}/\delta n$. If the difference $\Sigma_{xc} - \delta E_{xc}/\delta n$ is regarded as a perturbation on the Kohn-Sham equation (6), the correction C to the exact Kohn-Sham gap may be found by perturbation theory. To first order,

$$\begin{aligned} C &= [I(J) - A(J)] - [\varepsilon_{J+1}(J-\delta) - \varepsilon_J(J-\delta)] \\ &= \langle\psi_{J+1}| \left\{ \Sigma_{xc} - \left.\frac{\delta E_{xc}}{\delta n}\right|_{J-\delta} \right\} |\psi_{J+1}\rangle - \langle\psi_J| \left\{ \Sigma_{xc} - \left.\frac{\delta E_{xc}}{\delta n}\right|_{J-\delta} \right\} |\psi_J\rangle . \end{aligned} \qquad (127)$$

To leading order in e^2, Eq. (127) reduces to the exchange-only derivative discontinuity of Eqs. (122) and (123).

Eq. (127) is not quite exact, because there are terms of higher order in $\Sigma_{xc} - \delta E_{xc}/\delta n$, which have been formally summed by Sham and Schlüter.[20] These higher-order terms could be very small. To leading order in e^2, they correspond to the difference between the Hartree-Fock orbitals and the exact Kohn-Sham exchange-only orbitals, a difference which is almost negligible in atomic calculations.[46]

Barring unlikely accidents, the correction C of Eq. (127) is evidently non-zero in any crystal with a non-zero Kohn-Sham gap $\varepsilon_{J+1}(J) - \varepsilon_J(J)$, since the Kohn-Sham orbital $\psi_{J+1}(r)$ then differs nontrivially from $\psi_J(r)$. This conclusion is manifestly valid to lowest order in e^2 (since the exchange potential Σ_x is the same nonlocal operator for every orbital), and the conclusion can hardly be changed by terms of higher order in e^2. If the Kohn-Sham gap is zero, then the kinetic and exchange energies display no derivative discontinuity, by Eqs. (113), (122) and (123). However, the correlation energy can still manifest a derivative discontinuity, and does so[16] in a Mott insulator,[69,70] for which C of Eq. (127) is presumably non-zero due to a discontinuous energy-dependence of Σ_{xc}. (Note, however, that a non-zero Kohn-Sham gap for a Mott-insulator arises in LSD calculations,[71] which break the spin symmetry of the ground state.)

While the existence of a finite correction to the exact Kohn-Sham gap is unsurprising from the viewpoint of traditional many-body theory,[9,56] it was by no means obvious from the viewpoint of the exact density functional theory[1,2] until the discovery[16] of the derivative discontinuity of $E_{xc}[n]$. As a difference of ground-state energies for the (J-1)-, J-and (J+1)- electron systems, the physical gap certainly falls within the reach of the exact density functional theory. Thus Williams and von Barth[21,41] argued that the physical gap in a semiconductor or insulator would be predicted exactly by the gap in the exact Kohn-Sham band structure. Their argument was essentially the one preceding Eqs. (30) and (31) of this manuscript, but they assumed that $\delta E_{xc}/\delta n(\underset{\sim}{r})$ was continuous (i.e., C = 0).

The Williams-von Barth conclusion[21,41] is presumably still valid within approximations like LDA which display no derivative discontinuity of $E_{xc}[n]$. More precisely, the gap in the LDA band-structure for a semiconductor should equal the LDA total energy difference E(J-1) - 2E(J) + E(J+1). It is instructive to consider why even this is not true for a wide-gap insulator like Li F. The LDA, LSD and SIC approximations like to break symmetries.[62,72,73] In particular, when an electron is removed from the top of the valence (F^-) band in Li F, the hole in LDA localizes[63] around one of the lattice sites. The resulting density change is not infinitesimal, so the arguments preceding Eqs. (30) and (31) do not go through. However, this sort of symmetry-breaking is less likely to occur for the more extended orbitals at the top of the valence band in a semiconductor.

How important is the band-gap correction C of Eq. (127) in

Table 4. Fundamental gaps in the insulator band structure. (eV)

	Ne[a]	NaCℓ[a]	Ar[b]	LiCℓ[b]
LDA	11.5	5.6	7.9	5.8
LM	12.7	6.0	-	-
SIC	20.2	9.2	13.9	10.9
experiment	21.4	9.0	14.2	9.9

[a]From Ref. 61 [b]From Ref. 76.

real insulators and semiconductors? Tentative answers may be found by comparing the Kohn-Sham band gaps, calculated in some approximation such as LDA, to experiment. It is often found that the LDA gaps are 40 or 50% too narrow[7,8,74,75] in comparison with the measured gaps.

Table 4 illustrates the LDA underestimation of the fundamental gaps in the insulators Ne, Na Cℓ, Ar and Li Cℓ. The LM approximation of Eq. (96) improves the gaps slightly in the insulators Ne and Na Cℓ[61] (Table 4), and in the semiconductor silicon.[77] Table 3 shows that the LM approximation underestimates the exact Kohn-Sham "gap" in atomic neon by about 1.7eV, so the exact Kohn-Sham gap in crystalline neon is probably about 14.4eV, still considerably less than the measured gap of 21.4eV.

The SIC potential, as a result of orbital-dependence not present in the Kohn-Sham potential, gives a good account of the gaps in wide-gap insulators.[7,61,76] Table 4 shows the SIC gaps calculated in two different ways: (1) Heaton, Harrison and Lin[76] transform back and forth between delocalized Bloch and localized Wannier orbitals. Their SIC orbital energies reduce exactly to those of Eq. (100) in the separated-atom limit. (2) Norman and Perdew[61] use a simplified self-interaction correction[48] which can be applied directly in the Bloch representation. Their SIC orbital energies reduce only approximately to those of Eq. (100) in the separated-atom limit.

As discussed at the end of section 5A, the orbital-dependent SIC potential may be regarded as a rough approximation to the real part of the self-energy. More direct approximations to the self-energy Σ_{xc} have also produced an accurate gap in diamond,[59] and a significant improvement over LDA in silicon.[8,78] The next funda-

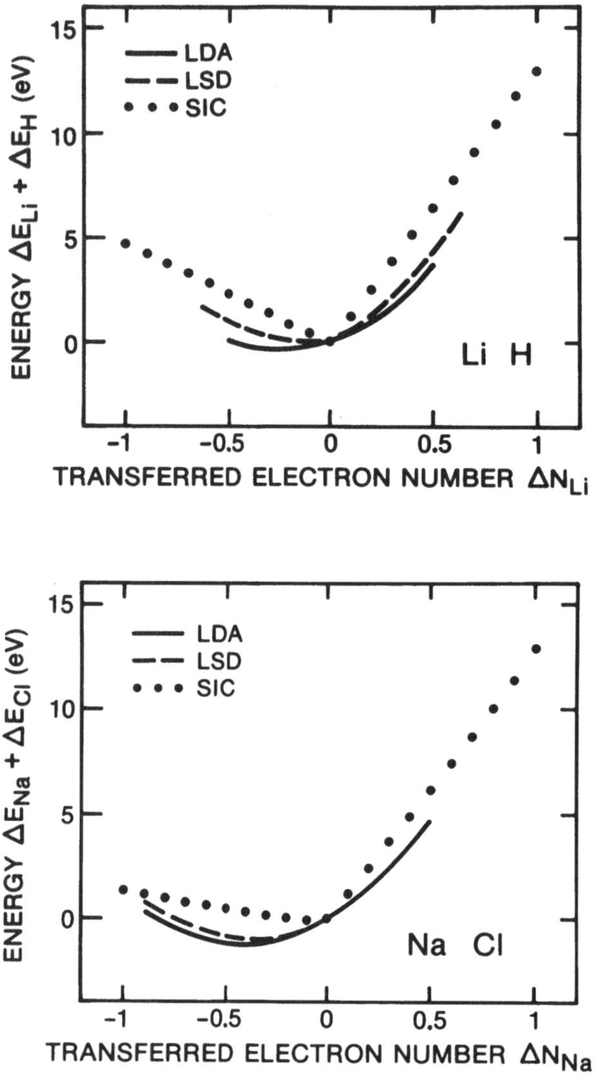

Fig. 6. Total energy of a separated diatomic system $\alpha\beta$, as a function of the number of electrons transferred from atom β to atom α.

mental advance in band theory may be the a priori construction of simple but accurate approximations for the self-energy.

C. The Separated-Atom Limit

The exact dissociation limit was discussed in section 2C. Figure 1 of that section shows how an open-shell atom dissociates from a metallic substrate, with a proper integer charge in the exact theory but with an improper fractional charge in the LSD approximation. The LSD energies were calculated self-consistently, with fractional occupation numbers, for the atom and for the substrate separately, and then added to form the LSD total energy. It may be worthwhile to explain here why[22] fractional occupation numbers can be used for this problem: In LSD the density around a chemisorbed atom will arise partly from localized orbitals (core levels) and partly from resonances[4,79,80] of the extended substrate orbitals. As the atom is moved away from the substrate, the resonances narrow down to sharp "atomic levels", and the density around the atom approaches a limit which can be thought of most simply as if it arose entirely from localized orbitals, some of which might be fractionally occupied.

Figure 6 shows how several approximations describe the total energy of a separated diatomic system $\alpha\beta$, as a function of the number of electrons transferred from atom β to atom α. The total energies were obtained by adding the results of self-consistent calculations, with fractional occupation numbers, for atom α and atom β separately. In the systems chosen, Li H and Na $C\ell$, the LDA displays no derivative discontinuity, while the LSD approximation displays a small one. The LSD energy minimizes properly at the neutral-atoms configuration for Li H, but improperly at the fractional-charge configuration $Na^{+0.4}$ $C\ell^{-0.4}$ for Na $C\ell$. It was in fact Slater's observation,[3] that diatomic molecules dissociate to fractional charge in the Xα approximation, which gave the original impetus to the investigations of this manuscript. Of the three approximations considered here, only SIC mimics the exact straight-lines behavior of section 2C.

Consider the LSD description of separated Na $C\ell$ at the energy-minimizing configuration $Na^{+0.4}$ $C\ell^{-0.4}$. Here the derivative of the total energy, with respect to the number of electrons transferred to the Na atom, is zero. By Janak's theorem (10), this implies equalization of the orbital energies of the Na 3s↑ level (occupied by 0.6 electrons) and the $C\ell$ 3p↑ level (occupied by 2.4 electrons). In the LSD description of the combined Na $C\ell$ system at infinite separation, these degenerate levels mix to form a common ↑ "molecular level" occupied by 3 electrons.

In the LSD description of separated Na $C\ell$, the chemical potentials of Na and $C\ell$ equalize via the transfer of fractional charge. In the exact Kohn-Sham-Mermin description of separated Na $C\ell$, the chemical potentials equalize with no transfer of charge

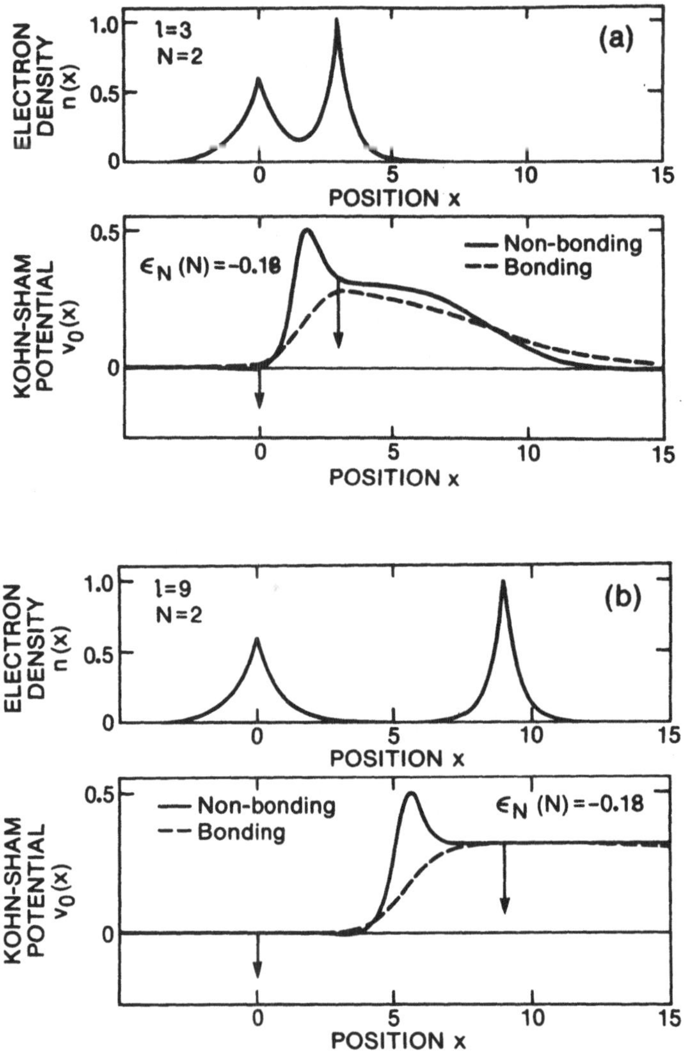

Fig. 7. Electron density and Kohn-Sham potential for a model of two "one-dimensional one-electron atoms" at separation ℓ. a) ℓ = 3 bohr, b) ℓ = 9 bohr.

(i.e., the Na and the Cℓ each remain neutral). The fact which permits this is precisely the indeterminacy of the Kohn-Sham-Mermin orbital energies for integer electron number J, i.e.,

$$- I(J) \leq \varepsilon_J(J) \leq \varepsilon_{J+1}(J) \leq - A(J) \ . \tag{128}$$

Equalization of the exact Kohn-Sham-Mermin orbital energies $\varepsilon_{3s}(Na)$ and $\varepsilon_{3p}(C\ell)$ is accomplished via the formation around the neutral $C\ell$ atom of a region of positive, "constant" exchange-correlation potential.[81]

These remarks are best illustrated by a simple model, in the spirit of the model used in Figure 4. Consider two "one-dimensional one-electron atoms" at finite separation ℓ, as shown in Figure 7. The "external potential" is $v(x) = -a\delta(x) - \delta(x-\ell)$ where $a = 0.6$ hartree, and there are two electrons, one bound to each site. The total density n(x) for the system has been constructed in two different schemes:

$$\text{non-bonding } n(x) = ae^{-2a|x|} + e^{-2|x-\ell|} \tag{129}$$

$$\text{bonding } n(x) = c \left[\sqrt{a}\ e^{-a|x|} + e^{-|x-\ell|}\right]^2 , \tag{130}$$

where c normalizes n(x) to 2 electrons. The Kohn-Sham-Mermin molecular orbital $\psi(x)$ is constructed as $[n(x)/2]^{\frac{1}{2}}$, and the potential $v_0(x)$ is then found from the Sternheimer construction of Eq. (91), with $\varepsilon = - a^2/2$.

Figure 7 shows that, although the Kohn-Sham-Mermin potential $v_0(x)$ does tend to zero as $x \to \pm \infty$, there is a region of positive, "constant" exchange-correlation potential which develops around the "atom" with the tighter density distribution. The resulting barrier between the two "atoms" keeps electron density from flowing toward the one with the tighter density distribution. This barrier, which appears in the region of maximal overlap between the two "atomic" densities, has a shape and width independent of the separation ℓ in the limit $\ell \to \infty$. Clearly, approximations will be hard-pressed to reproduce this behavior.

6. ACKNOWLEDGMENTS

Many people have contributed to the development of this work, including Jose Balduz, Mel Levy, Mike Norman and Bob Parr. Special thanks to Dave Langreth and Walter Kohn for constant encouragement. Much of the work was completed during a sabbatical visit to the Institute for Theoretical Physics, Santa Barbara, California, and presented at the NATO Advanced Studies Institute "Density Functional Methods in Physics," Alcabideche, Portugal, September 1983; thanks to the organizers and participants of both programs. Thanks also to Ann McKay for flawless typing on short notice.

This work was supported in part by the National Science Foundation under grants DMR 80-16117 and PHY 77-27084 (supplemented by funds from NASA).

REFERENCES

1. P. Hohenberg and W. Kohn, Inhomogeneous Electron Gas, Phys. Rev. 136: B864 (1964).
2. W. Kohn and L.J. Sham, Self-Consistent Equations Including Exchange and Correlation, Phys. Rev. 140: A1133 (1965).
3. J.C. Slater, "The Self-Consistent Field for Molecules and Solids", McGraw-Hill, N.Y. (1974).
4. N.D. Lang, Density Functional Approach to the Electronic Structure of Metal Surfaces and Metal-Adsorbate Systems, in "Theory of the Inhomogeneous Electron Gas", S. Lundqvist and N.H. March, eds., Plenum, N.Y. (1983); see also this volume.
5. J.P. Perdew, D.C. Langreth and V. Sahni, Corrections to the Local Density Approximation: Gradient Expansion vs. Wave-Vector Analysis for the Metallic Surface Problem, Phys. Rev. Lett. 38: 1030 (1977).
6. D.D. Koelling, The Band Model for d- and f- Metals, lecture notes for the Gent Summer School 1982, Plenum.
7. J.P. Perdew and A. Zunger, Self-interaction Correction to Density-Functional Approximations for Many-Electron Systems, Phys. Rev. B23: 5048 (1981).
8. C.S. Wang and W.E. Pickett, Density Functional Theory of Excitation Spectra of Semiconductors and Application to Silicon, Phys. Rev. Lett. 51: 597 (1983).
9. L.J. Sham and W. Kohn, One-Particle Properties of an Inhomogeneous Interacting Electron Gas, Phys. Rev. 145: 561 (1966).
10. J.F. Janak, Proof that $\partial E/\partial n_i = \varepsilon_i$ in Density-Functional Theory, Phys. Rev. B18: 7165 (1978).
11. M. Levy, Universal Variational Functionals of Electron Densities, First-Order Density Matrices, and Natural Spin Orbitals and Solution of the v-Representability Problem, Proc. Nat. Acad. Sci. USA 76: 6062 (1979); see also this volume.
12. M. Levy, Electron Densities in Search of Hamiltonians, Phys. Rev. A 26: 1200 (1982).
13. E.H. Lieb, Density Functionals for Coulomb Systems, Int. J. Quantum Chem. 24: 243 (1983); see also this volume.
14. W. Kohn, v-Representability and Density Functional Theory, Phys. Rev. Lett. 51: 1596 (1983); see also this volume.
15. N.D. Mermin, Thermal Properties of the Inhomogeneous Electron Gas, Phys. Rev. 137: A1441 (1965).
16. J.P. Perdew, R.G. Parr, M. Levy and J.L. Balduz, Density Functional Theory for Fractional Particle Number: Derivative Discontinuities of the Energy, Phys. Rev. Lett. 49: 1691

(1982). As L.J. Sham has pointed out, the first half of the last sentence of the abstract should be qualified with the phrase "for open-shell systems."
17. J.P. Perdew and M. Levy, Density Functional Theory for Open Systems, in "Many-Body Phenomena at Surfaces", D.C. Langreth and H. Suhl, eds., Academic.
18. R.G. Parr and L.J. Bartolotti, Some Remarks on the Density Functional Theory of Few Electron Systems, J. Phys. Chem. 87: 2810 (1983).
19. J.P. Perdew and M. Levy, Physical Content of the Exact Kohn-Sham Orbital Energies: Band Gaps and Derivative Discontinuities, Phys. Rev. Lett. 51: 1884(1983)
20. L.J. Sham and M. Schlüter, Density Functional Theory of the Energy Gap, Phys. Rev. Lett. 51: 1888(1983)
21. U. von Barth, Density Functional Theory for Solids, lecture notes for the Gent Summer School 1982, Plenum; see also this volume.
22. J.P. Perdew and J.R. Smith, Can Desorption be Described by the Local Density Formalism?, unpublished manuscript (1983).
23. J. Eichler and T.S. Ho, Collective Effects in Atomic Collisions I. Constrained Hartree-Fock Approach for Diatomic Systems, Z. Phys. A 311: 19 (1983).
24. R.G. Parr, R.A. Donnelly, M. Levy and W.E. Palke, Electronegativity: The Density Functional Viewpoint, J. Chem. Phys. 68: 3801 (1978).
25. E.P. Gyftopoulos and G.N. Hatsopoulos, Quantum-Thermodynamic Definition of Electronegativity, Proc. Nat. Acad. Sci. USA 60: 786 (1968).
26. R.G. Parr and R.G. Pearson, Absolute Hardness: Companion Parameter to Absolute Electronegativity, J.A.C.S. (to appear 1983); see also this volume.
27. S.M. Valone, A One-to-One Mapping between 1-Particle Densities and Some N-Particle Ensembles, J. Chem. Phys. 73: 4653 (1980).
28. U. von Barth and L. Hedin, A Local Exchange-Correlation Potential for the Spin-Polarized Case, J. Phys. C 5: 1629 (1972).
29. A.K. Rajagopal and J. Callaway, Inhomogeneous Electron Gas, Phys. Rev. B 7: 1912 (1973).
30. J. Harris and R.O. Jones, The Surface Energy of a Bounded Electron Gas, J. Phys. F 4: 1170 (1974).
31. D.C. Langreth and J.P. Perdew, The Exchange-Correlation Energy of a Metallic Surface, Solid State Commun. 17: 1425 (1975); Exchange-Correlation Energy of a Metallic Surface: Wave-Vector Analysis, Phys. Rev. B 15: 2884 (1977).
32. O. Gunnarsson and B.I. Lundqvist, Exchange and Correlation in Atoms, Molecules and Solids by the Spin-Density Functional Formalism, Phys. Rev. B 13: 4274 (1976).
33. M.M. Morrell, R.G. Parr and M. Levy, Calculation of Ioniza-

tion Potentials from Density Matrices and Natural Functions, and the Long-Range Behavior of Natural Orbitals and Electron Density, J. Chem. Phys. 62: 549 (1975).

34. M. Levy, On Long-Range Behavior and Ionization Potentials, technical report, U. of North Carolina (1975).
35. D.W. Smith, S. Jagannathan and G.S. Handler, Density Functional Theory of Atomic Structure. I. Exchange and Correlation Potentials for Two-Electron Atoms, Int. J. Quantum Chem. S13: 103 (1979).
36. J. Katriel and E.R. Davidson, Asymptotic Behavior of Atomic and Molecular Wave Functions, Proc. Nat. Acad. Sci. USA 77: 4403 (1980).
37. C.-O. Almbladh and U. von Barth, Exact Results for the Charge and Spin Densities, Exchange-Correlation Potentials, and Density-Functional Eigenvalues, unpublished manuscript (1983); see also this volume.
38. M. Levy, J.P. Perdew and V. Sahni, Exact Differential Equation for the Density of a Many-Particle System, unpublished manuscript (1983); see also this volume.
39. C.-O. Almbladh and A.C. Pedroza, Density-Functional Exchange-Correlation Potentials and Orbital Eigenvalues for Light Atoms, unpublished manuscript (1983).
40. U. Gupta and A.K. Rajagopal, Exchange-Correlation Potential for Inhomogeneous Electron Systems at Finite Temperatures, Phys. Rev. A 22: 2792 (1980); see also this volume.
41. A.R. Williams and U. von Barth, Applications of Density Functional Theory to Atoms, Molecules and Solids, in "Theory of the Inhomogeneous Electron Gas", S. Lundqvist and N.H. March, eds., Plenum, N.Y. (1983).
42. D.C. Langreth and M.J. Mehl, Beyond the Local Density Approximation in Calculations of Ground-State Electronic Properties, Phys. Rev. B 28: 1809 (1983).
43. D.C. Langreth and J.P. Perdew, Theory of Nonuniform Electronic Systems. I. Analysis of the Gradient Approximation and a Generalization that Works, Phys. Rev. B 21: 5469 (1980).
44. J.P. Perdew, Self-Interaction Correction, in "Local Densities in Quantum Chemistry and Solid State Physics", J. Avery and J.P. Dahl, eds., Academic.
45. H. Stoll and A. Savin, Density Functionals for Correlation Energies of Atoms and Molecules, this volume.
46. J.D. Talman and W.F. Shadwick, Optimized Effective Atomic Central Potential, Phys. Rev. A 14: 36 (1976).
47. V. Sahni, J. Gruenebaum and J.P. Perdew, Study of the Density-Gradient Expansion for the Exchange Energy, Phys. Rev. B 26: 4371 (1982).
48. J.P. Perdew and M.R. Norman, Electron Removal Energies in Kohn-Sham Density-Functional Theory, Phys. Rev. B 26: 5445 (1982).

49. M.R. Norman and D.D. Koelling, A Kohn-Sham Functional for Atoms, unpublished manuscript (1983).
50. H.F. King and R.E. Stanton, Multiple Solutions to the Hartree-Fock Problem. II. Molecular Wavefunctions in the Limit of Infinite Internuclear Separation, J. Chem. Phys. 50: 3789 (1969).
51. L.A. Cole and J.P. Perdew, Calculated Electron Affinities of the Elements, Phys. Rev. A 25: 1265 (1982).
52. C.E. Moore, Ionization Potentials and Ionization Limits Derived from the Analyses of Optical Spectra, Nat. Stand. Ref. Data Ser., Nat. Bur. Stand. (U.S.) 34: 1 (1970).
53. H. Hotop and W.C. Lineberger, Binding Energies in Atomic Negative Ions, J. Phys. Chem. Ref. Data 4: 539 (1975).
54. K.D. Sevier, Atomic Electron Binding Energies, At. Data Nucl. Data Tables 24: 323 (1979).
55. K. Nuroh, M.J. Stott and E. Zaremba, Calculation of the 4d Subshell Photoabsorption Spectra of Ba, Ba^{+} and Ba^{++}, Phys. Rev. Lett. 49: 862 (1982).
56. L. Hedin and S. Lundqvist, Effects of Electron-Electron and Electron-Phonon Interactions on the One-Electron States of Solids, Solid State Physics 23: 1 (1969).
57. M. Rasolt and S.H. Vosko, Investigations of Nonlocal Exchange and Correlation Effects in Metals via the Density-Functional Formalism, Phys. Rev. B 10: 4195 (1974).
58. A.H. MacDonald, M.W.C. Dharma-wardana and D.J.W. Geldart, Density Functional Approximation for the Quasiparticle Properties of Simple Metals: I. Theory and Electron Gas Calculations, J. Phys. F 10: 1719 (1980).
59. G. Strinati, H.J. Mattausch and W. Hanke, Dynamical Aspects of Correlation Corrections, Phys. Rev. B 25: 2867 (1982).
60. F.Sacchetti, Electron-Electron Interaction and Single-Particle Properties in Copper, J. Phys. F 12: 281 (1982).
61. M.R. Norman and J.P. Perdew, Simplified Self-Interaction Correction Applied to the Energy Bands of Neon and Sodium Chloride, Phys. Rev. B 28: 2135 (1983).
62. O. Gunnarsson, B.I. Lundqvist and J.W. Wilkins, Contribution to the Cohesive Energy of Simple Metals: Spin-Dependent Effect, Phys. Rev. B 10: 1319 (1974).
63. A. Zunger and A.J. Freeman, Ground- and Excited-State Properties of Li F in the Local-Density Approximation, Phys. Rev. B 16: 2901 (1977).
64. N.D. Lang and A.R. Williams, Core Holes in Chemisorbed Atoms, Phys. Rev. B 16: 2408 (1977).
65. J. Harris and R.O. Jones, Density Functional Theory of 3d-Transition Element Atoms, J. Chem. Phys. 68: 3316 (1978).
66. O. Gunnarsson and R.O. Jones, Density Functional Calculations for Atoms, Molecules and Clusters, Phys. Scripta 21: 394 (1980).
67. L. Wilk and S.H. Vosko, Estimates of Non-Local Corrections

to Total, Ionization, and Single-Particle Energies, J. Phys. C 15: 2139 (1982).
68. J.G. Harrison, Density Functional Calculations for Atoms in the First Transition Series, J. Chem. Phys. 78: 4562 (1983).
69. N.F. Mott, "Metal-Insulator Transitions", Barnes and Noble, N.Y. (1974).
70. B.H. Brandow, Electronic Structure of Mott Insulators, Adv. in Phys. 26: 651 (1977).
71. L.M. Sander, H.B. Shore and J.H. Rose, Self-Consistent Band Structure Theory of the Metal-Insulator Transition, Phys. Rev. B 24: 4879 (1981).
72. A. Zunger, One-Electron Broken-Symmetry Approach to the Core-Hole Spectra of Semiconductors, Phys. Rev. Lett. 50: 1215 (1983).
73. D.M. Bylander and L. Kleinman, Broken Symmetry in the Local Spin Density Approximation and its Application to Relativistic Atoms, Phys. Rev. Lett. 51: 889 (1983).
74. S.B. Trickey, A.K. Ray and J.P. Worth, Adequacy of Local-Exchange Excitation Hamiltonians in Insulators, Phys. Stat. Sol. (b) 106: 613 (1981).
75. A.E. Carlsson and J.W. Wilkins, Band-Overlap Metallization of Ba S, Ba Se and Ba Te, unpublished manuscript (1983).
76. R.A. Heaton, J.G. Harrison and C.C. Lin, Self-Interaction Correction for Density-Functional Theory of Electronic Energy Bands of Solids, Phys. Rev. B 28: 5992(1983)
77. U. von Barth and R. Car, unpublished calculation referred to in Ref. 21.
78. W. Hanke, Exchange-Correlation Potential for One-Electron Excitations in a Semiconductor, unpublished manuscript (1983).
79. J.R. Smith, ed., "Theory of Chemisorption", Springer-Verlag, Berlin (1980).
80. N.D. Lang and A.R. Williams, Theory of Atomic Chemisorption on Simple Metals, Phys. Rev. B 18: 616 (1978).
81. M. Levy and J.P. Perdew, The Exact Exchange-Correlation Potential at Infinite Atomic Separation, Bull. Am. Phys. Soc. 28: 266 (1983).

Note added in proof: For recent work on the Kohn-Sham gap in Ga As, see F. Manghi, G. Riegler, C.M. Bertoni, C. Calandra and G.B. Bachelet, Nonlocal Exchange and Correlation and Semiconductor Band Structure, Phys. Rev. B 28: 6157(1983).

HADRONIC DENSITY OF STATES

Rajat K. Bhaduri

Physics Department, McMaster University
Hamilton, Ontario
Canada L8S 4M1

INTRODUCTION

I shall talk on hadronic density of states and its implications in elementary particle physics and nuclear physics. These lectures will have very little to do with the density functional formalism, but I hope that they would be of sufficient interest to hold your attention.

What is hadronic density of states? First, to introduce the subject, let me remind you of nuclear density of states in a nucleus, with which we are more familiar. Energy can be imparted to a nucleus by inelastic scattering by, for example, neutrons, electrons etc, and the transitions to excited states depend on the density of final states available to the system. From such reactions, one can deduce experimentally the number of states $\rho(E)dE$ that are available to the nucleus between excitation energy of E and E+dE. It is found that $\rho(E)$ grows very rapidly with the excitation energy, and reasonable fits can be obtained with a form like[1]

$$\rho(E) \sim E^{-2}\exp(2\sqrt{aE}) \tag{1}$$

where the constant a is related to the density of the single particle states at the fermi energy, and may be taken to be $a \approx A/8\ \mathrm{MeV}^{-1}$, where A is the nucleon number. This is shown in Figs. 1 and 2. Note that $\rho(E)$ in Eq (1) refers to a single nucleus of mass number A.

By hadrons we refer to strongly interacting mesons and baryons. The lightest hadron is the pion, which comes in three charge states. Then there are the other pseudoscalar and vector mesons like η, K, ρ, K*, ω, ϕ etc, and their excited states. The lightest baryon is

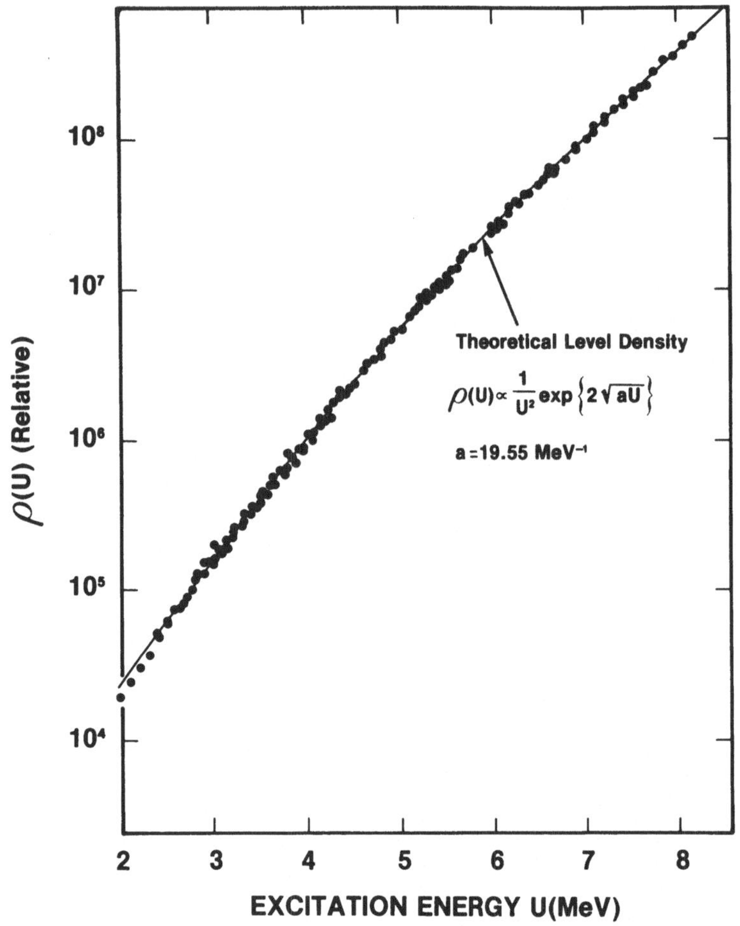

Fig. 1 Nuclear density ρ(U), deduced from inelastic neutron scattering on Ag. The data are taken from K. Tsukada, S. Tanaka, M. Maruyama and Y. Tomita, Nucl. Phys. 78, 369 (1966).

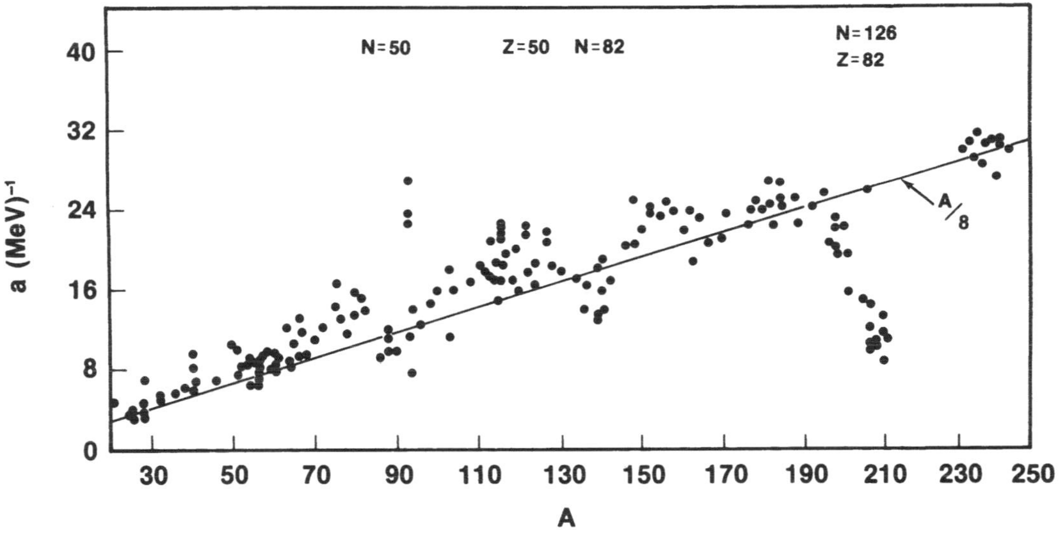

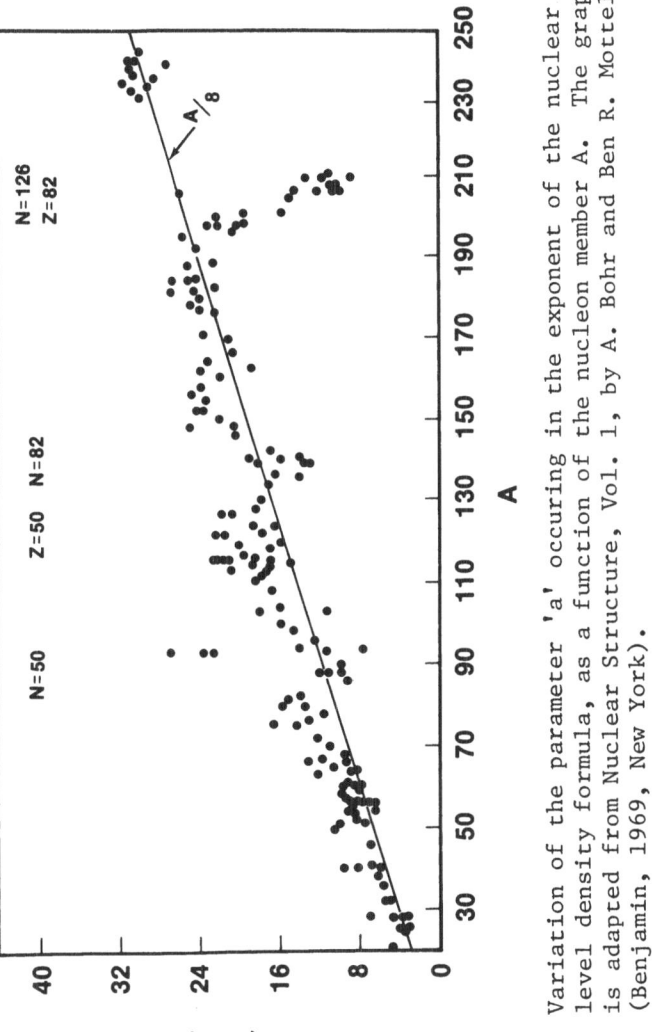

Fig. 2 Variation of the parameter 'a' occuring in the exponent of the nuclear level density formula, as a function of the nucleon member A. The graph is adapted from Nuclear Structure, Vol. 1, by A. Bohr and Ben R. Mottelson (Benjamin, 1969, New York).

the proton. Resonances in the nucleon and delta may be studied by analysing the 'bumps' in the pion-nucleon scattering cross sections. The resonances may also be produced by photo-excitation, $\gamma+N\to N^*$, which subsequently decay by strong interaction $N^*\to N+\pi$, or $N^*\to\Delta+\pi\to N+\pi+\pi$. These resonances have typically widths of 150-400 MeV, and are assigned quantum numbers like spin, i-spin, parity etc. Each such resonance will be considered to be a hadronic state with multiplicity $(2J+1)(2I+1)$, where J is the spin and I the i-spin quantum number. Starting from $E \sim 140$ MeV, where there are three states, the density of states increases rapidly with energy, as shown in Fig. 3. We see from this figure that beyond $E \sim 2.5$ GeV, the density starts falling off. This is only because at high excitations the resonances are very many and with large widths, so that there is too much overlapping, and it is very difficult to assign quantum numbers and count them individually. Actually, the hadronic states seem to grow even faster with energy than the nuclear case. Theoretically, one expects[2] that for large excitation energy E, the hadronic density of states $\rho(E)$ should go like

$$\rho(E) \sim E^a \exp(bE) \tag{2}$$

where a is a negative number like -3, and b is positive. This expectation is not contradicted by the data in the limited energy range, but we cannot expect to confirm this simply by counting new resonances with increasing energy. Indication from the limited data is that b^{-1} is in the range 160-230 MeV.

We first ask the question: why is the hadronic density of states of any interest? Can one, from experimental data, determine if it really grows exponentially with energy? We believe that hadrons may be described in the quark-model, with mesons as $q\bar{q}$ pairs and baryons as qqq. These are the active or "valance" quarks; in addition there may be virtual $q\bar{q}$ pairs, and the interaction between the quarks is mediated by gluons. With increasing excitation energy, the role of these may become very important. All these coloured quarks and gluons are confined in each hadron which itself is a colour singlet, or colour neutral. This is in analogy with atoms which contain charged protons and electrons, but are electrically neutral. If you take such a gas of atoms and heat it to a high enough temperature, the electrons would dissociate more and more from the atomic nuclei with increasing temperature. At very high temperature, you would end up with a plasma of bare nuclei and an electron gas. If, however, you consider a gas of hadrons, and pump in more and more energy, the quarks do not dissociate from the hadrons like electrons do, only more hadrons are created. It is believed[3], though, that at a temperature in the range $\tau \sim 150$-200 MeV, there is a phase transition from the hadronic phase to a quark-gluon plasma. In this phase of deconfinement, the quarks and gluons are not confined in individual colourless hadronic clusters. We may ask the question whether the hadronic density of states has anything to do with the mechanism of

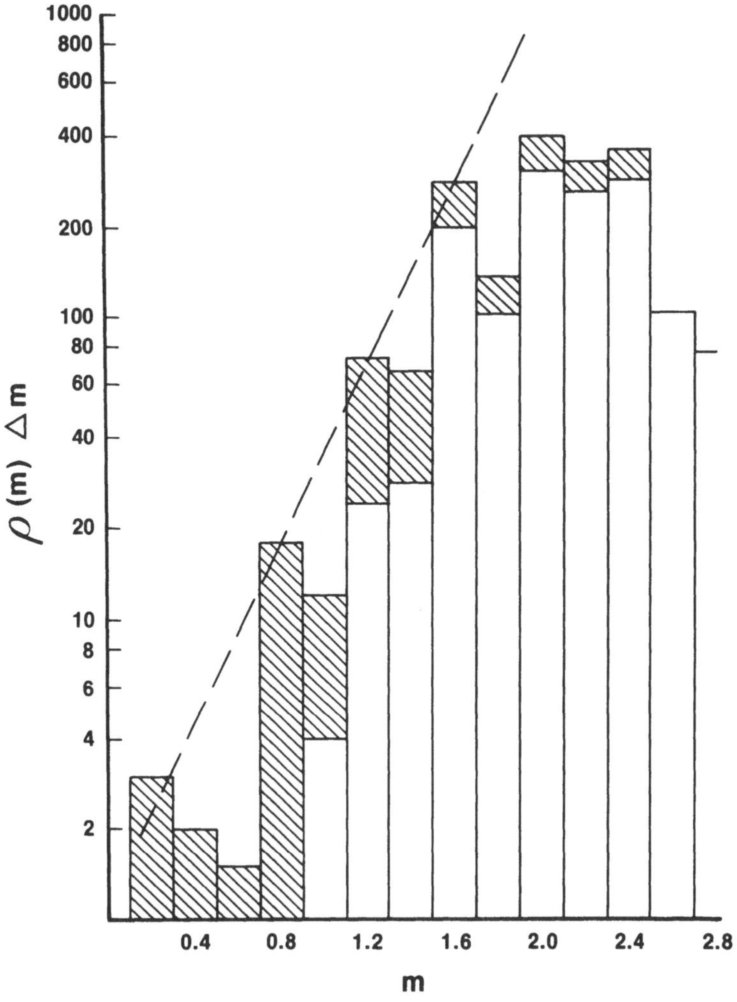

Fig. 3 The hadronic density of states $\rho(m)$. Plotted is the number of hadronic states detected experimentally (as reported in Particle Data Table, Phys. Lett. 111B, 1982) between m and $m+\Delta m$, with $\Delta m = 200$ MeV. The shaded area denotes the mesonic and the unshaded the baryonic states.

confinement, and if it can give any clue about the phase transition.

HEAVY-ION COLLISIONS

In the laboratory, high hadronic temperatures may be attained in high energy heavy-ion collisions, using machines like the Berkeley Bevelac. Experiments have been done with beams of protons, α-particles, C, Ne and Ar at energies in the range 0.4 GeV/A to 2.1 GeV/A, incident on suitable targets, and angular distribution of light fragments and their energies measured. These include p, $\pi^{\pm}$, d, ^{3}He and ^{3}H and ^{4}He. Large amounts of data exist[4], especially for inclusive cross sections (like Ne+NaF $\rightarrow$ p+anything). The data is generally analysed either in terms of the thermal (or fireball) model[5], or the N-N collision (or cascade) model[6]. The first model, in its simplest form, assumes that there is thermalization of the available energy (after possible particle production like $\pi^{\pm}$, K, etc) in a reaction region, which can be treated as a hot hadron gas at temperature T. This picture would be valid if the mean free path $\lambda << R$, where R is a measure of the size of the fireball. If, on the other hand, $\lambda >> R$, then single NN collision model[7], which is the other extreme, would suffice. Probably the thermal model description is not bad, because in a linear cascade model calculation with row-on-row geometry[8], it was found that thermalization of a nucleon takes place for more than two collisions. We shall display some data on the inclusive cross-section of protons and pions at $\theta_{CM} = 90°$, and interpret them in the naive thermal model. Since the particles are detected between momenta $\underset{\sim}{p}$ and $\underset{\sim}{p}+d\underset{\sim}{p}$, the results can be expressed in terms of the cross-section

$$\frac{d^3\sigma}{dp_x dp_y dp_z} .$$

It is more convenient to express this in terms of the Lorentz-invariant form

$$\frac{E}{p^2}\frac{d^2\sigma}{dp d\Omega} = \sigma_I .$$

In Fig 4, we display σ_I for the inclusive production of protons and pions at $E_{lab} = 0.8$ GeV/A as a function of the kinetic energy E_k^* in the N-N centre-of-mass frame. These particles are detected at $\theta_{CM} = 90°$. It will be seen that except for the low energy shoulder, the data may be fitted by

$$\sigma_I \propto \exp(-E_k^*/E_0) , \qquad (3)$$

where $E_0 \sim 70$-80 MeV for protons and 60-70 MeV for pions. In the thermal model, E_0 may be regarded as the effective temperature of the gas. In Fig 5, we show some proton data that demonstrate that this effective temperature keeps increasing with increasing beam energy,

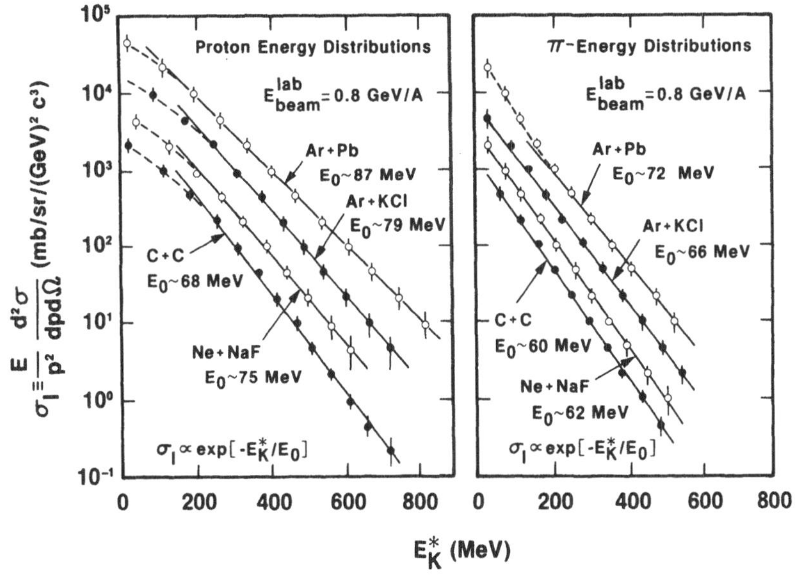

Fig. 4 Inclusive cross sections for production of protons and pions as a function of the kinetic energy (in C.M. frame) in heavy-ion collisions. The beam energy is 0.8 GeV/nucleon, and the particles are detected at $\theta_{CM} = 90^o$. The data are taken from Ref. 4.

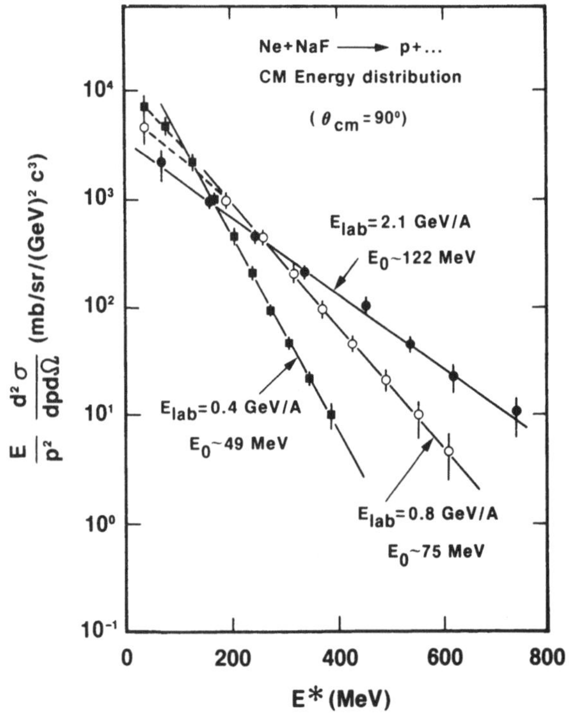

Fig. 5 Variation of the effective temperature E_0 with increasing beam energy in fits of inclusive proton production cross-section. The data is taken from Ref. 4.

reaching $E_0 \sim 122$ MeV for $E_{lab} = 2.1$ GeV/A. One could wonder whether it is possible to demonstrate the approach to a limiting hadronic temperature from these experiments by using higher and higher beam energy. In thermal models, one always finds that the equilibrium temperature for pions is less than that of protons. It is thought that pions, having a smaller mean free path, remain trapped in the reaction region more than the protons, leaving the reaction region at a later stage. The initial hot fireball expands and cools, and so the detected pions show a lower temperature[9]. By contrast, photons, leptons or even K^+ escape from the fireball earlier than the protons, and should sample a higher temperature. This is perhaps only part of the story, other explanations like radial collective flow of matter in the fireball may also contribute to the differences in temperature of the different species.

Indirectly, one could hope to learn more about the hadronic density of states by studying the decay products from the fireball. In its hot early stage, the composition should depend sensitively on $\rho(E)$, a situation analogous to the expanding hot universe. If $\rho(E)$ increased very rapidly with E, then, with increasing energy, it is more efficient to increase entropy by creating heavy hadrons than by channeling the energy into thermal motion. The outgoing detected particles at a later stage, will of course, also depend on the dynamical evolution and the reactions that take place from its inception, but may throw some light on the form of $\rho(E)$. Some preliminary studies on this aspect have been made by Glendenning and Karant[10].

PHASE TRANSITION

Very qualitatively, let me discuss why one may suspect that there may be, at a high enough temperature, a phase transition from the hadronic phase to a quark-gluon plasma. Consider a 'gas' of hadrons in a large "box", like nucleons in a large nucleus. If a large amount of energy is pumped into this box, more hadrons are created, the number of which depend on the assumed hadronic density of states. Each hadron has a size, and in most models it is believed that the size of a hadron grows with excitation energy. If the density of hadrons is high, then with their growing size there should be more and more overlapping of these hadrons. Initially, the quarks and gluons were confined in each individual hadron, but with more and more overlapping of the hadrons, the individual identity of the hadrons are lost, and the quarks and gluons form a plasma in the box. Although this is a picturesque description, it is not clear here why there should be a sudden transition from a hadronic to plasma phase from this kind of description. But it is clear that the extent of overlapping of the hadrons should depend sensitively on the hadronic density of states and on the rate of growth of their size. Of course all this presumes that experimentally a box-like region can be maintained for some length of time during which the hadronic density grows.

One can take another scenario to estimate, very crudely, the temperature at which a phase transition from the hadronic phase may take place. The nucleon and the delta and (the non-strange mesons) are made of u and d quarks which, in quantum chromodynamics, are believed to be nearly massless. These 'current' quarks are confined in a hadron, but because of their confinement, they acquire an effective mass, which is called the constituent mass. It is found that the relativistic motion of the current quarks in a hadron may approximately be mocked up by a nonrelativistic description of the constituent quarks[11], at least for low-energy spectroscopy. Let us stretch this point of view, and assume that as long as the quarks are confined in a hadron, we can describe their motion nonrelativistically with a constituent mass. Consider a hot hadron at a high temperature in which these point-like constituent quarks and $q\bar{q}$ pairs form a gas. Let us also assume that the temperature and density are such that we may assume it to be a perfect classical gas, with pressure $P = \rho\tau$, where ρ is the number density of quarks plus antiquarks per unit volume, and τ the temperature in units of the Boltzmann constant. The mass density ρ_m is then $m_q\rho$, where m_q is the constituent mass of a quark (or antiquark). The velocity of sound u is then given by

$$u^2 = \left(\frac{\partial P}{\partial \rho_m}\right)_s = c^2\left(\frac{\gamma\tau}{m_q c^2}\right) , \qquad (4)$$

where we make the standard assumption that the density compressions are adiabatic. The ratio of specific heats, $c_P/c_V = \gamma = 5/3$ for point nonrelativistic particles. We see that the velocity of sound u will approach the velocity of light c when $5/3\ \tau = m_q c^2$, and if we take the usual value of ~ 300 MeV for $m_q c^2$, $\tau = 180$ MeV. At this temperature, the concept of the constituent mass m_q should break down. Since this concept is related to confinement, maybe deconfinement of quarks takes place at this temperature with the accompanying phase transition. Our arguments here have been very qualitative, and should not be taken too seriously. Lattice QCD calculations[12] do show, however, that deconfinement may be taking place around this temperature.

DERIVATION OF THE HADRONIC DENSITY OF STATES

We had mentioned earlier that one expects $\rho(E)$ to grow exponentially with energy (or mass) of the hadron. Let us see why. Consider a system of volume V with eigen modes E_i's. Then the canonical partition function of the system is

$$Z = \sum_i e^{-\beta E_i} = \int_0^\infty \rho(E) e^{-\beta E} dE \qquad (5)$$

where

$$\rho(E) = \sum_i \delta(E-E_i)$$

is the density of states. Thus $Z(\beta)$ is the Laplace transform of $\rho(E)$. Inverting the equation, we get

$$\rho(E) = \frac{1}{2\pi i}\int_{-i\infty}^{+i\infty} Z(\beta)e^{\beta E}d\beta = \frac{1}{2\pi i}\int_{c-i\infty}^{c+i\infty} e^{S(\beta)}d\beta \ , \qquad (6)$$

where

$$S(\beta) = \ell n Z(\beta) + \beta E \ . \qquad (7)$$

Note that the integration in Eq (6) is along the imaginary axis, and for complex β the function $S(\beta)$ is complex. This results in the integrand being oscillatory, with most of the contribution coming from a point β_0 on the real axis at which $S(\beta)$ is stationary. For a given E, we see that β_0 can be found from Eq (7)

$$\left.\frac{\partial S}{\partial \beta}\right|_{\beta_0} = 0 \quad , \quad \text{so } E = -\left.\frac{\partial}{\partial \beta}\ell n Z(\beta)\right|_{\beta_0} \ . \qquad (8)$$

This is just the thermodynamic definition of the average energy for an equilibrium temperature β_0^{-1}. Expanding $S(\beta)$ about β_0, we get

$$S(\beta) = S(\beta_0) + \frac{1}{2!}(\beta-\beta_0)^2 \left.\frac{\partial^2}{\partial \beta^2}\ell n Z(\beta)\right|_{\beta_0} + \ldots\ldots \ . \qquad (9)$$

Substituting this in Eq (6), we get

$$\rho(E) = \frac{e^{S(\beta_0)}}{2\pi i}\int_{c-i\infty}^{c+i\infty} e^{1/2(\beta-\beta_0)^2\alpha^2} d\beta \ ,$$

where we have put

$$\left.\frac{\partial^2}{\partial \beta^2}\ell n Z(\beta)\right|_{\beta_0} = \alpha^2 \ .$$

The integral can now be done by putting $\beta-\beta_0 = ix$, where x is real. Thus, the intercept c on the real axis is identified with the real saddle point β_0. The result is

$$\rho(E) = \exp S(\beta_0)/(2\pi\alpha^2)^{1/2} \quad . \tag{10}$$

We may identify $S(\beta_0)$ as the entropy of the system. As a simple example, take the system under consideration to be photons in a cavity ($\varepsilon = \mu = 1$) which is embedded in a perfect conductor. For the system of photons at temperature τ, the entropy is $S \sim V\tau^3$, and the energy is $E \sim V\tau^4$, the latter being often known as Stefan's law. We then see that $S \sim V^{1/4}E^{3/4}$, so $\rho(E) \propto \exp(\text{const. } V^{1/4}E^{3/4})$. Of course, there is also an energy dependent denominator coming from α in Eq (10), which is a slowly varying function. If, however, the volume of the cavity itself grew linearly with energy, then the density of states of this system would grow as exp(const.E).

Much of the mathematics of the above example can be taken over to the problem of gluons confined in a hadron. To appreciate this, let us discuss the meaning of the dielectric constant ε and the permeability μ in nonrelativistic physics, and then their relation to the nature of the vacuum[13] in QED and QCD. In a polarizable medium, the potential energy $U(r)$ between two test charges q_1, q_2 is given by

$$U(r) = q_1 q_2 / 4\pi\varepsilon r \ , \tag{11}$$

where ε is the dielectric constant and r the distance between the charges. In vacuo, $\varepsilon = 1$. Due to the polarizability of the medium, there is screening of the charges, and $\varepsilon > 1$. Also, if the medium contains magnetic dipoles, then the energy density E of the medium in the presence of an external magnetic field $\underset{\sim}{H}$ is given by

$$E = -\frac{1}{2}\, 4\pi\chi H^2 \ , \tag{12}$$

and since the magnetic dipoles align with the external field and lower the energy, the susceptibility $\chi > 0$. The magnetic permeability μ is defined in terms of χ:

$$\mu = 1 + 4\pi\chi \ , \tag{13}$$

so that for a paramagnetic substance $\mu > 1$. In solid-state physics, it is possible to have both ε and μ to be greater than unity, as in a free electron gas at zero temperature.

In a relativistic field theory, the vacuum in many ways behaves like a polarizable medium, but there Lorentz invariance demands that $\mu\varepsilon = 1$. This means that screening between charges (in QED) can take place due to the creation of virtual electron-positron pairs, and $\varepsilon > 1$, $\mu < 1$. However, when the test charges are very close to each other, this screening breaks down, and $\varepsilon \to 1$. Thus, in QED vacuum, the dielectric constant ε is a function of r, and the interaction strength increases as $r \to 0$. Note that the interaction between the charges is brought about by the exchange of photons which carry spin. However, since the photons have no electric charge, they have no contribution to μ. In QCD, instead of two electrical charges, we have, let us say, two quarks with colour, and these exchange gluons. In contrast to photons, gluons themselves carry colour charge, and of course they have spin like the photons. So they make the QCD vacuum behave like a paramagnetic medium, with $\mu > 1$. Lorentz invariance then demands that $\varepsilon < 1$, so that the QCD vacuum is "antiscreening". With increasing distance between the quarks, $\mu(r)$ becomes greater and greater, and $\varepsilon(r) \to 0$ for r very large. From Eq (11), it follows that the interaction between quarks decreases as $r \to 0$ and $\varepsilon \to 1$, and this is called asymptotic freedom. Moreover, we can make the coloured quarks and gluons confined in a volume V by assuming that inside the volume, $\varepsilon = \mu = 1$, but outside $\varepsilon = 0$ and $\mu = \infty$. This is, of course, in idealization, but it takes care of both asymptotic freedom and confinement in a simple model, which is called the Bag model.

Now we can go back to the example in electrodynamics where photons are confined in a cavity with $\varepsilon = 1$, surrounded by a perfect electric conductor with $\varepsilon = \infty$ (see Fig. 6a). The corresponding situation for gluons in a bag is shown in Fig. 6b, where the bag is surrounded by the QCD vacuum with $\mu = \infty$. Inside the bag, if we take the gluons to be noninteracting as a first approximation, then the colour electric and colour magnetic fields obey just the Maxwell equations. The boundary conditions of the bag problem are the same as the photons in the cavity if $\underset{\sim}{E}$ and $\underset{\sim}{B}$ are interchanged, and the two problems become mathematically identical. Note that inside the bag, because of asymptotic freedom, interactions can be taken care of, hopefully, perturbatively. But the QCD vacuum is nonperturbative.

At this point, let us recall that for photons in a cavity, we had assumed that the energy $E \propto V\tau^4$ and the entropy $S \propto V\tau^3$ to obtain the expression for the density of states. The energy of a photon (or gluon) with momentum $\underset{\sim}{p}$ is $\varepsilon_p = pc = \hbar ck$, where k is the wave number. Note that earlier we had denoted the dielectric constant by ε, but for want of a better symbol, in subsequent discussions ε will denote the photon energy. Moreover, we shall, following the usual practice, set $\hbar = c = 1$. The number of states between $\underset{\sim}{p}$ and $\underset{\sim}{p} + d\underset{\sim}{p}$ that are available to a photon in a large volume V is Vd^3p/h^3 times the spin degeneracy factor of 2, there being no longitudinal photons. For gluons, this should be further multiplied by a factor of 8, since there are eight different types of gluons. Denoting the density of states of a photon between ε and $\varepsilon + d\varepsilon$ by $g(\varepsilon)d\varepsilon$, the above argument yields

$$g(\varepsilon) = \frac{\eta V}{2\pi^2}\,\varepsilon^2 , \qquad (14)$$

where η is the degeneracy factor. The energy of the photons in the

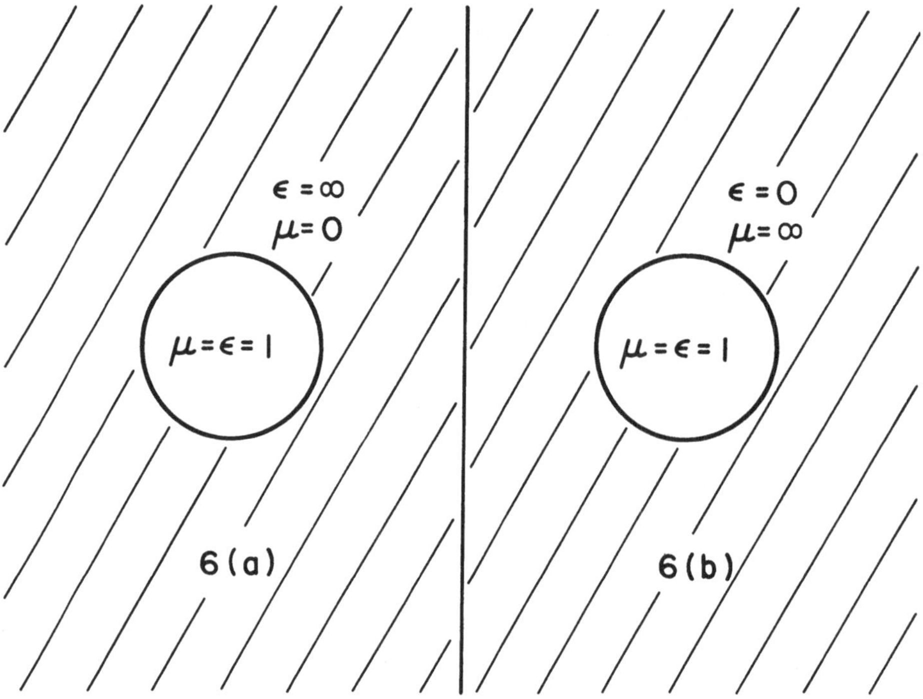

Fig. 6 (a) Protons confined in a cavity surrounded by a perfect electric conductor (shaded) with $\varepsilon = \infty$.
(b) Gluons in a bag, surrounded by a perfect paramagnetic medium. The unshaded area is the hadron, and the shaded area an idealized model of the QCD vacuum.

cavity at temperature β^{-1} is then

$$E = \int_0^\infty \frac{g(\varepsilon)\varepsilon d\varepsilon}{(e^{\beta\varepsilon}-1)} , \tag{15}$$

and the entropy S, to within a constant, is $S = \int dE/\tau$. From Eqs (14) and (15), we immediately obtain

$$E = \frac{\eta V}{2\pi^2} \frac{\pi^4}{15} \tau^4 ,$$

and the entropy

$$S = \frac{\eta V}{2\pi^2} \cdot \frac{4\pi^4}{45} \tau^3 .$$

In the bag model[14], the energy E is not quite the same as the mass m of the hadron. This is because we assume that the perturbative vacuum of the hadronic phase has a higher energy density than the surrounding QCD vacuum by an amount "B" which is a constant. This means that the mass 'm' of a hadron is given by the sum of the energy E of the gluons and quarks and the positive definite term BV:

$$m = E + BV . \tag{16}$$

This extra term also implies that the vacuum is exerting an inward pressure $P = -B$ on the bag, which is balanced by the outward pressure generated by the motion of the quark and gluons in the bag at equilibrium. Since for a zero mass (noninteracting) system $PV = 1/3\ E$, it follows that at equilibrium

$$m = 4BV , \tag{17}$$

so that in this model the volume of the hadron increases linearly with mass. Taking the mass corresponding to the nucleon, and taking a radius $R \sim 1$ fm, we see from the above equation that $B \sim 60$ MeV/fm^3. Using natural units $\hbar = c = 1$, we may write the same numbers as $B^{1/4} \approx 147$ MeV. Actually, estimates of B vary considerably, in the range $B^{1/4} \sim 120$-250 MeV. We had seen earlier that the entropy $S \propto V^{1/4}E^{3/4}$. Using Eq (17), it follows that $S \propto m$. The details are easy to work out, and it can be shown[15,16] that in this case (taking only noninteracting gluons in the bag),

$$\rho(m) = 1.59\ \eta\tau_0^2 m^{-3} \exp\left(\frac{m}{\tau_0}\right) , \tag{18}$$

which is of the same form as Eq (2) given at the beginning. Here τ_0

is the limiting temperature, given in this case by

$$\tau_0 = \left(\frac{90B}{\eta\pi^2}\right)^{1/4} .$$

Since we are only considering gluons here, $\eta = 16$.

At this point, let us go back to Fig. (3), where we have plotted the experimentally detected hadronic states in mass intervals of 200 MeV as a function of the mass. In this low mass range, we do not really expect the continuous asymptotic formula for $\rho(m)$ to hold due to quantum fluctuations. From the Figure, we can see that there seems to be bunching of states at intervals of 0.4 GeV, starting at 0.8 GeV. Beyond m = 2 GeV, so many states are missed that the density starts to fall off. The peaks in the distribution can be neatly joined by a straight line in the semilog plot, indicating an exponential increase in the density $\rho(m)$. This straight line behaviour may be reproduced by assuming

$$\rho(m) = C_0\tau_0^{-1}\exp\left(\frac{m}{\tau_0}\right) , \tag{19}$$

where $C_0 = 1.5$, $\tau_0 = .285$ GeV. On the other hand, the experimental data are also consistent with a form like Eq (18),

$$\rho(m) = C_3\tau_0^2 m^{-3}\exp\left(\frac{m}{\tau_0}\right) , \tag{20}$$

if one takes $C_3 = 20$, $\tau_0 = 0.175$ GeV. Thus we cannot really find τ_0 by fitting the experimentally detected states in a limiting range. Actually, the data in Fig. (3) do not necessarily imply a limiting temperature τ_0, since we can fit the data equally well by a form

$$\rho(m) = 1.25\exp(5m^{3/4})\ \text{GeV}^{-1} \tag{21}$$

where m is in GeV. This is the form one would expect if the volume V of the hadron remained fixed with mass, since the entropy $S \propto V^{1/4}E^{3/4}$. Since the thermodynamic temperature τ is defined through the relation

$$\frac{1}{\tau} = \left(\frac{\partial S}{\partial E}\right) ,$$

it follows that in this case the temperature would continue to rise with increasing energy E. In Eq (21), the preexponential factor should be m-dependent, but we have neglected this for simplicity.

Are there any experimental data which can differentiate between forms of $\rho(m)$ that lead to a limiting temperature and other forms like (21) that do not? Recently, Margolis and his collaborators[17] have examined the production of hadrons in very high energy pp and $p\bar{p}$ collisions. The energy of collision in the CM frame for pp is in the range $\sqrt{s}$ = 19.5 to 63 GeV, and for $p\bar{p}$ at $\sqrt{s}$ = 540 GeV. At a fixed energy, the cross section for hadron production of mass m falls off rapidly with increasing mass. The cross section $\sigma(m,s)$ for pp collision ranges from $\sim$ 100 mb for pion production down to 100 nb for ψ-production and less for heavier particles. Margolis et al attempt to explain the data in a statistical parton model in which a hadron is emitted centrally from a fireball, leaving a residual fireball which continues to emit particles. They claim that the data, specially for large m's, can only be fitted where the density of states in the fireball has no limiting temperature. In fact they get good fits to the data by assuming a form (21) for the density of states, with a coefficient 5.3 in the exponent. Their results depend, however, on a number of other assumptions in the model, and should not be taken as definitive. But their calculations do suggest that the bag model assumption of the hadronic volume increasing linearly with mass m is suspect. This may have interesting implications on the phenomenology of hadronic$\rightarrow$quark-gluon phase transition.

INCLUSION OF QUARK-ANTIQUARK DEGREES OF FREEDOM

Till now, in deriving equations for $\rho(m)$, we have only considered gluons. Even if, for simplicity, we take the baryon number to be zero, the hadron will contain $q\bar{q}$ pairs, which may be created in increasing numbers with increasing energy, just as gluons. Results are easy to derive to include these degrees of freedom. The energy of zero mass fermions is given by

$$E = \int_0^\infty \frac{g(\varepsilon)\varepsilon d\varepsilon}{(e^{\beta\varepsilon}+1)} \tag{22}$$

where $g(\varepsilon)$ is given by Eq (14), with an appropriate degeneracy factor η. If we take the number of flavours to be n_f, then for quarks $\eta = 3\times 2\times n_f$, with a colour factor of 3 and spin factor of 2. This should be further multiplied by a factor of 2, since for every quark there is an antiquark. It is then easy to show that the total energy (of gluons + $q\bar{q}$ pairs) is

$$E = \frac{8\pi^2}{15} V\tau^4(1 + \frac{21}{32} n_f) \tag{23}$$

and the entropy S is

$$S = \frac{32\pi^2}{45} V\tau^3(1 + \frac{21}{32} n_f) \ . \tag{24}$$

Eliminating τ between Eqs (23) and (24), we get

$$S = \frac{4}{3}\left[\frac{8\pi^2}{15}\left(1 + \frac{21}{32} n_f\right)\right]^{1/4} V^{1/4} E^{3/4} . \tag{25}$$

If we take $n_f = 3$, corresponding to the three light mass quark flavours u, d and s, then the numerical coefficient in Eq (25) is known, and we obtain

$$S = 2.65 V^{1/4} E^{3/4} , \tag{26}$$

where the volume V is in $(\text{GeV})^{-3}$. If we assume that the volume V of the hadron is fixed, then the hadronic density of states is of the form

$$\rho(E) \propto \frac{1}{V^{1/2} E^{5/2}} \exp(2.65 V^{1/4} E^{3/4}) . \tag{27}$$

It is worth mentioning that the preexponential factor in the above equation is not what one would obtain from our simple formula (10), but includes the correction factor due to the elimination of the centre-of-mass spurious states[15]. In fact, as Kapusta[20] has demonstrated, the preexponential factor depends rather sensitively on what constraints are applied (e.g. that the hadron should be a colour singlet).

Instead of the fixed volume hadron, if we take the volume of the bag to be proportional to its mass m, using Eq (17), we obtain

$$\rho(m) \propto \tau_0^2 m^{-3} \exp\left(\frac{m}{\tau_0}\right) ,$$

the same form as Eq (20). But now the limiting temperature τ_0 is given by

$$\tau_0 = \left[\frac{45B}{8\pi^2\left(1 + \frac{21}{32} n_f\right)}\right]^{1/4} = 0.66\ B^{1/4} \quad \text{for} \quad n_f = 3 . \tag{28}$$

From τ_0 = 175 MeV, we should then have $B^{1/4}$ = 265 MeV.

THE EQUATION OF STATE

Let us make a very simple model to determine how the pressure varies with temperature for a system of hadrons that are contained in a large volume V. For simplicity, assume that the hadrons are nonrelativistic ($3/2\tau \ll m$) and form a noninteracting classical gas obeying Maxwellian distribution. The pressure $P(m,\tau)$ due to hadrons of mass m at temperature τ is

$$P(m,\tau) = \int \frac{d^3p}{(2\pi)^3} \frac{p^2}{3m} \exp[-(\frac{p^2}{2m} + m)/\tau]$$

or

$$P(m,\tau) = \frac{\tau^{5/2}}{(2\pi)^{3/2}} m^{3/2} e^{-m/\tau} .$$

But the number of hadrons between mass m and m+dm is given by $\rho(m)dm$. So the total pressure $P(\tau)$ at temperature τ is

$$P(\tau) = \int^{\infty} P(m,\tau)\rho(m)dm$$
$$= \frac{\tau^{5/2}}{(2\pi)^{3/2}} \int^{\infty} \rho(m)m^{3/2} e^{-m/\tau} dm . \qquad (29)$$

If we take the form given by Eq (20) for $\rho(m)$,

$$\rho(m) = C_3\tau_0^2 m^{-3} \exp(\frac{m}{\tau_0}) , \qquad (20)$$

then a little algebra shows that

$$\frac{P(\tau)}{\tau_0^4} = \frac{C_3}{(2\pi)^{3/2}} (\frac{\tau}{\tau_0})^{5/2} \int_{x_0}^{\infty} x^{-3/2} e^{-x(\frac{\tau_0}{\tau} - 1)} dx \qquad (30)$$

where the lower limit $x_0 = m_0/\tau_0$, m_0 being the cutoff at the lower end. Actually it is more reasonable to take the discrete contributions of the lowest mass hadrons separately upto some chosen m_0, and then use Eq (30) for the excited states. So

$$\frac{P(\tau)}{\tau_0^4} = \sum_{m_i<m_0} \frac{1}{(2\pi)^{3/2}} (\frac{\tau}{\tau_0})^{5/2} (\frac{m_i}{\tau_0})^{3/2} e^{-m_i/\tau_0}$$
$$+ \text{ the contribution from Eq (30)} . \qquad (31)$$

We plot this as a function of τ/τ_0 in Fig. 7. Note that the pressure reaches a limiting value at $\tau = \tau_0$, rising steeply in the range $0 \leq \tau \leq \tau_0$. We have taken $x_0 = 7$, corresponding to $m_0 = 1232$ MeV for $\tau_0 = 175$ MeV.

The behaviour of $P(\tau)$ is very different, on the other hand, if we make the assumption that the volume of the hadron does not grow with energy, or grows less slowly than m. If we take, for simplicity,

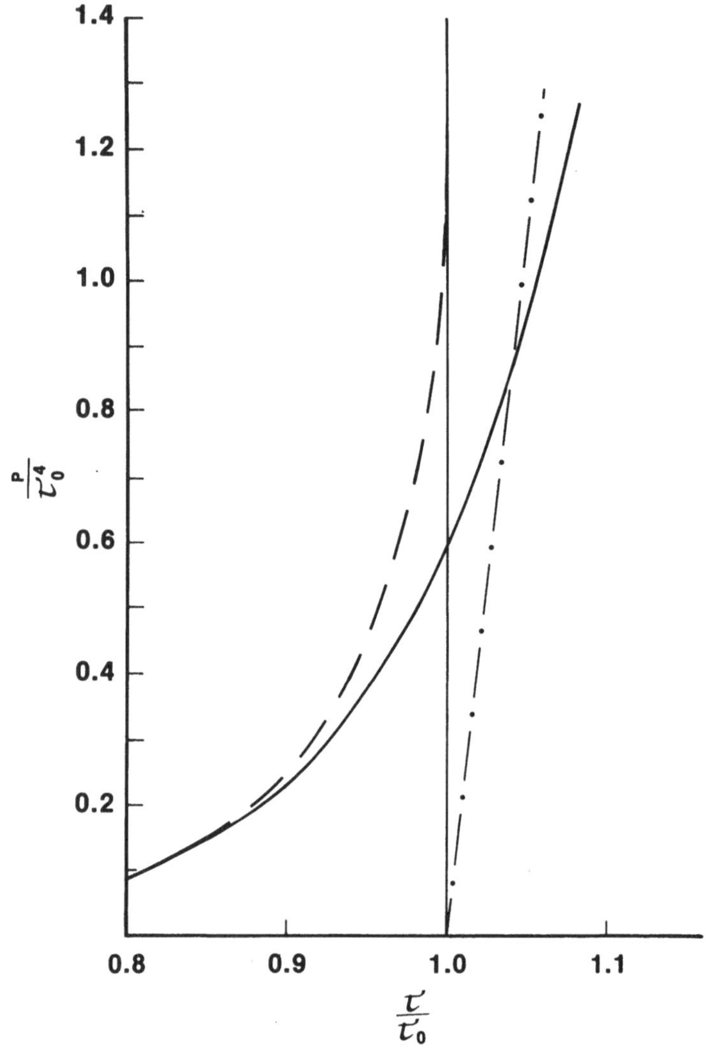

Fig. 7 The pressure P as a function of temperature τ in the hadronic and quark-gluon phase. There is a limiting temperature τ_0 when the hadronic density $\rho(m)$ grows exponentially with m (see Eq (20)). This is shown by the dashed curve. The solid curve shows the variation of P with τ when $\ell n\rho \sim m^{3/4}$, and this intersects the dot-dashed curve of the quark-gluon phase at $\tau/\tau_0 \gtrsim 1.04$. The temperature τ_0, defined by Eq (28), is taken to be 175 MeV.

the simplest empirical form (21),

$$\rho(m) = 1.25 \exp(5m^{3/4}) \text{ GeV}^{-1} ,$$

we can again calculate $P(\tau)$ as before. This is also plotted in Fig. 7, with τ_0 = 175 MeV and m_0 = 1232 MeV. Now τ_0 is just a parameter and does not have the significance of the limiting temperature.

Till now we have calculated the pressure of the system in the hadronic phase as a function of the temperature. If we assume that there is also a quark-gluon phase, then its pressure curve can also be found easily. To be consistent with our earlier approach, we again take it to be noninteracting. Then the pressure of this gas is given by

$$P(\tau) = \frac{8\pi^2}{45} (1 + \frac{21}{32} n_f)\tau^4 - B .$$

Using Eq (28), this may be written as

$$\frac{P(\tau)}{\tau_0^4} = \frac{B}{\tau_0^4} [(\frac{\tau}{\tau_0})^4 - 1] , \tag{32}$$

where, for $n_f = 3$, $B/\tau_0^4 \approx (3/2)^4$ from Eq (28). We also plot this function in Fig. 7 for comparison. Note that in the conventional bag model, the pressure curves in the two phases do not intersect, and this poses a difficulty that has been discussed by Kapusta[15]. If we assume, on the other hand, that there is no limiting hadronic temperature, then there is an intersection of the two phases which may be the transition temperature. From Fig. 7, we see that this takes place at $\tau \approx 1.04\ \tau_0$.

Our model has been very simple in the above discussion. In quark-gluon phase, the effect of interactions could be added perturbatively[19]. In the hadronic phase, a more careful treatment of finite volume effect[16] in the density of states could be done. All these are refinements to the simple model. The question remains open as to whether the logarithm of the hadronic density of states grows lineary with m or less rapidly. This question has not been resolved, but it is quite possible that the hadronic size does not grow as rapidly as the naive bag model prediction, and a phase transition to the quark-gluon plasma takes place around $\tau \sim 200$ MeV.

REFERENCES

1. H. A. Bethe, Rev. Mod. Phys. 9:69 (1937).
   A. Bohr and B. R. Mottelson, "Nuclear Structure", Vol. 1, p. 281, Benjamin, New York (1969).
2. R. Hagedorn, Nuovo Cimento Suppl. 3:147 (1965).
   S. Frautschi, Phys. Rev. D3:2821 (1971).

3. R. Hagedorn and J. Refelski, From Hadron Gas to Quark Matter, I and II, in: "Statistical Mechanics of Quarks and Hadron", J. Satz, ed., North Holland, Amsterdam (1981).
E. V. Shuryak, Phys. Rep. 61:71 (1980).
H. Satz, Phys. Rep. 88:321 (1982).
4. S. Nagamiya, M.-C. Lemaire, E. Moeller, S. Schnetzer, G. Shapiro, H. Steiner and I. Tanihata, Phys. Rev. C24:971 (1981).
5. J. Gosset, H. H. Gutbrod, W. G. Meyer, A. M. Poskanzer, A. Sandoval, R. Stock and G. D. Westfall, Phys. Rev. C16:629 (1977).
S. Das Gupta and A. Z. Mekjian, Phys. Rep. 72:131 (1981).
6. A. A. Amsden, J. N. Ginnochio, F. H. Harlow, J. R. Nix, M. Danos, E. C. Halbert and R. K. Smith, Phys. Rev. Lett. 38:1055 (1977).
7. R. L. Hatch and S. E. Koonin, Phys. Lett. 81B:1 (1978).
8. J. Hüfner and J. Knoll, Nucl. Phys. A290:460 (1977).
9. S. Nagamiya, Phys. Rev. Lett. 49:1383 (1982).
10. N. K. Glendenning and Y. J. Karant, Phys. Rev. C21:1501 (1980).
11. R. K. Bhaduri and M. Brack, Phys. Rev. D25:1443 (1982).
12. H. Satz, Phys. Rep. 88:321 (1982)
13. N. K. Nielsen, Am. J. Phys. 49:1171 (1981).
T. D. Lee, "Particle Physics and Introduction to Field Theory", Chapters 16 & 17, Harwood Academic Publishers, New York (1981).
14. A. Chodos, R. L. Jaffe, K. Johnson and V. F. Weisskopf, Phys. Rev. D9:3471 (1974).
15. J. I. Kapusta, Phys. Rev. D23:2444 (1981).
16. B. K. Jennings and R. K. Bhaduri, Phys. Rev. D26:1750 (1982).
17. Y. Afek, B. Margolis and L. Polvani, Phys. Rev. D25:1833 (1982).
P. L'Heureux and B. Margolis, Phys. Rev. D28:242 (1983).
18. J. I. Kapusta, Nucl. Phys. B196:1 (1982).
19. Carl-G. Källman, Phys. Lett. 126B:366 (1983).

SEMICLASSICAL DESCRIPTION OF NUCLEAR BULK PROPERTIES

Matthias Brack
Institut für Theoretische Physik
Universität Regensburg
D-8400 Regensburg, W. Germany

1. INTRODUCTION

In these lectures we shall discuss the use of density functionals for calculating static nuclear bulk properties such as average binding energies, density distributions and their moments, and deformation energies.

The idea of expressing the total energy of a nucleus as a functional of the local density $\rho(r)$ and to formulate with it a variational principle

$$\frac{\delta}{\delta\rho}\int d^3r\left\{\mathcal{E}[\rho(\vec{r})]-\lambda\rho(\vec{r})\right\}=0 \qquad (1.1)$$

has been used as early in the history of nuclear physics as 50 years ago, namely in the pioneering work which led to the famous semi-empirical Bethe-Weizsäcker mass formula.[1,2] Sophistication of the energy density functional $\mathcal{E}[\rho]$ was developed along with the understanding of the nuclear force[3-5] and led to the so-called energy density formalism.[6,7] The theoretical justification of the variational approach eq. (1.1) came from outside nuclear physics in form of the now well-known theorem by Hohenberg and Kohn.[8]

Whereas the main difficulty of density functional calculations in solid state physics and quantum chemistry lies in the development of sufficiently accurate exchange and correlation energy functionals, their applications in nuclear physics are further strongly handicapped by the fact that the basic nucleon-nucleon interaction is only partially known and, due to its repulsive core, cannot be used directly in a perturbation expansion. We refer to the literature for comprehensive discussions of our present knowledge of the nucleon-nucleon interaction[9] and its use in Brückner theory[10,11] calculations for nuclear matter and finite nuclei.[12,13]

The energy density variational calculations performed until the late sixties, in which mostly the Brückner G-matrix in the local density approximation[11,12] was used, have been reviewed by Lombard.[7] Typically, the experimental binding energies of spherical nuclei could be reproduced to within $\sim$ (1-10) MeV and their radii within $\sim$ (1-4) %. (The shell effects, which cannot be included in this formalism, contribute about $\pm$ (1-15) MeV to the total energy and less than 1 % to the radii.) The density profiles obtained with these calculations were as a rule rather poor. This deficiency can be traced back mainly to the use of an insufficient kinetic energy density functional: mostly, the Thomas-Fermi relation $\tau = \kappa\rho^{5/3}$ was used, sometimes a gradient correction with adjustable coefficients was added. The corresponding large errors in the kinetic energies were partially made up by readjustments of the nuclear force parameters, but this could not help to improve the resulting density profiles.

Several recent developments which took place over the last 10-15 years allow to reassess now the energy density formalism in a much more rigorous and quantitative way.[15,16] The developments are the following:

1. Phenomenological effective nucleon-nucleon interactions, which may be understood as mathematically simple parametrisations of a density-dependent effective G-matrix, can be constructed and used in the Hartree-Fock (HF) approximation to reproduce surprisingly well nuclear ground-state energies, densities, radii, deformation energies (in particular also fission barriers of heavy nuclei[15,16]) and some properties of highly collective excitations such as the nuclear giant resonances.[17,18] (See ref. 19 for a review of such effective forces and their applications in HF (plus RPA) calculations.) In particular, the Skyrme type effective interactions [3,20] allow to write the nuclear part of the HF energy as a functional of local one-body densities only, which makes them especially well-suited for the use in density functional methods.
2. The Strutinsky method[21] not only proved to be an efficient phenomenological tool for fission barrier calculations (see, e.g.ref.22), but it also provides a quantitative way[23] to extract an average part of the HF energy which is semiclassical in its nature and can be calculated by density functional methods. This allows to avoid the very difficult problem of describing the shell effects by density functionals; it was demonstrated[15,23] that the shell

effects can be included perturbatively at the end of a self-consistent semiclassical calculation without any significant loss of accuracy compared to an exact HF calculation.

3. The extended Thomas-Fermi (ETF) model, based on a semiclassical $\hbar$ expansion of the partition function or the Bloch density,[24] was reintroduced into nuclear physics[25,26] and successfully used to calculate the average energy of nucleons in realistic potentials.[27] It was, in particular, also shown[26-28] that this average energy is identical to that obtained with the microscopical Strutinsky averaging method. From the same ETF model, density gradient expansions of the kinetic energy density functional $\tau[\rho]$ and a spin-orbit density functional $\vec{J}[\rho]$ can be derived; they have recently been extended to include contributions from nonlocalities of the average nuclear potential such as variable effective nucleon masses and spin-orbit potentials.[29,30] The ETF functional for $\tau[\rho]$ was furthermore demonstrated in microscopical test calculations[29,30] to reproduce very accurately the average kinetic energy of finite nuclei.

The strategy of the density functional method discussed in these lectures is thus the following: We use effective Skyrme type forces, as they are determined in HF calculations, without touching their parameters. We then use the density functionals $\tau[\rho]$ and $\vec{J}[\rho]$ determined from the ETF model once for all, without readjusting any of their coefficients, in variational calculations for the average nuclear properties of interest. In this way, the semiclassical reslts can at any time be tested against microscopically averaged HF results and possible deficiencies of the density functionals can be disentangled from possible deficiencies of the Skyrme forces themselves. The shell effects, wherever they are of importance, are added perturbatively at the end of the semiclassical variational calculation in terms of the corresponding average mean fields.

These lectures will be structured as follows: In section 2 we discuss in some more detail the above mentioned newer developments, in order to provide the basic justification of the semiclassical variational method. In section 3, we shall present - after a discussion of the ETF-Euler variational equations - the results for static nuclear bulk properties obtained in semiclassical calculations with a restricted, but flexible variational space of trial nuclear densities. We shall also shortly discuss there the expansion of the semiclassical nuclear binding energies in a liquid drop model (LDM) type series, which allows to link the phenomenological LDM or droplet model parameters back to those of the Skyrme force. In section 4, we shall finally discuss some extensions of the semiclassical density functional method.

2. FOUNDATION OF THE SEMICLASSICAL VARIATIONAL METHOD FROM THE SKYRME-HF FORMALISM

2.1. The Skyrme-HF Energy Density

We shall briefly outline here the structure of the energy density obtained with effective forces of the Skyrme type.[3] They have mathematically a zero range; however, velocity dependent terms mock up the finite (but short) range of the nuclear force. This allows to write the nuclear part of the HF energy as a functional of local one-body densities only. Correspondingly, the total HF energy is written in the form

$$E_{HF} = \int d^3r \, [\mathcal{E}_{Sky}(\vec{r}) + \mathcal{E}_{Coul}(\vec{r})] \, . \tag{2.1}$$

The nuclear (Skyrme) part

$$\mathcal{E}_{Sky}(\vec{r}) = \mathcal{E}_{Sky}[\rho_q(\vec{r}), \tau_q(\vec{r}), \vec{J}_q(\vec{r})] \tag{2.2}$$

is a simple functional of the local nucleon densities $\rho_q(\vec{r})$, kinetic energy densities $\tau_q(\vec{r})$ and spin-orbit densities $\vec{J}_q(\vec{r})$ (q = n, p for neutrons and protons, respectively) defined by

$$\rho_q(\vec{r}) = \sum_{\nu,s} |\varphi_\nu(\vec{r},s,q)|^2 \, n_\nu^q \; , \tag{2.3}$$

$$\tau_q(\vec{r}) = \sum_{\nu,s} |\vec{\nabla}\varphi_\nu(\vec{r},s,q)|^2 \, n_\nu^q \; , \tag{2.4}$$

$$\vec{J}_q(\vec{r}) = (-i) \sum_{\nu,s,s'} \varphi_\nu^*(\vec{r},s',q) \vec{\nabla}\varphi_\nu(\vec{r},s,q) \times \langle s'|\vec{\sigma}|s\rangle \, n_\nu^q \; , \tag{2.5}$$

where $\varphi_\nu(\vec{r},s,q)$ are the single-particle wave functions with orbital and spin quantum numbers ν and s, respectively, and n_ν^q are the occupation numbers (equal to 1 or 0 in the pure HF case, or v_ν^2 if pairing correlations are included in the BCS approximation[32]). The Coulomb energy density is the sum of the direct term and the exchange term, the latter taken in the well-known Slater approximation which has proved sufficiently accurate for all practical purposes[33]:

$$\mathcal{E}_{Coul}(\vec{r}) = e^2 \rho_p(\vec{r}) \frac{1}{2} \int d^3r' \frac{\rho_p(\vec{r}')}{|\vec{r}-\vec{r}'|} - \frac{3}{4} e^2 \left(\frac{3}{\pi}\right)^{1/3} \rho_p^{4/3}(\vec{r}) \; . \tag{2.6}$$

We refer to the original paper of Vautherin and Brink[20] for the derivation and the exact form of the functional $\mathcal{E}_{Sky}(\vec{r})$. (For the extended form of Skyrme forces where the density dependent term contains a variable power of ρ, see e.g. ref. 34.)

As an illustration we give here the expression of $\mathcal{E}_{Sky}(\vec{r})$ for the case of a symmetric nucleus with $\rho_n = \rho_p = \rho/2$ etc.:

$$\mathcal{E}_{Sky}(\vec{r}) = \frac{\hbar^2}{2m}\tau + \frac{3}{8}t_0\rho^2 + \frac{1}{16}(3t_1+5t_2)\tau\rho + \frac{1}{16}t_3\rho^{2+\alpha} + \frac{1}{64}(9t_1-5t_2)(\vec{\nabla}\rho)^2 + \frac{3}{4}W_0\,\vec{J}\cdot\vec{\nabla}\rho. \tag{2.7}$$

The HF equations, obtained by varying the wave functions $\varphi_\nu^q = \varphi_\nu(\vec{r},s,q)$, take the form of Schrödinger equations with variable effective nucleon masses and spin-orbit potentials:

$$\hat{H}_{HF}^q\varphi_\nu^q = \left[-\vec{\nabla}\cdot\frac{\hbar^2}{2m_q^*(\vec{r})}\vec{\nabla} + V_q(\vec{r}) - i\vec{W}_q(\vec{r})\cdot(\vec{\nabla}\times\vec{\sigma})\right]\varphi_\nu^q = \varepsilon_\nu^q\varphi_\nu^q\ . \tag{2.8}$$

The local potentials $V_q(\vec{r})$, effective masses $m_q^*(\vec{r})$ and spin-orbit potentials $\vec{W}_q(\vec{r})$ are given by the relations

$$V_q(\vec{r}) = \frac{\delta\mathcal{E}(\vec{r})}{\delta\rho_q(\vec{r})} \equiv \frac{\partial\mathcal{E}}{\partial\rho_q} - \vec{\nabla}\cdot\frac{\partial\mathcal{E}}{\partial(\vec{\nabla}\rho_q)} + \Delta\frac{\partial\mathcal{E}}{\partial(\Delta\rho_q)}\ , \tag{2.9}$$

$$\frac{\hbar^2}{2m_q^*(\vec{r})} = \frac{\partial\mathcal{E}(\vec{r})}{\partial\tau_q(\vec{r})}\ , \tag{2.10}$$

$$\vec{W}_q(\vec{r}) = \frac{\partial\mathcal{E}(\vec{r})}{\partial\vec{J}_q(\vec{r})}\ . \tag{2.11}$$

where $\mathcal{E}(\vec{r})$ is the sum of the nuclear and the Coulomb energy density.

Usually, the force parameters $t_0, t_1, t_2, t_3, \alpha$ etc. are determined by fits of experimental groundstate properties of a series of (mostly spherical) nuclei. However, most of them are related to each other, and restricted in their range of values, by imposing the more or less well established saturation properties of infinite nuclear matter, such as the binding energy per nucleon E/A (i.e the volume energy of the mass formula), the saturation density ρ_∞, the effective mass m_∞^* or the nuclear matter incompressibility K_∞. Imposing their empirical values, the choice of the force parameters is greatly restricted, although still innumberable parameter sets can be found in the literature.[19,35] The parameter α of the density dependent term in the Skyrme functional eq. (2.7) is rather strongly restricted by the values of K_∞ and m_∞^*. In fact, if values in the ranges

$$\begin{aligned} 210 \text{ MeV} \lesssim K_\infty \lesssim 240 \text{ MeV} \\ 0.7 \lesssim m_\infty^*/m \lesssim 0.8 \end{aligned} \tag{2.12}$$

are imposed, as they are required in order to fit the giant monopole and quadrupole resonances by RPA calculations,[17,18] one finds that α must be of the order

$$1/6 \lesssim \alpha \lesssim 1/3 \; . \tag{2.13}$$

Having imposed "reasonable" nuclear matter properties alone guarantees, of course, in no way that a force will have good surface properties of finite nuclei, which then are adjusted by actual HF calculations and fits to experimental data. Even more it must be considered a great success that good fits to many data were obtained, considering the fact that the nuclear matter properties fix already five combinations of the typically 7-8 Skyrme parameters. For detailed comparisons of HF (+BCS) results to experimental data, we can only refer here to the abundant literature.[17-19,35-37]

It might be worth spending a few words on the nature of this HF + Skyrme formalism. Although it formally is a Hartree-Fock procedure, it may well go beyond this framework what the physics is concerned. Due to the fact that the Skyrme force is a parametrized G-matrix (and can be derived qualitatively from a Brückner G-matrix[13]), short-range correlations are built into it from the very beginning. But also long-range correlations can be contained in what above is called the HF energy, because the HF equations (2.8) can be understood as Kohn-Sham equations,[38] generalized to include nonlocal parts of the potential. Noting that, in fact, the mean fields in eq. (2.8) are nothing but functional derivatives of a parametrized energy density, one recognizes that due to the Hohenberg-Kohn theorem[8] all kinds of correlation energies may be contained in the energy E_{HF} eq. (2.1)

2.2 Separation of Shell Effects

The direct application of the Skyrme energy functional eq. (2.2) to the density variational method is handicapped by the presence of the kinetic energy and spin-orbit densities $\tau_q(\vec{r})$ and $\vec{J}_q(\vec{r})$. In principle, we know from the Hohenberg-Kohn theorem[8] that there exist unique functionals $\tau[\rho]$and $\vec{J}[\rho]$ which allow to express these densities in terms of the local nucleon densities $\rho_q(\vec{r})$. However we do not know these functionals and there is little chance to determine them exactly. They certainly must be nonlocal, since the shell effects contained in $\tau_q(\vec{r})$ and $\vec{J}_q(\vec{r})$ are not local, but global properties of the nucleus.[39,40]

This problem can be overcome by averaging out the shell effects and expressing the average of the energy by a functional of the average densities $\rho_q(r)$. This can be justified by means of Strutinsky's energy averaging method[21] which, in fact, allows to decompose the exact HF energy in a rather unique way into an average and a fluctuating ("shell-correction") part[21-23]:

$$E_{HF} \simeq \tilde{E}_{HF} + \delta_1 E_n + \delta_1 E_p \quad . \tag{2.14}$$

Hereby the average energy E_{HF} is practically calculated in the same way as the exact energy E_{HF} through eqs. (2.1) - (2.6), but replacing the quantum mechanical densities eqs. (2.3) - (2.5) by the averaged densities obtained by means of the Strutinsky averaging occupation numbers $\tilde{n}^q_\nu$,[22,28]

$$\tilde{\rho}_q(\vec{r}) = \sum_{\nu,s} |\varphi_\nu(\vec{r},s,q)|^2 \, \tilde{n}^q_\nu \quad , \tag{2.15}$$

etc. The shell-correction energy $\delta_1 E_q$ in eq. (2.14) is defined by

$$\delta_1 E_q = \sum_\nu \hat{\varepsilon}^q_\nu (n^q_\nu - \tilde{n}^q_\nu) \; , \tag{2.16}$$

where $\hat{\varepsilon}^q_\nu$ are the eigenvalues of the average HF Hamiltonians $\tilde{H}^q_{HF}$ defined through eqs. (2.8) - (2.11) in terms of the averaged densities, i.e.

$$\tilde{H}^q_{HF}\,\hat{\varphi}^q_\nu = \hat{H}_{HF}[\tilde{\rho}_q, \tilde{\tau}_q, \tilde{\vec{J}}_q]\,\hat{\varphi}^q_\nu = \hat{\varepsilon}^q_\nu\,\hat{\varphi}^q_\nu \quad . \tag{2.17}$$

Formally, eq. (2.14) just represents the lowest two terms of a Taylor expansion of the HF energy around the average parts of the densities. (See ref. 23 for a discussion and further literature on this subject.) In extended numerical calculations[23] it has been checked that the missing higher order terms in eq. (2.14) are negligible for all practical purposes. In particular if the averaging by means of the $\tilde{n}^q_\nu$ is done selfconsistently (see also the next subsection), the two sides of eq. (2.14) are equal to within less than $\sim$ 0.5 MeV even in heavy, strongly deformed nuclei (corresponding to an ccuracy of better than 10^{-3}).

Two important conclusions could be drawn from these numerical results[23]:

1) The averaged HF energy $\tilde{E}_{HF}$ has all the properties of a LDM type, semiclassical energy.
2) The selfconsistency is only important for the average quantities ($\tilde{E}_{HF}$, $\tilde{H}^q_{HF}$, $\tilde{\rho}_q$, etc); the shell effects can, in fact, be added perturbatively.

This provides us with a strong motivation to replace the above sketched microscopical selfconsistent calculations of $\tilde{E}_{HF}$ by a semiclassical calculation. For its realization, it was important to quantitatively secure the equivalence of the Strutinsky averaging procedure with a semiclassical expansion of the energy, as will be discussed in the following subsection.

2.3 Strutinsky Averaging as a Microscopical Link to the ETF Model

Strutinsky[21] and Tyapin[41] surmised that the numerically Strutinsky-averaged energies not only correspond to those obtained in the Fermi gas theory, but that they contain also inhomogeneity corrections such as they are obtained in the so-called extended Thomas-Fermi (ETF) model.[42]

Bhaduri and Ross[25] proposed to calculate the average energy of nucleons in various model potentials by employing a $\hbar$-expansion of the partition function, which actually had been developed long ago by Wigner and Kirkwood,[24] and demonstrated the closeness of their results to those of a numerical Strutinsky averaging. (We shall discuss the Wigner-Kirkwood expansion and the ETF relations derived from it in section 2.4.) For harmonic oscillator potentials, the exact equivalence of the Strutinsky averaging method and the semiclassical $\hbar$-expansion was proved analytically.[26,28] For realistic, deformed Woods-Saxon type potentials including spin orbit fields, the two methods were shown numerically[27] to yield identical energies to within $\sim$ 1 - 1.5 MeV (out of several GeV), which is roughly the uncertainty in either method.

It is thus well established that - at least as energies are concerned and with the numerical accuracy practically required - the microscopical Strutinsky averaging procedure is equivalent to a semiclassical $\hbar$-expansion. Therefore it seems natural to use the ETF functionals $\tau[\rho]$ and $\vec{J}[\rho]$ obtained from the same $\hbar$-expansion (see next section) in order to calculate the average HF energy $\tilde{E}_{HF}$ in a semiclassical, and thus much more economical way.

That the energy $\tilde{E}_{HF}$ - which was obtained microscopically in ref. 23, as explained in sect 2.2 - can be expressed as a functional of the average densities $\tilde{\rho}_q(\vec{r})$ eq. (2.15) is again a consequence of the Hohenberg-Kohn theorem. The iterative inclusion of the Strutinsky occupation numbers $\tilde{n}^q_\nu$ in the HF cycle has, in fact, been formulated in a strictly variational way,[23] including a proper constraint in the energy to be made stationary (and found to be minimized in actual calculations). The Hohenberg-Kohn theorem[8] applies therefore to this variational averaged system as well as it applies to any variational system of Fermions interacting through a 2-body force.

2.4 The ETF Model and its Density Functionals

We shall in the following sketch the semiclassical $\hbar$-expansion developed by Wigner and Kirkwood,[24] which provides a convenient tool to derive the ETF functionals $\tau[\rho]$ and $\vec{J}[\rho]$ which we are interested in. For the sake of a simple notation, we shall presently restrict ourselves to the case of N nucleons (one kind only) in a given local (HF) potential $V(\vec{r})$. Let φ_ν and ε_ν be the eigenfunctions and eigenvalues of the corresponding Schrödinger equation:

$$\hat{H}\varphi_\nu = [\hat{T} + V(\vec{r})]\varphi_\nu = \varepsilon_\nu \varphi_\nu \ . \tag{2.18}$$

Next we define the Bloch density matrix

$$C(\vec{r},\vec{r}\,';\beta) = \sum_\nu \varphi_\nu^*(\vec{r}\,')\varphi_\nu(\vec{r})e^{-\beta\varepsilon_\nu} , \tag{2.19}$$

where the sum goes over the complete spectrum (including an integral over the continuum, if present). From C, we obtain by an inverse Laplace transform the usual density matrix

$$\begin{aligned} \rho(\vec{r},\vec{r}\,') &= L_\lambda^{-1}\left[\frac{1}{\beta}C(\vec{r},\vec{r}\,';\beta)\right] \\ &= \frac{1}{2\pi i}\int_{c-i\infty}^{c+i\infty} d\beta\, e^{\lambda\beta}\frac{1}{\beta}C(\vec{r},\vec{r}\,';\beta) \ , \end{aligned} \tag{2.20}$$

from which in turn, the local densities $\rho(\vec{r})$ and $\tau(\vec{r})$ can be determined

$$\rho(\vec{r}) = \sum_{\nu=1}^{N} |\varphi_\nu(\vec{r})|^2 = \rho(\vec{r},\vec{r}\,') \ , \tag{2.21}$$

$$\tau(\vec{r}) = \sum_{\nu=1}^{N} |\vec{\nabla}\varphi_\nu(\vec{r})|^2 = \vec{\nabla}_r \cdot \vec{\nabla}_{r'}\,\rho(\vec{r},\vec{r}\,')\big|_{\vec{r}=\vec{r}\,'} \ . \tag{2.22}$$

In eq. (2.20), λ is the Fermi energy which is fixed by the particle number conservation

$$\int \rho(\vec{r})d^3r = N \tag{2.23}$$

The idea of Wigner and Kirkwood was to expand $C(\vec{r},\vec{r}';\beta)$ around its value obtained in the Thomas-Fermi approximation:

$$C_{TF}(\vec{r},\vec{r}';\beta) = \left(\frac{m}{2\pi\hbar^2\beta}\right)^{3/2} e^{-\beta V\left(\frac{\vec{r}+\vec{r}'}{2}\right)} e^{-\frac{m}{2\hbar^2\beta}(\vec{r}-\vec{r}')^2} \tag{2.24}$$

One makes the ansatz

$$C(\vec{r},\vec{r}';\beta) = C_{TF}(\vec{r},\vec{r}';\beta) \times \{1 + \hbar\chi_1 + \hbar^2\chi_2 + \dots\} , \tag{2.25}$$

thus expanding the ratio of the exact to the TF Bloch function in powers of $\hbar$. The χ_n are functions of $\vec{r},\vec{r}'$ and β which contain combinations of n gradients acting on $V(\vec{r})$. Uhlenbeck and Beth[42] worked out a recursive scheme to obtain the χ_n successively (see also ref. 43). By Laplace-inverting the series eq. (2.25) back term by term, one obtains an expansion of the density matrix eq. (2.20) and thus of $\rho(\vec{r})$ and $\tau(\vec{r})$, to which only even powers of $\hbar$ (i.e. χ_n with even n) contribute. We quote here the results up to order $\hbar^2$

$$\begin{aligned}\rho_{ETF}(\vec{r}) = \frac{1}{3\pi^2}\left(\frac{2m}{\hbar^2}\right)^{3/2}(\lambda - V(\vec{r}))^{3/2} \times \Theta(\lambda - V(\vec{r})) \times \\ \times \left\{1 - \frac{1}{8}\frac{\hbar^2}{2m}[\Delta V(\lambda - V)^{-2} + \frac{1}{4}(\vec{\nabla}V)^2(\lambda - V)^{-3}]\right\} ,\end{aligned} \tag{2.26}$$

$$\begin{aligned}\tau_{ETF}(\vec{r}) = \frac{1}{5\pi^2}\left(\frac{2m}{\hbar^2}\right)^{5/2}(\lambda - V(\vec{r}))^{5/2} \times \Theta(\lambda - V(\vec{r})) \times \\ \times \left\{1 - \frac{5}{8}\frac{\hbar^2}{2m}[\frac{5}{3}\Delta V(\lambda - V)^{-2} - \frac{3}{4}(\vec{\nabla}V)^2(\lambda - V)^{-3}]\right\} .\end{aligned} \tag{2.27}$$

In the lowest order terms we recognize the TF expressions; the $\hbar^2$-corrections lead to the well-known divergencies at the classical turning points $\vec{r}_\lambda$ given by $\lambda = V(\vec{r}_\lambda)$. (Due to the step functions, both densities are identically zero outside the classically allowed region.)

In spite of their turning point divergencies, the densities eqs. (2.26), (2.27) can be shown[43] to lead to finite energies and particle numbers, even if the $\hbar^4$ terms are included. This shows that the ETF densities are rather to be understood as distributions with well-defined integrals and moments (see also ref. 44). The energies so obtained form a rapidly converging asymptotic series

$$E_{ETF} = E_{TF} + E_2 + E_4 + \ldots \tag{2.28}$$

The sum of the first three terms (i.e. up to order $\hbar^4$) converges typically to within $\sim$ 1 MeV and agrees, as mentioned in section 2.3 above, with the energy obtained by Strutinsky averaging:

$$E_{ETF} \simeq \tilde{E}_{Str} = \sum_\nu \varepsilon_\nu \tilde{n}_\nu \, . \tag{2.29}$$

We shall not discuss here the technicalities of including effective mass and spin-orbit contributions, which can be done starting from a Hamiltonian of Skyrme type eq. (2.8); they can be found in the literature.[27,43]

Before coming to the construction of the ETF density functionals, we mention that a way of removing the turning point divergencies in $\rho_{ETF}(\vec{r})$ and $\tau_{ETF}(\vec{r})$ by partially resumming the Wigner-Kirkwood series eq. (2.25) will be discussed in sect. 4.2 below.

2.4. a) The functional $\tau[\rho]$ for a local potential

From eqs. (2.26) and (2.27) it is possible to eliminate algebraically the Fermi energy λ, the potential $V(\vec{r})$ and its derivatives, hereby consistently retaining all terms of order $\hbar^2$ and neglecting those of higher orders in $\hbar$. The result is (for one kind of nucleons)

$$\tau[\rho] = \tau_{TF}[\rho] + \tau_2[\rho] \tag{2.30}$$

with the well-known Thomas-Fermi relation

$$\tau_{TF}[\rho] = \kappa\rho^{5/3} \, , \qquad \kappa = \frac{3}{5}(3\pi^2)^{2/3} \tag{2.31}$$

and the second order gradient correction

$$\tau_2[\rho] = \frac{1}{36}\frac{(\vec{\nabla}\rho)^2}{\rho} + \frac{1}{3}\Delta\rho \quad . \tag{2.32}$$

The first term in $\tau_2[\rho]$ is the so-called Weizsäcker correction; this author[1] derived it in a somewhat ad hoc manner and obtained it with a 9 times larger coefficient. This coefficient has subsequently given rise to a lot of discussion.[4] By now it is clear that various alternative semiclassical expansion procedures[30,41,42] lead to exactly the same relations and coefficients. (For a recent review in which these alternative expansions are discussed and related, see ref. 45.) The coefficient 1/36 of the Weizsäcker term is thus well established in the framework of semiclassical expansions (and for smooth potentials $V(\vec{r})$). The second term in eq. (2.32) does not contribute to the integrated kinetic energy and has therefore often been ignored; it does however contribute to the total Skyrme energy through the terms containing $\tau\rho$, see eq. (2.7).

Going up to order $\hbar^4$ in the expansion of ρ_{ETF} and τ_{ETF} and proceeding in the same way, one obtains the next correction $\tau_4[\rho]$ to the functional, containing up to fourth derivatives of ρ. The somewhat lengthy expression for $\tau_4[\rho]$ is given in refs.[29,31]. When integrating over the whole space, the fourth and third derivatives of ρ can be eliminated by partial integration, and the expression simplifies to

$$\int\tau_4[\rho]d^3r = \frac{1}{6480}(3\pi^2)^{-2/3}\int\rho^{1/3}\left[8\left(\frac{\vec{\nabla}\rho}{\rho}\right)^4 - 27\left(\frac{\vec{\nabla}\rho}{\rho}\right)^2\frac{\Delta\rho}{\rho} + 24\left(\frac{\Delta\rho}{\rho}\right)^2\right]d^3r. \tag{2.33}$$

Similarly, one obtains

$$\int\rho\tau_4[\rho]d^3r = \frac{1}{3240}(3\pi^2)^{-2/3}\int\rho^{4/3}\left[-7\left(\frac{\vec{\nabla}\rho}{\rho}\right)^4 - 3\left(\frac{\vec{\nabla}\rho}{\rho}\right)^2\frac{\Delta\rho}{\rho} + 30\left(\frac{\Delta\rho}{\rho}\right)^2\right]d^3r. \tag{2.34}$$

This procedure can in principle be continued ad libitum, including higher and higher gradient corrections. However, the terms $\tau_n[\rho]$ with $n \geq 6$ diverge for densities which decay exponentially in the tail region. Therefore, the terms up to fourth order must be considered as the converging part of an <u>asymptotic series</u> for $\tau[\rho]$; we shall denote this part by $\tau_{ETF}[\rho]$:

$$\tau_{ETF}[\rho] = \tau_{TF}[\rho] + \tau_2[\rho] + \tau_4[\rho] \quad . \tag{2.35}$$

The above derivation of the functional $\tau_{ETF}[\rho]$ is strictly speaking not allowed at the classical turning points, where $\tau_{ETF}(\vec{r})$ and $\rho_{ETF}(\vec{r})$ are singular. It holds, however, at any other point. (In the classically forbidden region, $\tau_{ETF}[\rho]$ holds trivially since τ_{ETF} and ρ_{ETF} there are identically zero!) One may therefore hope to be able to use $\tau_{ETF}[\rho]$ everywhere in space by analytical continuation.

The functional $\tau_{ETF}[\rho]$ given by eqs. (2.31)-(2.35) has been tested numerically with the help of microscopically Strutinsky averaged densities $\tilde{\rho}(\vec{r})$ and $\tilde{\tau}(\vec{r})$, defined as in eq. (2.15), for different spherical and deformed potentials.[29,31] The results of these tests may be summarized as follows (for a more detailed discussion, see ref. 31):

1) The functional $\tau_{ETF}[\rho]$ eq. (2.35) reproduces the total Strutinsky averaged kinetic energy within less than $\sim$ 1.5 MeV, corresponding to a few parts in 10^4 for heavy nuclei. This holds independently of the radial shape of the potential, of its deformation and of the particle number, as it should be expected from the Hohenberg-Kohn theorem.
2) The functional also reproduces the integral $G\int\rho\tau d^3r$, as it occurs in the Skyrme energy, within less than 1 MeV (using realistic Skyrme parameters to determine G).
3) The terms due to $\tau_4[\rho]$ are essential for obtaining the correct deformation energies, in particular the fission barriers.

The points 1 and 3 are illustrated in figure 1 (taken from ref.31). It shows the kinetic energy for 112 particles in a deformed harmonic oscillator potential as a function of the deformation parameter $q = \omega_{\perp}/\omega_z$ which measures the frequency ratio. The different curves are obtained in terms of the Strutinsky averaged density $\tilde{\rho}(\vec{r})$ eq. (2.15) through the ETF functional eq. (2.35)

$$\tilde{T}_n[\tilde{\rho}] = \frac{\hbar^2}{2m}\int d^3r\ \tau\,[\tilde{\rho}(\vec{r})] \quad , \qquad (2.36)$$

whereby the index n shows where the functional (2.35) has been truncated (e.g. n = 2 means TF plus 2nd order gradients included). The reference quantity of the test is the microscopically Strutinsky averaged kinetic energy $\tilde{T}$ defined by

$$\tilde{T} = \frac{\hbar^2}{2m}\int\tilde{\tau}(\vec{r})\,d^3r \quad , \qquad (2.37)$$

whereby $\tilde{\tau}(\vec{r})$ has been averaged analogously to $\tilde{\rho}(\vec{r})$ eq. (2.15). It is seen that the full functional $\tau_{ETF}[\rho]$ up to 4th order reproduces the energy $\tilde{T}$ exactly within the accuracy of the drawing; the energies

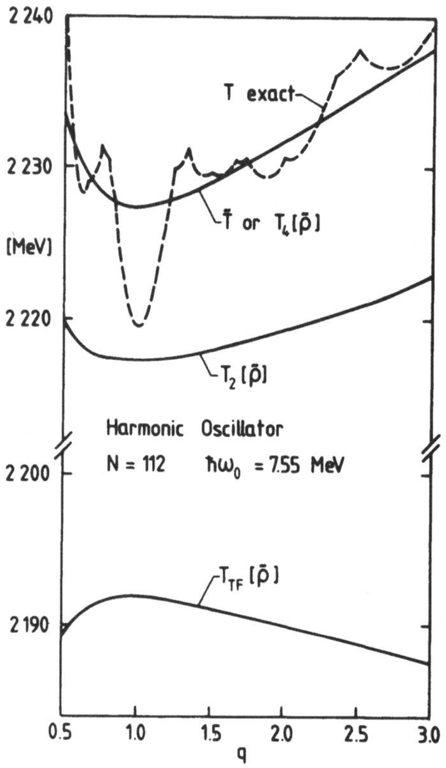

Fig. 1. Kinetic energy of 112 particles in axially deformed harmonic oscillator potential with frequency ratio $q = \omega_1/\omega_2$, obtained with the ETF functional $\tau[\rho]$ up to various orders of gradient corrections (see text). $\tilde{T}$ is the microscopically Strutinsky-averaged kinetic energy; the dashed curve shows the exact kinetic energy T which includes shell fluctuations.

$\tilde{T}$ and $T_4[\tilde{\rho}]$ agree in fact within less than 0.1 MeV at all deformations. The 4th order terms $\tau_4[\rho]$ still contribute 10 - 15 MeV to the total kinetic energy and are seen to be important for obtaining the correct deformation dependence. Fig. 1 also contains the exact kinetic energy T (shown by a dashed curve) which contains shell effects. It would of course be hopeless to try to reproduce this exact energy by a local gradient expanded functional, even if the exact quantum mechanical density $\rho(\vec{r})$ is put into the functional $\tau_{ETF}[\rho]$.[31]

We might add here some remarks concerning simplified functionals $\tau[\rho]$ of the form

$$\tau[\rho] = \kappa\rho^{5/3} + \eta \frac{1}{36} \frac{(\vec{\nabla}\rho)^2}{\rho} \tag{2.38}$$

with an adjustable coefficient η, as they have repeatedly been used in the literature.[4,46-48] It may be hoped, indeed, to mock up the 4th order terms $\tau_4[\rho]$ by choosing $\eta > 1$ such as to fit the functional eq. (2.38) to the correct total kinetic energy. However, it is not obvious, then, that the same value of η can be used for all potentials, all deformations and all particle numbers.

In order to illustrate this, we have calculated the quantity

$$\eta = \frac{\int d^3r \{\tau_2[\tilde{\rho}] + \tau_4[\tilde{\rho}]\}}{\int d^3r\, \tau_2[\tilde{\rho}]} \tag{2.39}$$

from the results obtained in ref. 31 for the harmonic oscillator potential. In figure 2, the number η eq. (2.39) is plotted against the particle number N (crosses) and against the deformation parameter q (circles). It is seen to be rather constant with values $\eta \simeq 1.4 - 1.5$ for not too small particle numbers. Similar values are also obtained for a deformed Woods-Saxon potential. The value $\eta \simeq 1.4 - 1.5$ is, however, about three times smaller than what typically has been used[46-48]; the reasons for this will be discussed in sect. 3.1 below.

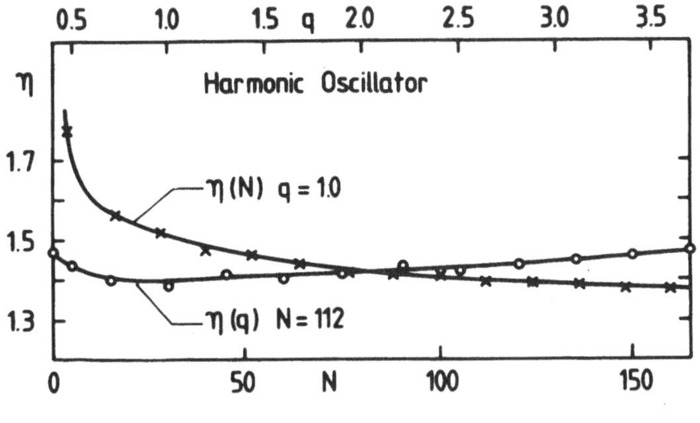

Fig. 2. The parameter η eq. (2.39) obtained for deformed harmonic oscillator potentials is plotted against particle number N (crosses, values on lower axis; evaluated for q = 1) and against deformation q (circles, values on upper axis, evaluated for N = 112 particles).

This result should be used with caution. We cannot expect this procedure of mocking up the 4th order contributions by a single parameter to work for nonlocal potentials. (Indeed, η is dependent on the effective mass of the force.[46]) Furthermore, problems arise with the surface of the densities, if such adjusted functionals as eq. (2.38) are used in variational calculations (see sect. 3.1).

2.4.b) The functionals $\tau[\rho]$ and $\vec{J}[\rho]$ for Skyrme-type nonlocal potentials

For velocity-dependent Skyrme forces, one has to generalize the functional $\tau_{ETF}[\rho]$, since it receives explicit contributions from the nonlocal parts of the HF-potential. Rewriting the Skyrme-HF Hamiltonian (see eq. (2.8)) in the form

$$H_{Sky} = -\frac{\hbar^2}{2m}\vec{\nabla}\cdot f(\vec{r})\vec{\nabla} + V(\vec{r}) - i\vec{W}(\vec{r})\cdot(\vec{\nabla}\times\vec{\sigma}) , \tag{2.40}$$

where $f(\vec{r}) = m/m^*(\vec{r})$, the Wigner-Kirkwood expansion eq. (2.25) can be readily obtained. (The Bloch density C is in this case a 2 x 2 matrix, the χ_n with n > 1 containing the Pauli matrices σ_i.) The second-order contribution to the kinetic energy density functional then becomes

$$\tau_2[\rho] = \frac{1}{36}\frac{(\vec{\nabla}\rho)^2}{\rho} + \frac{1}{3}\Delta\rho + \frac{1}{6}\frac{(\vec{\nabla}\rho\cdot\vec{\nabla}f)}{f} + \frac{1}{6}\rho\frac{\Delta f}{f} - \frac{1}{12}\rho\left(\frac{\vec{\nabla}f}{f}\right)^2 + \frac{1}{2}\left(\frac{2m}{\hbar^2}\right)^2\rho\left(\frac{\vec{W}}{f}\right)^2 . \tag{2.41}$$

The spin orbit density only gets contributions from the $\hbar^2$ and higher terms. The lowest-order expression is

$$-\vec{J}_2[\rho] = \left(\frac{2m}{\hbar^2}\right)\frac{1}{f}\rho\vec{W} = \left(\frac{2m^*}{\hbar^2}\right)\rho\vec{W} . \tag{2.42}$$

(A semiclassical spin-orbit correction equivalent to eq. (2.42) for $m = m^*$ has been derived earlier by Stocker et al.[49])

Carrying through the expansion to 4th order with effective mass and spin-orbit is extremely tedious. It has been done with an algebraic computer code by Grammaticos and Voros[30]; we refer to their papers for the explicit expressions for $\tau_4[\rho]$ and $\vec{J}_4[\rho]$. Again, after suitable partial integrations, the relevant contributions to the total energy only contain first and second derivatives of the densities $\rho_q(\vec{r})$. The corresponding expressions are given in ref. 15.

Note that for Skyrme forces $f(\vec{r}) = 1 + \beta\rho(\vec{r})$, and $\vec{W}(\vec{r})$ is proportional to $\vec{\nabla}\rho(\vec{r})$, so that the functionals $\tau[\rho]$ and $\vec{J}[\rho]$ ultimately only contain the density ρ and its gradients. We also recall to the reader that the equations in this section hold for <u>either proton or neutron</u> densities and <u>not</u> for the total densities $\tau = \tau_n + \tau_p$, $\rho = \rho_n + \rho_p$ and $\vec{J} = \vec{J}_n + \vec{J}_p$.

2.5 Summary

Let us summarize at this point the main steps of the derivation and justification of the semiclassical variational method.

1) HF calculations with effective Skyrme interactions allow to calculate a vast amount of nuclear ground-state properties, deformation energies and (with RPA) giant resonances to a satisfactory degree.
2) The HF energy can be split, by means of the Strutinsky averaging procedure, in a selfconsistent average part $\tilde{E}_{HF}$ and a shell-correction part, see eq. (2.14).
3) The averaged energy $\tilde{E}_{HF}$ and the corresponding selfconsistent average densities $\tilde{\rho}_q(\vec{r})$ can be obtained in a strictly variational way.[23] Therefore, by virtue of the Hohenberg-Kohn theorem, $\tilde{E}_{HF}$ and thus $\tilde{\tau}_q(\vec{r})$ and $\vec{J}(\vec{r})$ are unique functionals of $\tilde{\rho}_q(\vec{r})$.
4) The Strutinsky averaging method is practically equivalent to a semiclassical $\hbar$-expansion of the energy,[27] which in turn leads to the ETF density functionals $\tau[\rho]$ and $\vec{J}[\rho]$.
5) The ETF functional $\tau[\rho]$ with gradient corrections up to fourth order reproduces with high accuracy the average kinetic energy of nucleons in realistic potentials.[29,31]
6) Combining 3) and 4) allows to express $\tilde{E}_{HF}$ in terms of $\tilde{\rho}_q(\vec{r})$ only by means of the ETF-functionals $\tau[\rho]$ and $\vec{J}[\rho]$ and to perform semiclassical density variational calculations in order to optimize $\tilde{\rho}_q(\vec{r})$.
7) After selfconsistency has been reached for $\tilde{E}_{HF}$ and $\tilde{\rho}_q(\vec{r})$, the average mean fields eqs. (2.9) - (2.11) can be used to calculate the shell-correction energies $\delta_1 E_q$ (2.16) by solving once the Schrödinger equation (2.17). Adding $\delta_1 E_q$ to $\tilde{E}_{HF}$, thus incorporating the shell effects perturbatively, allows to recover the (exact) HF energy with sufficient accuracy.[15,23]
8) In the case of purely local potentials, the contribution to the total kinetic energy coming from the 4th order correction term $\tau_4[\rho]$ may be simulated by multiplying the Weizsäcker term in $\tau_2[\rho]$ by a factor $\eta \approx 1.4 - 1.5$. However, this procedure does not work for nonlocal Skyrme potentials (where η depends on the effective mass m^*_∞); it also leads to unphysical variational densities, as discussed in sect. 3.1 below.

3. SEMICLASSICAL VARIATIONAL CALCULATIONS

Inserting the functionals $\tau_{ETF}[\rho]$ and $\vec{J}_{ETF}[\rho]$ in the Skyrme energy density eq. (2.2) and making use of the variational definitions of $f_q = m/m_q^*$ and $\vec{W}_q$ by eqs. (2.10), (2.11), we can now express the total average energy of the nucleus as a functional of the spatial densities ρ_q only. The idea then is, as discussed in the introduction, to perform a variational calculation on the densities ρ_q, including Lagrange multipliers λ_q to ensure the correct particle numbers (N and Z):

$$\delta \int d^3r \left\{ \mathcal{E}[\rho_n, \rho_p] - \lambda_n \rho_n(\vec{r}) - \lambda_p \rho_p(\vec{r}) \right\} = 0 \quad . \tag{3.1}$$

(Here $\mathcal{E}[\rho_n, \rho_p]$ contains both the nuclear and the Coulomb parts.) In the following we shall discuss what happens if the variation is done exactly, i.e. if the corresponding Euler-Lagrange equations are solved.

3.1. Discussion of the ETF-Euler Variational Equations

In order to simplify the presentation, we shall again assume only one kind of particles - realistically, one will obtain two coupled differential equations for ρ_n and ρ_p - and leave out the effective mass and spin-orbit contributions (i.e. put f = 1 and $\vec{W} = 0$). These restrictions do not affect the conclusions drawn below.

The Euler-Lagrange equation then becomes

$$\frac{\hbar^2}{2m} \left\{ \frac{5}{3} \kappa \rho^{2/3} + D_2[\rho] + D_4[\rho] \right\} + V[\rho] = \lambda \quad , \tag{3.2}$$

where the term in curly brackets comes from the variation of the kinetic energy and the potential is

$$V[\rho] = \frac{\delta \mathcal{E}_{pot}}{\delta \rho} \quad , \tag{3.3}$$

cf. eq. (2.9). The second-order kinetic term is

$$D_2[\rho] = \frac{1}{36} \left[\frac{(\vec{\nabla}\rho)^2}{\rho^2} - 2 \frac{\Delta \rho}{\rho} \right] = \frac{\delta \tau_2[\rho]}{\delta \rho} \quad . \tag{3.4}$$

The term $D_4[\rho]$ correspondingly contains 7 contributions with up to 4th order derivatives of ρ.[29,31] The equation (3.4) can in principle only be solved numerically. However, it is possible to determine rather easily the asymptotic behavior of the solution both inside the nucleus and in the outer surface.

3.1 a) Asymptotic behavior in the outer surface

The fall-off of the density $\rho(r)$ at large distance r (we shall for simplicity assume spherical symmetry) is completely determined by the gradient corrections in the kinetic energy functional $\tau[\rho]$, if they are included at all. We shall accordingly discuss it in three steps.

1- Using $\tau_{TF}[\rho]$ only: If only $\tau_{TF}[\rho]$ is used, eq. (3.2) reduces to

$$\frac{\hbar^2}{2m} \cdot \frac{5}{3}\kappa\rho^{2/3} + V[\rho] = \lambda \quad . \tag{3.5}$$

If the potential $V[\rho]$ contains only powers of ρ and no gradients, the only solution of eq. (3.5) is $\rho(\vec{r}) = \rho_0$ and one obtains thus a liquid drop model type constant density with a sharp cut-off at the surface.

For Skyrme-like forces with a term $b(\vec{\nabla}\rho)^2$ in the potential energy, eq. (3.5) leads to a density profile which near the surface goes like[3]

$$\rho(r) \propto \quad \mathrm{Tgh}^2\left(\frac{r-R_0}{\alpha}\right) \tag{3.6}$$

for spherical nuclei, where α is essentially determined by the constant b in front of the $(\vec{\nabla}\rho)^2$ term. This density thus has to be cut off at a finite radius $r = R_0$ and put equal to zero outside, and is therefore not very physical. It leads to the deficiencies of the calculations reported in ref. 7 which we have already mentioned in the introduction.

2- Using $\tau_{TF}[\rho] + \tau_2[\rho]$: Berg and Wilets[4] pointed out that the inclusion of a Weizsäcker term in the variational equation eq. (3.2) (with $D_4 = 0$) leads to an asymptotic fall-off of the density with the correct exponential form (in the spherical case):

$$\rho(r) \xrightarrow[r\to\infty]{} \frac{1}{r^2} e^{-r/a} \tag{3.7}$$

The range a is given by the Fermi energy λ (which is always negative) and the coefficient of the Weizsäcker term:

$$a = \sqrt{-\frac{1}{36}\frac{\hbar^2}{2m}\cdot\frac{1}{\lambda}} \quad . \tag{3.8}$$

Unfortunately, this range is too small by a factor $\sim$ 2-3 compared with realistic nuclear surfaces. Consequently, the variational densities fall off too quickly in the outer surface and lead to an overestimation of the kinetic energy (which is partially compensated by an overestimation of the potential energy). This was confirmed in numerical calculations by Bohigas et al.,[50] who solved the Euler equations using the local functional $\tau_{TF}[\rho] + \tau_2[\rho]$ eqs. (2.31), (2.32) for a Skyrme force with $m^*_\infty/m \simeq 0.95$ and without spin-orbit force. The semiclassical energies obtained in this way differed from the exact HF energies by $\sim$ 0.4 - 0.6 MeV per nucleon, thus by far more than the order of magnitude of the shell corrections.

To overcome this defect - still in an attempt to solve the relatively easy second order differential equation - several authors used functionals of the type

$$\tau[\rho] = \alpha\rho^{5/3} + \eta\cdot\frac{1}{36}\frac{(\vec{\nabla}\rho)^2}{\rho} + \frac{1}{3}\Delta\rho \quad , \tag{3.9}$$

where α and η were adjustable parameters.[4,46-48,51] In particular in the so-called MTF-functional,[46] η was chosen to be $\sim$ 4-5, in order to obtain realistic tails of the densities, see eq. (3.8). This leads, however, to a drastic overestimation of the kinetic energy - in particular its surface contributions - which was compensated in ref. 46 by reducing the coefficient of the TF term (i.e. $\alpha < \kappa$). In this way it was possible to fit the kinetic energies of spherical nuclei quite well (see also ref. 47). However, the price to be paid for this is that α and β depend on the nucleon number and on the force (in particular on m^*_∞). The latter is obvious since the explicit effective mass and spin-orbit contributions in $\tau_2[\rho]$, shown in eq. (2.41), are ignored in eq. (3.9). Moreover, the MTF-functional[46] completely fails to give reasonable deformation energies due to a drastic overestimation of the surface energy contributions (see the next section).

Treiner and Krivine[48] recently used another functional of the type of eq. (3.9) with the original coefficient of the TF term (i.e. $\alpha = \kappa$) and $\eta = 2$, and added the correct second-order spin-orbit terms (see eqs. (2.41), (2.42)). This functional still slightly overestimates the surface energy, leading to a too high fission barrier as compared to the one obtained with the full, unchanged functional $\tau_{ETF}[\rho]$ including the 4th order contributions (see sect. 3.2). In fact, we have seen in fig. 2 above that a factor of $\eta \simeq 1.4$ to 1.5 would lead to reasonable deformation energies if spin-orbit and eff. mass contributions are neglected. However, the tails of the density distributions then become unrealistically steep, as seen from eq. (3.8).

One faces thus a basic dilemma when using adjustable functionals of the type of eq. (3.9): If one wants to obtain densities with good tails, one needs $\eta \simeq 4\text{-}5$; if one wants to obtain good energies, and in particular deformation energies, one needs $\eta \simeq 1.4 - 1.5$. (A similar dilemma exists also in atomic physics in the so-called Thomas-Fermi Weizsäcker theory.[52]) We shall see in the next section that this dilemma can be satisfactorily resolved by using the full, unchanged functional $\tau_{ETF}[\rho]$.

3- Using $\tau_{ETF}[\rho]$ up to 4th order: The full fourth order equation (3.2) was discussed in ref. 31. In this case the spherical solution of $\rho(r)$ falls off like

$$\rho(r) \xrightarrow[r\to\infty]{} \frac{c}{r^6} \quad ; \tag{3.10}$$

the coefficient c is given by

$$c = \left[-\frac{\hbar^2}{2m}(3\pi^2)^{-2/3} \cdot \frac{13}{45} \cdot \frac{1}{\lambda}\right]^{3/2} \quad . \tag{3.11}$$

This result at first looks rather discouraging, since eq. (3.10) is not the behavior we would like to expect from a nice density. However, we do not know at which distance from the nuclear surface the behavior r^{-6} will be assumed. In order to investigate this, let us take $\lambda \simeq -7$ MeV. We then find from eq. (3.11) that $c \simeq 0.03$ fm^3. If eq. (3.10) were to be true at a distance of $r = 10$ fm in ^{208}Pb, the density then would be 3×10^{-8} fm^{-3} at that point which is 4 orders of magnitude smaller than what it would be for a Fermi function type density. This indicates that eq. (3.10) is a purely mathematical result which is reached so far outside the nuclear surface that it will have no physical meaning. This is illustrated also in the semi-infinite nuclear matter calculations presented in ref. 15.

Unfortunately, the highly non-linear, fourth order differential equation (3.2) seems unaccessible to numerical solutions. Even in the semi-infinite case, where it can be integrated once analytically, we did not succeed in solving numerically the resulting third order equation. However, the results obtained with a restricted variational space for the densities $\rho_q(\vec{r})$ presented below are satisfactory enough, so that it is not necessary to solve eq. (3.2) exactly.

3.1.b) Asymptotic behavior inside the nucleus

The onset of the surface region, i.e. the deviation of the density from a constant value in the interior of a heavy nucleus, can also be estimated qualitatively without exactly solving the Euler equation. For simplicity we shall ignore the Coulomb interaction and the curvature effects, i.e. take the limit of a very large nucleus. Since in the inner region the density is very near its saturation value, we shall - following Skyrme,[3] and Strutinsky and Tyapin[53] who developed in this way a precursor of the droplet model - replace the Skyrme energy density by a schematic one which, however, preserves the correct saturation properties.

We thus write

$$\mathcal{E}[\rho] = \rho\tilde{e}_\infty(\rho) + a(\vec{\nabla}\rho)^2 + \frac{\hbar^2}{2m}(\tau_2[\rho]+\tau_4[\rho]) \qquad (3.12)$$

where the "volume part" $\tilde{e}_\infty(\rho)$ is taken to be

$$\tilde{e}_\infty(\rho) = a_v^\infty + \frac{K_\infty}{18\rho_\infty^2}(\rho-\rho_\infty)^2 \quad . \qquad (3.13)$$

This corresponds to a parabolic approximation of the saturation curve near the saturation density ρ_∞, which certainly is good enough for the following estimations. Writing

$$\rho(r) = \rho_\infty y(r) \, , \qquad (3.14)$$

the Euler equation then becomes (neglecting the curvature contribution)

$$\frac{K_\infty}{18}(3y^2 - 4y + 1) - 2a\rho_\infty y'' + \frac{\hbar^2}{2m}(D_2[y] + D_4[y]) = 0 \, , \qquad (3.15)$$

where $y'' = d^2y/dr^2$. The kinetic terms D_2 and D_4 play a minor role in the following development and we shall therefore drop $D_4[\rho]$ immediately. To arrive at eq. (3.15) we have also neglected the fact that the central density in finite nuclei is in general different from the saturation density ρ_∞ [54]; this has little bearing on the following argument and shall shortly be discussed in sect. 3.3.

We now write

$$y(r) = 1 - e^{(r-R)/\alpha} = 1 - \varepsilon(r) \quad . \qquad (3.16)$$

Inside the nucleus, $\varepsilon(r) \ll 1$ and we can expand eq. (3.15) in powers of ε. Keeping the linear terms in ε, we obtain an equation for α:

$$\alpha = \sqrt{\frac{18}{K_\infty}\left(a\rho_\infty + \frac{1}{36}\frac{\hbar^2}{2m}\right)} \quad . \tag{3.17}$$

For realistic Skyrme forces, $b\rho_\infty \simeq (10 - 13)$ MeV fm^2, so that α turns out to be of the order of ~ 1 fm. The Weizsäcker correction (the second term in the brackets in eq. (3.17)) only contributes ~ 3 % to this result; the term $D_4[\rho]$ in eq. (3.15) would have contributed even far less.

We learn from this that the asymptotic <u>inner</u> part of the nuclear surface is mainly determined by the "surface term" $a(\vec{\nabla}\rho)^2$ of the Skyrme energy density and by the incompressibility K_∞; the kinetic energy gradient terms play only a minor role here. The range α of the inner surface part is ~ 1 fm and thus about twice larger than the typical value of the diffuseness parameter of the density when parametrized by a Fermi function. This tends to make the realistic densities <u>asymmetric</u> around the half-density distance; the "shoulder" of the surface is broader than the tail of the surface. This asymmetry is, indeed, seen in the results discussed in sect. 3.2 below.

In order to summarize this section, let us repeat the main conclusions we have reached.

1- If the functional $\tau_{ETF}[\rho]$ is used to solve the Euler-Lagrange variational equation for the density, the gradient corrections to $\tau_{ETF}[\rho]$ completely determine the asymptotic fall-off of the density in the extreme surface. In <u>no</u> order of its gradient expansion can $\tau_{ETF}[\rho]$ give a realistic exponential fall-off. In particular with the gradient terms kept up to 4th order, one obtains a fall-off of the form $1/r^6$.

2- This latter result need not be in contradiction with the positive numerical results quoted in sect. 2.4 in the sense that the mathematical fall-off $\sim 1/r^6$ is only assumed at far distances outside the nucleus which play no physical role, whereas a realistic surface region is compatible with the 4th order functional $\tau_{ETF}[\rho]$.

3- Heuristic functionals $\tau[\rho]$ with only second order corrections and adjustable coefficients can either reproduce energies or density profiles, but not both at the same time.

4- Practically independently of the gradient corrections to $\tau[\rho]$, the inner asymptotic part of the surface is essentially determined by a balance between the gradient term $\sim(\nabla\rho)^2$ of the potential energy and the incompressibility K_∞. As a consequence, the density profile is in general asymmetric around its inflection point.

3.2 Variational Calculations for Finite Nuclei

In the following we shall present some selective results of variational semiclassical calculations[15] obtained with a restricted, but flexible variational space of trial nuclear densities. We shall be brief and refer the interested reader to the review article[15] in which these calculations have been discussed in detail.

For spherical nuclei we chose the radial densities to have the form (see also ref. 55):

$$\rho_q(r) = \frac{\rho_{0q}\,[1 + \rho_{1q}\,\exp(-r^2/\beta_q^2 R_q^2)]}{\left[1 + \exp\left(\frac{r - R_q}{\alpha_q}\right)\right]^{\gamma_q}} \quad . \qquad (q = p,n) \qquad (3.18)$$

We have thus 10 independent variational parameters ρ_{0q}, ρ_{1q}, β_q, α_q, and γ_q, with respect to which the total energy is minimized; the radius constants R_q are always determined to fix the nucleon numbers Z and N. For $\rho_{1q} = 0$ and $\gamma_q = 1$ we have the familiar Fermi functions with central density ρ_{0q} and surface diffuseness α_q. For $\gamma_{1q} \neq 1$, the surface is asymmetric; as we have discussed in sect. 3.1, we expect $\gamma_q > 1$ in realistic cases. For $\rho_{1q} \neq 0$, the Gaussian factor in eq. (3.18) allows for a depression or an enhancement in the central region measured by β_q; for physical reasons we are interested only in values $0.3 \lesssim \beta_q < 1.0$. (In fact, the energy was found to be stationary for $\beta_q \simeq 0.5$ in all cases where $\rho_{1q} \neq 0$ was favored at all.[15,55])

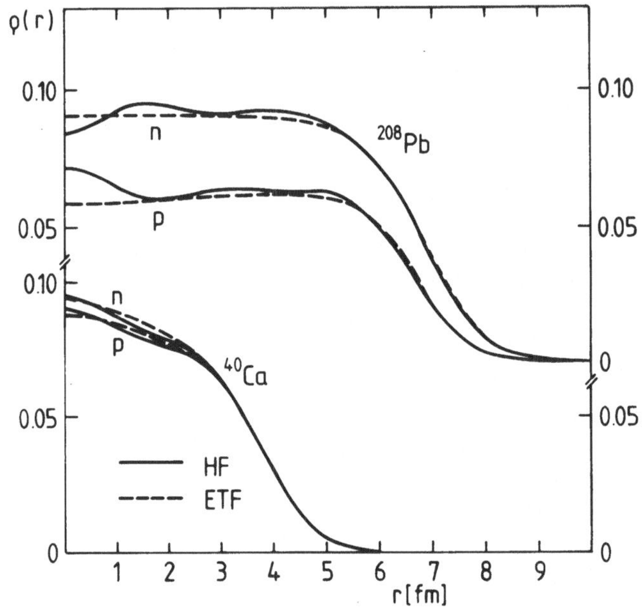

Fig. 3. Comparison of the microscopical HF densities and the variational ETF densities for ^{40}Ca and ^{208}Pb, obtained with the Skyrme force SkM*.

In figure 3 we compare the density profiles obtained for ^{40}Ca and ^{208}Pb with microscopical HF results, both calculated with the Skyrme force SkM*.[15,16] An almost perfect agreement is obtained in the surface and the tail region. In the interior part the ETF densities reproduce nicely the average trend of the HF results. In fact the possibility to build up a bump or a dip near the center, although it does not affect the binding energies by more than a few hundred keV, is important for obtaining this agreement. In particular for ^{40}Ca, we see that the central densities are enhanced by $\sim$ 20 %. It is worth underlining that this is not just a shell effect, but it must be understood as a bulk effect which results from the compression of the nucleus by the surface tension. In heavy nuclei such as ^{208}Pb, this compression effect is overpowered by the Coulomb repulsion between the protons, which leads to a slight depression at the center ($\sim$ 8 % for the proton and $\sim$ 2 % for the neutron density of ^{208}Pb).

In table 1 we present the binding energies of a series of spherical nuclei (all in MeV). B_{exp} are the experimental values; B_{HF} and B_{ETF} the HF and the ETF results (with SkM*). (In both calculations, a 1-body c.m. energy correction [16] has been included; it is not included in all the other results presented below.) Note the nice agreement between B_{HF} and B_{exp} especially for the ß-stable nuclei. The semiclassical energies B_{ETF}, which of course do not contain

Table 1

	B_{exp}	B_{HF}	B_{ETF}	B_{EVM}
^{16}O	127.6	127.7	128.0	127.4
^{40}Ca	342.1	341.1	345.9	340.4
^{48}Ca	416.0	420.1	421.8	418.4
^{56}Ni	484.0	485.4	483.9	483.1
^{90}Zr	783.9	784.5	786.6	782.7
^{114}Sn	971.6	969.2	976.0	967.9
^{132}Sn	1102.7	1110.7	1101.5	1108.3
^{140}Ce	1172.7	1173.9	1174.5	1171.6
^{208}Pb	1636.5	1636.4	1627.0	1633.7

the shell effects, are larger than the averaged HF energies by ∿ 3-8 MeV. This effect of a slight overbinding was observed earlier with other Skyrme forces[14] - it is larger by a factor of roughly 2 for the SIII force,[35] presumably due to its larger incompressibility - and must be considered as a slight defect of the ETF functionals. Although the variational principle holds strictly for the "ideal" (but unknown) exact functional $\mathcal{E}[\rho]$, the use of approximate functionals can lead to violations of the variational principle and thus to such overbinding effects. This slight deficiency of B_{ETF} is, however, healed after inclusion of the shell effects by the "expectation value method" (EVM),[56,57] which corresponds to performing a single HF iteration using the variational ETF densities as an input. The so obtained energies are shown in the last column of table 1 and are seen to reproduce the HF energies to within less than ∿ 1 MeV (^{16}O, ^{40}Ca) to ∿ 3 MeV (^{208}Pb).

In refs. 15,16 it was shown that also the HF neutron and proton r.m.s. radii - and in particular their difference, the so-called "neutron skin" - are also very accurately reproduced by the variational ETF calculations (the shell effects are practically negligible here).

This excellent agreement between the ETF and the (averaged) HF results for both energies and densities demonstrates the powerfulness of the 4th-order corrected ETF functionals; it cannot be obtained leaving out the $\tau_4[\rho]$ term, as discussed above.

In figure 4 we compare the variational ETF charge densities of 5 spherical nuclei to the experimental ones deduced from electron scattering experiments. A very good agreement is found for the average trends in all cases; the remaining differences are the typical shell fluctuations. (These are overestimated in HF calculations with most effective forces; see, however ref. 37 for a recent discussion of this effect.)

As already indicated above, the deviations of constant densities in the nuclear interior - governed by the parameters ρ_{1q} and β_q in eq. (3.18) - have very little influence on the total energy of the nucleus. This is demonstrated in table 2, where we list all the density parameters according to eq. (3.18) together with the minimized energies E_{ETF} of the 5 nuclei shown also in fig. 4. For ^{40}Ca and ^{208}Pb we also give the results obtained when the densities were restricted to pure Fermi functions (imposing $\gamma_q = 1$, $\rho_{1q} = 0$) or asymmetric Fermi functions (with $\gamma_q \neq 1$, but $\rho_{1q} = 0$). It is interesting to note that the 10 parameter variation lowers the total energy by only 2.3 MeV in ^{40}Ca (i.e. ∿ 0.7 %) and by 5.1 MeV in ^{208}Pb (i.e. ∿ 0.03 %) compared to the 4 parameter variation with pure Fermi functions. Furthermore, almost all of this gain in energy is already obtained with flat densities ($\rho_{1q} = 0$) with an asymmetric surface ($\gamma_q \neq 1$). As long as one is interested in binding or deformation energies alone, it is thus

Table 2

	ρ_{op}	ρ_{on}	α_p	α_n	γ_p	γ_n	ρ_{1p}	ρ_{1n}	$\beta_n=\beta_p$	R_p	R_n	E_{ETF}
^{40}Ca	0.0776	0.0804	0.472	0.464	<u>1.0</u>	<u>1.0</u>	<u>0.0</u>	<u>0.0</u>	-	3.761	3.720	-327.46
	0.0804	0.0834	0.577	0.580	1.43	1.49	<u>0.0</u>	<u>0.0</u>	-	3.999	3.984	-329.35
	0.0768	0.0777	0.566	0.553	1.42	1.42	0.136	0.199	0.5	4.027	4.001	-329.71
^{208}Pb	0.0610	0.0887	0.440	0.530	<u>1.0</u>	<u>1.0</u>	<u>0.0</u>	<u>0.0</u>	-	6.753	6.839	-1604.0
	0.0622	0.0911	0.532	0.662	1.42	1.47	<u>0.0</u>	<u>0.0</u>	-	6.981	7.202	-1608.9
	0.0639	0.0904	0.557	0.646	1.45	1.50	-0.086	-0.004	0.5	6.976	7.125	-1609.1
^{58}Ni	0.0756	0.0800	0.561	0.557	1.42	1.45	0.107	0.168	0.5	4.550	4.552	-491.6
^{116}Sn	0.0697	0.0870	0.559	0.605	1.48	1.50	0.019	0.078	0.5	5.734	5.854	-976.6
^{124}Sn	0.0664	0.0902	0.551	0.630	1.48	1.50	0.017	0.061	0.5	5.828	2.995	-1037.1

Parameters of the variational ETF densities for 5 spherical nuclei (obtained with SkM*). ρ_{0q} in fm^{-3} α_q and R_q in fm. The last column shows the total energy in MeV. The underlined values have been imposed in the variational calculations.

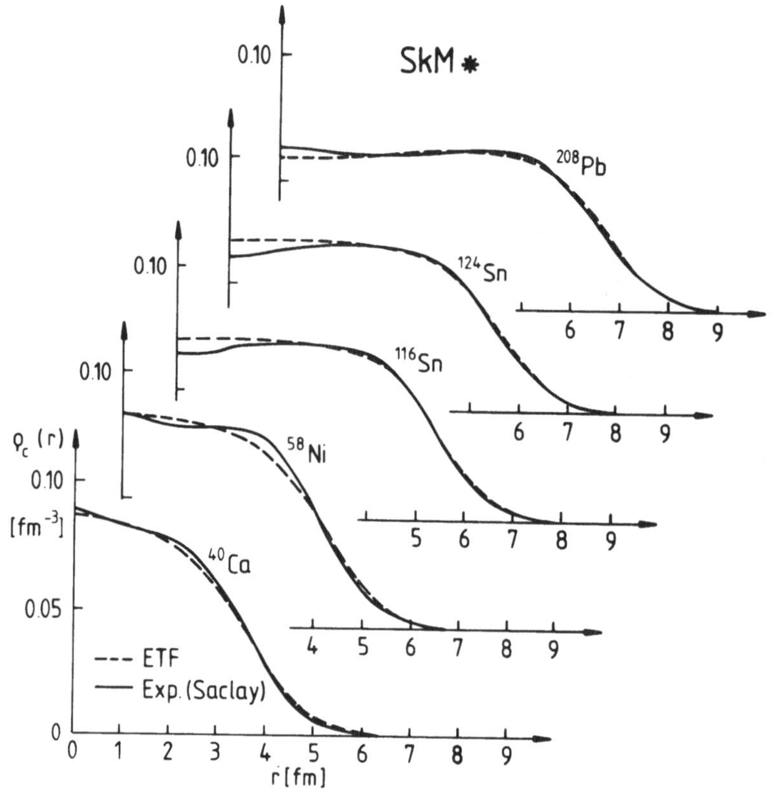

Fig. 4. Comparison of semiclassical ETF charge densities (including a proton form factor of 0.64 fm^2) with the experimental distributions extracted from electron scattering data [58] for five spherical nuclei.

perfectly sufficient to use 3-parameter densities (i.e. Fermi functions to the power γ_q) with a flat interior. The values of γ_q vary only a little, from $\sim$ 1.4 in light to $\sim$ 1.5 in heavy nuclei. For forces with larger K_∞ , the γ_q become smaller, as can easily be understood on the basis of the discussion in sect. 3.1 above. (For the SIII force, e.g., $\gamma_q \sim 1.2$.) When the $\tau_4[\rho]$ gradient corrections are omitted,[48,55] unphysically large values $\gamma_q \sim 2$-3 are obtained.

In order to describe deformed nuclei, we have to use a constraint since in a semiclassical model all nuclei are spherical in their ground states. In ref. 15 the constraint was introduced by starting from a deformed LDM "generating surface" with sharp edges, such as it has been used in shell-correction calculations for fission barriers.[22] It was then assumed that the diffuse densities have a constant surface thickness, so that they can be described simply by replacing $(r-R_q)$ in eq. (3.18) by the normal distance from the generating LDM surface (see the details in ref. 15); hereby the density profiles of eq. (3.18) were, for the reasons just given above, restricted to asymmetric Fermi functions ($\gamma_q \neq 1$ but $\rho_{1q} = 0$).

In figure 5 we present the semiclassical fission barriers of ^{204}Pu obtained for four different Skyrme forces. The (c,h) family of shapes[22] was used for which c is the main elongation parameter and h is a "necking" parameter. The cross indicates the location of the empirical LDM saddle point as it is known from shell-correction calculations.[22] We see that the forces SIII[35] and Ska[59] give too high fission barriers by a factor of $\sim$ 2. For the SIII force, this had been known from constrained HF calculations.[60]

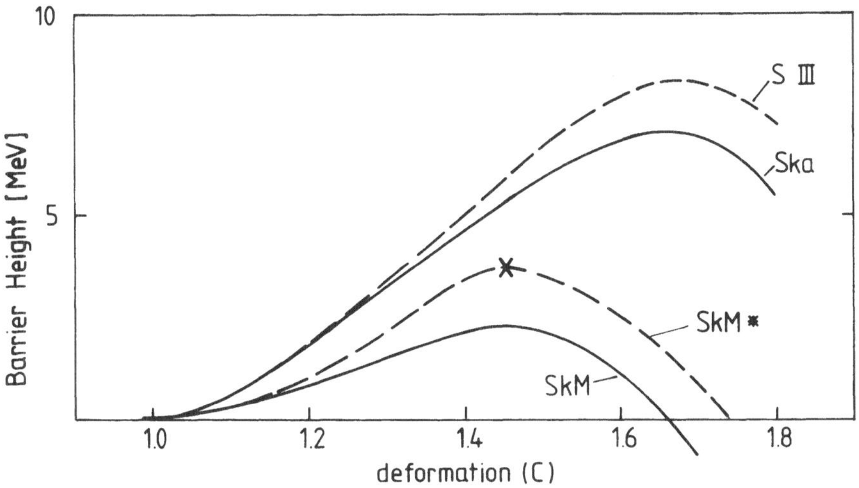

Fig. 5. Semiclassical fission barriers for ^{204}Pu. For each elongation c, the energies are minimized with respect to the neck parameter h. Four different Skyrme forces were used; the cross indicates the empirical LDM saddle point.

In fact, it was a puzzle for quite some time that HF calculations consistently led to too high fission barriers even with effective forces[60,61] which otherwise gave good results for ground state properties of both spherical and deformed nuclei (see a review article[62] on the status of fission barrier calculations up to 1979). Due to the excessive computer times required by the constrained HF calculations for heavy nuclei, it was practically not possible to refit the forces taking explicitly the fission barriers into account. This became, however, possible[63,14,15] with the semiclassical method described above which is more than 10^3 times faster if one is not interested in the shell effects. As we see from fig. 5, we can well distinguish the average barriers predicted with the different forces.

The Skyrme force SkM which was fitted to reproduce the giant nuclear monopole and quadrupole resonances[51] and therefore has an incompressibility K_∞ of 216 MeV, compatible with eq. (2.12), gives a somewhat too low barrier. (The forces SIII and Ska have higher values of K_∞; which leads to stiffer surfaces and thus to higher surface energies.) The force SkM* was explicitly adjusted with semi-classical calculations to reproduce the LDM saddle point energy[15]; it was shown at the same time to yield excellent binding energies and radii for stable spherical nuclei in HF calculations[16] (see the results shown in figs. 3,4 and table 1 above).

In fig. 6 we present a microscopical test of the semiclassical results. The corresponding HF calculations were done in ref.[16] The figure shows the full HF result, obtained with the SkM force, with

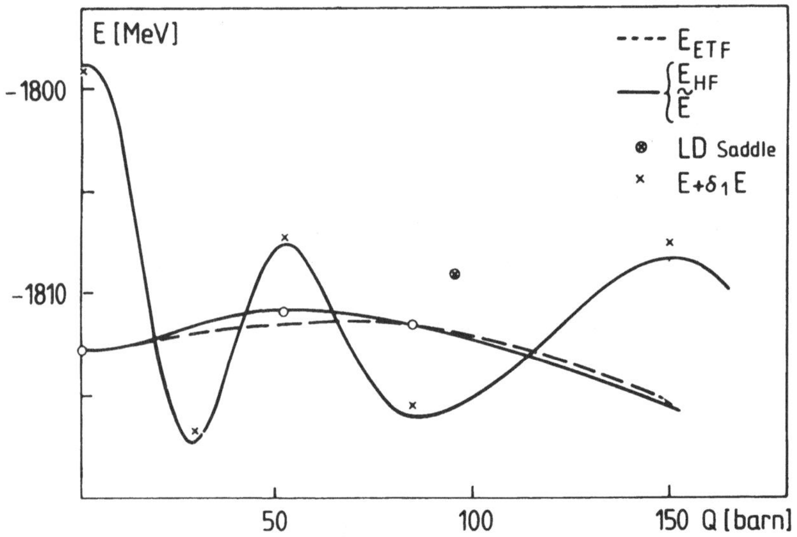

Fig. 6. Fission barrier of ^{204}Pu, calculated with the SkM force. The exact and Strutinsky-averaged HF results[16] are shown along with the semiclassical ETF result.[15] Q is the total quadrupole moment. The cross in a circle indicates the LDM saddle point. The crosses show the results after inclusion of the shell-correction energy $\delta_1 E$.

2 minima and 2 maxima. Also shown is the selfconsistently Strutinsky smoothed HF energy, calculated as discussed in sect. 2.3. The semi-classical ETF result is shown by the dashed line (adjusted at Q = 0).

The agreement of the two average curves is better than 1 MeV at all deformations included. This gives once again a nice confirmation of the semiclassical method. It shows in particular also that the slight overbinding of the ETF results discussed above (∿ 8 MeV in this nucleus) does not affect the deformation energies noticably. The crosses in fig. 6 show the results obtained after adding the shell-correction energy $\delta_1 E$ to the average curves; they reproduce the exact HF values within less than 0.5 MeV.

An interesting result is that in the semiclassical variational calculations, the density parameters ρ_{oq}, α_q, γ_q and R_q found for the spherical shape vary only very little with deformation; in fact, only an error of ∿ 0.5 MeV would be made for the realistic force SkM* if they were kept constant.[15] The influence of the asymmetry of the surface, governed by the parameters γ_q, on the fission barrier is shown in fig. 7, where the barrier of ^{204}Pu has been calculated once with $\gamma_q = 1$ and once with the variational values $\gamma_q \neq 1$. The difference is seen to be ∿ - 0.8 MeV at the saddle, corresponding to a decrease

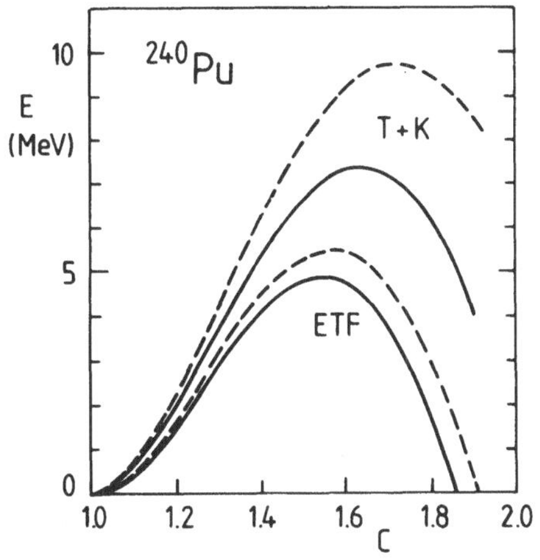

Fig. 7. The same as fig. 5 with force SkM*. The curves ETF show results obtained with the full 4th order functional $\tau[\rho]$ the curves "T + K" those with the phenomenological functional of Treiner and Krivine.[48] Dashed curves are obtained with Fermi function densities ($\gamma_q = 1$), solid curves with asymmetric density profiles ($\gamma_q \neq 1$).

of ∿ 0.3 MeV of the surface energy (see sect. 3.3). This difference is typical for forces which give approximately correct average fission barrieres.

We also show in fig. 7 the results obtained for the same force SkM*, but with the simplified functional $\tau[\rho]$ of Treiner and Krivine[48] where τ_4 was omitted and the Weizsäcker coefficient was multiplied by two. It leads to an overestimation of the barrier height.

The corresponding variational values of γ_q were found to be $\gamma_p \simeq 3.2$ and $\gamma_u \simeq 2.3$, reflecting a too steep tail of the densities. A similar calculation with the MTF functional[46] (in which the Weizsäcker term is multiplied by ∿4) gives a fission barrier of over 30 MeV for ^{204}Pu. This illustrates the problem discussed above in sect. 3.1 with readjusted functionals $\tau[\rho]$ without 4th order gradient terms.

3.3 Calculation of LDM parameters for Effective Forces

An interesting application of the variational ETF calculations with parametrized densities is the determination of the LDM parameters for a given effective force. Nuclei with A $\gtrsim$ 40 are "leptodermous",[54,63] i.e. the ratio of the surface diffuseness α to the bulk radius R is small:

$$\alpha/R \ll 1 \, . \tag{3.19}$$

The expansion of the nuclear binding energy in powers of α/R is the underlying technique of the liquid drop model.[63] If asymmetry and compression effects are taken into account by further expansions in powers of the small parameters δ and ε, defined by

$$\delta = \frac{\rho_{0n} - \rho_{0p}}{\rho_0} \, , \qquad (\rho_0 = \rho_{0n} + \rho_{0p}) \tag{3.20}$$

$$\varepsilon = -\frac{1}{3} \frac{(\rho_0 - \rho_\infty)}{\rho_\infty} \, , \tag{3.21}$$

one obtains the droplet model.[53,54]

The "leptodermous expansion" in powers of α/R was recently adapted to the total energy of an arbitrarily deformed nucleus within the Skyrme-ETF fomalism.[15,64] We refer to the recent review article[15] for the details and quote here just some of the main results. (For earlier calculations of surface energies from ETF results see ref. 30; similar analyses based on semiclassical[48,55] and HF calculations[65-67] may also be compared.)

For symmetric nuclei ($N=Z;\delta=0$) the expansion in powers of α/R leads to

$$E = a_v A + a_s A^{2/3} + a_c A^{1/3} + a_0 + \ldots \tag{3.22}$$

The dependence of the a_n on the deformation and the force parameters can be exactly separated. For their spherical part, one then expands

$$a_v = a_v^\infty + \frac{1}{2} K_\infty \varepsilon^2 + \ldots \tag{3.23}$$

$$a_s = a_s^\infty - 3\dot{a}_s \varepsilon + \frac{9}{2} \ddot{a}_s \varepsilon^2 + \ldots \tag{3.24}$$

$$a_c = a_c^\infty - 3\dot{a}_c \varepsilon + \frac{9}{2} \ddot{a}_c \varepsilon^2 + \ldots \tag{3.25}$$

etc. Minimizing the energy with respect to ε (fixing the surface parameters α and γ of the densities, which vary only very little for finite nuclei), one obtains the smooth variation of ε with A which reflects the effect of compression of the nucleus by the surface tension:

$$\varepsilon(A) = \frac{3\dot{a}_s A^{-1/3} + 3\dot{a}_c A^{-2/3}}{K_\infty + 9\ddot{a}_s A^{-1/3} + 9\ddot{a}_c A^{-2/3}} \quad . \tag{3.26}$$

It was shown[15] that for the realistic force SkM*, eq. (3.26) is needed to describe the A-dependence of the central density ρ_0 correctly, whereas the droplet model expression[54]

$$\varepsilon_{DM}(A) = (3\dot{a}_s / K_\infty) A^{-1/3} \quad , \tag{3.27}$$

which just contains the leading term of eq. (3.26) leads to large overestimations of ε in particular for medium and lighter nuclei. The surface compressibility parameter $\ddot{a}_s$, which is neglected in the droplet model, is known to play an important role for the compressibility of finite nuclei[18,36,68] which nowadays is known from the measurement of the nuclear breathing mode.

In table 3 we list the coefficients of the expansions in eqs. (3.24) and (3.25) (all in MeV), obtained from semi-infinite (symmetric) nuclear matter calculations with the variational ETF method (asymmetric Fermi profiles with $\gamma_q \neq 1$ were used).[15] The various Skyrme forces already mentioned above were used as well as the energy density of Tondeur[69] which is very similar to that of a Skyrme force. Table 3 also contains the effective curvature energy a_c defined by

$$a_c^* = a_c^\infty - \frac{9\dot{a}_s^2}{2K_\infty} \tag{3.28}$$

which is obtained[54] if the lowest order contribution from ε is included in eq. (3.22).

In the realistic case one has to include also the Coulomb energy and to expand everything also in powers of the asymmetry parameter δ eq. (3.20). We refer to the droplet model of Myers and Swiatecki[54] for this procedure. In table 4 we list the coefficients (in MeV) of the surface energy which are obtained if it is expanded to second order in the asymmetry parameters (for a fixed value of ε)[54]

$$a_s = a_s^\infty + H\tau^2 + 2P\tau\delta - G\delta^2 \quad ; \tag{3.29}$$

hereby τ is the so-called "neutron skin" parameter:

$$\tau = \frac{R_n - R_p}{r_0} \simeq \frac{2}{3}(I - \delta)A^{1/3} + O(A^{-1/3}) \tag{3.30}$$

with $I = (N-Z)/A$. We also give in table 4 the volume asymmetry energy J and the "surface stiffness coefficient" Q of the droplet model, defined by

$$Q = \frac{H}{\left(1 - \frac{2}{3}\frac{P}{J}\right)} \quad , \tag{3.31}$$

as well as the quantity $\tilde{J}$

$$\tilde{J} = \frac{2}{3}\left(P + \frac{GH}{P}\right) \tag{3.32}$$

of which a theorem derived by Myers and Swiatecki[54] tells that it should be equal to J. The same 5 effective forces as in table 3 were used; on the top line (quoted "DM") we also give the droplet model values. We see that for all forces, the theorem $\tilde{J} \equiv J$ is fulfilled within less than 3 % which may be considered as a test of the numerical calculations. (To obtain the results in table 4, pure Fermi functions with $\gamma_q = 1$ were used, because the above droplet model relations do not apply to density profiles with asymmetric surfaces.[15])

In summary it can be said that the variational ETF calculations can be used to justify and test the droplet model or similar extensions of the simple LDM. Some of the shortcomings of the droplet model have been discussed and some extensions and improvements have been proposed.[15] The main conclusion is that the variational ETF formalism with its 8 - 10 Skyrme force parameters is more powerful than the droplet model, even if the latter is extended to include some 20 or more phenomenological parameters.

Table 3

Force	a_s^∞	$\dot{a}_s$	$\ddot{a}_s$	a_c^∞	$\dot{a}_c$	$\ddot{a}_c$	a_o^∞	a_c^*
SIII	18.04	-12.40	-88.11	9.52	26.02	49.32	- 8.66	7.57
Ska	18.52	-11.60	-71.29	12.15	30.20	47.28	-13.88	9.86
SkM	16.60	-11.06	-58.64	12.19	27.26	37.27	-12.31	9.65
SkM*	17.22	-11.15	-60.37	12.82	29.27	40.76	-14.13	10.24
Tond.	18.11	-11.30	-65.19	12.74	30.25	44.23	-14.73	10.32

Table 4

force	a_s^∞	H	P	G	J	Q	$\tilde{J}$
DM	18.56	9.42	17.55	45.4	28.06	16.1	27.94
SIII	18.13	13.55	31.76	26.52	28.16	54.6	28.72
Ska	18.79	10.66	31.69	47.47	32.91	29.8	31.77
SkM	16.85	11.11	31.99	38.94	30.75	36.3	30.34
SkM*	17.51	10.56	31.61	39.04	30.03	35.4	29.77
Tondeur	18.41	11.46	33.24	39.83	32.12	37.0	31.32

4. EXTENSIONS OF THE SEMICLASSICAL METHOD

4.1. Partial Resummation of the Wigner-Kirkwood $\hbar$ Expansion

One of the unpleasant features of the $\hbar$-expanded densities $\rho_{ETF}(\vec{r})$ and $\tau_{ETF}(\vec{r})$ eqs. (2.26), (2.27) is their divergence at the classical turning points. It is the reason why they cannot be used directly in an iterative procedure to calculate the selfconsistent average HF potential. One way to circumvent this problem is the construction of the ETF functional $\tau_{ETF}[\rho]$ and its use in a density variational calculation, as we have discussed it extensively above.

Another way to solve the turning point problem is the use of partial resummations of the Wigner-Kirkwood expansion of the Bloch density matrix C eq. (2.25). Bhaduri[70] noticed that all terms which contain powers of the first gradient of the potential, $\vec{\nabla}V(\vec{r})$, can be summed up to infinite order in $\hbar$. In this way one obtains for the local Bloch density

$$C(\vec{r},\vec{r}';\beta) = C_{TF}(\vec{r},\vec{r}';\beta)\, e^{\frac{\hbar^2}{24m}\beta^3(\vec{\nabla}V)^2}\{1+\hbar^2\eta_2+\hbar^4\eta_4+\dots\}\,, \tag{4.1}$$

where C_{TF} is given by eq. (2.24) and the η_n contain second and higher order gradients of $V(\vec{r})$. The nice feature of eq. (4.1) is that it leads, after Laplace inversion, to densities $\rho(\vec{r})$ and $\tau(\vec{r})$ which are well-behaved everywhere in space, being in particular finite at the classical turning points and falling rapidly to zero in the classically forbidden region. Noting that the exponential factor appearing in eq. (4.1), in fact, is the Laplace transform of the Airy function, we see that in the lowest order in eq. (4.1) (i.e. neglecting η_2, η_4, etc.) one obtains by eq. (2.20) the folding product of the TF density matrix with an Airy function

$$\rho(\vec{r},\vec{r}') = \sigma(\vec{R})\int_{-\infty}^{\lambda}\rho_{TF}(\vec{r},\vec{r}';\lambda-E)\,\mathrm{Ai}[-\sigma(\vec{R})E]\,dE\,, \tag{4.2}$$

where

$$\sigma(\vec{R}) = \sigma\left(\frac{\vec{r}+\vec{r}'}{2}\right) = \left(\frac{8m}{\hbar^2}\right)^{1/3}\{[\vec{\nabla}V(\vec{R})]^2\}^{-1/3}. \tag{4.3}$$

It can be shown[71] that this result eq. (4.2) is identical to that of a <u>locally linear approximation</u> to the potential V.

The above procedure can be extended to sum up also all terms containing <u>second</u> order gradients to all powers; this corresponds to a locally <u>harmonic</u> approximation to $V(\vec{r})$.[71] (Nonlocal potentials can be treated in the same way.)

The densities obtained after these partial resummations have some unphysical oscillations in the interior part of the nucleus. They can be damped out if the Laplace inversion eq. (2.20) is not done analytically, but with the saddle point method, hereby using only the saddle point $\beta_o > 0$ on the real β axis.[70] This implicitly is a semiclassical approximation; as shown by Jennings,[26,43] the average (or ETF) part of the densities $\rho(\vec{r})$, $\tau(\vec{r})$ comes from contributions in the inverse Laplace transform from the region around $\beta = 0$, whereas poles (or saddle points) of $C(r,r';\beta)$ far from the real β axis - they usually ly on or near the imaginary β axis - lead to the fluctuating part (shell effect).

The combined method of partial resummation of $C(r,r';\beta)$ and using the saddle point method (with real $\beta_o > 0$) for the Laplace inversion leads thus to well-behaved semiclassical densities $\rho(\vec{r})$ and $\tau(\vec{r})$ (see ref. 72 for a discussion of some technical details and model examples). These densities can be used directly to calculate the average HF-Skyrme potentials eqs. (2.9) - (2.11) and thus, in an iterative cycle, to reach selfconsistency.

Compared to the variational ETF method discussed in the main parts of these lectures, the present method has the advantage that one does not need to know the functional $\tau[\rho]$. Numerically, the densities tend to become unstable since hihger and higher gradients of the potential are implicitly taken during the iterative cycle. They therefore have to be regularized e.g. by a fit to smooth parametrized densities.[73] It was found that when the same form of the densities was used as in eq. (3.18) above, the partial resummation method leads to very similar results as the variational ETF method using the functionals $\tau_{ETF}[\rho]$ and $\vec{J}_{ETF}[\rho]$; in particular the LDM and droplet model parameters reported in tables 3,4 above are closely reproduced,[74] thus implicitly providing a quantitative confirmation of the ETF functionals.

4.2. Semiclassical Description of Hot Nuclei and Nuclear Matter

Excited nuclear systems with temperatures larger than $\sim$ 3 MeV contain no shell effects and are therefore ideal objects for semiclassical investigations. Such hot compound nuclei can be produced in heavy ion and high-energy hadron induced reactions.[75] In astrophysics, there has recently been an increased interest in the equation of state of hot nuclear matter.[76,77]

The microscopical mean field (HF) theory can easily be generalized to finite temperatures in the statistical approximation.[78] Here one minimizes no longer the total intrinsic energy E, but the Helmholtz free energy F

$$F = E - TS \ , \tag{4.4}$$

where the entropy is given by

$$S = -\sum_{q\nu} [n_\nu^q \ln n_\nu^q + (1 - n_\nu^q) \ln(1 - n_\nu^q)] \tag{4.5}$$

in terms of the occupation numbers

$$n_\nu^q = \frac{1}{1+\exp\left(\frac{\varepsilon_\nu^q - \lambda_q}{T}\right)} \ . \tag{4.6}$$

(We put the Boltzmann constant $k \equiv 1$ and measure the temperature T in units of MeV.)

HF calculations at finite temperature are relatively easy to perform; it is sufficient to replace the HF occupation numbers n_ν^q in eqs. (2.3) - (2.5) by the occupation numbers eq. (4.6). Such calculations have been performed with Skyrme forces by different groups.[77,79-81] Hereby it must be assumed that the parameters of the Skyrme force themselves do not depend on T. This could in principle be checked by performing a Brückner G-matrix calculation at finite temperature; this has, however, not been endeavoured so far.

A well-known effect of the smoothing of the Fermi surface brought about by the occupation numbers eq. (4.6) is the washing out of the shell effects; the above mentioned HF results showed that beyond a critical temperature $T_c \simeq (2.5\text{-}3)$ MeV (which is roughly the same for all systems) the shell effects have disappeared. Systems at such temperatures are thus ideal objects for studies within a semi-classical framework. It is therefore obvious to try to apply the methods developed above to nuclei at $T > 0$. Thomas-Fermi calculations at finite temperature are by now standard.[76,82,83] However, we shall see in the following that it is not easy to construct the appropriate ETF functionals for $T > 0$.

The Wigner-Kirkwood expansion discussed in sect. 2.4 can easily be extended to finite temperatures. To do so, it is sufficient to know that the inclusion of the Fermi occupation numbers eq. (4.6) in the HF case is identical to a convolution of the spectral density with the function $f_T(E) = \frac{1}{4} \cosh^{-2}(E/2T)$.[23] Thus, due to the convolution theorem, the Bloch density eq. (2.19) is multiplied for $T > 0$ with the Laplace transform of $f_T(E)$

$$C_T(\vec{r},\vec{r}\,';\beta) = C(\vec{r},\vec{r}\,';\beta)\frac{\pi\beta T}{\sin(\pi\beta T)} \ . \tag{4.7}$$

Note that this result is still exact within the HF framework. Proceeding now as in the $T = 0$ case, i.e. replacing the "cold" Bloch density C by its Wigner-Kirkwood expansion eq. (2.25) and doing the inverse Laplace transforms term by term, one finds the expressions for the densities $\rho_{ETF}(\vec{r})$ and $\tau_{ETF}(\vec{r})$ at $T > 0$ and that of the entropy density $\sigma(\vec{r})$ defined by

$$S = \int d^3r \, \sigma(\vec{r}) = - \frac{\partial F}{\partial T} \quad . \tag{4.8}$$

The resulting expressions are up to order $\hbar^2$ (for a local potential, with effective mass m*),[84]

$$\rho^T_{ETF}(\vec{r}) = \frac{1}{2\pi^2} \left(\frac{2m^*}{\hbar^2}\right)^{3/2} \times \Big\{ T^{3/2} J_{1/2}(\eta) + \tag{4.9}$$
$$+ \frac{1}{24} \frac{\hbar^2}{2m^*} \left[\frac{3}{4} T^{-3/2} J_{-5/2}(\eta) (\vec{\nabla} V)^2 + T^{-1/2} J_{-3/2}(\eta) \Delta V \right] \Big\} \quad ,$$

$$\tau^T_{ETF}(\vec{r}) = \frac{1}{2\pi^2} \left(\frac{2m^*}{\hbar^2}\right)^{5/2} \Big\{ T^{5/2} J_{3/2}(\eta) - \tag{4.10}$$
$$- \frac{1}{4} \frac{\hbar^2}{2m^*} \left[\frac{3}{8} T^{-1/2} J_{-3/2}(\eta) (\vec{\nabla} V)^2 + \frac{5}{6} T^{1/2} J_{-1/2}(\eta) \Delta V \right] \Big\} \quad ,$$

$$\sigma_{ETF}(\vec{r}) = \frac{1}{2\pi^2} \left(\frac{2m^*}{\hbar^2}\right)^{3/2} T^{3/2} \Big\{ \frac{5}{3} J_{3/2}(\eta) - \eta J_{1/2}(\eta) + \tag{4.11}$$
$$+ \frac{1}{24} \frac{\hbar^2}{2m^*} \left[\frac{1}{4} T^{-3} J_{-3/2}(\eta) (\vec{\nabla} V)^2 - T^{-2} J_{-1/2}(\eta) \Delta V \right] \Big\} \quad ,$$

$$\eta = \frac{\lambda - V(\vec{r})}{T} \tag{4.12}$$

and $J_\nu(\eta)$ are the Fermi integrals

$$J_\nu(\eta) = \int_0^\infty \frac{x^\nu}{1 + \exp(x - \eta)} dx \quad . \tag{4.13}$$

To lowest order in eqs. (4.9) - (4.11) we recognize the well-known TF expressions. At this order it is possible to eliminate the quantity η numerically from the above densities; this defines the exact TF functionals at $T > 0$:

$$\tau^{T>0}_{TF}[\rho] = \tau_{TF}[\eta(\rho)] \quad , \tag{4.14}$$

$$\sigma_{TF}[\rho] = \sigma_{TF}[\eta(\rho)] \quad , \tag{4.15}$$

where $\eta(\rho)$ is obtained from inverting the function $J_{1/2}(\eta)$ in the leading (TF) term of eq. (4.9). Unfortunately, this procedure cannot be extended in an obvious way to include correctly all the $\hbar^2$-corrections. We are thus forced to make further approximations. Two possible ways shall be discussed in the following.

4.2.a) Low temperature expansion

In the limit $\eta \gg 1$, i.e. for $T \ll (\lambda - V)$, the Fermi integrals can be expanded in a series of decreasing powers of η [85]

$$J_\nu(\eta) = \frac{\eta^{\nu+1}}{\nu+1}\left[1 + \nu(\nu+1)\frac{\pi^2}{6}\eta^{-2} + \dots\right] \quad . \qquad (4.16)$$

The leading terms of the $J_\nu(\eta)$ give then just the old expressions $\rho_{ETF}(\vec{r})$ and $\tau_{ETF}(\vec{r})$ at T = O, eqs. (2.26), (2.27); the next terms give corrections of order T^2. From these expressions one obtains the corrected functionals:

$$\tau^T_{ETF}[\rho] = \tau_{ETF}[\rho] + \frac{2m^*}{\hbar^2}\alpha(\rho)T^2, \qquad (4.17)$$

$$\sigma_{ETF}[\rho] = 2\alpha(\rho)T \quad , \qquad (4.18)$$

where

$$\alpha(\rho) = \frac{1}{12}(3\pi^2)^{1/3}\left(\frac{2m^*}{\hbar^2}\right)\rho^{1/3} \quad . \qquad (4.19)$$

As in the T = O case, higher order corrections would contain inverse powers of ρ and must therefore be left out.

The total free energy density then becomes

$$\mathcal{F}(\vec{r}) = \mathcal{F}[\rho] = \mathcal{E}[\rho] - \alpha(\rho)T^2, \qquad (4.20)$$

where $\varepsilon[\rho]$ is the full ETF energy density functional described in sect. 2 for T = O. Note that the spatial integral of $\alpha(\rho)$ is nothing but the TF approximation to the well-known level density parameter a_0

$$a_0 = \frac{\pi^2}{6}\tilde{g}(\lambda) , \qquad (4.21)$$

where $\tilde{g}(\lambda)$ is the average single-particle level density (of one kind of particles). The functional eqs. (4.19), (4.20) has been used by several autors[86,87] to discuss thermal properties of nuclei. In the case of a variable effective mass $m^*(\vec{r}) = m/f(\vec{r})$, two correction terms to eq. (4.19) arise which remain finite; they have been shown, however, not to modify the numerical results very much.[15]

The problem with the above relations is that the low-temperature expansion $T \ll (\lambda-V)$ is only justified in the interior part of the nucleus (or in infinite nuclear matter), where λ-V is of the order of 30 - 40 MeV and the approximation holds up to fairly high temperatures. In the nuclear surface, however, λ-V quickly becomes smaller, going through zero at the classical turning points which still are in the surface region where the density is a few percent of its saturation value. Thus in the very region where one is interested in going beyond the TF approximation, namely in the surface region, the low-T expansion breaks down. It is thus not surprising that unsatisfactory results have been obtained with the functional (4.20).[88,89]

4.2.b) Gradient-corrected finite T functional

Since the low-T expansion breaks down in the surface, one might try to use the exact relations (valid for all T) at least in the TF approximation given above, and to add the gradient correction terms $\tau_2[\rho]$ and $\tau_4[\rho]$ known from the T = 0 case in an ad hoc manner. This leads to the functional

$$\tau_{ETF*}[\rho] = \tau_{TF}^{T>0}[\rho] + \tau_2[\rho] + \tau_4[\rho] \tag{4.22}$$

where $\tau_{TF}^{T>0}[\rho]$ is the exact finite T functional in eq. (4.14). Since we cannot know any gradient corrections to $\sigma[\rho]$ at T = 0 we will use $\sigma_{TF}[\rho]$ eq. (4.15) along with $\tau_{ETF*}[\rho]$. This procedure has been proposed by Barranco and Treiner[88,89]; they used, however, a readjusted Weizsäcker term in $\tau_2[\rho]$ and omitted $\tau_4[\rho]$ which, as we have seen in sect. 3, is to be used very cautiously.

4.2.c) Comparison of numerical results

We shall in the following be using both approximate ETF functionals, eqs. (4.20) and (4.22), including in all cases the full, unrenormalized "cold" correction terms $\tau_2[\rho]$, $\tau_4[\rho]$, as well as $\vec{J}_2[\rho]$ and $\vec{J}_4[\rho]$ discussed in sect. 2. We also shall quote results obtained with the partial resummation method described in sect. 4.1 which can be generalized to finite temperatures without difficulties.[74,84] In fact, for that purpose it is sufficient to replace the exact Bloch density C in eq. (4.7) by that obtained with the partial resummation method. Since the Laplace inversion there is made numerically by the saddle point method, it causes no problem to take into account the temperature dependent factor in eq. (4.7) exactly (i.e. without low-T expansion). We shall first test the different approximations using the force SIII for which HF calculations at T > 0 have been performed[79] and can be used for comparison.

In figure 9 we plot for ^{208}Pb the "effective level density parameter" a_{eff} defined by

$$a_{eff} = \frac{S^2}{4E^*} \tag{4.23}$$

versus the excitation energy E*

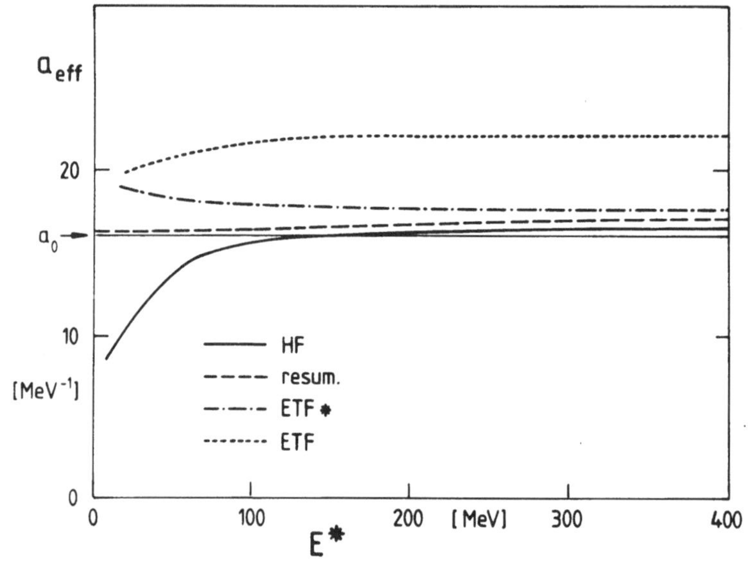

Fig. 8. Effective level density parameter a_{eff} eq. (4.23) for ^{208}Pb versus excitation energy E* for ^{208}Pb (SIII force used). The various approximations are: HF (full line), partial resummation method (dashed line), modified ETF* functional with exact T dependence in the TF terms, eq. (4.22) (dashed-dotted line), and low-T-expanded ETF functional eq. (4.20) (dotted line). a_0 is the level density parameter eq. (4.21).

$$E^* = E(T) - E(0) . \tag{4.24}$$

The relation (4.23) is that of the Fermi gas theory which is reached when the shell effects are washed out,[79] so that in this limit a_{eff} tends to the level density parameter a_0 defined in eq. (4.21). We see in fig. 8, indeed, that the curves $a_{eff}(T)$ are approximately constant for $E^* \gtrsim 150$ MeV (corresponding to $T \gtrsim 3$ MeV).

Whereas the HF result approaches the correct value a_0 eq. (4.21) - for the slight variation at $E^* \gtrsim 200$ MeV see the discussion in ref. 79 - , the low-T expanded functional (ETF) leads to a value which is more than 30 % too high. This is the well-known failure of this approximation.[89] The modified functional eq. (4.22) (ETF*) gives an asymptotic value of a_{eff} only $\sim$ 7 % higher than the HF result, which is a considerable improvement. The result of the partial re-summation method, in which the temperature dependence is treated exactly, comes closest to the HF result and clearly is an excellent approximation above $E^* \simeq 100$ MeV where the shell effects have disappeared.

In refs. 15,84 it was also shown that the temperature dependence of the r.m.s. radii obtained with HF is very well reproduced by both the ETF* functional eq. (4.22) and the resummation method, whereas with the low-T expanded functional eq. (4.20) it is strongly exaggerated above $T \simeq 3$ MeV.

A question which has been much discussed in the literature is how the fission barriers depend on temperature.[81,86,87,90] The fission of an excited nucleus is usually thought to be an isothermal process; therefore one has to look at the deformation behavior of the free energy F. Due to the well-known decrease of the free surface energy, the fission barriers also decrease with increasing temperature. (The variation of the Coulomb energy with temperature is not very important.) This was shown by explicit calculations of fission barriers with the variational ETF method at $T > 0$.[15]

In table 5 we list as a function of temperature the free surface energy a_s^∞ obtained with the three above methods for the SkM* force. It is clearly seen that the low-T expansion leads to an exaggeration

Table 5

T	a_s^∞ ETF	a_s^∞ ETF*	a_s^∞ resum.	ETF* a_c^*	ETF* Q	ETF* k_s	J
0	17.51	17.51	17.63	10.3	35.4	-57.3	30.03
1	17.30	17.33	17.53	10.0	35.3	-57.4	30.00
2	16.64	16.85	17.22	9.6	35.0	-57.5	29.91
3	15.50	16.08	16.70	8.7	34.4	-57.9	29.76
4	13.78	15.08	15.70	7.7	33.5	-58.6	29.54

LDM and droplet model parameters (all in MeV) for the force SkM* as functions of temperature T (in MeV). The free surface energy a_s^∞ is obtained in three approximations discussed in the text; the parameters a_c^*, Q and k_s are obtained with the ETF* functional eq. (4.22). For the volume asymmetry energy J, all approximations give the same result.

of the temperature dependence of a_s^∞. The partial resummation method (3rd column) reproduces the T-dependence found in HF calculations at $T \gtrsim 2$ MeV; the corrected functional $\tau_{ETF^*}[\rho]$ eq. (4.22) comes rather close to it, althoug the decrease of a_s^∞ with T here also is somewhat too strong. We also give in table 5 the effective curvature energy a_c^* eq. (3.28), the surface stiffness parameter Q and the surface asymmetry energy k_s defined by[15,54]

$$K_s = -\frac{9}{4}\frac{J^2}{Q} , \tag{4.25}$$

all evaluated for semi-infinite nuclear matter with Fermi function profiles. It is interesting to note that the absolute value of k_s increases with T due to the inverse dependence of Q which decreases faster with T than the volume asymmetry energy J (given in the last column of tab. 5).

We learn from these results that the temperature dependence of surface properties depend rather crucially on the approximations made. In particular, the low-T expansion leads to rather bad results which strongly exaggerate the T dependence. The best agreement with finite-T HF results is obtained with the partial resummation method, and reasonable agreement with the corrected ETF* functional eq. (4.22) in which the exact T dependence is contained in the TF expressions for $\tau[\rho]$ and $\sigma[\rho]$. In the context of density functional theory there remains, however, still a challenge to find better functionals $\tau[\rho]$ and $\sigma[\rho]$ in which the correct T dependence is contained also in the gradient corrections.

4.3. Application of the ETF Method to the Nuclear Breathing Mode [68]

We finally want to mention briefly an application of the variational ETF method to the calculation of the nuclear breathing mode energies.[68] We refer to the lectures of Holzwarth[91] and Treiner[92] for detailed discussions of the nuclear giant resonances of which the breathing mode, corresponding to density compressional vibrations, has only recently been established experimentally.

Starting from spherical nuclear ground-state densities described by simple Fermi functions, we can introduce compression modes by writing

$$\rho_q(r,t) = \frac{\rho^c_{0q}(t)}{1+\exp\left(\frac{r-R^c_q(t)}{\alpha^c_q(t)}\right)} , \tag{4.26}$$

where the density parameters now are supposed to be periodically time dependent functions:

$$\rho^c_{0q}(t) = \rho_{0q} + \delta\rho_{0q}\sin(\omega t) , \tag{4.27}$$

etc. We shall define two independent dimensionless (isoscalar) collective degrees of freedom by

$$q_\rho(t) = \frac{\rho^c_{0q}(t)}{\rho_{0q}} = 1 + \delta q_\rho(t) , \tag{4.28}$$

$$q_\alpha(t) = \frac{\alpha_q^c(t)}{\alpha_q} = 1 + \delta q_\alpha(t) \; ; \tag{4.29}$$

the radii parameters $R_q^c(t)$ shall for each set of values q_ρ, q_α be determined by the conservation of particle numbers. The variables q_ρ and q_α define a two-dimensional collective Hamiltonian $(i,j = \rho,\alpha)$

$$H_{coll} = \frac{1}{2}\sum_{i,j} B_{ij}\,\dot{q}_i\dot{q}_j + \frac{1}{2}\sum_{i,j} K_{ij}\,(q_i - 1)(q_j - 1) + E_0 \; ; \tag{4.30}$$

we have assumed small amplitude oscillations ($\delta q_i << 1$) and therefore used a quadratic approximation of the potential energy part. The compressibility modulus K_{ij} can easily be determined[68] from the variational ETF ground-state energies discussed in sect. 3.2, by

$$K_{ij} = 9\,\frac{\partial^2}{\partial q_i \partial q_j}\left(\frac{E}{A}\right) \; ; \tag{4.31}$$

shell effects in the K_{ij} are small (of the order of ~ 1 %) and can therefore be safely neglected. The inertial tensor B_{ij} can be obtained from classical hydrodynamics (which is allowed for the 0^+ mode[91]) in terms of the velocity fields $v_i(r)$:

$$B_{ij} = \frac{9}{A} m \int \rho\, v_i\, v_j\, d^3r \; ; \tag{4.32}$$

the latter can be found from solving the continuity equations

$$\frac{\partial \rho}{\partial q_i} + \vec{\nabla}\cdot(\rho \vec{v}_i) = 0 \; ; \qquad \vec{v}_i(r) = \frac{\vec{r}}{r} v_i(r) \tag{4.33}$$

(here $\rho = \rho_n + \rho_p$). Eq. (4.30) is that of two coupled harmonic oscillators (taking B_{ij} to be constant at $q_i = 1$); it is solved by diagonalizing the secular matrix $K_{ij} - \omega^2 B_{ij}$. Of the two resulting frequencies ω_1, ω_2 we can identify the lower with the experimentally known breathing mode energy

$$\hbar\omega_1 = E_{0^+} = E_{GMR} \; ; \tag{4.34}$$

the second corresponds to a higher mode (still to be found).

In figure 9 we show the results of the semiclassical calculations obtained in this way with the SkM* force; they are seen to reproduce perfectly the experimental peak energies within their error bars.

This result illustrates, as an example, the usefulness of the variational ETF approach also in dynamical applications. In fact, the breathing mode energies shown in fig. 9 are practically identi-

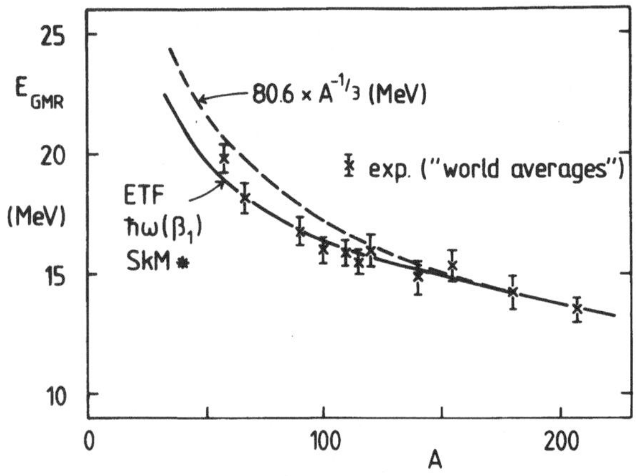

Fig. 9. Breathing mode (giant monopole resonance, GMR) energies versus nucleon number. Crosses are experimental peak energies with error bars, taken from ref. 36. The solid line shows the energies $\hbar\omega(\beta_1) = \hbar\omega_1$ found from the ETF model calculations[68] with the SkM* force.

cal with those which are obtained in microscopical RPA calculations at much higher cost. Furthermore, the interpretation in terms of oscillating parameters of simple trial densities gives a rather nice physical insight into the role of the couplings of surface and bulk contributions to the breathing mode.[68]

5. ACKNOWLEDGEMENTS

The author is indebted to C. Guet and H.-B. Håkansson, with whom most of the ETF calculations were performed in a fruitful collaboration, and to J. Bartel and W. Stocker for their contributions to the work discussed in the last section.

Note added in proof:

The correct functionals $\tau[\rho]$ and $\sigma[\rho]$ up to second order with temperature-dependent coefficients have meanwhile been derived in ref.[93].

REFERENCES

1. C.F. v. Weizsäcker, Z. Phys. 96: 431 (1935)
2. H.A. Bethe and F. Bacher, Rev. Mod. Phys. 8: 82 (1936)
3. T.H.R.Skyrme, Phil. Mag. 1: 1043 (1956); Nucl. Phys. 9: 615 (1959)
4. R.A. Berg and L. Wilets, Phys. Rev. 101: 201 (1956); L. Wilets, Rev. Mod. Phys. 30: 542 (1958)
5. K. Kumar and R.K. Bhaduri, Phys. Rev. 122: 1926 (1961)
6. K.A. Brückner, J.R. Buchler, S. Jorna and R.L. Lombard, Phys. Rev. 171: 1188 (1968); H.A. Bethe, Phys. Rev. 167: 879 (1973)
7. R.J. Lombard, Ann. of Phys. 77: 380 (1973)
8. J.P. Hohenberg and W. Kohn, Phys. Rev. 136: B 864 (1964)
9. G.E. Brown and A.D. Jackson, The Nucleon-Nucleon-Interaction (North-Holland/Amer. Elsevier, New York, 1976)
10. K.A. Brückner, Phys. Rev. 97: 1353 (1955)
11. K.A. Brückner, J.L. Gammel and H. Weitzner, Phys. Rev. 110: 431 (1958)
12. H.A. Bethe, Ann. Rev. Nucl. Sci. 21: 93 (1972)
13. J.W. Negele, Phys. Rev. C1: 1260 (1970); J.W. Negele and D. Vautherin, Phys. Rev. C5: 1472 (1972); C11: 1031 (1975)
14. C. Guet, H.-B. Håkansson and M. Brack, Phys. Lett. 97 B: 7 (1980)
15. M. Brack, C. Guet and H.-B. Håkansson, Regensburg Preprint TPR-83-16, submitted to Physics Reports
16. J. Bartel, P. Quentin, M. Brack, C. Guet and H.-B. Håkansson, Nucl. Phys. A 386: 79 (1982)
17. O. Bohigas, A.M. Lane and J. Martorell, Phys. Reports 51: 267 Phys. Reports 51: 267 (1979)
18. J.P. Blaizot, Phys. Reports 64: 171 (1980)
19. P. Quentin and H. Flocard, Ann.Rev.Nucl.Part.Sci. 28: 523 (1978)
20. D. Vautherin and D.M. Brink, Phys. Rev. C 5: 626 (1972)
21. V.M. Strutinsky, Nucl. Phys. A 95: 420 (1967); A 122: 1 (1968)
22. M. Brack, J. Damgård, A.S. Jensen, H.C. Pauli, V.M. Strutinsky and C.Y. Wong, Rev. Mod. Phys. 44: 320 (1972)
23. M. Brack and P. Quentin, Nucl. Phys. A 361: 35 (1981)
24. E.P. Wigner, Phys. Rev. 40: 749 (1932); J. G. Kirkwood, Phys. Rev. 44: 31 (1933)
25. R.K. Bhaduri and C.K. Ross, Phys. Rev. Lett. 27: 606 (1971)
26. B.K. Jennings, Nucl. Phys. A 207: 538 (1973)
27. B.K. Jennings, R.K. Bhaduri and M. Brack, Nucl. Phys. A 253: 29 (1975) and references quoted therein
28. M. Brack and H.C. Pauli, Nucl. Phys. A 207: 401 (1973)
29. M. Brack, B.K. Jennings and Y.H. Chu, Phys. Lett. 65 B: 1 (1976)
30. B. Grammaticos and A. Voros, Ann. of Phys. 123: 359 (1979); 129: 153 (1980)
31. C. Guet and M. Brack, Z. Phys. A 297: 247 (1980)
32. D. Vautherin, Phys. Rev. C 7: 296 (1973)
33. C. Titin-Schnaider and P. Quentin, Phys. Lett. 49 B: 397 (1974)

34. M.J. Giannoni and P. Quentin, Phys. Rev. C 21: 2076 (1980)
35. M. Beiner, H. Flocard, N.V. Giai and P. Quentin, Nucl. Phys. A 238: 29 (1975)
36. J. Treiner, H. Krivine, O. Bohigas and J. Martorell, Nucl. Phys. A 371: 253 (1981)
37. F. Tondeur, Phys. Lett. 123 B: 139 (1983); F. Tondeur, M. Brack, M. Farine and J.M. Pearson, Preprint 1983
38. W. Kohn and L.J. Sham, Phys. Rev. 137: A 1697 (1965); 140: A 1133 (1965)
39. R. Balian and C. Bloch, Ann. of Phys. 69: 76 (1972) and earlier references quoted therein
40. V.M. Strutinsky, A.G. Magner, S.R. Ofengenden and T. Døssing, Z. Phys. A 283: 269 (1977)
41. A.S. Tyapin. Sov. J. Nucl. Phys. 11: 401 (1970); 14: 50 (1972)
42. G.E. Uhlenbeck and E. Beth, Physica 3: 729 (1936)
43. B.K. Jennings, Ph. D. Thesis, McMaster University, 1976 (unpublished)
44. A. Voros, Thèse d'Etat, Paris University, Orsay, 1977 (unpublished)
45. N.L. Balazs and B.K. Jennings, preprint TRI-PP-83-55, 1983, to be published in Physics Reports
46. H. Krivine and J. Treiner, Phys. Lett. 88 B: 212 (1979)
47. X. Campi and S. Stringari, Nucl. Phys. A 337: 313 (1980)
48. J. Treiner and H. Krivine, Preprint Orsay IPNO/TH 82-18, 1982.
49. W. Stocker, G. Süssmann and S. Knaak, Nucl. Phys. A 187: 38 (1972)
50. O. Bohigas, X. Campi, H. Krivine and J. Treiner, Phys. Lett. 64 B: 381 (1979)
51. H. Krivine, J. Treiner and O. Bohigas, Nucl. Phys. A 366: 155 (1980)
52. E. Lieb, Rev. Mod. Phys. 53: 603 (1981)
53. V.M. Strutinsky and A.S. Tyapin. Sov. Phys. JETP 18: 664 (1964)
54. W.D. Myers and W.J. Swiatecki, Ann. of Phys. 55: 395 (1969); 84: 186 (1974)
55. Y.H. Chu, B.K. Jennings and M. Brack, Phys. Lett. 68 B :407 (1977); see also Y.H. Chu, Ph.D. Thesis, Stony Brook 1977, unpubl.
56. C.M. Ko, H.C. Pauli, M. Brack and G.E. Brown, Nucl. Phys. A 236: 269 (1974)
57. M. Brack, Phys. Lett. 81 B: 239 (1977)
58. J.B. Bellicard et al., Saclay progress report CEA-N-2207: 81(1981)
59. S. Köhler, Nucl. Phys. A 258: 301 (1976)
60. H. Flocard, P. Quentin, A.K. Kerman and D. Vautherin, Nucl. Phys. A 203: 433 (1973)
61. D. Gogny, Nucl. Phys. A 237: 399 (1975)
62. M. Brack, in "Physics and Chemistry of Fission 1979", Jülich (IAEA Vienna, 1980) Vol. I, p. 227
63. W.D. Myers and W.J. Swiatecki, Nucl. Phys. 81: 1 (1966)
64. M. Brack, C. Guet, H.-B. Håkansson, A. Magner and V.M. Strutinsky, 4th Int. Conf. on nuclei far from stability,

Helsingør 1981 (CERN 81-09, Geneva) p. 65
65. M. Farine, J. Côté and J.M. Pearson, Phys. Rev. C 24: 303 (1981), and earlier references therein
66. M. Pearson, Nucl. Phys. A 376: 507 (1982)
67. F. Tondeur, J.M. Pearson and M. Farine, Nucl. Phys. A 394: 462 (1983)
68. M. Brack and W. Stocker, Nucl. Phys. A 388: 230 (1982); A 406: 413 (1983)
69. F. Tondeur, Nucl. Phys. A 315: 353 (1978)
70. R.K. Bhaduri, Phys. Rev. Lett. 39: 329 (1977)
71. M. Durand, M. Brack and P. Schuck, Z. Phys. A 286: 381 (1978); A 296: 87 (1980)
72. J. Bartel, M. Durand and M. Brack, Z. Phys. A 315: 341 (1984);
73. J. Bartel and M. Vallieres, Phys. Lett. 114 B: 303 (1982)
74. J. Bartel, PhD Thesis, University of Regensburg, 1984; and to be published
75. M. Lefort, Nucl. Phys. A 387: 3c (1982), and references quoted therein
76. see, e.g. J.M. Lattimer, Ann. Rev. Nucl. Part. Sci. 31: 337 (1981), and references quoted therein
77. P. Bonche and D. Vautherin, Nucl. Phys. A 372: 496 (1981); Astron. Astrophys. 112: 168 (1982)
78. see, e.g. J. Des Cloiseaux in "Many Body Physics", Les Houches 1967 (Gordon and Breach, New York, 1968) p. 1
79. M. Brack and P. Quentin, Phys. Lett. 52 B: 159 (1974); Physica Scripta A 10: 163 (1974); see also ref. 23
80. U. Mosel, P. Zint and K.H. Passler, Nucl. Phys. A 236: 252 (1974)
81. G. Sauer, H. Chandra and U. Mosel, Nucl. Phys. A 264: 221 (1976)
82. M. Barranco and J.R. Buchler, Phys. Rev. C 22: 1729 (1980); C 24: 1191 (1981);
D.Q. Lamb, J.M. Lattimer, C.J. Pethick and D.G. Ravenhall, Nucl. Phys. A 360: 459 (1981)
83. D.G. Ravenhall, C.J. Pethick and J.M. Lattimer, Nucl. Phys. A 407: 572 (1983)
84. J. Bartel, M. Brack and C. Guet, Phys. Lett. B, in print
85. see, e.g., E.C. Stoner, Phil. Mag. 28: 257 (1939)
86. X. Campi and S. Stringari, Z. Phys. A 309: 239(1983)
87. M. Barranco, M. Pi and X. Viñas, Phys. Lett. 124 B: 131 (1983), and references quoted therein
88. M. Barranco and J. Treiner, Nucl. Phys. A 351: 269 (1981)
89. J. Treiner, preprint Orsay IPNO/TH 83-26
90. W. Stocker and J. Burzlaff, Nucl. Phys. A 202: 265 (1973); R.W. Hasse and W. Stocker, Phys. Lett. 44 B: 26 (1973)
91. G. Holzwarth, these proceedings
92. J. Treiner, these proceedings
93. M. Brack, Workshop on semiclassical methods in nuclear physics, Grenoble, 1984, to be published in Journal de Physique (Paris).

THE SCALING APPROACH TO NUCLEAR GIANT MULTIPOLE RESONANCES

Gottfried Holzwarth

Universität Siegen, FB 7

59 Siegen 21, West Germany

INTRODUCTION

In the following lectures we shall discuss an example where straight application of density-functional methods fails to give a useful approximation to the dynamical behaviour of a many-body system. We shall show, however, that a rather simple extension of the functional method allows for a reliable description of many dynamical features of collective motion in Fermi fluids.

Specifically we shall be concerned with the "Giant Multipole Resonances" (GMR)[1] observed in almost all atomic nuclei which may be interpreted as collective oscillations of the nucleon fluid. Many different types of GMR have been theoretically suggested, distinguished by their isospin T (protons and neutrons moving in phase (T=0), or in opposite phase (T=1)), their parity Π ("Electric modes" with $\Pi = (-)^L$, "Magnetic modes" with $\Pi = (-)^{1+L}$, where L characterizes the multipolarity of the GMR), "surface modes" characterized by a flow-velocity field $\vec{v}$ satisfying $\vec{\nabla}\cdot\vec{v} \equiv 0$, or "compressional" modes which involve density changes in the nuclear interior. Quite a number of these have been experimentally identified and successfully interpreted in the microscopic formalism of the "Random Phase Approximation" (RPA). The outstanding feature of GMR is that they exhaust a large fraction of corresponding sum rules, which means that many nucleons move in a coherent way. This leads to the expectation that it should be possible to describe their basic properties in terms of simple field variables, like the density $\rho(\vec{x},t)$, or the velocity $\vec{v}(\vec{x},t)$.

Especially after the successful application of energy-density functionals

$$E[\rho] = \int(\tau[\rho] + V[\rho])\, d^3x \tag{1}$$

for deriving static nuclear properties[2] through the variational equation

$$\frac{\delta}{\delta\rho} E[\rho] - \lambda_F = 0 \tag{2}$$

one tried to use the same functional (1) also in dynamical calculations. For the potential part $V[\rho]$ in (1) the Skyrme parametrization[3] has proven very useful, we shall, however, for simplicity discuss here a simplified version

$$V[\rho] = \sum_\sigma a_\sigma \rho^\sigma + a_s(\vec{\nabla}\rho)^2 \tag{3}$$

which consists of a volume part $\Sigma a_\sigma \rho^\sigma$ and a surface part $a_s(\vec{\nabla}\rho)^2$ with constants a_σ and a_s. The local density $\rho(\vec{x},t)$ is the sum of proton and neutron single-particle densities as we shall consider here mainly T=O states of spin-saturated systems.

In order to obtain also the kinetic energy density τ as a functional of the local density ρ it is necessary to make assumptions about the distribution of particle momenta $\vec{p}$ in phase space. Assuming isotropic momentum distribution and the occupation of available states up to $|\vec{p}| = P_F$, i.e.

$$f_o(\vec{x},\vec{p}) = f_o(\vec{x},p) \propto \Theta(P_F - p) \tag{4}$$

for the groundstate distribution function f_o, leads to the kinetic energy functional approximated by[4]

$$\tau[\rho] = \frac{\hbar^2}{2m}\left(\frac{3}{5}\left(\frac{3}{2}\pi^2\right)^{2/3}\rho^{5/3} + \frac{1}{4}\eta\frac{(\nabla\rho)^2}{\rho} + \zeta\Delta\rho\right). \tag{5}$$

The volume term $\rho^{5/3}$ is the usual Thomas Fermi expression. Together with the surface terms with parameters η and ζ we shall refer to (5) as the "Extended Thomas Fermi" (ETF) approximation.

Once we accept the equation of state (1) to hold also for time-dependent densities, conservation of mass and momentum flow leads to the coupled set of fluid-dynamical equations

$$\frac{\partial}{\partial t}\rho + \vec{\nabla}\cdot\vec{j} = 0 \tag{6}$$

$$\frac{\partial}{\partial t}\vec{j} = -\frac{1}{m}\rho\vec{\nabla}\frac{\delta}{\delta\rho}E[\rho] \; . \tag{7}$$

In the Euler equation (7) we have already linearized the left-hand side as we shall consider only small oscillations. The current $\vec{j}$ is connected with the velocity field $\vec{v}(\vec{x},t)$ as usual

$$\vec{j} = \rho\vec{v}. \tag{8}$$

We shall refer to the set (6), (7) as "hydrodynamics" or "first sound" propagation. It may now be easily seen that this approach must fail to reproduce essential features of surface modes (satisfying $\vec{\nabla}\cdot\vec{v} \equiv 0$) encountered in droplets of Fermi fluids:

Frequencies ω for a velocity field $\vec{v}(\vec{x},t) = \dot{\alpha}(t)\,\vec{u}(\vec{x})$ are given by[5 6]

$$\omega^2 = C/B \tag{9}$$

with the inertia parameter

$$B = \frac{1}{2} m \int \rho_0\, u^2\, d^3x \tag{10}$$

and the force constant

$$C = \frac{1}{2}\int \left(\frac{\delta^2}{\delta\rho^2} E[\rho]\right)_{\rho=\rho_0} (\delta\rho)^2\, d^3x = \frac{1}{2}\frac{\partial^2}{\partial\alpha^2} E[\rho] \; . \tag{11}$$

where

$$\rho = \rho_0 + \alpha\,\delta\rho \quad , \quad \delta\rho = -\vec{\nabla}\cdot(\rho_0\vec{u}) \; .$$

Evidently, volume parts in $E[\rho]$ cannot contribute to C for a velocity field $\vec{u}$ satisfying $\vec{\nabla}\cdot\vec{u} = 0$, because for the contributions of the volume part E_{vol} of E to C we have

$$C = \frac{1}{2}\int\left(\frac{\partial^2}{\partial\rho_0^2} E_{vol}[\rho_0]\right)(\vec{u}\cdot\vec{\nabla}\rho_0)^2 d^3x = \frac{1}{2}\int(\vec{u}\cdot\vec{\nabla}\rho_0)\left(\vec{u}\cdot\vec{\nabla}\frac{\partial}{\partial\rho_0}E_{vol}[\rho_0]\right)d^3x.$$

According to (2) $\partial/\partial\rho_0\, E_{vol}\,[\rho_0]$ may be completely expressed through surface terms and the constant λ_F therefore C can be expressed through surface terms alone. As an example we consider the Giant Quadrupole Resonance (GQR) with

$$\vec{u} = \vec{\nabla}Q(\vec{x}), \qquad Q(\vec{x}) = r^2 Y_{20} \tag{12}$$

which leads to[6]

$$\omega^2 = \frac{4\, E_{sur}[\rho_0]}{m\, A\, \langle r^2 \rangle} \quad . \tag{13}$$

The surface part of the groundstate energy $E_{sur}[\rho_0]$ in the numerator and the mean square radius $\langle r^2 \rangle$ in the denominator are both proportional to $A^{2/3}$, therefore we obtain $\omega \propto A^{-1/2}$. This result is in serious conflict with the experimentally observed GQR which by its formfactors is clearly identified as a surface mode but occurs near $\hbar\omega \approx 60\ A^{-1/3}$ MeV[1].

Quantum mechanical sum rules[7] may be used to estimate frequencies of modes which exhaust large portions of the sums:

$$\hbar^2\omega^2 \approx m_3/m_1 \tag{14}$$

with

$$m_1 = \sum_n \hbar\omega_n |\langle n|Q|o\rangle|^2 = \frac{1}{2} \langle o|\, [Q, [H,Q]]\, |o\rangle$$

$$m_3 = \sum_n \hbar^3\omega_n^3 |\langle n|Q|o\rangle|^2 = -\frac{1}{2} \langle o|\, [[Q,H], [H, [Q,H]]]\, |0\rangle \ .$$

Using (12) it turns out that the numerator of (14) contains also the total ground-state kinetic energy $\langle T \rangle$ together with the surface part of the potential

$$m_3/m_1 = \frac{4(\langle T \rangle + \langle V \rangle_{sur})}{m\, A\, \langle r^2 \rangle / \hbar^2} \ . \tag{15}$$

This expression gives the right $A^{-1/3}$ dependence for the frequency because the volume part of $\langle T \rangle$ dominates in the numerator. Comparison of (13) and (15) gives clear evidence that the kinetic energy is not treated correctly in the energy-density formalism if we use the static equation of state (1) with (5) in studying the dynamical behaviour of the system.

THE SCALING APPROXIMATION

In the following we shall discuss a simple extension of the set (6), (7) of the form[8 9]

$$\rho \frac{\partial}{\partial t} \vec{s} + \vec{j} = 0 \tag{16}$$

$$\frac{\partial}{\partial t} \vec{j} = \frac{1}{m} \frac{\delta}{\delta \vec{s}} E[\vec{s}] \ . \tag{17}$$

We shall call this the "Scaling Approximation" (SCA) because the modified Euler equation (17) results from the quantum-mechanical time-dependent variational principle

$$\delta \int \langle\psi | i\partial_t - H | \psi\rangle \, dt = 0$$

if the deviations of the time-even part $\hat{\rho}_+$ of the single-particle density matrix $\hat{\rho}$ from its ground-state values $\hat{\rho}_0$ are restricted to the form

$$\hat{\rho}_+(\vec{x},\vec{x}',t) = \exp(\tfrac{1}{2}(\vec{s}(\vec{x},t)\cdot\vec{\nabla} + \vec{\nabla}\cdot\vec{s}(\vec{x},t))) \times$$

$$\times \exp(\tfrac{1}{2}(\vec{s}(\vec{x}',t)\cdot\vec{\nabla}' + \vec{\nabla}'\cdot\vec{s}(\vec{x}',t))) \, \hat{\rho}_0(\vec{x},\vec{x}') . \quad (18)$$

This means that the time-dependence is introduced into $\hat{\rho}_+$ by scaling the arguments $\vec{x}$ and $\vec{x}'$ of $\hat{\rho}_0$ with an arbitrary (variational) time-dependent scaling field $\vec{s}(\vec{x},t)$. If we let $\vec{x}=\vec{x}'$ in (18) we obtain for the local density

$$\rho(\vec{x},t) = \exp(\vec{\nabla}\cdot\vec{s}(\vec{x},t))\,\rho_0(\vec{x}) \ . \quad (19)$$

(In (18) and (19) the gradients are understood to act on all functions standing to the right of the $\vec{\nabla}$). Equation (16) defines the velocity in terms of the scaling field

$$\vec{v} = -\frac{\partial}{\partial t}\vec{s} \ . \quad (20)$$

From (19) follows $\dot{\rho} = \vec{\nabla}\cdot(\rho\dot{\vec{s}})$ therefore the continuity equation (6) is satisfied. It should, however, be noted that (6) equates only the longitudinal parts of velocity and scaling field, while (16) requires also the transverse parts to be equal.

Through (18) the kinetic energy part in E becomes a functional of $\vec{s}$ which to any desired order is obtained from

$$\tau[\vec{s}] = \frac{\hbar^2}{2m}(-\Delta_y \hat{\rho}_+(\vec{x},\vec{x}',t))_{\vec{y}=0} \quad , \quad (\vec{y} = \vec{x} - \vec{x}') \ . \quad (21)$$

As we shall be concerned here with the harmonic approximation we give only the second order term[8] $E^{(2)}[\vec{s}]$:

$$E^{(2)}[\vec{s}] = \frac{1}{2}\int \tau^{(o)}_{ik}(s_{ij}s_{jk}+s_{ij}s_{kj}-s_{\ell}s_{ik\ell})d^3x + \frac{\hbar^2}{8m}\int s_{ik\ell}s_{\ell ki}\rho_0 d^3x$$

$$+ \int V^{(2)}[\vec{s}]d^3x \quad (22)$$

where

$$s_{ijk..} = .. \frac{\partial}{\partial x_k} \frac{\partial}{\partial x_j} s_i \,,$$

$$\tau^{(o)}_{ik} = - \frac{\hbar^2}{m} \left(\frac{\partial}{\partial y_i} \frac{\partial}{\partial y_k} \hat{\rho}_0(\vec{x},\vec{x}')\right)_{\vec{y}=0} \tag{23}$$

and $V^{(2)}$ is the second order term in an expansion of (19) and (3) in terms of $\vec{s}$. It is this functional (22) which now enters into the equation of motion (17).

It is easily seen that the functional (22) resolves the problem encountered in the first sound dynamics. In order to show this it is sufficient to consider only the volume terms in (22) and to omit also the quantum correction term containing the square of second order derivatives $s_{ik\ell}s_{\ell ki}$. The volume terms in the static quantity $\tau^{(o)}_{ik}$ must be diagonal ($\tau^{(o)}_{ik} = \delta_{ik} \frac{2}{3} \tau^{(o)}_{vol}$) therefore we have from (22)

$$E_{vol}[\vec{s}] = \int \{ \frac{1}{3}\tau^{(o)}_{vol}(s_{ij}s_{ji} + s_{ij}s_{ij} + s_{ii}s_{jj})$$
$$+ \frac{1}{2}(\frac{\partial^2}{\partial \rho_0^2} V_{vol}(\rho_0))\, \rho_0^2\, s_{ii}s_{jj} \} d^3x.$$

For $\vec{s} = \alpha(t)\, \vec{\nabla} r^2 Y_{20}$ we obtain

$$E_{vol} = \alpha^2 \frac{5}{4\pi} \cdot 4 \int \tau^{(o)}_{vol}\, d^3x, \qquad B = \frac{1}{2}\, m \frac{5}{4\pi}\, 2 \int \rho_0 r^2\, d^3x \,.$$

Therefore we get for the frequency (with $C = \frac{1}{2}(\partial^2/\partial\alpha^2 E)$)

$$\omega^2 = \frac{C}{B} = 4\int\tau^{(o)}_{vol}\, d^3x \,/\, m\, A \langle r^2\rangle$$

in agreement with the sum-rule value (15). The physical reason for this result is that through (18) we have allowed for dynamical distortions of the local momentum distribution which can be easily demonstrated by evaluating the Wigner transform of the scaled density matrix (18)[8].

We have obtained the above results without really solving the dynamical problem (16), (17) but just by making use of the energy functionals (1) or (22) together with an imposed flow field (12). This is not unphysical but rather corresponds to the physical situation where a certain flow pattern is forced on the system through a given excitation mechanism. However, for studying the detailed response of the system to a given external field we must solve the set (16), (17). After variation of (22) with respect to $\vec{s}$ and making use of the equilibrium condition

$$\partial_k \tau_{ik}^{(o)} + \partial_i \sum_\sigma a_\sigma (\sigma-1)\, \rho_o^\sigma - 2a_s\, \rho_o\, \partial_i\, \Delta\rho_o = 0 \qquad (24)$$

the basic fluid-dynamical equation of motion in SCA reads[8]

$$- m\omega^2 \rho_o s_i = \partial_j (\tau_{ik}^{(o)} (s_{jk}+s_{kj})) + s_{jki}\tau_{jk}^{(o)} - \frac{\hbar^2}{4m} \partial_k \partial_\ell (s_{\ell ki}\rho_o)$$
$$+ \rho_o \partial_i (\rho_o s_{jj} \frac{\partial^2}{\partial\rho_o^2} \sum_\sigma a_\sigma \rho_o^\sigma)$$
$$+ 2a_s \rho_o \partial_i (\Delta\partial_j (s_j \rho_o) - s_j \Delta\partial_j \rho_o) \quad . \qquad (25)$$

The quantum correction and the surface term of the potential make it a fourth-order differential equation for the scaling field $\vec{s}$. It should be stressed that in contrast to the density-functional method (6), (7) (where we are forced to use an approximation like ETF for τ (5)) the static input functions ρ_o and $\tau_{ik}^{(o)}$ in (25) can be calculated in any desired approximation. A very good choice would be to take ρ_o and $\tau_{ik}^{(o)}$ from a Hartree-Fock calculation with the self-consistent field $w[\rho]$ given by the functional derivative

$$w[\rho] = \frac{\delta}{\delta\rho} \int V[\rho]\, d^3x = \sum_\sigma \sigma\, a_\sigma \rho^{\sigma-1} - 2\, a_s \Delta\rho \; . \qquad (26)$$

Then ρ_o and $\tau_{ik}^{(o)}$ would reflect the detailed single-particle structure of the groundstate which, as we shall see, is especially important for the asymptotic ($r \to \infty$) behaviour of the solutions of (25). Much more crude and much simpler would be to take $\tau_{ik}^{(o)}$ in the ETF approximation and to obtain ρ_o from the variational problem (2). One might even obtain analytical solutions of (25) by using a simple square density $\rho_o(r) = \rho_{oo}\Theta(R-r)$ and take the TF value for $\tau_{ik}^{(o)}$

$$(\tau_{ik}^{(o)})_{TF} = \delta_{ik} \frac{1}{5} \rho_{oo} \frac{p_F^2}{m} \; . \qquad (27)$$

Only in this last case it will be necessary to explicitly supply boundary conditions for $\vec{s}$ at the nuclear surface r=R. In all other cases the normalization conditions

$$\int \rho_o\, \vec{s}^{(n)} \cdot \vec{s}^{(m)}\, d^3x = \delta_{nm} \qquad (28)$$

for solutions with discrete eigenfrequencies $\omega_n^2 \neq \omega_m^2$ or

$$\int \rho_o\, \vec{s}^{(\omega)} \cdot \vec{s}^{(\omega')}\, d^3x = \delta(\omega-\omega') \qquad (29)$$

for solutions with continuous eigenvalues prove to be sufficient to completely determine the solutions of (25)[10]. The orthogonality of solutions with different eigenfrequencies is due to the fact that (25) may be written in the form

$$m\omega^2 \rho_0 \vec{s}(\vec{x}) = \vec{\mathcal{L}}\,[\vec{s}(\vec{x})]$$

where the differential (vector) operator

$$\vec{\mathcal{L}}[\vec{s}] \equiv \frac{\delta}{\delta \vec{s}} E[\vec{s}]$$

is hermitian for any two vector fields $\vec{\varphi}$ and $\vec{\psi}$ for which the integrals exist:

$$\int \vec{\varphi} \cdot \vec{\mathcal{L}}[\vec{\psi}]\, d^3x = \int \vec{\psi} \cdot \vec{\mathcal{L}}[\vec{\varphi}]\, d^3x \quad .$$

After these general remarks about the SCA we shall discuss at first a few analytical properties of (25) for the case of a simple square density.

SQUARE DENSITY MODEL[11 12]

We omit surface terms from (3) and introduce the abbreviation

$$v(\rho_0) = \sum_\sigma a_\sigma \rho_0^\sigma \qquad (\rho_0 \equiv \rho_{00} = \text{const.})$$

and the Landauparameter

$$F_o = \frac{3m}{P_F^2} \rho_0 \frac{\partial^2}{\partial \rho_0^2} v(\rho_0) \quad . \tag{30}$$

First we notice that (25) implies a quite different sound speed for compressional modes as compared to the first sound propagation. From (6), (7) we obtain for the density change $\delta\rho = \rho - \rho_0$

$$-\, m\omega^2 \delta\rho = \rho_0 \left(\frac{\partial^2}{\partial \rho_0^2} E(\rho_0)\right) \Delta\delta\rho = \frac{P_F^2}{3m}(1+F_o)\ \Delta\delta\rho, \tag{31}$$

i.e., the first sound speed is $c_L^2 = v_F^2(1+F_o)/3$. For the square density (25) simplifies (with (27)) to

$$-\, m\omega^2 \vec{s} = \frac{P_F^2}{m}\left(\frac{2}{5} + \frac{F_o}{3}\right) \vec{\nabla}(\vec{\nabla}\cdot s) + \frac{P_F^2}{5m} \Delta\vec{s} \quad . \tag{32}$$

Therefore the longitudinal part of $\vec{s}$ is determined by

$$- m\omega^2 \vec{\nabla}\cdot\vec{s} = \frac{P_F^2}{m}\left(\frac{3}{5} + \frac{F_o}{3}\right) \Delta\, (\vec{\nabla}\cdot\vec{s}) \tag{33}$$

i.e. the longitudinal scaling sound speed is $c_L^2 = v_F^2(3/5+F_o/3)$. We see that the contribution from the kinetic energy to the longitudinal sound speed is almost twice as large as in the first sound result. Interestingly (33) allows for transverse modes with a transverse sound speed $c_T^2 = v_F^2/5$ while in first sound propagation such modes are excluded.

Another decisive difference concerns the boundary conditions which solutions of (31) or (32) should satisfy at the nuclear surface r=R. The physical requirement is that no forces should be exerted on the freely moving surface. In the density-functional method the pressure is, of course, proportional to the density change $\delta\rho$, i.e. we must require for solutions of (31) (which are of the form $\delta\rho_L = j_L(kr)Y_{LM}$)

$$j_L(kR) = 0. \tag{34}$$

The pressure tensor underlying the dynamics (32), however, is

$$P_{ik} = \frac{P_F^2}{m}\left(\frac{F_o}{3} + \frac{1}{5}\right) \delta_{ik}\, s_{jj} + \frac{P_F^2}{5m}(s_{ik} + s_{ki}) \tag{35}$$

which can be seen by writing (32) in the Euler form

$$\frac{\partial}{\partial t} j_i = - \frac{1}{m} \partial_k P_{ik} \quad , \quad \text{with } P_{ik} = P_{ki} \quad . \tag{36}$$

On the free surface we therefore must require

$$x_k P_{ik}\big|_R = 0. \tag{37}$$

This condition is in general much more complicated than (34) because it mixes the transverse and longitudinal components in the solutions of (32) which are of the form (for Electric modes)

$$\vec{s}_L = \alpha_L \vec{\nabla} (j_L(k_{||}r)Y_{LM}) + \beta_L \vec{\nabla} \times (j_L(k_{\perp}r)\vec{Y}_{LLM}) \ . \tag{38}$$

As we have seen, in first sound dynamics the restoring forces for surface modes originate in the surface energy alone, therefore the square-density model contains no first sound surface modes. The scaling mechanism, however, does provide for the volume terms which comprise the essential part of the restoring force for surface modes. Therefore one might wonder to which extent the solutions (38) may represent surface modes: In fact, the lowest energy eigenvalue obtained from (37), (38) for L=2 is $\hbar\omega \approx 56\ A^{-1/3}$ MeV in rather close agreement with our previous result. On the other hand, it is clear that (38) cannot really represent pure surface modes unless $k_{||}R \to 0$, which is not possible because $k_{||}$ is tied to the frequency through the fixed sound speed. The actual values of $k_{||}R$ turn out to be of the order of 1 (or 2 for $k_{\perp}R$) so that in the nuclear interior the Bessel-functions in (38) are still well approximated by powers r^L. In this limit there are no transverse components in (38) (because of the identity $\vec{\nabla} \times r^L\vec{Y}_{LLM} \propto \vec{\nabla}\, r^L Y_{LM}$), but the deviations of $j_L(k_{\perp}r)$ from $(k_{\perp}r)^L$ near the nuclear surface represent genuine transverse components. It is striking to notice that for smooth self-consistent densities $\rho_0(r)$ the combination of longitudinal and transverse components acts to reestablish the Tassie transition density

$$\delta\rho_L = \vec{\nabla} \cdot (\rho_0(r)\vec{\nabla} r^L Y_{Lo}) \tag{39}$$

throughout the nuclear surface with very good accuracy. On the other hand one may notice that (33) does allow for pure surface modes if we restrict $\vec{s}$ from the outset to be irrotational

$$\vec{s} = \vec{\nabla}\,\phi \tag{40}$$

and solve (33) for the displacement potential ϕ:

$$\phi_L = (j_L(kr) + \alpha_L r^L)\ Y_{LM}\ . \tag{41}$$

In this case the transverse part of the boundary condition (37) is replaced by another condition[13] due to additional surface terms which arise from the partial integration of the gradient in (40), such that both constants k and α_L in (41) are uniquely determined.

Eq. (32) contains also purely transverse "Magnetic" modes

$$\vec{s}_L = j_L(k_{\perp}r)\ \vec{Y}_{LLM}\ . \tag{42}$$

Writing the excitation operator $B_L^+ = \vec{s}_L \cdot \vec{\nabla}$ with (42) in the form

$$B_L^+ = j_L(k_{\perp}r)\vec{Y}_{LLO} \cdot \vec{\nabla} = \frac{1}{r} j_L(k_{\perp}r) F_L(\Omega)\hat{L}_z \quad , \quad (\hat{L}_z = i\vec{x} \times \vec{\nabla})$$

gives a very intuitive picture of these Magnetic modes: They are rotations around the z-axis by coordinate-dependent angles[8]. A most interesting type is the Magnetic Quadrupole mode (L=2) for which one has $F_2(\Omega) = \cos\Theta$, therefore (for $k_\perp R \approx 1$)

$$B_2^+ \approx z\hat{L}_z = [\vec{x} \otimes \vec{L}]_{20} \; . \tag{43}$$

Evidently, the 2^- mode represents a twisting motion of the nucleus where the top is rotated in opposite phase to the bottom. In the square density model the twist frequency occurs near $\hbar\omega \approx 50 A^{-1/3}$MeV (depending on ρ_0). For general smooth surface densities $\rho_0(r)$ it is, however, quite sensitive to the surface profile and lies near $40\ A^{-1/3}$ for realistic nuclei. This is in good agreement with experimental results for the center of the 2^- strength distribution. Microscopically (43) shows that B_2^+ involves $1\ \hbar\omega_0$ (one shell distance) transitions, therefore we expect this mode to be strongly mixed with the $1\ \hbar\omega_0$ spin-flip mode $[\vec{x} \otimes \vec{\sigma}]_2$ in a realistic nucleus. This makes a clear experimental identification of the twist components difficult which, however, would be highly desirable because the simple density-functional method does not contain these modes.

Before we present some detailed results of the scaling approach in more realistic cases we shall discuss its connection with the RPA in a classical analogue. This will show how the scaling concept (18) emerges as a step in a systematic expansion and allows us to go beyond it in order to check its validity.

CONNECTION WITH RPA IN THE CLASSICAL LIMIT

In the microscopic RPA method the normal coordinates $\hat{P}, \hat{Q}$ for collective modes are obtained by solving the variational problem

$$\delta \langle 0 | [\hat{P}, \hat{H}] - \hbar^2\omega^2 \frac{M}{i\hbar} \hat{Q} | 0 \rangle = 0, \tag{44a}$$

$$\delta \langle 0 | [\hat{Q}, \hat{H}] - \frac{i\hbar}{M} \hat{P} | 0 \rangle = 0 \; . \tag{44b}$$

Variation with respect to the ground-state determinant $|0\rangle$ leads to the RPA equations for the particle-hole and hole-particle matrix elements of $\hat{P}$ and $\hat{Q}$. (M is the collective inertia parameter). In the classical limit we replace the commutators by Poisson brackets $[\ ,\ ] \rightarrow i\hbar \{\ ,\ \}$, operators $\hat{A}$ by their Wigner transforms $A(\vec{x},\vec{p})$, and ground-state expectation values by averaging in phase space over the single-particle distribution function $f(\vec{x},\vec{p})$:

$$\delta \int (\{P(\vec{x},\vec{p}),\ h(\vec{x},\vec{p})\} + M\omega^2\ Q(\vec{x},\vec{p}))\ f(\vec{x},\vec{p})\ d^3x\ d^3p = 0, \tag{45a}$$

$$\delta \int (\{Q(\vec{x},\vec{p}),h(\vec{x},\vec{p})\} - \frac{1}{M} P(\vec{x},\vec{p}))f(\vec{x},\vec{p})\, d^3x\, d^3p = 0 . \qquad (45b)$$

We define local tensor fields $\chi(\vec{x})$, $s_\alpha(\vec{x})$, $\phi_{\alpha\beta}(\vec{x})$, $\Theta_{\alpha\beta\gamma}(\vec{x})$, etc. by expanding the time-even Q and time-odd P in powers of the momentum variable $\vec{p}$:

$$Q(\vec{x},\vec{p}) = \chi(\vec{x}) + \frac{1}{2} p_\alpha p_\beta\, \phi_{\alpha\beta}(\vec{x}) + \ldots \qquad (46)$$

$$P(\vec{x},\vec{p}) = s_\alpha(\vec{x})\, p_\alpha + \frac{1}{6}\, \Theta_{\alpha\beta\gamma}(\vec{x})\, p_\alpha p_\beta p_\gamma + \ldots \quad . \qquad (47)$$

and use for h the self-consistent single-particle Hamiltonian

$$h(\vec{x},\vec{p}) = \frac{p^2}{2m} + \frac{\partial}{\partial\rho} v(\rho(\vec{x})) \quad . \qquad (48)$$

The coupled equations for P and Q are then obtained by considering variations $\delta f = f - f_o$, where δf is restricted to the Fermi surface, i.e.

$$\delta f(\vec{x},\vec{p}) = \delta(p - p_F) \sum_{\kappa\mu} \nu_{\kappa\mu}(\vec{x}) Y_{\kappa\mu}(\hat{p}) \qquad (49)$$

with arbitrary functions $\nu_{\kappa\mu}(\vec{x})$.

For an infinite system this would corresponds to solving the Landau-Vlassov equation[14]. The fact that in a finite system boundary conditions have to be specified for each of the tensor fields χ, s_α, $\phi_{\alpha\beta}$, ... makes it necessary to truncate the expansions (46), (47). To the order indicated in (46), (47) one obtains the following equations of motion:

$$\bar{\chi} = -\frac{1}{M\omega^2} \frac{p_F^2}{3m} (1 + F_o)(\vec{\nabla}\cdot\vec{s} + \frac{p_F^2}{10} \partial_\alpha \Theta_{\alpha\gamma\gamma}) \qquad (50a)$$

$$s_\mu = \frac{M}{m} (\partial_\mu \bar{\chi} + \frac{p_F^2}{5} \partial_\gamma \bar{\phi}_{\gamma\mu}) - \frac{p_F^2}{10} \Theta_{\mu\gamma\gamma} \qquad (50b)$$

$$\bar{\phi}_{\mu\nu} = \delta_{\mu\nu} \frac{2}{3Mm\omega^2} (\vec{\nabla}\cdot\vec{s} + \frac{p_F^2}{7} \partial_\alpha \Theta_{\alpha\gamma\gamma}) \qquad (50c)$$

$$- \frac{1}{Mm\omega^2} (\partial_\mu (s_\nu + \frac{p_F^2}{14} \Theta_{\nu\alpha\alpha}) + \partial_\nu (s_\mu + \frac{p_F^2}{14} \Theta_{\mu\alpha\alpha}) + \frac{p_F^2}{7} \partial_\alpha \Theta_{\alpha\mu\nu})$$

$$\Theta_{\mu\nu\rho} = -\frac{2}{P_F^2}\,(\delta_{\mu\nu}(s_\rho - \frac{M}{m}\partial_\rho\overline{\chi}) + \delta_{\nu\rho}(s_\mu - \frac{M}{m}\partial_\mu\overline{\chi}) + \delta_{\mu\rho}(s_\nu - \frac{M}{m}\partial_\nu\overline{\chi}))$$

$$+ \frac{M}{m}(\partial_\mu\overline{\phi}_{\nu\rho} + \partial_\nu\overline{\phi}_{\rho\mu} + \partial_\rho\overline{\phi}_{\mu\nu}) \qquad (50d)$$

with

$$\overline{\chi} = \chi + \frac{P_F^2}{6}\,\phi_{\alpha\alpha}\ ,$$

and

$$\overline{\phi}_{\mu\nu} = \phi_{\mu\nu} - \frac{1}{3}\,\delta_{\mu\nu}\,\phi_{\alpha\alpha}\ .$$

In the square-density model the free variation of the nuclear surface at r=R in the coordinate space integration in (45) gives the scalar, vector, and tensor boundary conditions to be satisfied by the solutions of (50):

$$(M\omega^2\,\frac{P_F^2}{15}\,\phi_{\alpha\alpha} + \frac{P_F^2}{3m}(1+F_o)\,\partial_\alpha s_\alpha + \frac{P_F^4}{30\,m}(\frac{4}{7}+K_o)\partial_\alpha\Theta_{\alpha\gamma\gamma})_{r=R} = 0, \qquad (51a)$$

$$(x_\alpha\phi_{\alpha\beta})_{r=R} = 0, \qquad (51b)$$

$$(x_\alpha(\Theta_{\alpha\beta\gamma} - \frac{1}{3}\,\delta_{\beta\gamma}\Theta_{\alpha\nu\nu}))_{r=R} = 0\ . \qquad (51c)$$

Like the Landauparameter F_o, K_o is another constant depending on the form of $v(\rho)$.

It is easy to see that (50) contains the first sound propagation as the lowest approximation: Keeping only χ and $\vec{s}$ in (50a,b) yields

$$- m\omega^2\chi = \frac{P_F^2}{3m}\,(1 + F_o)\,\Delta\chi$$

in agreement with (31). The scalar boundary condition (51a) reduces to $(\vec{\nabla}\cdot\vec{s})_{r=R} = 0$ or with (50a) to $\chi|_R = 0$ in agreement with (34).

We shall now show that inclusion of the tensor $\phi_{\mu\nu}$ in (50a, b,c) is equivalent to the scaling approximation. Inserting (50a) and (50c) into (50b) yields

$$- m\,\omega^2 s_\mu = \frac{P_F^2}{m}\,(\frac{2}{5} + \frac{F_o}{3})\,\partial_\mu\vec{\nabla}\cdot\vec{s} + \frac{P_F^2}{5m}\,\Delta s_\mu$$

in agreement with (32). Inserting (50c) into the vector boundary condition (51b) yields

$$x_\nu \phi_{\nu\mu}\big|_R = x_\mu\left(\frac{1}{3}\phi_{\alpha\alpha} + \frac{2}{3Mm\omega^2}\vec{\nabla}\cdot\vec{s}\right) - \frac{1}{Mm\omega^2} x_\nu(\partial_\nu s_\mu + \partial_\mu s_\nu)\big|_R = 0 \ .$$

Eliminating the trace $\phi_{\alpha\alpha}$ through the scalar boundary condition (51a) and comparing with the pressure tensor (35) we have

$$x_\nu \phi_{\nu\mu}\big|_R = -\frac{5}{Mm\omega^2} x_\nu P_{\nu\mu}\big|_R \ .$$

Thus we have obtained equation of motion and boundary condition of the SCA. The correspondence goes still deeper: Evaluation of the current

$$j_\alpha = i\omega M \int \{f_o, Q\} \frac{p_\alpha}{m} d^3p = i\omega \frac{M}{m} \rho_0(\partial_\alpha \bar{\chi} + \frac{p_F^2}{5} \partial_\beta \bar{\phi}_{\beta\alpha}) \qquad (52)$$

and a comparison with (50b) shows that (in the approximation scheme where $\Theta_{\alpha\beta\gamma} \equiv 0$) we have direct proportionality between $\vec{j}$ and the scaling field $\vec{s}$ as it is required by eq. (16). In the SCA this relation is postulated, therefore no more assumptions about the time-odd part of $\hat{\rho}$ are necessary once the time-even part $\hat{\rho}_+$ is fixed through (18) (which corresponds to $P = s_\alpha p_\alpha$).

Including finally $\Theta_{\alpha\beta\gamma}$ in (50 a-d) allows to investigate the validity of the SCA by going one step further. The details of such an investigation are somewhat involved[15] and we shall present only some general results here.

One obtains two values for the longitudinal sound speed

$$c_L^2 = v_F^2 \frac{1}{2}\left(\frac{6}{7} + \frac{F_o}{3} \pm \left(\left(\frac{F_o}{3}\right)^2 + \frac{1}{35}\left(8F_o + \frac{96}{7}\right)\right)^{1/2}\right) \ . \qquad (53)$$

This allows to describe in addition to the giant electric modes also low-lying collective modes. This is an especially interesting possibility for octupole (E3) modes where the low-lying component is well known. It is also helpful in the case of quadrupole modes where the existence of a low-lying collective E2 state depends on the specific nucleus. However, for a very large nucleus (which the square density model corresponds to) there will always be a low-lying state and therefore inclusion of $\Theta_{\alpha\beta\gamma}$ allows to separate it from the giant state.

The transverse sound speed is also quite different from its value in the SCA: One obtains

$$c_T^2 = v_F^2 \cdot \frac{3}{7} \quad . \tag{54}$$

Together with these changes the modified boundary conditions (51) act, however, in such a way as to reestablish the results of the SCA for frequencies, transition densities and flow patterns to a remarkable degree for strongly collective giant states in all cases where low-lying collective states of the same multipolarity do not exist (i.e. for Magnetic and Isovector Electric modes). For Isoscalar Electric surface modes the quality of agreement is increased if the SCA is restricted to a purely longitudinal scaling field in order to exclude the transverse components of low-lying states from the giant states which the SCA with unrestricted $\vec{s}$ does not separate.

UNBOUND RESONANCES IN THE SCALING APPROACH

For existing nuclei the GMR lie generally above the neutron-emission threshold and should therefore be considered as bound states embedded in the particle continuum. It is a very appealing feature of the SCA equations (25) that they naturally yield the GMR as resonances embedded in a continuum of solutions if one uses as an input into (25) smooth selfconsistent ground-state densities. At this point, however, it again turns out that one has to be extremely cautious with the use of density functionals:

As we have said the input for (25) (namely $\rho_0(r)$ and $\tau_{ik}^{(o)}(r)$) may be taken from any feasible model (HF, ETF,..). Any selfconsistent theory will lead to an exponentially decreasing ground-state density

$$\rho_0(r \to \infty) \to e^{-\mu r}/r^2 \quad . \tag{55}$$

In HF-theory μ reflects the binding energy λ_F of the last bound particle, i.e.

$$\mu = \frac{2}{\hbar}(2m|\lambda_F|)^{1/2} \quad ,$$

while in ETF (with the kinetic energy density approximated by (5)) the self-consistent solution of (2) decreases like (55) with

$$\mu = \frac{2}{\hbar}(2m|\lambda_F|/\eta)^{1/2}$$

as if the last particle was bound with λ_F/η instead of λ_F. For static properties this difference is quite unimportant because it appears only for r much larger than the nuclear radius. For the dynamical calculations considered here the difference in the asymptotic behaviour of ρ_0 is, however, crucial because it determines the asymtotic form of the equation of motion (25) which reads in the asymptotic limit[11]

$$- m\omega^2\vec{s} = - \frac{\hbar^2}{4m} (\partial_r^2 - \mu\partial_r)^2 \vec{s} .$$

With exp(γr) for the asymptotic radial part of $\vec{s}$ we have

$$(\gamma^2 - \mu\gamma) = \pm 2m\omega/\hbar$$

or $\hbar\omega < |\lambda_F|$ for HF , (56a)

$\hbar\omega < |\lambda_F|/\eta$ for ETF (56b)

for bound discrete solutions. For ω values above these critical limits the basic equation (25) has a continuous spectrum, corresponding to single-particle escape. Evidently, the threshold for the onset of the continuum is correctly reproduced with a HF-density $\rho_0(r)$, while in the ETF case it is shifted to much higher values (commonly used values for η are 1/9 or 4/9) corresponding to the seemingly stronger bound last particle. It turns out that this difference is not very important for calculating the position of a specific resonance, nor for the form of the corresponding transition densities or currents. However, obviously, if we calculate the width (= escape width) of a specific resonance it will make a decisive difference how far above threshold the peak of the resonance occurs. If we denote the thresholds (56) as $\hbar\omega_{crit}$ one finds approximately

$$\Gamma^{\uparrow} \propto (\omega - \omega_{crit})^{5/3} .$$

For instance, in ETF with the value η = 4/9 it will frequently happen that a GMR appears as a discrete bound state, while with the HF-input one has to calculate a strength function

$$S_Q(\omega) = (\frac{\hbar}{2M\omega}) (\int \delta\rho_\omega(\vec{x})Q(\vec{x})\ d^3x)^2 \qquad (57)$$

which characterizes the dynamical response of the system to an external field $Q(\vec{x})$, as a function of the continuous variable ω. The strength function (57) will then display a resonance structure around the energy value obtained in ETF as discrete eigenvalue. This structure may be directly compared with corresponding RPA results calculated with an identical potential-energy function. A few examples of such comparisons are shown in the figures[16]. We conclude that it is crucial for the calculation of strength functions to use an HF-input for the scaling functional. But then it is very gratifying to see how well the simple SCA reproduces essential features of transition densities, currents and strength distributions which otherwise may only be obtained through microscopic continuum RPA calculations which also yield a lot of intermediate (and often rather arbitrary and accidental) single-particle structure in the functions $S(\omega)$.

We may therefore conclude that for low-multipolarity Giant Resonances in finite Fermi systems the successful use of functionals of the local density for the potential part of the total energy may be complemented by the scaling functional (22) for the kinetic part to obtain a fluid-dynamical set of equations which determine the main features of physically relevant field variables and strength functions.

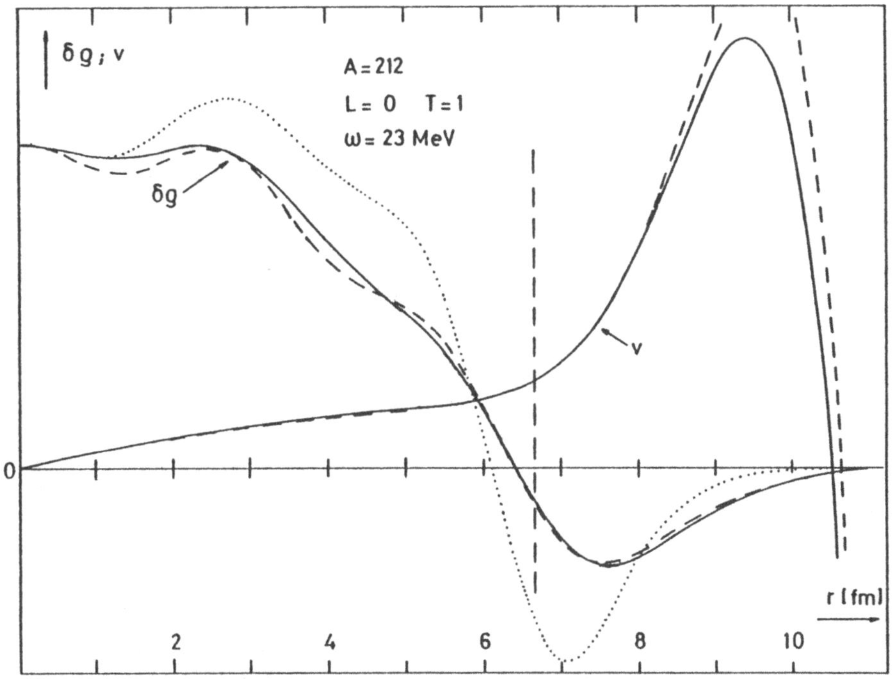

Fig. 1. Comparison of the transition density $\delta\rho(r)$ and radial flow velocity $v(r)$ for the isovector monopole resonance in an N=Z nucleus with A=212 (without Coulomb forces). The full curves are results of a microscopic RPA calculation, the dashed curves are SCA results. The dotted curve is the Tassie transition density (i.e. it corresponds to $v(r) \equiv r$). The vertical dashed line indicates the half-density radius. The energy for this comparison is chosen near the peak of the resonance at $\hbar\omega$ = 23 MeV (cf. Fig. 3).

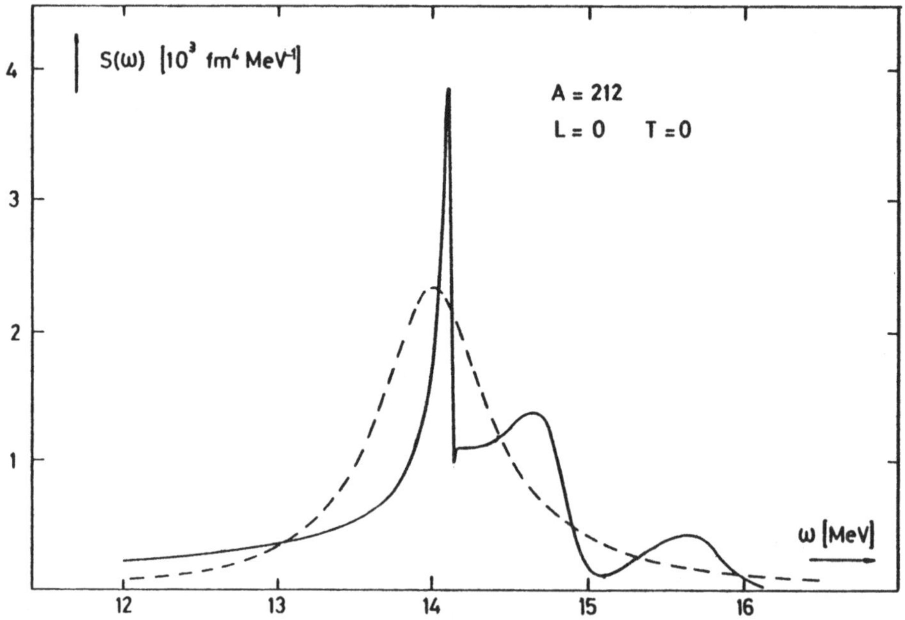

Fig. 2. Strengtfunction S(ω) for isoscalar monopole excitation (full line: RPA, dashed line: SCA)

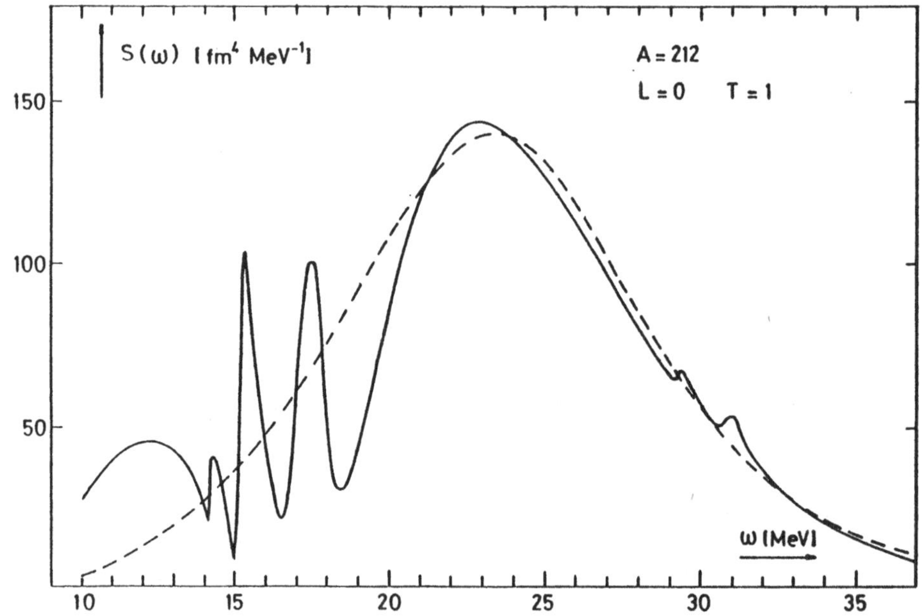

Fig. 3. Strengthfunction S(ω) for isovector monopole excitation (full line: RPA, dashed line: SCA)

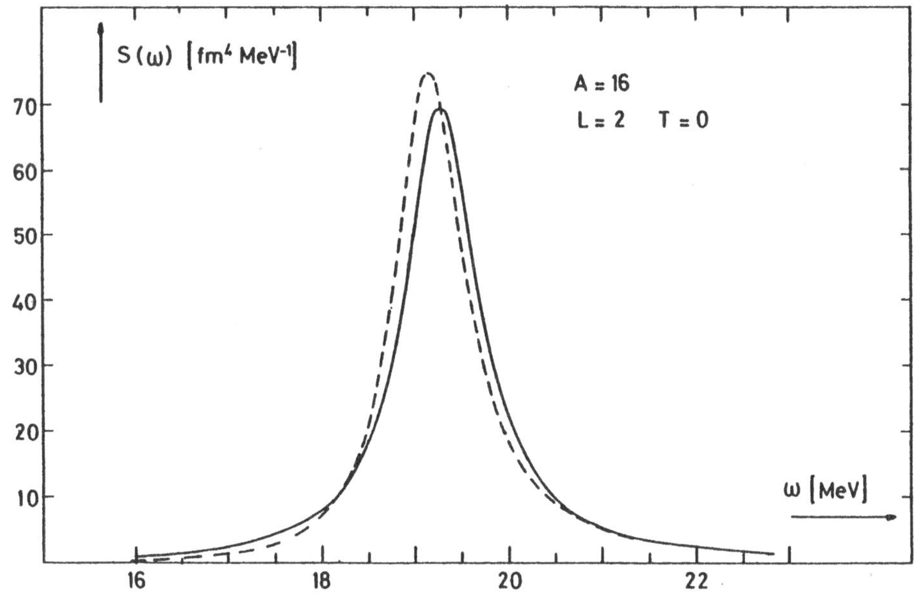

Fig. 4. Strengthfunction S(ω) for isoscalar quadrupole excitation (full line: RPA, dashed line: Irrotational SCA)

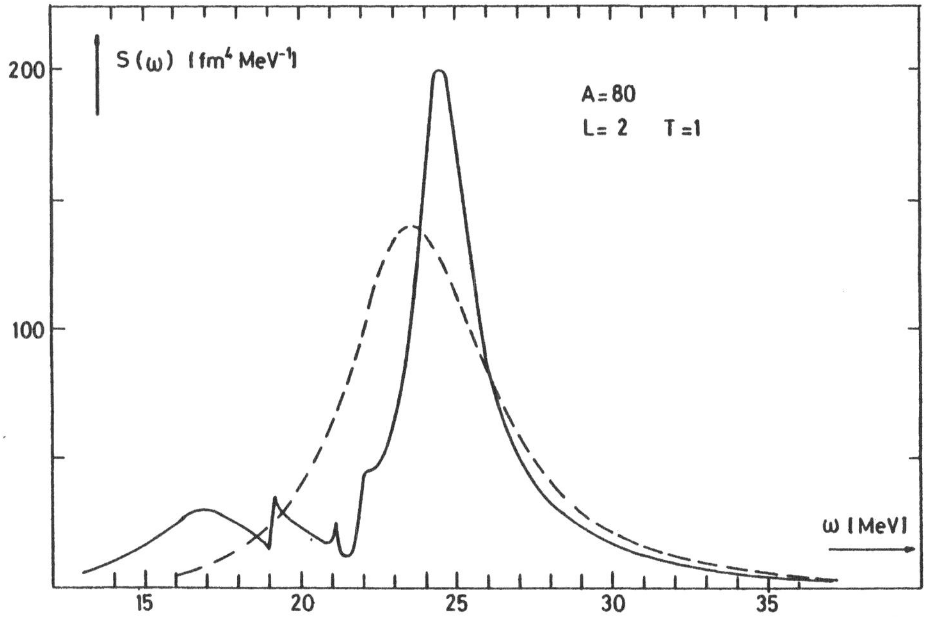

Fig. 5. Strengthfunction S(ω) for isovector quadrupole excitation (full line: RPA, dashed line: Irrotational SCA).

REFERENCES

1. F. E. Bertrand (Editor), "Giant Multipole Resonances", Nuclear Science Research Conference Series, Vol. 1 Harwood Academic Publishers 1980.
   J. Speth and A. van der Woude, Rep. Prog. Phys. 44:719 (1981).
2. O. Bohigas, X. Campi, H. Krivine and J. Treiner, Phys. Letters 64 B:381(1976).
   G. Eckart and G. Holzwarth, Z. Phys. A 281:385(1977).
3. M. Beiner, H. Flocard, N. Van Giai and P. Quentin, Nucl. Phys. A 238:29(1975).
   H. Krivine, J. Treiner and O. Bohigas, Nucl. Phys. A 336: 155(1980).
4. D. A. Kirshnits, "Field theoretical methods in many-body systems", Pergamon, London 1967.
   M. Brack, B. K. Jennings and Y. H. Chu, Phys. Letters 65 B: 1(1976).
   H. Krivine and J. Treiner, Phys. Letters 88 B:212(1979).
5. A. Bohr and B. R. Mottelson, "Nuclear Structure", Vol. 2, Ch. 6 A, Benjamin, Reading 1975.
6. H. Sagawa and G. Holzwarth, Prog. Theor. Phys. 59:1213 (1978).
7. J. Martorell, O. Bohigas, S. Fallieros and A. M. Lane, Phys. Letters 60 B:313(1976).
   O. Bohigas, A. M. Lane and J. Martorell, Phys. Rep. 51: 267(1979).
8. G. Holzwarth and G. Eckart, Nucl. Phys. A 325:1(1979).
9. S. Stringari, Nucl. Phys. A 279:454(1977).
10. G. Eckart and G. Holzwarth, Phys. Letters 118 B:9(1982).
11. G. Eckart, G. Holzwarth and J. P. da Providencia, Nucl. Phys. A 364:1(1981).
12. F. E. Serr, Phys. Letters 97 B:180(1980).
13. K. Andō and S. Nishizaki, Prog. Theor. Phys. 68:1196(1982).
14. T. Yukawa and G. Holzwarth, Nucl. Phys. A 364:29(1981).
15. G. Holzwarth and H. Thorn, Siegen University 1983, to be publ.
16. K. Andō and G. Eckart, Siegen University 1983, to be publ.

DENSITY FUNCTIONALS IN HIGH-ENERGY HEAVY-ION COLLISIONS

J.A. Maruhn

Institut für Theoretische Physik
Universität Frankfurt
D-6000 Frankfurt 1, W. Germany

1. Introduction

High energy heavy-ion collisions provide a possibly unique opportunity for the study of nuclear matter under extreme conditions, such as occur elswhere only in neutron stars or black holes, or in the early stages of the big bang.

In these experiments a beam of heavy ions, typically Ne or Ar, with an energy of 200-2100 MeV per nucleon is made to collide with a target containing nuclei of arbitrary mass number. The relative speed of the two nuclei is much larger than the estimated speed of sound in nuclear matter (between 0.1 and 0.2 c), so that compression effects can be expected[1].

Whether they really do occur, however, is much less certain than for macroscopic materials under similar conditions. Since the radius of even a very heavy nucleus is only about 7-8 fm while estimates[2] for the mean free path of a nucleon inside a nucleus amount to >1.5 fm (because of the Pauli principle, the mean free path becomes very large at lower energies and temperatures, but in the regime considered here this should not be an important consideration), nuclei could show a considerable degree of transparency in such collisions, which would make the desired production of excited and compressed matter much less likely.

A number of models have been proposed to investigate this problem more closely (for a review see ref. 3). None of these is felt to come close to an "exact" description of these processes, even if one neglects all quantum-mechanical effects and treats the nuclei as classical many-body systems consisting of nucleons as elementary constituents. The two models which give a relatively comprehensive treatment of the dynamics of such a reaction and have been widely used for trying to understand the data are the hydrodynamic model[1,4,5] and the cascade model[6,7,8].

The hydrodynamic model assumes that even in such a nuclear collision the nucleons populate a local equilibrium distribution instantaneously, so that the reaction is determined by the equation of state of nuclear matter - corresponding to a special form of density functional. In the intranuclear cascade, on the other hand, the reaction is described in terms of collisions between the individual nucleons that make up the nuclei. At first sight the cascade model thus seems to be better justified, since no assumptions about local equilibrium have to be made. However, it has its own problem in that the treatment of the reaction by collisions between two nucleons breaks down at higher densities. To quote a macroscopic analogue: while water can be described by a hydrodynamic theory, its molecules are too closely packed to allow a description in terms of two-body interactions.

Lacking a more general theoretical approach to the problem, the present attitude is to let experiments decide which of the two models is closer to the truth. In this sense we can say that it has to be found out experimentally whether density functionals play any role in high-energy heavy ion collisions.

In the following we will review two possible derivations of the hydrodynamical equations, one based on the Boltzmann equation and one starting from TDHF. In both cases the establishement of local equilibrium cannot be proved but has to be assumed. Nevertheless, these derivations give considerable insight into the theoretical problems. The lecture then continues with a short review of hydrodynamic model calculations and the present status of comparison with experiment.

2. The "Standard" Derivation of Hydrodynamics

In this chapter we examine the usual textbook derivation of hydrodynamic equations from the Boltzmann equation[9]. As we shall see, this derivation is somewhat too restricted for judging the applicability of hydrodynamics in a nuclear collision, but nevertheless allows us to identify the principal problems.

The Boltzmann equation (in a form without an external force)

$$\frac{\partial f}{\partial t} + \vec{v}\cdot\nabla f = S$$

describes the time evolution of the single-particle phase space density $f(\vec{r},\vec{v},t)$ through convection in phase space - the left hand side of the equation - and through two-body collisions, which are described by the collision term S, to be explained later. At this point it should be noted already that the Boltzmann equation constitutes a continuum approximation i.e. in principle there should be many particles in a phase space element of a size small compared to the range of characteristic variations in f. This condition seems hard to fulfil in a nuclear collision with a few hundred particles and short-range structure in the density, but, as will be seen in the next section, quantum mechanics may help in this problem.

The collision term S has the form

$$S = \int d^3v_1\, d^3v_2\, db\, d\varphi\, w(\vec{v}_1, \vec{v}_2 \rightarrow \vec{v}, \vec{v}'; b\varphi) \times$$
$$\times \left[f(\vec{r}, \vec{v}_1, t) f(\vec{r}, \vec{v}_2, t) - f(\vec{r}, \vec{v}, t) f(\vec{r}, \vec{v}', t) \right],$$

where w describes the probability for a collision of particles with velocities $\vec{v}_1, \vec{v}_2$ leading to velocities $\vec{v}, \vec{v}'$ at an impact parameter b and azimuthal angle ϕ. The negative contribution gives the probability for particles to be scattered out of $\vec{v}, \vec{v}'$, and the two pairs of velocities are related by the usual conversation laws.

The collision term introduces the following additional assumptions into the theory:

a) The system must be dilute, since only binary conditions are accounted for.
b) The collision mean free path must be short compared to the range of characteristic variation in f, since all quantities in S are evaluated at the same space point.
c) Particles are always uncorrelated - they forget the collisions instantanously (the celebrated "molecular chaos" assumption).

Note that a simulation of nuclear collisions with an intranuclear cascade model is subject to the assumption of diluteness but not the other two.

The derivation of hydrodynamic equations now proceeds relatively straightforwardly. Integrating the Boltzmann equation over the velocity $\vec{v}$ with a weight $\chi(\vec{v})$ yields

$$\frac{\partial}{\partial t}\langle \chi \rangle + \nabla \cdot \langle \chi \vec{v} \rangle = \int d^3v\, \chi(\vec{v}) S$$

where the brackets denote a velocity average:

$$\langle a \rangle = \frac{1}{\rho} \int d^3v\, a(\vec{v}) f(\vec{r}, \vec{v}, t), \quad \rho(\vec{r}, t) = \int d^3v\, f(\vec{r}, \vec{v}, t).$$

The crucial point then consists in inserting for $\chi(\vec{v})$ the collisional invariants, i.e. the quantities conserved in a microscopic collision, $1, \vec{v}, v^2$. For these one can show that the right-hand side vanishes because gain and loss cancel exactly in that case.

The three invariants thus lead to three conservation equations:

$$\frac{\partial \rho}{\partial t} + \nabla \cdot (\rho \vec{u}) = 0$$

$$\frac{\partial (\rho \vec{u})}{\partial t} + \nabla \cdot (\rho \vec{u} \vec{u}) = - \nabla \cdot \overleftrightarrow{p}$$

$$\frac{\partial (\rho \varepsilon)}{\partial t} + \nabla \cdot (\rho \varepsilon \vec{u}) = - \nabla \cdot \vec{q} - \nabla \cdot (\overleftrightarrow{p} \cdot \vec{u}),$$

where the following quantities are defined as averages: the mean velocity $\vec{u}=\langle\vec{v}\rangle$, the mean energy per particle $\varepsilon=m\langle|\vec{v}-\vec{u}|^2\rangle/2$, the stress tensor $\overleftrightarrow{p}=\rho\langle(\vec{v}-\vec{u})(\vec{v}-\vec{u})\rangle$, and the heat flux $\vec{q}=\rho\langle(\vec{v}-\vec{u})|\vec{v}-\vec{u}|^2\rangle$.

This set of equations becomes closed only if $\overleftrightarrow{p}$ and $\vec{q}$ can be expressed as functions of ρ, $\vec{u}$, and ε, which is true for example if local equilibrium holds, i.e. if the distribution function is given by a local Maxwell-Boltzmann distribution (or a Fermi-Dirac one, more generally). The pressure is then given by the usual expressions for an ideal gas.

How then can we find out whether local equilibrium really holds in a given situation? In this derivation the mechanism for equilibration, the two-particle collisions, has dropped out completely. It can be reintroduced in the linearized Boltzmann equation, whose study shows that deviations from local equilibrium decay exponentially, with the slowest decay in the long-range deviations which give rise to viscosity and thermoconductivity[9].

On the other hand, the very fact that the collisions drop out during the derivation already indicates that the hydrodynamic equations are of a more general character then this derivation would seem to imply. In fact the Boltzmann equation is not valid for liquids (because of the diluteness assumption), which are described perfectly well by hydrodynamics. In addition there may be other mechanisms for equilibration, such as single-particle friction, or fields in collisionless plasmas which may lead to local equilibrium but are not included in this framework.

The crucial question for nuclear collisions is thus whether there is a sufficiently strong mechanism for local equilibration. Before examining this problem in more detail, however, we want to present a different derivation of hydrodynamics which sheds some light on the continuum aspect.

3. The Quantum Aspect

The preceding "standard" derivation of hydrodynamics from the Boltzmann equation indicated two principal conditions of validity for the hydrodynamic approach, namely the requirement of a short mean free path leading to instantaneous local equilibrium and that of large particle numbers allowing a continuum treatment. On the other hand, we know that the density functional approach as presented in other lectures at this school is quite close to a hydrodynamic theory and works surprisingly well even for nuclear ground states, where the nucleon mean free path is very large, and for relatively small numbers of nucleons. The reasons for this become apparent in an alternative derivation[10] of the hydrodynamical equations which incorporates quantum mechanics.

In 1927 already Madelung[11] showed that even the one-particle Schrödinger equation

$$i\hbar\frac{\partial\psi}{\partial t}=-\frac{\hbar^2}{2m}\nabla^2\psi+V\psi,$$

can be written in a hydrodynamic form by decomposing the wave function as

$$\psi(\vec{r},t) = \varphi(\vec{r},t) \exp[im\, S(\vec{r},t)/\hbar]$$

with ϕ and S real fields. Inserting this into the Schrödinger equation and decomposing the result into real and imaginary parts yields

$$\frac{\partial \rho}{\partial t} + \nabla \cdot (\rho \vec{v}) = 0$$

$$\frac{\partial (\rho \vec{v})}{\partial t} + \nabla \cdot (\rho \vec{v} \vec{v}) = -\frac{1}{m} \rho \nabla V - \frac{1}{m} \rho \nabla \left(\frac{\tau}{\rho}\right)$$

Here the definitions of the probability density $\rho = \phi^2$, the flow velocity $\vec{v} = \nabla S$ (this definition is consistent in the sense that the probability current is given by $\rho \nabla S$),and the internal kinetic energy $\tau = \phi(-\hbar^2/2m)\nabla^2 \phi$ have been inserted. τ plays the role of a "quantum pressure" which prevents excessive localization of the wave function.

Clearly in this approach a one-particle system appears as a continuum by virtue of its quantum-mechanical probability distribution. To carry the derivation over to a many-particle system the following ideas are used.

a) The starting point is the TDHF-equations. The two-particle collisions are assumed to drop out effectively (like in the derivation from the Boltzmann equation) and to be incorporated by the very assumption of local equilibrium.
b) The definitions pf ρ and $\vec{v}$ should be based on the density matrix in order to avoid a dependence on arbitrary phases.
c) Temperature will be introduced by the introduction of a thermally excited local equilibrium.
d) The average field and "quantum pressure" terms have to be rewritten in terms of the standard pressure and potential terms, which will allow definition of a fluid with a definite equation of state and self-interaction, or, equivalently, a density functional.

The necessary derivations are given in the literature[10] and we only present a short discussion of the principal points here.

The one-particle density matrix is obtained by a summation over occupied states

$$\hat{\rho}(\vec{r},\vec{r}\,',t) = \sum_\alpha \psi_\alpha(\vec{r},t)\psi_\alpha^*(\vec{r}\,',t),$$

and the hydrodynamic fields can then be expressed as (quantities with index α as in the one-particle case)

$$\rho(\vec{r},t) = \hat{\rho}(\vec{r},\vec{r},t) = \sum_\alpha \rho_\alpha$$

$$\vec{u}(\vec{r},t) = \frac{\hbar}{2im\rho} \lim_{\vec{r}\,' \to \vec{r}} (\nabla - \nabla')\hat{\rho}(\vec{r},\vec{r}\,',t) = \frac{1}{\rho}\sum_\alpha \rho_\alpha \vec{v}_\alpha .$$

Note that while for a single particle the velocity field is always irrotational this is generally not true for many particles.

With these definitions the TDHF equations

$$i\hbar \frac{\partial}{\partial t} \hat{\rho}(\vec{r},\vec{r}',t) = -\frac{\hbar^2}{2m} (\nabla^2 - \nabla'^2) \hat{\rho}(\vec{r},\vec{r}',t) + \int d^3r'' \left[V(\vec{r}-\vec{r}'') - V(\vec{r}'-\vec{r}'') \right] \times \left[\hat{\rho}(\vec{r},\vec{r}',t) \hat{\rho}(\vec{r}'',\vec{r}'',t) - \hat{\rho}(\vec{r},\vec{r}'',t) \hat{\rho}(\vec{r}'',\vec{r}',t) \right]$$

lead to

$$\frac{\partial \rho}{\partial t} + \nabla \cdot (\rho \vec{u}) = 0$$

$$\frac{\partial}{\partial t} \rho \vec{u} + \nabla \cdot \left[\sum_\alpha \rho_\alpha \nabla S_\alpha \nabla S_\alpha \right] = -\frac{1}{m} \sum_\alpha \rho_\alpha \nabla \left(\frac{\tau_\alpha}{\rho_\alpha} \right) - \frac{\rho}{m} \nabla \int d^3r'' \, V(\vec{r}-\vec{r}'') \rho(\vec{r}'') + \frac{1}{m} \int d^3r'' \, \nabla V(\vec{r}-\vec{r}'') \, |\rho(\vec{r},\vec{r}'')|^2 \, .$$

The first equation already has the desired form while in the second one we have to transform the individual terms on the right hand side one by one.

As in kinetic theory, the first of these can be rewritten by introducing the average and fluctuating parts of the velocities:

$$\sum_\alpha \rho_\alpha \vec{v}_\alpha \vec{v}_\alpha = \rho \vec{u} \vec{u} + \frac{1}{m} \overleftrightarrow{p},$$

where p is the stress tensor

$$\overleftrightarrow{p} = m \sum_\alpha \rho_\alpha (\vec{v}_\alpha - \vec{u})(\vec{v}_\alpha - \vec{u})$$

which under the assumption of local equilibrium can be replaced by the local pressure p (a function of ρ and the temperature T), plus possibly the viscous stress tensor. In the same approximation the second term can also be written as

$$\frac{1}{m} \sum_\alpha \rho_\alpha \nabla \left(\frac{\tau_\alpha}{\rho_\alpha} \right) = \frac{\rho}{m} \nabla \left(\frac{\partial \tau}{\partial \rho} \right)$$

The interaction terms can be rewritten in a more useful way if one adopts the splitting up of the potential into a density-dependent part of zero range and a longer-range one (which can include the Coulomb potential):

$$V(\vec{r}-\vec{r}') = V_s[\rho(\vec{r})] \, \delta(\vec{r}-\vec{r}') + V_\ell(\vec{r}-\vec{r}')$$

V_s leads to a simple density-dependent contribution in both the direct and (using the Slater approximation) the exchange potentials, while the exchange part of the long-range interaction has to be dropped:

$$\int V(\vec{r}'-\vec{r})\,\rho(\vec{r}')\,d^3r' = \frac{\partial}{\partial\rho}\big(W_s(\rho)\rho\big) + \bar{V}_l(\vec{r}) \;.$$

Here the density-dependent parts have been written in terms of energy-per-nucleon functions $W_{s,x}$, which will later appear in the equation of state.

Collecting all of these results we find the following equation of motion

$$\frac{\partial(\rho\vec{u})}{\partial t} + \nabla\cdot(\rho\vec{u}\vec{u}) = \frac{1}{m}\nabla\cdot\overleftrightarrow{p} - \frac{1}{m}\rho\nabla\frac{\partial(W_0\rho)}{\partial\rho} - \frac{1}{m}\rho\nabla\bar{V}_l$$

with $W_0=\tau/\rho+W_s+W_x$.
Now it is only remains to unite the first two terms on the right hand side. If we neglect viscosity - it can easily be incorporated if desired - the stress tensor contains only the scalar pressure from thermal fluctuations, i.e. approximately the pressure of a Fermi gas in the local density approximation. The second term can be rewritten:

$$\rho\nabla\frac{\partial(W_0\rho)}{\partial\rho} = \nabla\left(\rho^2\frac{\partial W_0}{\partial\rho}\right) ,$$

and if we also express the pressure in terms of a binding energy per nucleon $W_{th}(\rho,T)$ we obtain

$$p = \rho^2\frac{\partial}{\partial\rho}\left(W_0+W_{th}\right)\Big|_S ,$$

so that the equation of motion becomes

$$\frac{\partial(\rho\vec{u})}{\partial t} + \nabla\cdot(\rho\vec{u}\vec{u}) = -\frac{1}{m}\nabla p - \frac{1}{m}\rho\nabla\bar{V}_l \;.$$

The pressure has to be obtained by differentiation at constant entropy per nucleon S; this is a complication that effects only W_{th} (W_0 is only density-dependent) and can be handled by standard thermodynamic relations.

The introduction of the new field T implies that an additional equation of motion is needed; its derivation proceeds along the same lines so that we refer to ref.[10] for details. The final result is just the standard equation for energy conservation in classical hydrodynamics.

In practice one has to be careful with the separation of equation of state $W_0(\rho)$ and long-range interaction V_l. For a Yukawa-interaction, e.g., V_l also contributes to binding in infinite nuclear matter and W_0 alone will, in fact, not bind nuclear matter at all. For finite size nuclei the range of V_l determines the stability against break-up and there will be strong dispersion of sound waves.

Summing up the foregoing derivation, we have seen that

a) the continuum approximation may be made valid by quantum-mechanic effects;

b) the properties of nuclear matter enter the model through the zero-temperature binding energy as a function of density, a long-range interaction and a Fermi-gas expression for thermally excited material;
c) in this derivation also the validity of the local equilibrium approximation can be assumed only and not proven. We still need a strong mechanism for equilibration in a nucleus-nucleus collision.

Let us finally write down the connection to density functionals. At zero temperature the hydrostatic equilibrium condition

$$\nabla p = \rho \nabla \bar{V}_I \quad , \quad p = \rho^2 \frac{\partial W}{\partial \rho}$$

may be derived by minimizing the energy functional

$$E = \int d^3r \; W_0 \rho + \frac{1}{2} \int d^3r \int d^3r' \; V_I(\vec{r}-\vec{r}\,')\rho(\vec{r}\,')\rho(\vec{r}) .$$

4. Typical results

The hydrodynamic equations discussed in the previous chapters define the model to a large extent. They have to be complemented only by the equation of state, the viscosity, and the thermoconductivity.

The equation of state is usually given in terms of the binding energy per nucleon as a function of density ρ and specific entropy s, and is split up as

$$W(\rho, s) = W_0(\rho) + W_{th}(\rho, s) .$$

$W_0(\rho)$ represents the binding energy in the absence of thermal excitation and can be taken from nuclear matter calculations. Its behaviour and that of the pressure and speed of sound at zero thermal excitation is illustrated in fig. 1. The shape of these curves above equilibrium density is unexplored by conventional experiments (except for the region very close to equilibrium, which influences the giant monopole resonance) and thus largely open to speculation about possible phase transitions to exotic states at high density and/or temperature).

For the thermal part W_{th} all calculations up to now simply employed a Fermi gas expression. The pressure and temperature can be obtained by the standard thermodynamic relations

$$p = \rho^2 \frac{\partial W}{\partial \rho}\Big|_s \quad , \quad T = \frac{\partial W}{\partial s}\Big|_\rho .$$

Finally, the viscosity and thermoconductivity of nuclear matter are also not well known. Since the dominant part of the dissipation in this model happens in the shock fronts and is thus independent of the existence of explicit dissipative terms, most calculations, including those presented here, do not take them into account.

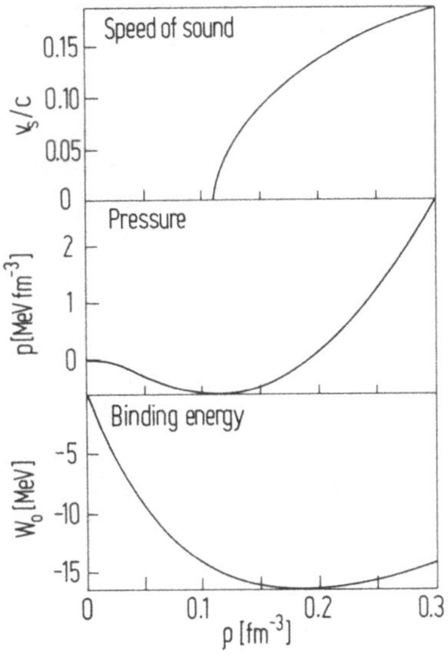

Fig. 1: Properties of nuclear matter assumed for the hydrodynamic model calculations. The binding energy, pressure and speed of sound are plotted as functions of the density at zero temperature. The behaviour of these curves at higher dendensities is practically unknown.

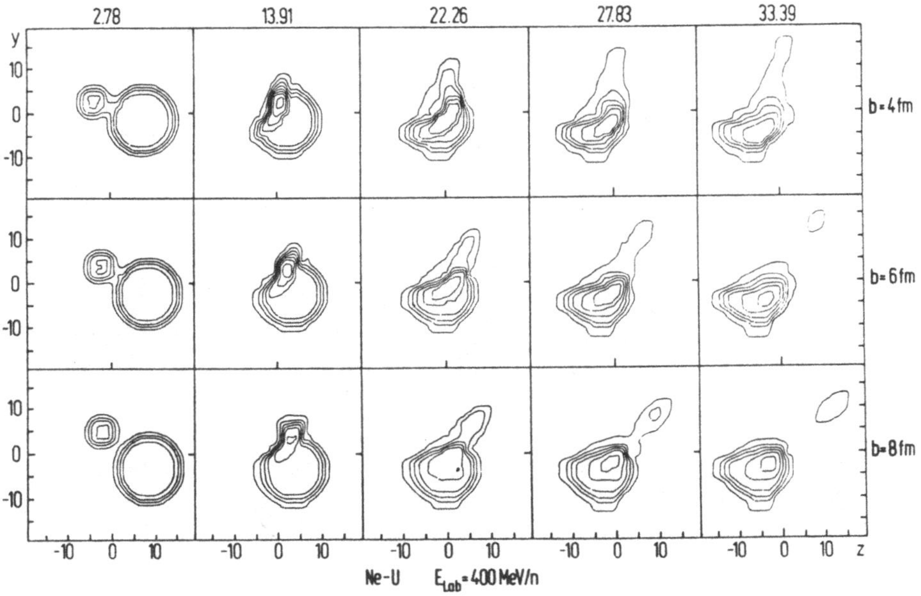

Fig. 2: Collisions of ^{20}Ne on ^{238}U at 400 Mev per nucleon in the hydrodynamic model. Plotted are density contur lines for the time into the collision given above the figures (in fm/c). The horizontal sequences correspond to different impact parameters. The formation of shock fronts, compressed matter and the "bounce-off" deflection of the projectile at an angle strongly dependent on impact parameter can be seen clearly.

The dynamical calculation itself is done numerically on a spatial grid of 64^3 points. The density distributions of the projectile and target nuclei are inserted into this grid and the velocity is set such that they approach each other at a given relative kinetic energy and impact parameter.

Fig. 2 shows a series of density profiles for collisions of Ne+U, at an energy of 400 MeV per nucleon. At all impact parameters a shock wave quickly develops as the projectile hits the target. For central collisions the projectile and target amalgamate completely to form a highly excited system in the final state, which still retains the characteristic curved shape of the initial shock and will evaporate many individual nucleons and light clusters. For more peripheral collisions, which, of course, take a much larger share of the cross section, there are also remainders of both target and projectile deflected at finite angles. This "bounce-off"- effect[12] is a characteristic prediction of the hydrodynamic model: all cascade calculations seem to yield an angular deflection of at most 3^0, while here we observe angles more than ten times larger.

The observation of such large deflection angles thus would be a good indication for an approach to local equilibrium and for the formation of compressed and excited nuclear matter. The problem experi entally is that the effects depend strongly on the impact parameter and are visible clearly in the reaction plane only,which cannot be identified directly. And indeed the experimental distributions averaged over azimuthal angles and impact parameter show smooth forward peaking only.

A selection in impact parameter can be tried for example by selecting especially "violent" events, e.g. those containing a very large number of particles in the final state, so that the nuclei have been broken up almost completely. Early experiments of this type by Schopper's group[13], were later supported by Stock et al.[14]. The latter data are shown in fig. 3 and clearly indicate that sidewards peaking of the distributions appears only when a selection for central events is made. Both of these experiments, however, are hampered by relatively large statistical uncertainties.

Recently a new type of experimental analysis, the so-called global analysis, was introduced in order to look for this sidewards deflection. The basic idea is to define quantities describing the average motion of particles in the final state for each event separately. In practice one needs a detector which measures the momentum vector $P_k^{(i)}$ for particle i (cartesian component k). By summing up over all particles a "flow tensor" is constructed:

$$F_{kl} = \sum_i p_k^{(i)} p_l^{(i)} / m_i$$

This tensor defines an ellipsoid in space whose principal axes and orientation angles can be determined. One may then examine the distribution of these quantities directly - chief among these the angle between the longest axis and the beam direction, as well as the "flow ratio" f_{13}, the ratio of largest to smallest

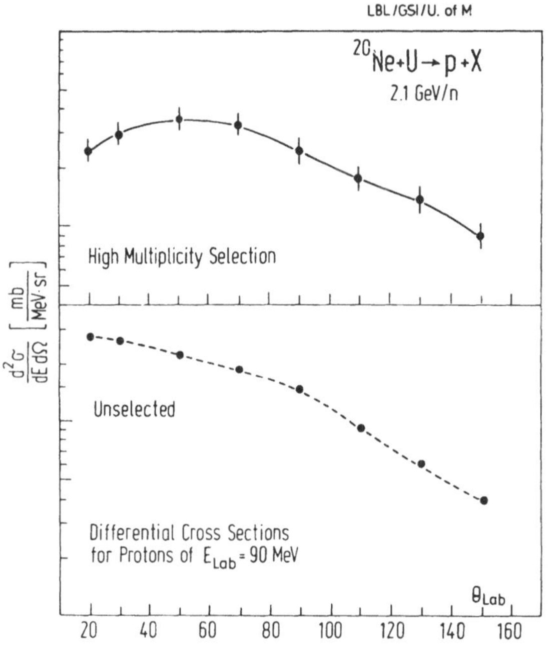

Fig. 3: Experimental data [14] on the angular distribution of protons in the reaction Ne+U at 2.1 GeV per nucleon. If the spectra are summed up over all events, forward peaking results, while an inclusion of events with a large multiplicity (central events) yields a strong depression, possibly a minimum, in the beam direction.

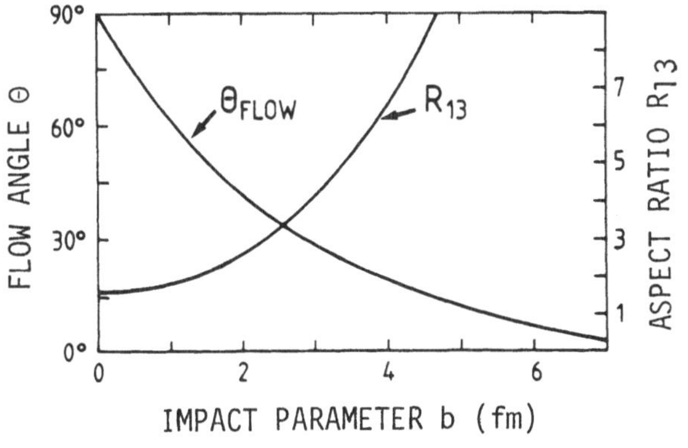

Fig. 4: Systematic behaviour of the flow ratio and angle in the hydrodynamic model. For small impact parameters, i.e. central wents, the flow ellipsoid nearly becomes a sphere while peripheral reactions lead to a stretched shape in the beam direction.

axis, which gives an idea of the deformation of the ellipsoid. The hydrodynamic predictions are shown in fig. 4. For large impact parameters the angle should be small and the flow ratio very large, as befits a peripheral reaction, while for more central collisions a large angle should reflect the bounce-off-effect. Experiments have been done to study these quantities, but their analysis has not been completed at the time of this writing, so that it can only be said that preliminary analysis seems to support the hydrodynamic predictions or lie midway between these and cascade calculations.

However, there is still the problem that hydrodynamics is a continuum theory which does not contain the statistical fluctuations in the flow tensor due to finite particle numbers. It is known that these cause systematic distortions in the flow quantities, so that a finite particle number analysis of hydrodynamics has to be performed. This problem is being investigated.

5. Outlook: nuclear matter at high density

While most of the current interest is focussed on the basic question whether the formation of high-density nuclear matter can be demonstrated clearly in these collisions, the problem of extracting the equation of state should not be lost from sight. In this context it is especially important to investigate the possible effects of phase transitions, such as pion condensation or the formation of a quark-gluon-plasma.

One piece of data that reflects the equation of state directly is the number of pions produced, if these are formed in thermal equilibrium. Fig. 5 shows that this number rises smoothly in hydrodynamics with energy and jumps relatively rapidly when a phase transition is reached, which makes more energy available for pion creation. Unfortunately it was found in these analyses that the thermal energy tends to dominate at these energies so that it becomes increasingly hard to identify the contribution W_0. This becomes apparent also in fig. 6, where the flow angle for a fixed impact parameter is plotted as a function of energy for various equations of state. K is the incompressibility; the higher the value of K the more energy is stored in compression. The "pion condensate" refers to an equation of state containing a phase transition to spin-isospin ordering. Apparently a lowering of K and a liberation of thermal energy through the phase transition both lead to a hotter, more equilibrated system with a smaller bounce-off-angle. The magnitude of the effect is much larger, however, at low energies and it becomes quite doubtful whether equation of state effects arising from non-thermal contributions really are measurable at very high energies.

The presently available data on pion production do not yet show indications for phase transitions.

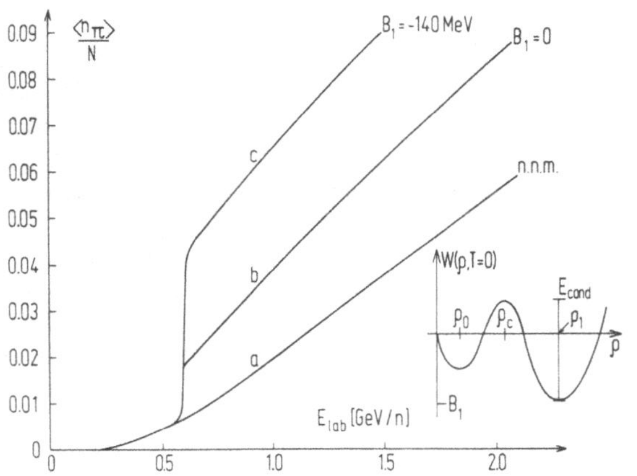

Fig. 5: Pion production as a function of the bombarding energy in the hydrodynamic model. This is computed as the number of pions expected in thermal and chemical equilibrium. The smooth curve labelled a corresponds to the standard equation of state while the inclusion of a phase transition leads to sudden jumps in the curve (depending on the heat liberated in the transition, curve b or c may be attained).

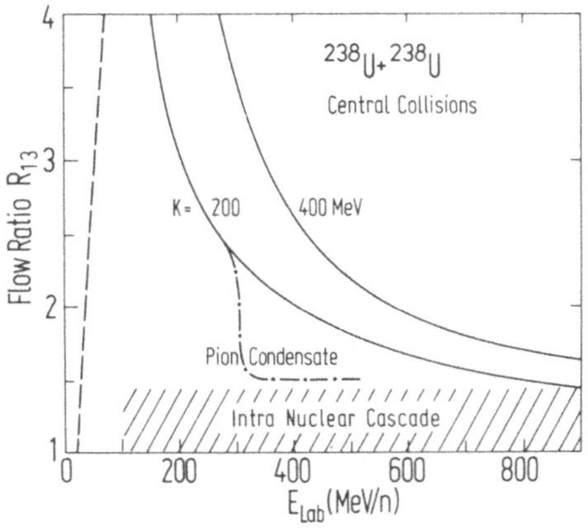

Fig. 6: The flow ratio as a function of bombarding energy also shows a dependence on the equation of state. The value K indicates the incompressibility of nuclear matter, while "pion condensate" indicates a phase transition at higher density. The intranuclear cascade model alway yields small ratios corresponding - infinite multiplicity distortions are accounted for - to isotropic behaviour. Note that in these flow tensor calculations only the particles in the hot interaction zone, the "participants" are taken into account (excluding the target and projectile remnants). Only these participants are sufficiently energetic to register inthe detectors.

6. Summary

As we have seen, the field of high-energy heavy-ion collisions is still in its initial stages. The basic reaction mechanism has not yet been established convincingly, although some tantalizing evidence for hydrodynamic behaviour exists. The collaboration between theorists and experimentalists is unusually stimulating in this field; new ways of looking at the data and demands for theoretical predictions push progress further regularly towards the goal of producing and identifying matter under extreme conditions hither to inaccessible to experimental investigation.

Acknowledgement:

The author would like to thank the members of the Frankfurt hydrodynamic theory group, notably W. Greiner, H. Stöcker, G. Buchwald, G. Graebner, J. Theis, L.P. Csernai, P.R. Subramanian, H. Kruse, D. Barthel, and T. Rentzsch for many years of fruitful collaboration.

References

1. W. Scheid, H. Müller, and W. Greiner, Phys. Rev. Lett. 32: 741 (1974).
2. A. Kind, Nuovo Cimento 10:176 (1953).
   A. Kind and P. Patergnani, Nuovo Cimento 10:1375 (1953).
   M. T. Collins and J. J. Griffin, Nucl. Phys. A348:63 (1980).
3. J. A. Maruhn and W. Greiner, "Relativistic Heavy Ion Reactions: Theoretical Models", to be published in "Heavy Ion Science", ed. D.A. Bromley, Plenum Press.
4. A. A. Amsden, G. F. Bertsch, F. H. Harlow, and J. R. Nix, Phys. Rev. Lett. 35:905 (1975).
   F. H. Harlow, A. A. Amsden, and J. R. Nix, J. Comp. Phys. 20:119(1976).
5. J. Hofmann, H. Stöcker, U. Heinz, W. Scheid, and W. Greiner, Phys. Rev. Lett. 36:88 (1976).
   H. Stöcker, J. A. Maruhn, and W. Greiner, Phys. Lett. 81B: 303 (1979).
   H. Stöcker, J. A. Maruhn, and W. Greiner, Phys. Rev. Lett. 44:725 (1980).
   H. Stöcker, G. Buchwald, L. P. Csernai, G. Graebner, J. A. Maruhn, and W. Greiner, Nucl. Phys. A387:205c (1982).
   G. Buchwald, G. Graebner, J. Theis, J. A. Maruhn, W. Greiner, and H. Stöcker, Phys. Rev. C28: 1119 (1983).
6. Y. Yariv and Z. Fraenkel, Phys. Rev. C20:2227 (1979).
7. K. K. Gudima and V. D. Toneev, Yad. Fiz. 27:658 (1978).
   K. K. Gudima and V. D. Toneev, Yad. Fiz. 31:1455 (1980).
8. J. Cugnon, T. Mizutani, and J. Vandermeulen, Nucl. Phys. A352:505 (1981).

9. Y. M. Lifschitz and L. P. Pitayevski, "Fizicheskaya Kinetika", (Nauka, Moscow 1979).
10. C. Y. Wong, J. A. Maruhn, and T. A. Welton, Nucl. Phys. A253:469 (1975).
11. E. Madelung, Z. Phys. 40:322 (1926).
K. K. Kan and J. J. Griffin, Phys. Rev. C15: 1126 (1977).
12. H. Stöcker, J. A. Maruhn, and W. Greiner, Z. Physik A293: 173 (1979).
13. H. G. Baumgardt, J. U. Schott, Y. Sakamoto, E. Schopper, H. Stöcker, J. Hofmann, W. Scheid, and W. Greiner, Z. Physik A237:241 (1975).
14. R. Stock, H. Gutbrod, W. G. Meyer, A. M. Poskanzer, A. Sandoval, J. Gosset, C. H. King, G. King, Ch. Lukner, Nguyen Van Sen, G. D. Westfall, and K. L. Wolf, Phys. Rev. Lett. 44:1243 (1980).

ON THE SEMICLASSICAL DESCRIPTION OF NUCLEAR FERMI LIQUID DROPS

Peter Schuck

Institut des Sciences Nucléaires

38026 Grenoble-Cédex, France

INTRODUCTION

Recent years have seen quite some developments in the understanding and application of semiclassical methods to gross properties of nuclei[1-4]. Talking about gross properties we mean in fact properties of the nucleus where in a systematic and well defined way the influence of individual shells has been averaged out. One typical quantity to be considered is the single particle level density where the average part of say a fully quantal distribution (corresponding e.g. to some sort of Wood-Saxon or H.F. potential) is easily conceivable. The average part of nuclear groundstate masses is another such quantity and in fact represented by the well known Bethe Weizsäcker mass formula. These are two very well known examples where the average nuclear properties have been studied since long and in very great detail. There are however many more quantities and properties whose average behavior could be investigated in the same way and what in fact is interesting to do. Among those we want to cite e.g. the moment of inertia of rotating nuclei, average nuclear pairing properties, average m-particle-n-hole level densities, average behavior of collective nuclear vibrations, average current distributions in rotating and vibrating nuclei, and many things more. In short, we would like to describe semiclassically all nuclear properties which survived could we artificially blow up nuclei to quasi macroscopic dimensions - like e.g. droplets of liquid He^3 - where it is clear that the continuum limit is reached, i.e. no shell effect present any more, but still all quantities depending on the size parameters like e.g. volume, surface, curvature, and deformation of the nuclear droplets. In this region we would like to establish the laws the different quantities obey as a function of these parameters which in most cases can be resumed in a power law dependence on the cubic root of the nucleon number A. These laws should then be taken and extrapolated back to the sizes of real nuclei which at the same time then also define their behavior on the average. We know by now that the well known Strutinsky averaging procedure [5] for nuclei of realistic sizes is exactly equivalent to this point of view for the purely theoretical approaches to the nucleus; but also on the experimental side a specific quantity measured as a function of a large number of nuclei allows to extract exactly the corresponding experimentally determined average of the same quantity. Agreement of average experimental and theoretical numbers then allows us to conclude about our understanding of the nucleus. A whole realm of nuclear properties is thus open to our semiclassical investigations.

One of the interests of the use of semiclassics then resides in the fact that on a microscopic (not phenomenological) level this method is by orders of magnitude easier to handle than the Strutinsky method* (e.g. a selfconsistent Hartree-Fock Strutinsky calculation). This therefore is by now increasingly practiced [1,6] for the microscopic calculation of average nuclear ground state properties starting from a (effective) nucleon-nucleon interaction. In this way for example the parameters of the Bethe Weizsäcker mass formula can be directly and very economically related to the parameters of the (effective) nucleon-nucleon force. The point we want to make here, and an essential part of this lecture shall be devoted to this aim, is that this procedure not only can be applied to groundstate properties of nuclei but to all quantities present in the afore mentioned quasi macroscopic nuclei. To give a specific example which on the theoretical side has not been explored in a satisfactory way in this context we want to mention the properties of nuclear vibrational states let us say the quadrupole vibrations; there does not exist at present a theory which allows to calculate the full spectrum of this type of vibrations for our quasi macroscopic nuclei which extrapolated back to real sizes would define the average vibrational behavior. A pertinent question in this context is for example whether the currents of the well known low lying 2^+ states in nuclei which certainly would be present in these quasi macroscopic nuclei as well, exhibit there a rotational or irrotational flow pattern. In realistic nuclei these states show strong shell fluctuations and theoretical macroscopic calculations (RPA) reveal that (for magic nuclei) the flow patterns have strong rotational components (vortices); but are these rota-

*This should not be understood as a criticism of the Strutinsky method which has been conceived as a very valuable phenomenological method!

tional components due to shell fluctuations or do they also persist in large nuclei where no shell effects are present ? This question illustrates the need for theories treating in a well defined way the average behaviour(i.e. exempt of shell effects) also of dynamical quantities*.

We insist on the fact that such a theory should contain an average of(and therefore comprise) spherical and open shell nuclei in one and thus give a common theory for all nuclei depending only in a smooth way on particle number**. The question raised above about the nature of the low lying 2^+ states also shows that for the study of some properties shell effects can actually be quite disturbing in revealing the true nature of a state or other quantities (another example will be given later on in the context of currents in rotating nuclei).

For the average vibrational theory we will sketch one possibility of approach below but the main emphasis of these lectures will be to lay the basis of a precise and selfcontained framework of (Extended) Thomas Fermi theory which is generalised in many different ways allowing for up to now unknown systematics and applications. The interest of such an endeavor is two fold: firstly, as we said already above, because of its equivalence to Strutinsky averaging, disturbing shell fluctuations are systematically eliminated revealing in a direct, simple, and transparent way basic physical features which in a purely quantum mechanical way should have been detected only at the cost of a laborious detour. The second interest lies in the fact that the numerical effort in this extended Thomas Fermi theory is almost reduced to the one of infinite matter calculations with its enormous advantage of translational invariance (Thomas Fermi = use of plane waves locally); one can then hope that many body theory for finite nuclei can be pushed within this formalism to degrees of sophistication where a full quantal solution is out of scope due to numerical limitations. More details of what we are thinking of will be given below.

Before ending this introduction we would like to point out that we believe that we also made some progress in the formal aspects of the semiclassical expansions in powers of $\hbar$. This concerns for instance the physical interpretation of the Wigner Kirkwood expansion of the Wigner transform of the density matrix into an asymptotic series in powers of $\hbar$. We will give arguments that this

---

*By this we do not mean quantities related to adiabatic motion which still can be incorporated into (constrained) groundstate properties.

**It seems unclear to us whether the scaling approach [7] to nuclear dynamics allows for such an interpretation.

type of expansion has (in a certain sense) by far better convergence properties and less restrictive critera to be obeyed than those usually advanced for the validity of Thomas Fermi theory ($\nabla V / (V k_F) \ll 1$). As a matter of fact we will elaborate on an argument which says that as long as the diffusivity of the Wigner phase space distribution around the local Fermi surface stays small compared to the Fermi energy the Wigner Kirkwood expansion converges essentially in a very rapid way to the exact average part of the considered quantity. In this context we will then also interprete the Wigner Kirkwood expansion as a leptodermous expansion in phase space in analogy to the leptodermous expansion in real space introduced by Swiatecki [8]. We thus hope to give with our semiclassical method a mathematically well defined tool to the practitioner whose handling he might feel more confident upon than this was the case hitherto and which will encourage for further applications and investigations.

THE WIGNER TRANSFORM OF THE DENSITY MATRIX

A basic quantity in nuclear physics where the mean field approach is believed to be a very valuable first order approximation is the single particle density matrix. In order to study this quantity semiclassically it is convenient to pass over to its Wigner transform and also to take a simple example not completely irrelevant to nuclear physics, i.e. the three-dimensional spherical harmonic oscillator. For closed shell nuclei it is then straightforward to show that [9]

$$f_{H.O.}(\vec{R},\vec{p}) = \int d^3 s\, e^{-\frac{i}{\hbar}\vec{p}\vec{s}} \langle \vec{R} + \frac{\vec{s}}{2} | \hat{\rho} | \vec{R} - \frac{\vec{s}}{2} \rangle = f\left(\frac{H_c}{\hbar\omega_0}\right) \quad (1)$$

where $H_c = p^2/2m + (m\omega_0^2) R^2/2$ is the classical energy and $\vec{R}$, $\vec{s}$ are the center of gravity and relative coordinates of the nonlocal density matrix $\langle \vec{r} | \hat{\rho} | \vec{r}\,' \rangle$. For a general potential the $\vec{R},\vec{p}$ dependence can of course not be lumped together in one variable but it will still be convenient to keep as one variable H_c which is a kind of polar coordinate in phase space, the corresponding angle being characterised for example by the ratio $m\,\omega_0 R/p$. Usually however f depends on three variables (for spherical potentials) R, $p_\perp$, $p_\parallel$ where $p_\perp$ and $p_\parallel$ are the momenta perpendicular and parallel to the nuclear surface respectively. In Fig. 1a we show one radial cut through the distribution $f_{H.O.}(R,p)$ for A = 224 [9]. We see that the exact distribution has strong shell fluctuations (becoming even negative) and as a matter of fact the Wigner transform of the density matrix is the quantity where the shell fluctuations are the most pronounced. The Strutinsky averaged [5] distribution on the contrary shows no wiggles in the interior any more and has there become identical to the pure Thomas Fermi distribution. It is interesting to note that as $A \to \infty$ the

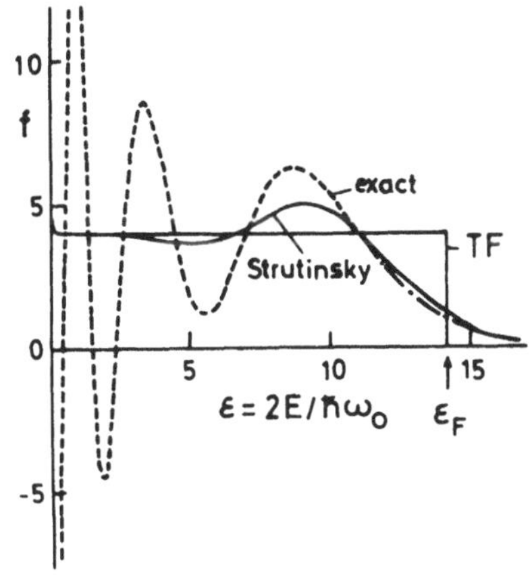

Fig. 1a . One of the radial cuts (all cuts are equal for the harmonic oscillator) through the Wigner distribution function for a harmonic potential with A =224 nucleons. The dashed, full and dashed dotted curve represent the exact quantum mechanical, the Strutinsky averaged, and the Thomas Fermi results respectively.

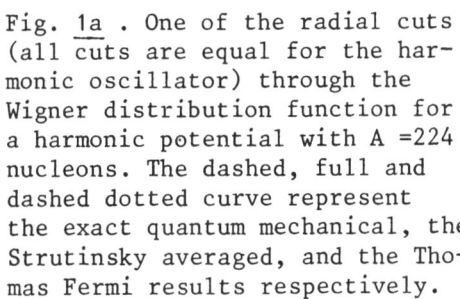

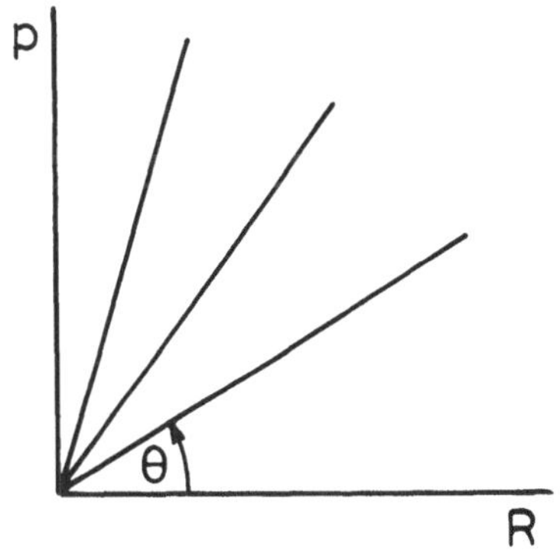

Fig. 1b . Schematic indication of the radial cuts through f.

exact Wigner function does not approach the Thomas Fermi distribution in an analytic way (because of the non analytic dependence of f on $\hbar$, see eq.(1))[10]. Instead the fluctuations become even wilder as A increases; they are however at the same time more and more pushed towards the origin, the remainder approaching the smooth Strutinsky distribution which itself reaches the step like Fermi function for $A = \infty$. For our quasi microscopic nuclei described in the introduction it suffices to put some little "dirt" into the system in order to wash out the infinitely dense oscillations at the origin; we then can extrapolate back to real nuclear sizes and will find total agreement with the displayed Strutinsky distribution. So both procedures lead to the same well defined elimination of shell fluctuations as already mentioned in the introduction. The smooth distribution nevertheless still exhibits quite interesting features : not all oscillations have been completely washed out and there remains for instance a last bump before the final descent. This is a typical edge effect still present in half infinite systems; (one can easily convince oneself of this fact in calculating the Wigner distribution for half infinite matter with a hard wall); in analogy to similar oscillations in the electron density close to the surface of a solid we would like to call this a "Friedel bump". Another remarkable feature is the very pronounced diffusivity of the surface region of f. (Fitted one in a rough way a Fermi function to the distribution of Fig.1a then the corresponding diffusivity parameter would be of the order of 4MeV, a rather high value indeed). This diffusivity stays almost unchanged whether we consider the quantum mechanical or the smooth distribution and has therefore, like the Friedel bump, nothing to do with the existence of shells but is rather due to the finiteness of the system.

LEPTODERMOUS EXPANSION IN PHASE SPACE

It is known since long that smooth functions like the Strutinsky curve in Fig. 1a where the surface thickness is small compared to the total length of the function allow for a very accurate asymptotic expansion in terms of distributions. A very well known example is the Sommerfeld low temperature expansion of the Fermi function[11] (obvious notation) :

$$F(\varepsilon) = (1+e^{\frac{\varepsilon-\mu}{T}})^{-1} \simeq \Theta(\mu-\varepsilon) + \frac{\pi^2}{6} T^2 \delta'(\mu-\varepsilon) + \ldots \qquad (2)$$

This is an asymptotic expansion since the Fermi function has an essential singularity in T and consequently does not allow for a Taylor series expansion in powers of T. Nevertheless the expansion (2) represents for $T \ll \mu$ an extremely good approximation and it is easy to see that the error in (2) is of the order $\exp(-\mu/T)$. In nuclear physics one often parametrises the nuclear density by a Fermi function and the corresponding expansion (2), very valid for medium and heavy nuclei, is called the leptodermous expansion[8]. For the smooth Wigner function of Fig. 1a or for the corresponding one using a Wood Saxon potential (see Fig. 2 and explanation how it has been obtained below) it is obvious that a leptodermous expansion also would be extremely accurate (width/length $\simeq 1/10$, i.e. the exponentially small error $\simeq 10^{-4} - 10^{-5}$). As a matter of fact such a leptodermous expansion in phase space for the average function exists since long very well known under the name Wigner Kirkwood expansion and reads (the derivation can be found in text books[12,13] for a general but local single particle potential $V(\vec{R})$ (the generalisation to non local potentials is straightforward) :

$$f(\vec{R},\vec{p}) = \Theta\,(\mu - H_c) - \frac{1}{4}\frac{\hbar^2}{2m}\,\Delta V\,\delta'(\mu - H_c) + \frac{1}{12}\frac{\hbar^2}{2m}[(\vec{\nabla}V)^2 + (\vec{p}.\vec{\nabla})^2 V/m)] \tag{3}$$

$$\delta''\,(\mu - H_c) + \ldots$$

We see that here $\hbar$ plays exactly the same role as T in (2), i.e. it triggers the surface thickness of the distribution and as a matter of fact it is well known that the exact Wigner transform (as well as the Strutinsky averaged one) has an essential singularity in $\hbar$ very similar to the one of T in (2) (see also (1)).

From (3) we also can make some more precise statements about the validity and convergence of such an expansion: the prefactors (more precisely something like the square root of their absolute values) to the δ' and δ'' functions (3) yield of course a measure of the magnitude of the surface width of the Wigner function (compare (2) for this); as long as these prefactors indicate a width of the distribution (which changes with angle θ , see Fig.2) which is everywhere much smaller than the Fermi energy the expansion (3) converges to the exact average result (with an exponentially small error, see above). We checked this criterion numerically [14] and found that for a Wood Saxon potential with a diffusivity parameter as small as $a \simeq 0.3$ fm this criterion is still very well fulfilled but that for values $a \lesssim 0.3$ fm the expansion (3) starts rapidly to diverge. It is clear that the value $a \simeq 0.3$ is well below the usual values $0.4 \lesssim a \lesssim 0.6$ fm encountered in nuclear physics and that before even for a Woods Saxon potential the above mentioned accuracy of $10^{-4} - 10^{-5}$ of the expansion holds. As a matter of fact extensive numerical studies comparing groundstate energies obtained from the Strutinsky method and the semiclassical method [13,15] fully confirm our above reasoning. On the other hand, this kind of accuracy is needed for such subtle quantities like e.g. nuclear fission barriers[9]. We want to mention that the above convergence and validity criteria are by far less restrictive than the usual criterion for the validity of Thomas Fermi theory [16]

$$\vec{\nabla}V\,/(Vk_F) \ll 1$$

which breaks down already very shortly after the point where the nuclear potential leaves its constant value of saturation. Our considerations explain why already to lowest order Thomas Fermi approximation the corresponding nuclear density globally very nicely represents the <u>average</u> nuclear density and that lowest order correction in $\hbar$ allows for a quite accurate reproduction of the whole tail region of the nuclear density [17]. In order to do that we need however some method to approximately reconstruct the actual average Wigner distribution to which expansion (3) is its asymptotic series. This can be done with the so-called partial $\hbar$ resummation technique which we do not want to explain here because of lack of space and because in fact we will not need it extensively for our further considerations. Besides it is well documented in the literature [6,13,17]. It will serve only to visualize how the corresponding distributions really look like (which is hardly feasible looking at (3))but as a mathematical tool (3) is perfectly capable to solve all problems one usually is interested in (including e.g. the Hartree-Fock problem as will be seen below). The reconstitution techniques (partial $\hbar$ summation) are clumsier to use and loose in accuracy and mathematical rigour so that we prefer to stick to the simplicity and accuracy of (3) usually very well converged after the $\hbar^2$-term. In order to demonstrate that in (3) all the information of the smooth Wigner distribution is contained we simply mention [18] that a skilled reconstitution of the true function using (3) yields a distribution which is indistinguishable within the thickness of the line from the smooth curve in Fig. 1a. The same technique has then been used to reconstruct the Wigner function for a Woods Saxon potential (Fig. 2). Compared to the harmonic oscillator case we see some differences : the distribution depends now on the angle θ of the cut in phase space (Fig. 1b). For cuts exploring more the interior of the nucleus the distribution is more infinite matter (i.e. steplike)becoming more diffuse for cuts involving the surface region; no corresponding Strutinsky calculation is available as yet but preliminary comparison with the exact Wigner function shows very reasonable agreement in the surface region indicating that also for general nuclear potentials our method yields very accurate average distributions. Again, we also see the Friedel bump more or less developed at all angles. The reproduction of this bump is we think a remarkable feature implying that information about it is contained in (3). As a matter of fact this bump is the reason why the distribution f (see Fig. 1 a) is very asymetric around the border μ of the corresponding Thomas Fermi step function (this is e.g. not the case for the example of eq. (2) where the step function cuts the Fermi function exactly at half height, i.e. symmetrically); indeed closer inspection of (3) (in broadening the δ-function somewhat) shows that the $\hbar^2$ - correction terms indeed "shuffle" distribution from the edge of the step function away more to the interior allowing also for the build up of the bump. Realising this we may gain an appreciation for the potential power and accuracy contained in (3).

We have been quite detailed and elaborate on certain points in this paragraph because we think that this constitutes the basis to what will follow where due to restricted space we will be forced to be much briefer. We also feel that these things have not been very much clarified neither in the nuclear physics literature nor are we aware of any detailed discussion elsewhere.

SEMICLASSICAL HARTREE-FOCK

The considerations in the preceding paragraph lead us directly to a very simple solution of the nuclear Hartree-Fock problem for the average selfconsistent Wigner transform of the density matrix : we assume that we can represent a distribution like the one shown in Fig. 2 by a superposition of known functions; one could for instance imagine an expansion in a harmonic oscillator basis but probably, in view of the special structure of the radial cuts in Fig. 2, the following procedure is more efficient. Let us try to find a good guess for a function which has enough freedom to represent any of the curves in Fig. 2 quite accurately in choosing appropriately some parameters. A relatively crude guess is a Fermi function but probably a superposition of a Fermi function and derivative of a Fermi function allows already a quite accurate representation; we shall call this function $F(\varepsilon)$. Since this function is supposed to be quite close to the seeked for function it can be used as a kind of "basis" to represent the true function and it is well known that the following procedure is convergent under certain conditions (we call $f_\theta(\varepsilon)$ the radial cuts in Fig.2) :

$$f_\theta(\varepsilon) = \left(1 + C_1(\theta)\frac{\partial}{\partial\varepsilon} + C_2(\theta)\frac{\partial^2}{\partial\varepsilon^2} + \dots\right) F_\theta(\varepsilon) \qquad (4)$$

To make this expansion plausible we consider the ratio $r(t) = f(t)/F(t)$ of the Fourier transforms $f(t)$ and $F(t)$ of $f(\varepsilon)$ and $F(\varepsilon)$ respectively; under the condition that $r(t)$ has a convergent Taylor series in powers of t eq.(4) is convergent as a direct Fourier transform of (4) shows. Whether our ansatz (4) works reasonably i.e. with only a few coefficients C_i depends of course on the quality of our guess for $F(\varepsilon)$. As it can be seen from Fig. 2 the θ- dependence of the parameters should be quite smooth and can certainly be fit by low order polynomials in θ . The anisotropy in the $\vec{p}$ dependence we talked about above can also be incorporated with one or two further parameters as we have a quite good knowledge of the angular dependence in $f(R,\vec{p})$ (see for this ref[4]). For most cases however the angle average $\tilde{f}(R,p)$ will be sufficient. We therefore suppose to have with (4) a quite accurate and flexible parametrisation of $f(R,p)$. The Wigner transform of the Hartree-Fock potential is then schematically given by

$$V^{H.F.}(R,p) = \int d^3r'\, v(\vec{R}-\vec{r}')\, \rho(\vec{r}') + \int d^3k\, v(\vec{p}-\vec{k}) f(R,k) \qquad (5)$$

where $\rho(r) = 2 \int d^3p\, f/(2\pi\hbar)^3$ is the density and $v(p)$ the Fourier transform of the two body force $v(r)$. We insert $V^{H.F.}$ of (5) into the Wigner Kirkwood expansion (3) generalised to non local potentials and find a Wigner Kirkwood distribution $f_{W.K.}(R,p)$ which thus depends on the say 6 to 8 parameters necessary to parametrise f(R,p) through (4) (or on any other expansion coefficients). These parameters can then be fixed by the condition that the expectation values of an equal number of operators be equal calculated in using f given through (4) or using f_{WK} of (3). Such operators can be low powers of r and p (the powers may be fractional or negative) or combination of powers or functions of r and p such as r.p or $H^{HF} = T + V^{HF}$. It is easy to construct in this way say around 10 operators $O_i(R,p)$. As we said the parameters in (4) can then be fixed by the condition

$$\int d^3R\, d^3p\, O_i\, f_{WK} = \int d^3 R\, d^3 p\, O_i\, f \qquad (6)$$

where the number of operators O_i must be equal to the number of parameters entering into f. Since as we have discussed above f_{WK} yields extremely accurate results for expectation values (error: $10^{-4} \ldots 10^{-5}$) this procedure should then allow also for a very accurate determination of f(R,p). Of course the result may depend on the specific choice of the set $\{O_i\}$ of operators and we should in each case check that our final result does not depend sensitively on the choice in enlarging the number of parameters and consequently the number of operators. The dependence on a specific set of operators may be significantly reduced by the possibility that the above proposed procedure is variational. So far this is an unproven conjecture but there are reasons which we do not want to dwell on here, to believe that this is in fact the case. In any case the numerical examples we will study now are very encouraging.

These examples[19] are very much simplified by the fact that we studied only zero range forces of the Skyrme type. Since the corresponding potential (5) is local i.e. p-independent the whole procedure is reduced to a parametrisation of ρ (r) only. We took as a parametrisation for ρ a superposition of the first and second power of a Fermi function which is equivalent to taking a Fermi function plus its first derivative. We thus had four parameters to be fixed by (6). For the O_i's we took the particle number, R^{-1}, R and R^2. We calculated for the Skyrme forces SIII and SVII for N = Z = 20 and 92 (no Coulomb field, no spin orbit) the total binding energies and the rms radii and compare these quantities to those obtained from selfconsistent Hartree-Fock-Strutinsky calculations. (see Table 1). In Fig. 3 we also show for the case SIII,Z,N=92 the comparison of the exact and semiclassical densities. We see that there is a very satisfactory overall agreement as well as for the binding energies, rms radii and the density itself. The semiclassical density is slightly too diffuse in the surface however which might be an indication that the chosen parametrisation is not yet optimal. In Table 1 we also give the relative weight factors with which the first and second powers of the Fermi function in $\rho_{s.c.}$ are represented; we see that always the second power is by far the most important one. This subtle feature

Table 1.

Comparison of total binding energies and rms radii using Hartree-Fock(HF), Hartree-Fock Strutinsky (HFS), and Wigner Kirkwood (WK) for the two Skyrme forces SIII and SVII. (We are grateful to Dr. J. Bartel for providing us with the HFS values). The ratio of the weights of Fermi function to the power one and two (see text) is given by α_1/α_2.

A		40	140	184
	HF	- 397.3	- 1649.6	- 2246.9
E_{tot} [MeV]	HFS	- 382.1		- 2230.7
S III	WK	- 384.2	- 1642.7	- 2225.3
$\langle r^2 \rangle$	HF	3.37	4.89	5.33
S III	WK	3.41	4.93	5.37
α_1/α_2	WK	0.11	0.015	0.003
	HF	- 397.1	- 1649.2	- 2243.4
E_{tot} [MeV]	HFS		- 1648.5	- 2238.4
S VII	WK	- 388.0	- 1651.1	- 2237.4
$\langle r^2 \rangle$	HF	3.38	4.90	5.35
S VII	WK	3.38	4.92	5.36
α_1/α_2	WK	0.10	0.004	0.01

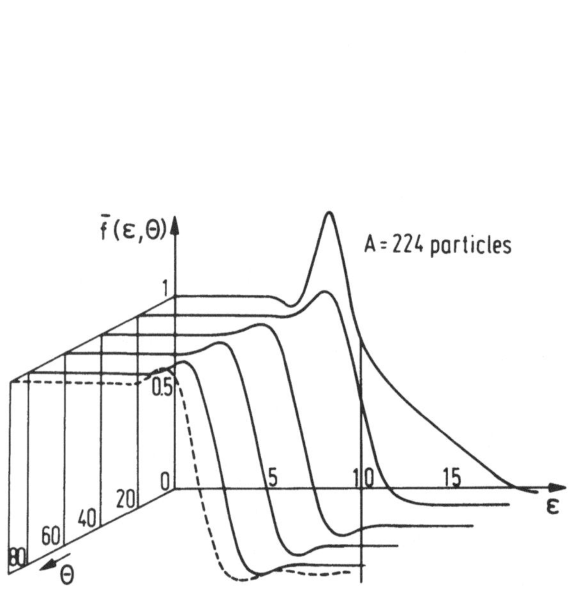

Fig. 2 Smooth Wigner distribution $\tilde{f}(\varepsilon,\theta)$, $\varepsilon = 2H_c^{Ho}/\hbar\omega_o$, $\theta =$ artg b/R for a Woods Saxon potential (V_o =-50MeV, R = 7.3 fm, a =o.6fm, A = 224, usual notation). The twiggle indicates that we averaged over all directions of $\vec{p}$ in order to obtain a convenient graphical representation as a function of two variables only.

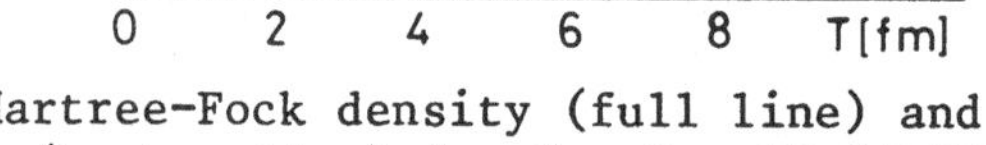

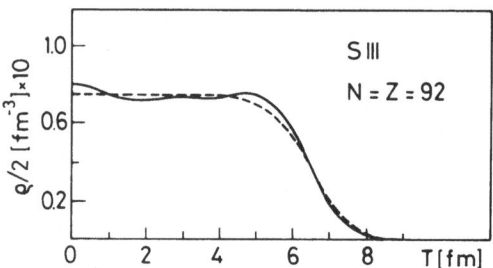

Fig. 3 Exact Hartree-Fock density (full line) and semiclassical density (broken line) for N = Z = 92 (SIII).

actually has been known before [1,19] revealing that the surface region of the nuclear density is asymmetric around the half density point. It is very satisfying that our semiclassical procedure reproduces this fact indicating once again that in the Wigner Kirkwood series (3) very fine details of the density or density matrix are contained. So far the method has not yet been checked in very great detail and it would for example be interesting if including a third power of the Fermi function improved the density. One could also think that taking as a first guess for the density a Fermi function to an arbitrary power where the power is an ajustable parameter would be helpful. A further indication that the above procedure is quite successful lies in the fact that the semiclassically calculated level density parameter agrees well with the Hartree-Fock Strutinsky result. We also calculated the static monopole polarizability in constraining the density by an external monopole field and found excellent agreement between semiclassical and Hertree-Fock results.

Another advantage of this method is that it is very easily generalised to finite temperatures. In fact we only have to replace in (3) the step function (and consequently the derivatives of δ-functions which are successive derivatives of the step function) by

$$\theta(\mu - H_c) \rightarrow F_T = \left(1 + \exp\left(\frac{H_c-\mu}{T}\right)\right)^{-1}$$

$$\delta'(\mu - H_c) \rightarrow \frac{\delta^2 f}{\delta\mu^2} \quad ;\text{etc.} \tag{7}$$

We also would like to mention that a "liquid drop expansion" in volume, surface, curvature can straightforwardly be achieved[14]. Since applications of semiclassical method for groundstate properties are quite numerous and well documented for instance in what concerns the calculation of fission barriers [20,21] we here want to content ourselves only with one further example namely the dependence of the height of the fission barrier on the compression or dilation of the nucleus. We calculated this result some years back[22] within the density functional formalism but there is no doubt that similar results could be reproduced with the method outlined in this paragraph. We show in Fig. 4 the fission barrier height for ^{240}Pu as a function of the central density ρ_0. We see that the barrier is lowered more in compressing it than it is raised in dilating it; one could speculate that in a breathing mode the nucleus sees on the average a reduced width favouring the fission process.

Ending this paragraph we would like to point out again the simplicity and theoretical well- roundedness of our appraoch; this in view of other existing methods to solve the same problem like the well known density functional formalism[9,20] or the recently developed $\hbar$ - resummation technique[6]. The density functional formalism suffers from several drawbacks : its use is difficult to justify beyond the classical turning point; the convergence in powers of $\hbar$ is quite a bit slower than the Wigner Kirkwood series and it has been convincingly shown [23] that one has to go to fourth order for convergence whereas in (3) second order is sufficient. Fourth order terms are however already quite complicated [20,24] becoming almost intractable in the most general cases including such features as finite range forces, spin orbit terms, rotational motion, finite temperature, and nuclear superfluidity. Generalisations of this kind are however still easily feasible with the Wigner-Kirkwood expansion as we will show below. Further difficulties arise with the calculation of the level density parameter at the Fermi surface where density functional methods seem to give 20% - 30% too high values compared to corresponding Hartree-Fock-Strutinsky values [25], not to talk about quantities which are impossible to calculate with density functional methods like e.g. the derivative of the single particle level density at the Fermi surface which, as can easily be checked, gives a divergent result whereas it is perfectly defined using the Wigner-Kirkwood expansion (3).

The other method named "partial $\hbar$ resummation" [6] has as we mentioned already other difficulties related for instance to inconsistencies in the $\hbar$ - expansion and to uncontrolled accuracy in a necessary saddle-point approximation; still it is very useful to reconstruct approximately the true function to which (3) is its asymptotic expansion.

SEMICLASSICAL TREATMENT OF NUCLER SUPERFLUIDITY

We will be extremely brief in this paragraph because the main results are already documented elsewhere [2,22,26] since quite some time, we will however indicate recent developments in this field.

It is well known that for superfluid nuclei the H.F. scheme has to be replaced by the H.F. + BCS or the Hartree-Fock Bogoliubov theory (HFB);in addition to the normal density ρ and the normal field V^{HF} additional quantities κ (anormal density) and Δ (pair potential) are introduced leading to generalised density and Hamiltonian :

$$\rho \to \mathcal{R} = \begin{pmatrix} \rho & \kappa \\ \kappa^+ & 1-\rho^* \end{pmatrix} ; \quad H^{H.F.} \to \mathcal{H} = \begin{pmatrix} H^{H.F.} - \lambda & \Delta \\ \Delta^+ & -H^{H.F.*} + \lambda \end{pmatrix} \tag{8}$$

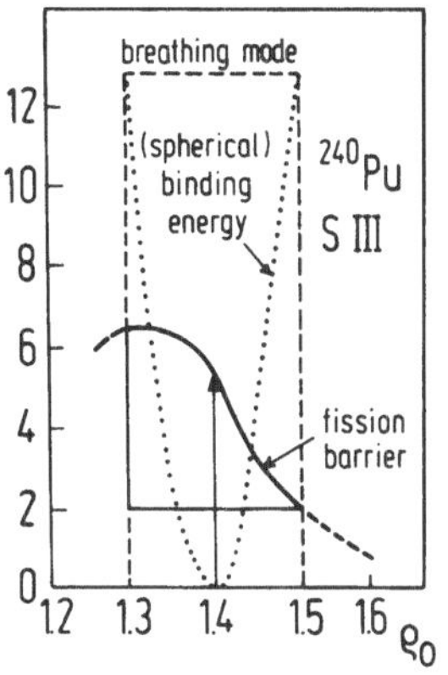

Fig. 4 Dependence of the fission barrier height on the central density ρ_o . The vertical broken lines indicate the extrema of ρ_o for the breathing mode while the base of the parabola shows the equlibrium density.

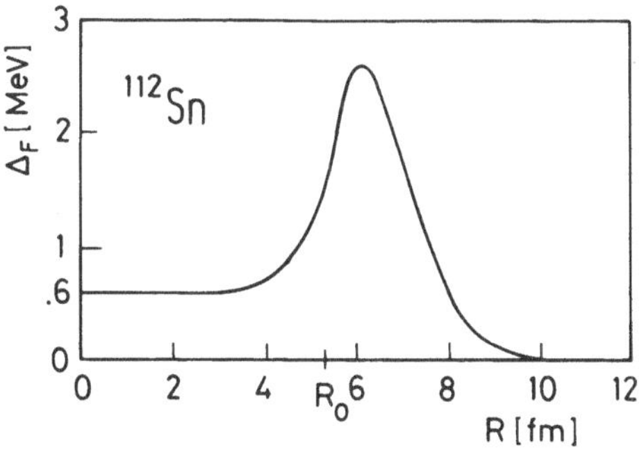

Fig. 5 Spatial dependence of $\Delta_F(R) = \Delta(R,p_F(R))$ for ^{112}Sn. R_o indicates the position of the half density.

Solving the HFB eqs $[\mathcal{R}, \mathcal{H}] = 0$; $\mathcal{R}^2 = \mathcal{R}$ to lowest order in $\hbar$ leads to the following solution:

$$f = \frac{1}{2} \{ 1 - (H_W^{H.F.} - \lambda)/E_W \}$$
$$\kappa = \frac{1}{2} \frac{\Delta_W}{E_W} \; ; \; E_W = \sqrt{(H_W^{H.F.} - \lambda)^2 + \Delta_W^2} \tag{9}$$

where the index W stands for Wigner transform. It is easy to see that for $\Delta \rightarrow 0$ we get back the usual lowest order Thomas Fermi theory of non superfluid nuclei. We solved the corresponding gap equation

$$\Delta_W = \int d^3k \; v(\vec{p}-\vec{k}) \; \kappa(\vec{R},\vec{k})$$

using the Gogny force and obtained for the gap at the locel Fermi surface

$$p = p_F(R) \propto \rho^{1/3}(R), \text{i.e. for } \Delta_F(R) = \Delta\,(R,p_F)$$

the result shown in Fig. 5 (we want to point out that we solved the gap equation numerically but a very accurate approximate solution can be obtained analytically using standard methods of the infinite matter case in solid state physics[26]). We remark from

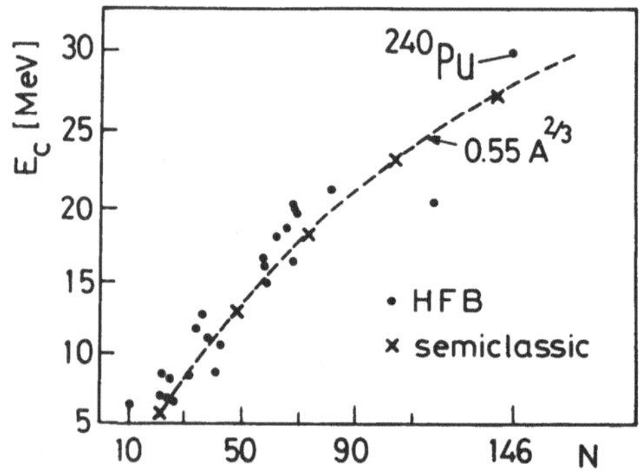

Fig. 6 Comparison of the exact quantal and semiclassical results for the condensation energy E_C for nuclei constrained to sphericity as a function of neutron number[27].

Fig. 5 that in the interior of the nucleus we obtain $\Delta_F \simeq 0.5$ MeV which at the same time represents the nuclear matter value of the nucler gap calculated with the Gogny force. We also see that Δ_F has an extreme surface variation shooting up to almost 3 MeV at a point where the density has dropped to a third of its saturation value. That this lowest order Thomas Fermi solution for the gap is not devoid of reality is shown in Fig. 6 where we compare the semi-classically calculated condensation energy

$$E_c = \frac{1}{2} \mathrm{Tr}\ (\kappa\Delta) = \frac{1}{2} \int d^3 R d^3 p\ \Delta_W \kappa_W / (2\pi\hbar)^3$$

as a function of neutron number with results obtained by an exact solution of the HFB eqs.[27] constraining the nuclei to sphericity. We see that the semiclassical values follow the average trend but passing through the maximal values rather than through the true average. In view of the extreme sensitivity of the solution on all parameters (exponential dependence on strength, range of the force, effective mass, density) we think however that this lowest order result can be considered as a success; after all it is a quite subtle question whether local density, i.e. Thomas-Fermi theory is applicable for a quantity like the pairing gap; we do not want to dwell on this question further here but rather answer it by the affirmative in view of our results. However, it is clear from the rapid spatial variation of Δ_F in the surface and the overshooting of E_c that $\hbar$, i.e; gradient-corrections to this lowest order result will be quite important. Since gradient corrections are inefficient in the interior of the nucleus this means that the very pronounced surface peak in Δ_F will most likely be quite substantially reduced, leading at the same time to a reduction of E_c.

Gradient corrections to the ($\hbar = 0$) result (9) can in fact be elaborated in a similar fashion as in the non superfluid case [28]. We briefly want to sketch this here : it is well known [13] that the normal density is related in the following way to the Bloch density

$$\rho = \mathcal{L}^{-1}\ C^\beta / \beta \text{ with } C^\beta = \exp\ (-\ \beta H)$$

and $\mathcal{L}^{-1}$ standing for inverse Laplace transformation. It is easy to relate in a similar way the density matrix $\mathcal{R}$ of eq. (8) to a generalised Boch density $\mathcal{C} = \exp\ (-\beta\ \mathcal{H})$ the lowest order (in $\hbar$) expression of which is obtained by the replacement $\mathcal{H} \rightarrow \mathcal{H}_W$; this leads of course again to eqs (9). Inserting the formal expansion

$$\mathcal{C} = \mathcal{C}_0 + \hbar\ \mathcal{C}_1 + \hbar^2\ \mathcal{C}_2 + \ldots$$

into the Wigner transform of the Bloch equation

$$\mathcal{C}_W + \frac{1}{2} \{\mathcal{H}\,\mathcal{C} + \mathcal{C}\,\mathcal{H}\}_W = 0$$

allows to evaluate the expressions for $\mathcal{C}_1, \mathcal{C}_2, \ldots$ and therefore the correction terms to (9). These results will be published in more detailed form elsewhere [28]. The inclusion of these correction terms will, however, as we said, be necessary to obtain precise quantitative agreement with the average of the fully quantal results. Also the well known empirical law for the even odd mass difference $\Delta_{eo} \simeq 12\, A^{-1/2}$ may then be explainable (this seems to be a quite subtle law to be derived semiclassically because of the necessity to specify to even and odd systems).

In conclusion, we can say that together with the method described in the previous paragraph we will be able to solve semiclassically the full HFB problem very economically for the average behavior of the normal and superfluid components of nuclei.

SLOW ROTATIONAL MOTION OF NORMAL AND SUPERFLUID NUCLEI

Besides for the description of purely static nuclei in their groundstate, the methods outlined in the preceding paragraphs can be equally applied to nuclei in slow (i.e. adiabatic) collective motion. A very well known example is slow nuclear rotation around an axis (say x-axis) perpendicular to the symmetry axis (say z-axis) of a deformed nucleus; we then have to add in the usual way an additional term $-\ \omega \ell_x$ to the single particle Hamiltonian. This can be studied for normal as well as for superfluid nuclei.

a) Non superfluid case

The Hamiltonian to be studied is now given by $H_\omega = H - \omega \ell_x$ where $\ell_x = y\, p_z - z\, p_y$ is the x- component of the angular momentum. H_ω can be treated exactly as H in the preceding paragraphs. The analogous expression to (3) is somewhat more complicated but straightforward to calculate. It is also straightforward to evaluate the moment of inertia in the $\hbar = 0$ limit; as is well known [29] it is equal to the one of rigid rotation. Including lowest order $\hbar$ - corrections reduces this value by about 4% if spin is neglected. It is interesting to study to what kind of flow of the nuclear fluid this reduction is due to. To this purpose we calculate the semiclassical current in the intrinsic frame. In the pure Thomas-Fermi limit this is zero (rigid rotation). Thus any finite current in the body fixed system must come from $\hbar$-corrections. Since these always involve derivatives on the mean field potential (see (3)) it is clear that the intrinsic current is a surface phenomenon. It is easy to calculate this residual intrinsic current say for a deformed Woods-Saxon potential within the Wigner Kirkwood expansion and we indeed find a surface current in the intrinsic frame which opposes the rotational motion reducing thus the moment of inertia. We show such a flow pattern in Fig. 7 where the pronounced surface peaking is clearly exhibited. This phenomenon is the analogue to the Bohr van Leeuwen theorem in diamagnetism which says that in the classical limit no diamagnetism exists and that diamagnetism only comes from the amperic current flowing in the surface layer of the sample [30]. Our nuclear Bohr van Leeuwen theorem reveals a quite small (as is diamagnetism) but interesting effect. It also demonstrates quite nicely that in looking at the average behavior using semiclassics subtleties can be evidenced straightforwardly which in purely quantum mechanical treatment would be completely hidden by predominating shell effects : calculating namely the intrinsic current quantum mechanically for a specific nucleus reveals strong vortex structures entirely due to shell effects. This structure only becomes washed out if the individual nucleus is heated (say in a heavy ion reaction) up to temperatures of the order 3-5 MeV. It can indeed be shown [31] that the intrinsic current given in Fig. 7 is up to $T \simeq 10$ MeV quite independent of the temperature so that Fig. 7 represents also the true picture for an individual but heated nucleus.

b) Superfluid case

It is very well known that the rotational motion of nuclei is most strongly influenced by superfluidity reducing for example the moment of inertia by a factor of roughly two. It is therefore interesting to see whether this phenomenon can be easily understood within our semiclassical framework. In the shell model basis {n} the moment of inertia for a superfluid nucleus is given in the approximation of Inglis, i.e. neglecting the readjustment of the mean field to the rotational motion by:

$$\mathcal{J}_B = \sum_{n_1 n_2} \langle n_1 | \ell_x | n_2 \rangle \; \delta\rho_{n_1 n_2} \tag{10}$$

with

$$\delta\rho_{n_1 n_2} = (U_{n_1} V_{n_2} - U_{n_2} V_{n_1})^2 \langle n_2 | \ell_x | n_1 \rangle / (E_{n_1} + E_{n_2}) \tag{11}$$

where $\delta\rho$ is called the static linear free response to the rotational field ℓ_x; U, V and E are the usual BCS occupation numbers and quasiparticle energies. In order to simplify things we consider here the gap to be a constant: then (11) can be written in the following form

$$\delta\rho_{n_1 n_2} = \langle n_2 | \ell_x | n_1 \rangle \langle n_1 | R_{12} | n_2 \rangle$$

$$R_{12} = \frac{(U_1 V_2 - U_2 V_1)^2}{(E_1 + E_2)} = \frac{E_1 E_2 - \tilde{H}_1 \tilde{H}_2 - \Delta^2}{2 E_1 E_2 (E_1 + E_2)} \tag{12}$$

where E_i ($i = 1,2$) are the quasiparticle energies ($E^2 = \tilde{H}^2 + \Delta^2$) and $\tilde{H}$ the usual single particle Hamiltonian relative to the Fermi energy ($\tilde{H} = H - \lambda$). It can be verified that (12) approaches $\delta\,(\lambda - (H_1 + H_2)/2)$ in the limit $\Delta \to 0$. Following Migdal [32] one observes that for finite values of Δ (12) still exhibits a sharp maximum as a function of the center of mass energy $H = (H_1 + H_2)/2$ for $H = \lambda$. We will approximate this function by a δ-function with however a Δ-dependent strength. To evaluate this strength we transform R_{12} to center of mass and relative energies H and $h = H_1 - H_2$ and integrate over H :

$$\int dH\, R(H,h) = 1 - g\left(\frac{h}{2\Delta}\right) \tag{13}$$

with

$$g(x) = \frac{\text{ar sinh}(x)}{x\sqrt{1 + x^2}} \tag{14}$$

The moment of inertia therefore can be written as :

$$\mathcal{J}_B = \mathcal{J}_{11} - \mathcal{J}_{12}$$

$$\mathcal{J}_{12} = \sum_{n_1 n_2} g((\varepsilon_{n_1} - \varepsilon_{n_2})/2\Delta)\, |\langle n_1 | \ell_x | n_2 \rangle|^2\, \delta(\lambda - (\varepsilon_{n_1} + \varepsilon_{n_2})/2) \tag{15}$$

where of course the ε_n's are the shell model single particle energies. Introducing the Fourier transform of $g(x)$ and $\delta(x)$ we can also write:

$$\mathcal{J}_{11} = \frac{1}{2\pi\hbar} \int_{-\infty}^{+\infty} dt\, e^{\frac{i}{\hbar}\lambda t}\, \text{Tr}\,\{C^{t/2}\, \ell_x\, C^{t/2}\, \ell_x\} \tag{16}$$

$$\mathcal{J}_{12} = \frac{1}{2\pi\hbar} \int_{-\infty}^{+\infty} d\tau\, g(\tau) \int_{-\infty}^{+\infty} dt\, e^{\frac{i}{\hbar}\lambda t}\, \text{Tr}\,\{C^{t/2}\, \ell_x C^{t/2} \ell_x(\tau)\} \tag{17}$$

$$C^t = \exp\left(-\frac{i}{\hbar} Ht\right)$$

where $\ell_x(\tau) = \exp(i \frac{H}{2\Delta} \tau)\ \ell_x \exp(-i \frac{H}{2\Delta} \tau)$. The expressions under the traces in (16,17) are the products of four single particle operators and it is well known[13] that to lowest order in $\hbar$ this is just the phase space integral over the ordinary product of the Wigner transforms of the different operators. Within this approximation it is easy to show that $\mathcal{J}_{11}$ is equal to the rigid body moment of inertia ($\mathcal{J}_{11} = \mathcal{J}_{rig}$). The expression for $\mathcal{J}_{12}$ is somewhat more involved since $\ell_x(\tau) = y(\tau)\, p_z(\tau) - z(\tau)\, p_y(\tau)$ ($\vec{r}(\tau)$, $\vec{p}(\tau)$ being the classical trajectories) is not known in analytical form for a general potential. Under certain conditions a power series expansion in $\hbar^2 / 4\Delta^2$ may be valid otherwise partial resummation of the series using Padé approximants may be adequate. For the harmonic oscillator the classical trajectories and therefore $\ell_x(\tau)$ are easily evaluated and of course well known in analytical form. Inserting this into (17) performing the time integrals and evaluating the phase space integral yields

$$\mathcal{J}_{12} = \frac{1}{12\hbar^3} \frac{\lambda^4}{\omega_2^3\, \omega_y^4} \frac{4\Delta^2}{\hbar^2} \frac{1}{2} (g_- \nu_+^2 + g_+ \nu_-^2)$$

$$\nu_\pm = \hbar \frac{\omega_z \pm \omega_y}{2\Delta} \tag{18}$$

with $g_\pm = g(\nu_\pm)$ and ω_y, ω_z being the harmonic oscillator frequencies in y- and z- direction (we recall that we consider an axially symmetric nucleus with prolate deformation rotating around the x-axis). Expression (18) is exactly equivalent to the one found by Migdal [32] using however quite a different procedure. From (18) we see that in fact in treating $\ell_x(\tau)$ to all orders in $\hbar^2/4\Delta^2$ a partial resummation of the $\hbar$ expansion for $\mathcal{J}_{12}$ has been achieved involving however the most subtle part since we see from it that the order of the limits $\Delta \to 0$ and $\hbar \to 0$ are not interchangeable. As a matter of fact $\Delta \to 0$ first and $\hbar \to 0$ second yields $\mathcal{J}_{12} = 0$ and therefore $\mathcal{J}_B = \mathcal{J}_{rig}$. On the other hand reversing the order yields $\mathcal{J}_{12} = \mathcal{J}_{rig}$ and therefore $\mathcal{J}_B = 0$. In Fig. 8 we show this behavior of $\mathcal{J}_B$ as a function of Δ for the model nucleus ^{168}Er for a deformation of $\varepsilon = 0.27$. We also show the exact quantum mechanical result[31] and see the close agreement between two approaches for finite values of Δ. It turns out that Δ is a very effective smearing parameter and that already for realistic values of Δ semiclassical and quantum mechanical results are almost indistinguishable. From Fig. 8 we also see that the reduction of the moment of inertia by a factor of two from its rigid value for $\Delta \simeq 1$ MeV is very well confirmed.

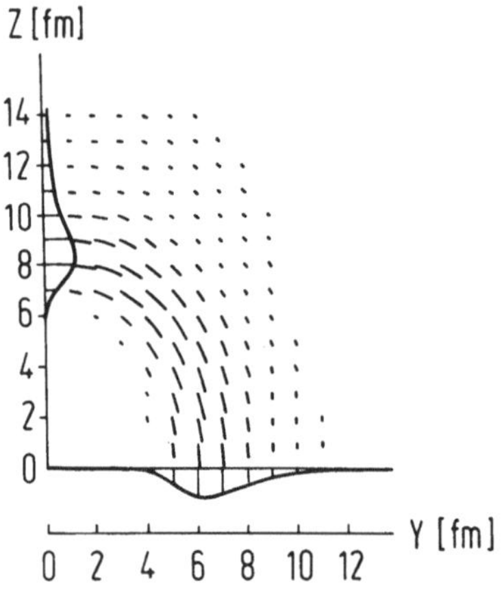

Fig. 7 Back flow in the body fixed frame due to ħ-corrections (Woods Saxon potential, A = 168 deformation, ε = 0.27 [22]).

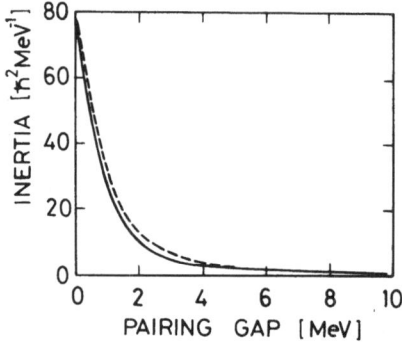

Fig. 8 Comparison of quantum mechanical (broken line) and semi-classical (full line) moment of inertia as a function of Δ.

Nature provides us not so very much with a variation of $\mathfrak{J}$ as a function of Δ for constant deformations ε but rather with the inverse situation, i.e. the moment of inertia for different deformations with a roughly constant gap can be measured. In Fig. 9 we show the experimental findings[33] together with the results of our expression [15,18]. We see that our formula yields for constant finite gap : $\mathfrak{J}_B(\varepsilon = 0) = 0$ and $\mathfrak{J}_B(\varepsilon \to \infty) = \mathfrak{J}_{rig.}$ in quite good agreement with the experimental data.

It should be mentioned that for large values of Δ the formula (10,11) becomes inadequate and the readjustement $\Delta^{(1)}$ of Δ to the rotational field should be taken into account. Also this can be treated in our framework and we find that the moment of inertia approaches the irrotational flow value as $\Delta \to \infty$. For

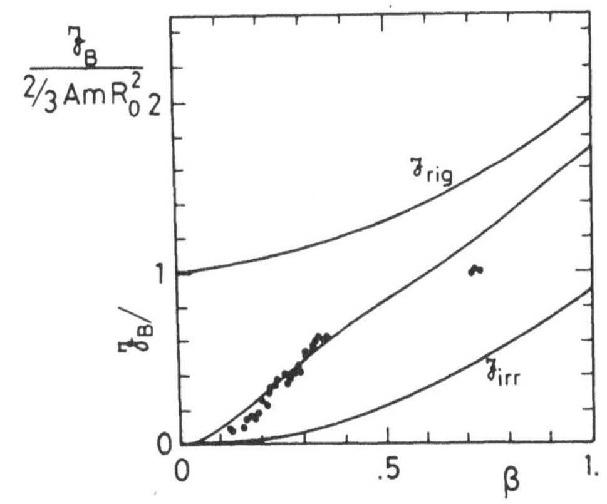

Fig. 9 Semiclassical moment of inertia as a function of deformation (Hill-Wheeler coordinats are used). The points at $\beta \simeq 0.7$ correspond to the shape isomers ($\Delta = 0.93$ MeV)

$\Delta \simeq 1$ MeV the influence of $\Delta^{(1)}$ is however not very important and we shall not discuss it further here.

Contrary to the procedure chosen by Migdal, we are perfectly capable to also calculate the current; in the body fixed frame we obtain for $\vec{j} = (j_x, j_y, j_z)$

$$
\begin{aligned}
j_x &= 0 \\
j_y &= - m\, \rho(\vec{R})\ \frac{R_z}{\omega_y} [(\omega_z + \omega_y)\, g_- - (\omega_z - \omega_y)\, g_+] \\
j_z &= m\, \rho\, (\vec{R})\ \frac{R_y}{\omega_z} [(\omega_z + \omega_y)\, g_- + (\omega_z - \omega_y)\, g_+]
\end{aligned}
\tag{19}
$$

Depending on the value of Δ we obtain from (19) a more or less strong backflow until for $\Delta \to \infty$, no matter flows in the laboratory frame while the potential is still turning. We checked this with quantum mechanical calculations and found close agreement with (19) down to almost realistic values of Δ. For very small values of Δ the afore mentioned vortices due to shell structure develop however quantum mechanically whereas the semiclassical treatment yields only the average part of it.

Concluding this paragraph, we can say that our semiclassical methods very satisfactorily allow a quantitative understanding of normal and superfluid nuclei in rotation and that for instance for superfluid nuclei semiclassical and fully quantal results are very close due to the smearing effect of Δ. Further studies refining the above treatment using for instance a coordinate dependent gap would be interesting.

ONE PARTICLE - ONE HOLE AND TWO PARTICLE-TWO HOLE LEVEL DENSITIES

The static linear responsefunction (12) of the last paragraph is in fact already a two body correlation function and in applying successfully our semiclassical methods implies that we know how to calculate more body correlation functions in the Thomas Fermi limit. We here want to make a more detailed study for the 1p-1h and 2p-2h level densities.

It is very well known since long how to calculate in Thomas-Fermi theory the single particle level density $g^E_{1p} = \sum_i \delta(E - \varepsilon_i) = \mathrm{Tr}\ \delta(E-H)$: the Hamiltonian has to be replaced by its classical counterpart ($H \to H_c$) and the trace must be converted into a phase space integral. For a three dimensional spherical harmonic oscillator potential we then find $\hbar\omega_o\ g^{TF}_{1p} = \varepsilon^2$ with $\varepsilon = E/\hbar\omega_o$. In order to proceed to more complicated level densities we therefore will try to express these also in a representation independent way using traces. It is easy to verify that we have:

$$g^E_{1p-1h} = \sum_{ph} \delta\,(E-\varepsilon_p+\varepsilon_h) = Tr_1\, Tr_2\{\theta\,(\varepsilon_F-H_2)\theta\,(H_1-\varepsilon_F)\delta(E-H_1+H_2)\} \tag{20}$$

$$g^E_{2p-2h} = \sum_{p_1<p_2,h_1<h_2} \delta(E-\varepsilon_{p_1}-\varepsilon_{p_2}+\varepsilon_{h_1}+\varepsilon_{h_2})$$

$$= Tr_1\, Tr_2\, Tr_3\, Tr_4\,\{\theta\,(\varepsilon_F-H_4)\,\theta\,(\varepsilon_F-H_3)\,\theta(H_2-\varepsilon_F)\,\theta(H_1-\varepsilon_F) \tag{21}$$

$$\delta(E-H_1-H_2+H_3+H_4)\}$$

-(exchange)

The unit step functions θ in (20,21) are the projectors on the particle or hole spaces. In (21) we do not specify the extra terms coming from the Pauli principle between the two particles and two holes[3] because we here only are interested in the principle of the procedure. From what we said above about the single particle level density we immediately find the way to pass to the semiclassical limit in (20,21) : we replace the H_i's by H_{ic} and the traces

$$T_{r_i} \text{ by } n_i \int d^3 r\, d^3 p_i \,/\, (2\pi\hbar)^3$$

with n_i the spin, isospin degeneracy. We thus obtain for (20) a double phase space integral and for (21) a four fold phase space integral. For the spherical harmonic oscillator all integrals can be performed analytically and we obtain for example for the 1p-1h level density ($\lambda = \varepsilon_F/\hbar\omega_o$) :

$$g_{1p-1h} = \frac{\lambda^4\varepsilon}{2} + \frac{\varepsilon^5}{60} - \frac{(\varepsilon-\lambda)^3}{6}\,\theta(\varepsilon-\lambda)\,[\lambda^2+\frac{\lambda}{2}(\varepsilon-\lambda)+\frac{1}{10}\,(\varepsilon-\lambda)^2] \tag{22}$$

and a similar but lengthier expression for g_{2p-2h}[3] .

In Fig. 10 we show the comparison of our semiclassical calculation with an exact quantum mechanical one. However in order to ease the comparison we display the integrated quantities i.e. the number of levels as a function of energy. For reasons of completeness we also show the single particle case. We see that the semiclassical results pass very nicely through the average quantal values; even such fine details as the very low level density below $2\hbar\omega_o$ for g_{2p-2h} are well accounted for. We thus have a very reliable theory to also calculate correlation functions semiclassically.

The above procedure to obtain the semiclassical limit for correlation functions is however not quite as trivial as our derivation might indicate. In fact this procedure does not give the complete $\hbar \to 0$ limit. Let us explain this in more detail: our prescription leads e.g. for the δ-function in (20) to $\delta\,(E - H_{1c} + H_{2c})$. The argument of this function contains the difference of the potential at two different points : $V(\vec{R}_1) - V(\vec{R}_2)$. Thomas Fermi theory however supposes that the potential is locally a constant, i.e. all gradients of the potential are supposed to be negligible; introducing center of gravity and relative coordinates $\vec{R}_1 = \vec{R} + \vec{s}/2$ and $\vec{R}_2 = \vec{R} - \vec{s}/2$ we see that to lowest order in an expansion in powers of $\vec{s}$ the potential terms drop out in the δ - function and to the same order the total expression in the curly brackets depends only on the single spatial coordinate $\vec{R}$. This represents the complete $\hbar \to 0$ limit and it is straightforward to show that going to higher powers in $\vec{s}$ introduces $\hbar$-corrections of the same order. We thus come to the conclusion that our procedure in retaining all powers in the relative ph-distance treats therefore the relative ph motion to all orders in $\hbar$ whereas the individual motion of the particle and the whole is treated to lowest order in $\hbar$ (i.e. Thomas Fermi) only. One also easily sees that the total $\hbar = 0$ limit ($\vec{s} = 0$) corresponds to what is usually called the local density approximation where in the infinite matter result λ is simply replaced by the local Fermi energy $\lambda\,(\vec{R}) = \lambda - V(\vec{R}) = p_F^2\,(\vec{R})\,/\,2m \propto \rho(\vec{R})^{2/3}$. It therefore has to be recognized that the local density approximation for the evaluation of correlation functions contains a very severe assumption namely that the correlation function does not depend on for example the relative p-h distance. It is however clear that the p-h correlation function has to go to zero at least for ph distances greater than the nuclear radius. In many physical examples the p-h responsefunction is however multiplied by operators which restrict the p-h distances to small values so that the local density approximation may still be valid. This is a quite subtle point and necessitates detailed investigation in each case; we will give an example of this below. We also immediately see that we run into troubles with the local density approximation if we wanted to use it for the evaluation of the p-h level density (20). Because of the absence of the relative p-h variable in this case the integral over this variable contained in the double

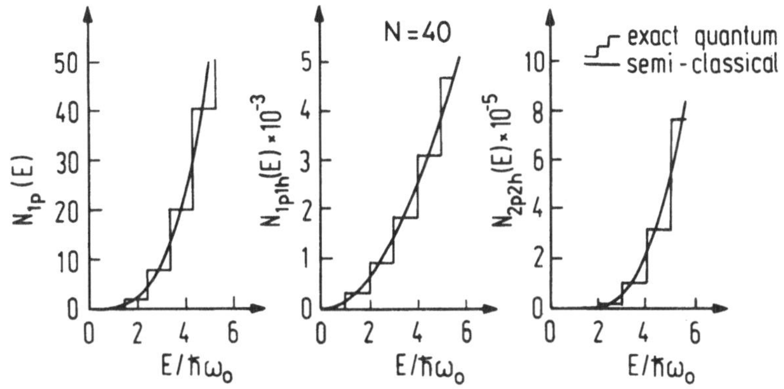

Fig. 10 Number of states for g_{1p}, g_{1p-1h}, g_{2p-2h}.

phase space integration yields a divergent contribution unless it is somewhat ad hoc restricted to the nuclear volume.

The method developed above to evaluate the p-h densities has of course a number of further applications. First of all we want to mention that, using the same method, it is easy to calculate any m-p n-h level density. Second lifetimes of quasiparticles and giant resonances can be directly related to the quantities involving 2p-1h and 2p-2h level densities, respectively. For example one can directly convince oneself that the imaginary part W of the optical potential for elastic nucleon scattering is to second order in the two body interaction given by (we use Fourier representation of the 2p-1h δ-function and a contact force) :

$$W(\vec{r},\vec{r}',E) = \frac{\pi}{2} V_o^2 \int \frac{dt}{2\pi\hbar} e^{\frac{i}{\hbar}Et} C_h^t(\vec{r},\vec{r}')\,(C_p^t(\vec{r},\vec{r}'))^2 \tag{23}$$

where $C_h^t(\vec{r},\vec{r}') = \int \frac{d^3p}{(2\pi\hbar)^3} e^{-\frac{i}{\hbar}\vec{p}\,\vec{s}}\,\theta(\lambda - H_c)\, e^{iH_c t}$

$$C_h^t(\vec{r}\vec{r}') = \int \frac{d^3p}{(2\pi\hbar)^3} e^{\frac{i}{\hbar}\vec{p}\,\vec{s}}\,\theta(H_c - \lambda)\, e^{-iH_c t} \tag{24}$$

This formula is straightforward to evaluate and we can in view of the accuracy of our results for the level densities have quite well founded confidence concerning the average behaviour of this quantity (i.e. without shell fluctuations) which yields the imaginary part of the optical potential completely off the energy shell.

In this particular example the contact force constrains the particle and holes to the same center of mass position, nevertheless the result (23) is not equal to the local density approximation where one takes the infinite matter result replacing simply ε_F by $\varepsilon_F(R)$ everywhere. Preliminary results [34] confirm the well known parabolic energy dependence $W \propto (E - \varepsilon_F)^2$ close to the Fermi surface but our expression is in no way only restricted to this energy region. Once the imaginary part W is known one can also calculate the corresponding real part U using well known dispersion relations relating both quantities. We thus are able to determine to lowest order perturbation theory an effective energy dependent potential M = U + i W usually called the mass operator[13]. Because of the relative simplicity of the semiclassical calculus we could then imagine that we add M to the usual Hartree-Fock potential and perform a semiclassical selfconsistent calculation including $V^{H.F.}$ and M using the procedure outlined above. On the contrary a full quantum mechanical treatment of this problem seems to be quite difficult. Corrections contained in M are presently attracting much interest because they are believed to cure a long-standing problem of Hartree-Fock theory : the single particle level density around the Fermi level turns always out to be roughly a factor of two too small. Including M the single particle level density at the Fermi energy is given by

$$g_{1p}(\varepsilon_F) = \frac{2}{(2\pi\hbar)^3} \frac{1}{\pi} \int d^3 R \, d^3 p \, \mathrm{Im}(E - H_W^{H.F.} - M_W^E)^{-1} \Big|_{E = \varepsilon_F} \tag{25}$$

For M = 0 eq. (25) yields of course the same result as discussed above but the influence of M^E due to its energy dependence seems likely to be important.

The other very interesting quantity closely related to (23) is the spreading width of giant resonances arising from the decay of the collective motion into incoherent 2p-2h states. Introducing an effective potential M^E_{ph} [35,36] which contains all effects besides the resummation of p-h bubbles (RPA) we can write for the exact linear response function R schematically

$$R^E = R^E_{RPA} + R^E_{RPA} \; M^E_{ph} \; R^E \tag{26}$$

Evaluating M_{ph} to second order of the two body interaction and sandwiching it between the transition density $\delta\rho$ of the collective mode yields for the width of the giant resonance μ

$$\Gamma_\mu = \delta\rho_\mu \, \mathrm{Im} \, M_{ph} \, \delta\rho_\mu$$

Using again a contact force we obtain to second order in the two body interaction

$$\Gamma_\mu = \mathrm{Im} \int d^3 r_1 \, d^3 r'_1 \, d^3 r_2 \, d^3 r'_2 \, \delta \rho_\mu(\vec{r}_1 \vec{r}'_1) \, M_{ph}(\vec{r}_1 \vec{r}'_1 \vec{r}_2 \vec{r}'_2) \, \delta \rho_\mu(\vec{r}_2 \vec{r}'_2) \tag{27}$$

with

$$\begin{aligned} \mathrm{Im}\, M_{ph}(\vec{r}_1\vec{r}'_1 \, \vec{r}_2\vec{r}'_2) = V_o^2 \int dt \; e^{iEt} \{ & - 2C_p^t(\vec{r}_1\vec{r}_2) \, C_p^t(\vec{r}'_1 \vec{r}_2) \, C_h^t(\vec{r}'_1 \vec{r}_2) \, C_h^t(\vec{r}'_1\vec{r}'_2) + \\ & + \left[C_p^t(\vec{r}_1\vec{r}_2) \, C_p^t(\vec{r}'_1\vec{r}'_2) - C_p^t(\vec{r}_1\vec{r}'_2) \, C_p^t(\vec{r}'_1\vec{r}_2) \right] C_h^2(\vec{r}'_1\vec{r}'_2) \\ & + \left[C_h(\vec{r}'_1\vec{r}'_2) \, C_h(\vec{r}_1\vec{r}_2) - C_h(\vec{r}'_1\vec{r}_2) \, C_h(\vec{r}_1\vec{r}'_2) \right] C_p^2(\vec{r}_1\vec{r}_2) \} \end{aligned} \tag{28}$$

Inserting expression (24) into (28) we see that eq. (28) is essentially the inverse Wigner transform of the expression under the traces of the 2p-2h level density (21). Assuming for the transition densities a simple analytic form like the ones given by the schematic model or the Tassie model we are capable to calculate the average spreading width as a function of particle number in a very reliable way. In order to give a graphical impression of processes eq. (28) accounts for we represent in Fig. 11 the corresponding Feynman graphs

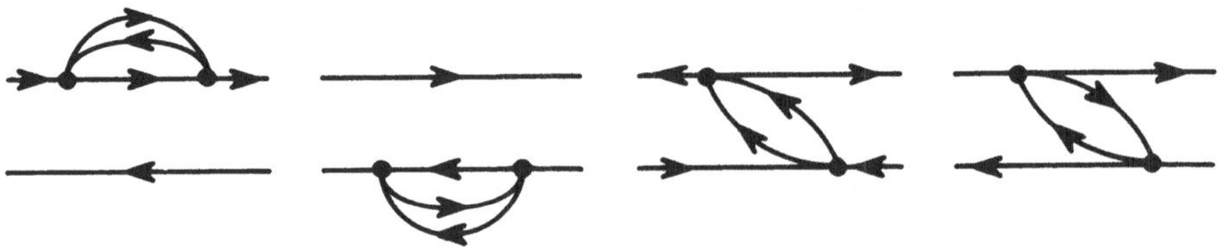

Fig. 11 Two particle - two hole contribution to the spreading width of giant resonances.

This example should be sufficient to illustrate the possible applications of our theory for m particle - n hole level densities.

LINEAR RESPONSE AND NUCLEAR STRUCTURE FUNCTIONS

It is clear that the p-h level density is intimately related to the nuclear response function and it would be interesting to see to which accuracy our method of the preceding paragraph allows us to calculate the linear nuclear response. Let us study to this purpose the so-called nuclear structure function [13,37].

$$S_o(E) = \sum_{p\,h} |\langle p|\ Q\ |\ h\ \rangle|^2\ \delta(E-\varepsilon_p+\varepsilon_h) \qquad (29)$$

where we put an index zero to indicate that in expression (29) residual p-h interactions have been neglected. Using our formalism (29) can be written with the same approximation which lead us from (20) to (22) :

$$S_o^{TF} = \frac{1}{2\pi\hbar} \int dt\ e^{\frac{i}{\hbar}Et} \int d^3r\ d^3r'\ Q(\vec{r}\ \vec{r}')\ Q^t_{ph}(\vec{r},\vec{r}') \qquad (30)$$

with

$$Q^t_{ph}(\vec{r},\vec{r}') = \int d^3r_1\ d^3r_2\ C^t_h\ (\vec{r}\ \vec{r}_1)\ Q(\vec{r}_1\ \vec{r}_2)\ C^t_p(\vec{r}_2\ \vec{r}') \qquad (31)$$

and C_h, C_p given by (24). A frequently encountered excitation operator (γ-rays, inelastic electron scattering etc) is the plane wave : $Q\ (\vec{r},\vec{r}') = e^{i\vec{q}\vec{r}}\ \delta(\vec{r}-\vec{r}')$. Inserting this operator in (30) and using (24) we obtain the structure function as a function of excitation energy E and momentum transfer $\vec{q}$ [38]

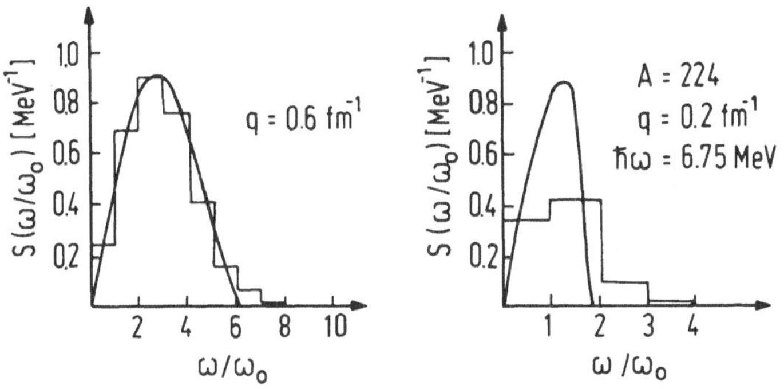

Fig. 12 Structure function in the local density approximation for two values of the momentum transfer. Also shown is the histogramme average of the exact result.

$$S_o^{TF}(E,\vec{q}) = \int \frac{d^3R d^3p}{(2\pi\hbar)^3}\, \theta(\lambda - H_c)\, \{\delta(E - \frac{1}{2m}(p^2 - (\vec{p}-\vec{q})^2) - (E \rightarrow -E))\} \quad (32)$$

We remark that in this example replacing λ by λ (R) in the infinite matter expression leads to the same result as (32) and (32) therefore coincides with the local density approximation. It is intuitatively clear that this approximation should become the better the larger the momentum transfer q since in this case only small portions of the nucleus treatable as local pieces of nuclear matter are being probed. This depends however on the resolution we are looking for. A very fine resolution would in fact reveal a lot of structure in which most of the time we are however not interested and thus a Thomas Fermi approach, which yields by definition only the average,adequate.

We want to investigate this in a model study for a non interacting nucleon gas in a spherical harmonic potential. In this case the structure function is easily evaluated exactly and to be compared with the semiclassical result obtained straightforwardly from (32)[39]

$$S(E,q) = \frac{1}{6\pi} \frac{\hbar}{m^2 \omega_o^3 q} \begin{cases} 0 & k_F < a \\ \frac{2}{5}(k_F^2 - a^2)^{5/2} & a < k_F < b \\ \frac{2}{5}(k_F^2 - a^2)^{5/2} - \frac{2}{5}(k_F^2 - b^2)^{5/2} & k < k_F \end{cases} \quad (33)$$

where $a = \frac{m\,E}{q\hbar^2} - \frac{q}{2}$; $b = \frac{m\,E}{q\hbar^2} + \frac{q}{2}$. In Fig. 12 we show this

result as a function of q for the harmonic oscillator together with the quantum mechanical one averaged over this a rectangular distribution $2\hbar\omega_o$ wide (a $2\hbar\omega_o$ wide distribution and not a $1\hbar\omega_o$ wide one is to be used because it can be shown that Thomas-Fermi theory averages individually over the positive and negative parity part of the spectrum).We see that already for $q = 0.6\text{fm}^{-1}$ we get a very reasonable agreement between quantum mechanics and semiclassics. This agreement improves for increasing values of q; below $q \simeq 0.6\ \text{fm}^{-1}$ however the detailed form of the semiclassical distribution starts to deviate from the quantum mechanical one and for $q \rightarrow 0$ we obtain a $\delta'(E)$ distribution whereas the quantum result has a δ-peak at $1\hbar\omega_o$ corresponding to the dipole excitation. The $\delta'(E)$ distribution has however such a strength that the well known dipole sum rule is still fulfilled [38]. One can easily convince oneself that in the $q \rightarrow 0$ limit taking for Q the angular momemtum operator ℓx and calculating the inverse energy weighted sum rule we again get the rigid body moment of inertia.

We thus see that this very simple theory yields quantitatively reliable results for moderate to high q-values and that even for small q-values energy integrated quantities can be accurately evaluated. It is clear that for $q \to 0$ the local density approximation must break down because in order to get the position of the dipole δ- peak in the $q \to 0$ limit correctly we must know the eigenfrequency which is a global quantity and can not be found by a local approach.

However in generalising somewhat our above approach we will indeed be able to get some information about the exact position of the resonance still using only Thomas-Fermi information. To this end we write eq. (29) in the following form:

$$S_o(E) = \frac{1}{2\pi\hbar} \int dt \; e^{\frac{i}{\hbar} Et} \; \mathrm{Tr} \, \{Q \; C_h^t \; Q^+ \; C_p^t\} \tag{34}$$

where $C_h = \rho \exp(\frac{i}{\hbar} H t)$ and $C_p = (1-\rho) \exp(-\frac{i}{\hbar} H t)$ and all quantities are to be understood as operators. Approximating the operators ρ and $\exp(\frac{i}{\hbar} H t)$ by their average values (by using the corresponding Strutinsky averaged expressions or - what amounts the same - the (Extended)Fermi expressions) but still using each average operator as a nondiagonal matrix and performing proper matrix multiplication we should simply transform the exact spectrum of S (E) into a broadened one reflecting the averaging procedure. Using for the operators their non local Thomas-Fermi approximation and again the harmonic oscillator potential we obtain for $Q = r^2 Y_{20}$ the following result

$$\tilde{S}_o(E) \propto \Omega \, (3 + 6\Omega + 8\Omega^2 + 8\Omega^3) e^{-2\Omega} ; \; \Omega = E/\hbar\omega_o \tag{35}$$

which is graphically displayed in Fig. 13*. The peak is somewhat shifted to the left of the exact peak position at Ω = 2 but the center of gravity of our distribution is very close to the value of 2. The width Γ of the peak has to be interpreted as always in Strutinsky averaging : it is a distributed (e.g. Gaussian) average of nuclei of different masses A (sizes) where the width ΔA corresponds approximately to the nucleon number of one shell. Linear response on an ensemble of nuclei of different sizes will of course give a broadened distribution. This example shows us that width's and therefore damping processes in general using (local) Fermi gas approximations can be easily overestimated by a large factor since of course for a nucleus with a <u>definite</u> <u>mass</u> the

* see end of text.

width of Fig. 13 is unphysical. We therefore should be very careful in using such kind of arguments in describing damping and dissipation processes.

DESCRIPTION OF THE COLLECTIVE NUCLEAR VIBRATIONS ON THE AVERAGE

In this last paragraph we would like to take up again the subject of the semiclassical description of collective nuclear vibrations about which we talked already somewhat in the introduction.

We refer again to our picture, tacitly present in all other paragraphs, of a quasi macroscopic nucleus where the continuum limit has been reached but surface, curvature ... still playing the role to define the droplet. What we have in mind is to calculate say the 2^+ states of such a nucleus, explore their A-dependence, and extrapolate back to real nuclei. This will then define the average (i.e. exempt of shell effects) of excitation energies, transition densities, currents, etc. Some states (e.g. giant resonances) will depend very little on the shell structure so that our procedure might almost coincide with a fully microscopic RPA calculation but other states (e.g. the low lying 2^+ states) are known to show strong shell fluctuations and thus a semiclassical treatment will reveal only the average structure. For example the flow pattern of the low lying states may individually exhibit vortices but on the average be a state in local equilibrium. It will then be a matter of taste whether we consider the average behavior or the individual one to be the true nature of the state. It is also clear that such a macroscopic nucleus is superfluid and this feature should be taken into account; it will however certainly influence more the low lying than the high lying part of the spectrum.

One approach which might eventually account, in some sense, for the situation we have been describing is the recently developed theory of fluid dynamics[7]. One there relates the time dependent density matrix ρ (t) by a unitary transformation to the static one (ρ_{st}) :

$$\rho(t) = e^{F} \rho_{st} e^{F^{+}} \tag{36}$$

and expands the Wigner transform of F in powers of $\vec{p}$ (the original version of this specific derivation of fluid dynamics can be found in ref.[13] Ch. 13) :

$$F_W = F_o(\vec{R},t) + \vec{F}_1(\vec{R}).\vec{p} + \ldots \qquad (37)$$

Using in addition for $(\rho_{st})_W$ the semiclassical one displayed in Fig. 2 could lead to the description of the average of the vibrational states. It is however unclear whether the expansion (37) converges and in which sense it converges (it could be an asymptotic expansion). In fact this kind of procedure which leads to equations similar to the ones of an elastic medium have been criticized recently by Bodell and Pethick for the infinite matter case [40]. Whether there are additional arguments in favour of (37) for finite systems seems to be an open question.

We here therefore want to briefly sketch an alternative approach which could be more systematic. The first observation is that for our quasi macroscopic nuclei which perform small amplitude vibrations around equilibrium the time dependent density

$$\rho(t) = \rho_{st} + \delta\rho(t) \qquad (38)$$

changes only around the local Fermi surface of $f=(\rho_{st})_W$ in Fig. 2. Therefore $(\delta\rho(t))_W = \delta f(\vec{R},\vec{p},t)$ is a time dependent wavepacket in phasespace whose domain in phasespace is roughly constrained to be of the order $\partial f/\partial\lambda$, i.e. it will be concentrated in the surface region of f with a width equal to the width of the surface in phasespace of f. This is actually a rather restricted domain (equal to a δ-shell in infinite matter) and it could very well be that this wavepacket can be represented quite accurately as a superposition of a small number of coherent states

$$\rho(t) = \sum_{\alpha\alpha'} |\alpha > \rho_{\alpha\alpha'} < \alpha| \qquad (39)$$

where the $|\alpha>$ are the coherent states (in one dimension) :

$$< x| \; x_c \; y_c; \alpha> = \left(\frac{2\alpha}{\pi}\right)^{1/4} \exp\left[-\alpha(x-x_c)^2 + ip_c x\right] \qquad (40)$$

$x_c(t)$, $p_c(t)$ being the classical trajectories and a complex time-dependent parameter. The coherent states (40) form a (over) complete basis and (39) is always possible. The hope would be that only a few well chosen terms in (39) will be sufficient. This kind of theory has recently been successfully developed and applied by E. Heller[41] . It is certainly a systematic theory in the sense the convergence properties of the expansion (39) can be studied as a function of the number of terms to be retained. The hope that a small number of terms might be sufficient in our case stems from the fact that the phasespace distribution of (40) corresponds to ellipsoids of any size and position in phasespace through proper choice of the parameters α, x_c and p_c and that the distribution in Fig. 2 is a relatively smooth function. The development should ideally cover two well known limits : i) the zero sound modes in infinite nuclear matter ($A\to\infty$) and ii) the case of a harmonic selfconsistent potential (light nuclei). The first one is of course the most critical one and detailed studies are necessary in this case. For the second case where one is in the situation of almost harmonic potentials the following procedure which is a special case of (38) could be successful . For our quasi macroscopic nuclei we can use the classical limit of the TDHF equation, i.e. the Vlasov equation :

$$\frac{\partial f}{\partial t} + \frac{\vec{p}}{m} \cdot \frac{\partial f}{\partial \vec{R}} - \frac{\partial V^H}{\partial \vec{R}} \cdot \frac{\partial f}{\partial \vec{p}} = 0 \qquad (41)$$

where for reasons of simplicity we restrict ourselves to the local Hartree-potential only (the generalisation to non local Hartree-Fock is straightforward). If the potential $V^H(\vec{R},t)$ is only slightly anharmonic in $\vec{R}$ a development of V^H around the center of mass $R_c(t) = \int d^3 R\, d^3 p\, \vec{R}\, f(\vec{R},\vec{p},t)/(2\pi\hbar)^3$ of the wavepacket f up to second order

$$V^H \simeq V^H_c + \sum_i V^H_{c_i} (\vec{R} - \vec{R}_c)_i + \frac{1}{2} \sum_{ij} V^H_{cij} (R_i - R_{ci})(R_j - R_{cj}) \qquad (42)$$

(the index c should indicate: to be evaluated at the classical trajec tory) may be sufficient. We remark that this procedure is still exact in the case of a separable Q.Q force. Inserting (41) into (40) yields the following equation if we assume to transform to locally harmonic coordinates :

$$\frac{\partial f}{\partial t} + \frac{\vec{p}}{m} \cdot \frac{\partial f}{\partial \vec{R}} - m \sum_i \Omega_i^2(t)\, R_i \frac{\partial f}{\partial p_i} = 0 \qquad (43)$$

where the time dependent frequencies are directly related to the first and second derivatives of the potential. Equation (42) is the one of a three dimensional harmonic oscillator and the solution can be given in analytic form [42] as one coherent state :

$$f(\vec{R},\vec{p},t) = F(\lambda - \tilde{H}(t)) \qquad (44)$$

where F is an arbitrary function and

$$\tilde{H} = \sum_i (\bar{p}_i^2/(2m) + m\, \Omega_i^2\, \bar{R}_i^2) \qquad (45)$$

$$\bar{p}_i = \xi_i (p_i - m u_i) \quad \bar{R}_i = R_i/\xi_i \quad u_i = \dot{\alpha}_i R_i \; ; \; \alpha_i = \ln \xi_i$$

The parameters $\xi_i(t)$ obey the (coupled) set of classical equations

$$\ddot{\xi}_i - \Omega_i^2(0)/\xi_i^3 + \Omega_i^2(t)\, \xi_i = 0 \quad ; \quad \xi_i(0) = 1; \quad \dot{\xi}_i(0) = 0 \qquad (46)$$

As we said for a Q.Q force the solution (43-45) is the exact solution of the (large amplitude) TDHF equation but for general potentials (41) restricts to small amplitude vibrations where the R_c explores only a small region of V^H around the equilibrium position and a locally harmonic approach is valid. This approach could well be sufficiently accurate for the description of the giant quadrupole resonance but probably for the low lying 2^+ states a superposition of more coherent packets as in (38) will be necessary.

CONCLUSIONS

In this series of lectures we aimed at presenting a self-contained semiclassical theory entirely based on the extended Thomas-Fermi or Wigner-Kirkwood $\hbar$ expansion in phasespace. We saw that not only the Wigner transform of the single particle density matrix can be understood and very accurately represented in this way but that also generalisations to correlation functions are straightforward. First, we demonstrated a generalisation to superfluid nuclei and to superfluid nuclei in slow rotation. The latter involves already the (static) particle-hole correlation function and we saw how e.g. the reduction of the moment of inertia by roughly a factor of two could be explained very easily in an analytic way. In a more obvious way m-p -n-h correlation functions are involved in the corresponding mp-nh level densities. We very clearly pointed out the necessity to treat particles (holes) individually in Thomas Fermi approximation. This means that the "Thomas Fermi particles (holes)" can move around independently in phasespace whereas the local density approximation is only valid if the particles (holes) are spatially close by. Indeed our 2p-2h level density passes nicely through the average of the corresponding quantum values. This can then be used to evaluate spreading width of giant resonances and the imaginary part of the nucleon-nucleus optical potential. A further very promising result is that the linear response function for transferred momenta $q > 0.6\ fm^{-1}$ can be very accurately represented in our p-h-Thomas Fermi approach. In the last paragraph we give somewhat speculative arguments that say the 2^+ states of quasi macroscopic Fermi Liquid Drops could be well calculated in expanding the time dependent density matrix on a set of coherent states and a simple example for nearly harmonic potentials is given.

We hope that we have given with these lectures an impression of how powerful a tool the correct application of $\hbar$-expansion techniques can be and we also hope that, what we mean by "correct application" of these methods, may have become more transparent.

ACKNOWLEDGMENTS

The scientific content of these lectures could not have been presentedin this form without the highly appreciated collaboration of the following colleagues (as a matter of fact quite a few subjects of the lectures are only in preparation for publication and I am grateful to my colleagues for allowing me to present results prior to publication): Dr. M.Durand , Dr. V.S. Ramamurthy (Wigner function), Dr. J. Treiner (semiclassical Hartree-Fock), Dr. R. Bengtsson (semiclassical pairing), Dr. M. Durand, Dr. J. Kunz (nuclear rotation), Dr. Hasse (level densities and optical potential), Dr. R. Hasse, Dr. U. Stroth (linear response).

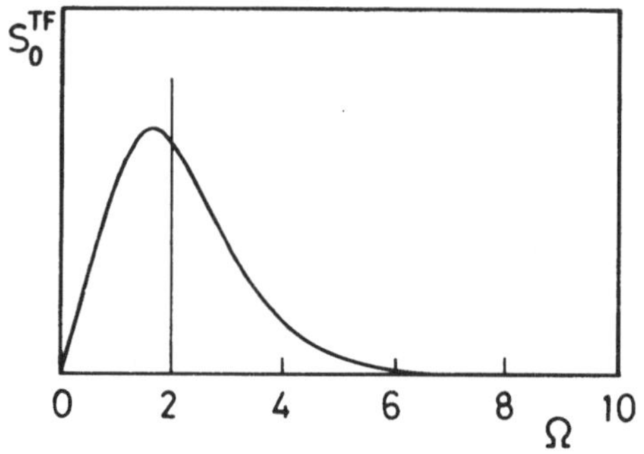

Fig. 13 Quadrupole transition strength calculated in the Thomas-Fermi limit (arbitrary units). The vertical line indicates the peak position in the exact case.

REFERENCES

1. R.K. Bhaduri, C.K. Ross, Phys. Rev. Lett.27:606 (1971), J. Treiner, H. Krivine, Nuc. Phys.A371:253 (1981), M. Brack, Contribution workshop on "Semiclassical Methods in Nuclear Physics", Proceedings Institute Laue Langevin, March 1980.

2. R. Bengtsson, P. Schuck, Phys. Lett. 89B:321 (1980).
3. G. Ghosh, R.W. Hasse, P. Schuck, J. Winter, Phys. Rev. Lett. 50:1250 (1983).
4. M. Durand, V.S. Ramamurthy, P. Schuck, Phys. Lett. 113B:116 (1982)
5. V.M. Strutinsky, F.A. Ivanjuk, Nucl.Phys. A255:405 (1975).
6. J. Bartel, M. Vallières, Phys. Lett. 114B:303 (1982).
7. G. Holzwarth, Lecture Notes to this conference.
8. J. Blocki, J. Randrup, W.J. Swiatecki, C.F. Tsang, Ann. Phys. 105:427 (1977).
9. M. Prakash, S. Shlomo, V.M. Kolomietz, Nucl. Phys. A370:30 (1981).
10. N. Rowley, P. Prakash, J. Phys. A : Math. Gen. 16:3219 (1983).
11. S. Chandrasekhar, An introduction to the study of stellar structure (Dover, N.Y. 1967) 2nd Edit. (1st Edit. 1939).
12. D.A. Kirzhnits : Field Theoretical Methods in Many Body Systems, Pergamon, Oxford 1967
13. P. Ring, P. Schuck, The Nuclear Many Body Problem, Springer-Verlag, 1980.
14. V.S. Ramamurthy, M. Ashgar, S.K. Kataria, Nucl.Phys. A398:544 (1983).
15. B.K.Jennings, Ph. D. Thesis, Mc Master University (1976).
16. S.E. Koonin, Ph. D. Thesis, MIT 1975.
17. M. Durand, M. Brack, P. Schuck, Z. Phys. A286:381 (1978).
18. M. Durand, V.S. Ramamurthy, P. Schuck, to be published.
19. J.Treiner, P. Schuck, to be published
J. Treiner, These d'Etat, Orsay 1981.
20. C. Guet, H.B. Hakansson, M. Brack, Phys. Lett.97B:7 (1980), J. Bartel, P. Quentin, M. Brack, C. Guet, H.B. Hakansson, Nucl. Phys. A385:79 (1982).
21. J. Treiner, R.W. Hasse, P. Schuck, J. Physique Lettres 44:L733 (1983).
22. P. Schuck, R. Bengtsson, M. Durand, J. Kunz, V.S. Ramamurthy, Proc. "Workshop on Nuclear Fission " (Bad Honnef, October 1981), Lecture Notes in Physics, Vol. 158 (Springer-Verlag, New York, 1982) p. 183.
23. C. Guet, M. Brack, Z. Physik A297:247 (1980).
24. B. Grammaticos, A. Voros, Ann. Phys.(NY) 123:359 (1979).
25. J. Treiner, private communication and J. Treiner, P. Schuck, to be published.

26. Proceedings of " International Workshop on Gross Properties of Nuclei and Nuclear Excitations VIII", p. 152, Institut für Kernphysik, Techn. Hochschule, Darmstadt, 1980.
27. M. Girod, D. Gogny, private communication.
28. D. Gogny, P. Schuck, to be published.
29. M. Brack, B.K. Jennings, Nucl. Phys. A258:264 (1976).
30. A.A. Lastney, B. Jancovici, Physica 102A:327 (1980).
31. M. Durand, J. Kunz, P. Schuck, to be published in Nucl. Phys.
32. A.B. Migdal, Nucl. Phys. 13:655 (1959).
33. R.M. Diamond, F.S. Stephenens, W.J. Swiatecki, Phys. Lett. 11:315 (1964).
34. R.W. Hasse, P. Schuck, contribution I11, Proceedings of the International Conference on Nuclears Physics, Florence 1983, Vol. 1.
35. P. Schuck, Z. Physik A279:31,(eq. (8)) (1976)
P. Schuck, Journal of Low Temperature Phys. 7:459 (eq.(19)) (1972).
36. C. Yannouleas, M. Dworzecka, J.J. Griffin, Nucl. Phys. A397:239 (1983).
37. P. Schuck, G. Ghosh, R.W. Hasse, Phys. Lett. 118B:237 (1982).
38. R. Rosenfelder, Ann. Phys. (NY) 128:188 (1979).
39. R.W. Hasse, P. Schuck, U. Stroth, to be published.
40. K. Bedell, C. Pethick, J. Low Temp. Phys. 49:213 (1982).
41. M.J. Davis, E.J. Heller, J. Chem. Phys. 71(8):3383 (1979).
42. P. Schuck, W. Brenig, Z. Phys. B - Condensed Matter 46:137 (1972).

AVERAGE NUCLEAR PROPERTIES FROM THE NUCLEAR EFFECTIVE INTERACTION

Jacques Treiner

Institut de Physique Nucleaire

B.P. n° 1 91406 Orsay Cedex, France

Atomic nuclei are small Fermi systems -even the heavy ones : the ratio of the number of nucleons in the surface region (~2.5 fm) to the number of nucleons in the bulk is ≃ 1 in the largest nuclei and of course greater in smaller ones. This characteristic implies that volume, surface, curvature...effects cannot be analyzed separately and are hard to disentangle from experimental data. Moreover in their ground state nuclei are close to being spherical and the symmetries associated with this shape produce quantum fluctuations which make the task of exhibiting the smooth trends of nuclear properties more difficult. In one of their reference paper on the Droplet Model[1] , Myers and Swiatecki point out " an analogy that serves to bring out this relation between average nuclear properties and shell-effect fluctuations [...] The topography of the earth, in which the average shape (a somewhat oblate spheroid) is modified by sharp local fluctuations (mountains). On the one hand the determination of the earth radius and oblateness from a few local triangulations, especially in mountainous country, would be wellnigh impossible. On the other hand, given methods that perform suitable averages over the sharp fluctuations of the earth's surface (such as experiment with artificial satellites), one is able to determine the shape of the earth with great accuracy, including higher harmonics. Thus the pear-shaped component of the earth's figure may be measured quite accurately even though its amplitude is much less than the amplitude of the fluctuations associated with local mountain ranges. "

In the macroscopic approach[1-3], the average trends of nuclear properties are described in terms of global quantities such as volume, surface and curvature energies, asymmetries energies associated to the neutron excess, Coulomb energy, incompressibility modulus, nuclear shapes...The model provides a good description of ground state masses and radii and a qualitative understanding of the fission process ; quantum effects can then be added in a perturbative way using the Strutinsky prescription [4], leading to the so-called macroscopic-microscopic model. Among the great successes of the method is of course the understanding of one of the important discoveries in nuclear physics in the past 20 years : the existence of fission isomers. However two features in this scheme seem unsatisfactory : i) the model is not self-consistent, i.e. shell effects are not treated self-consistently with the macroscopic part and ii) some of the coefficients entering the mass formula are not easy to extract directly from a macroscopic analysis of nuclear masses ; this is the case e.g. for the surface symmetry energy and for the curvature energy ; large uncertainties and strong correlations to the other coefficients are then obtained even when one incorporates more experimental information, such as fission barriers heights. For example the range of values found in the literature for the surface symmetry energy is (-25,-160)Mev and (0-10)Mev for the curvature energy.

The Hartree-Fock (HF) approximation has proven in the past 15 years to be a tractable alternative which overcomes at least one of the difficulties of the macroscopic-microscopic model : here average properties and shell effects are treated in a unique and self-consistent framework. The basic ingredient is the nuclear effective interaction, which can be calculated, within the local density approximation, from a given free nucleon-nucleon potential using the Bruckner-Goldstone expansion in infinite nuclear matter (see e.g. Brack's contribution to this school and references therein). Zero-range forces of Skyrme type [5] and finite range forces such as Gogny's [6] have been extensively used in the past decade, using the HF method or its time dependent version TDHF or RPA : nuclear masses and radii, fission barriers, thermal properties, giant resonances have been analyzed in a self-consistent scheme. A particularly interesting feature of the Skyrme parametrization of the effective interaction lies in the fact that the potential energy appears as a functional of the diagonal part of the one-body nucleon density matrix, thus reducing the HF equations to a set of differential equations which can be solved in coordinate space. If one makes use of the Thomas Fermi approximation for the kinetic energy density (or its various extensions [7]) one then has an Energy Density Functional (EDF) from which macroscopic properties can easily be calculated by solving the two coupled Euler equations describing the nucleus (one for neutrons and one for protons) [8-10].

Such a use of semi-classical methods is of course fully justified in situations where shell-effects are not important ;

for example, Extended Thomas Fermi (ETF) calculations of RPA sum-rules have significantly facilitate the discussion of the Giant Monopole Resonance and the Giant Dipole Resonance in relation with the nuclear incompressibility and the symmetry energies respectively[11-13,32,34]. It is also interesting, from a more practical point of view, in the process of improving existing effective interactions. One gain physical insight if one is able to characterize a given interaction by a set of physical quantities of global significance rather than by a set of parameters without any physical interpretation. The following lectures are oriented to such a practical aim : we shall see that ground state and thermal properties as well as Giant Resonances, in their smooth trends, can be analyzed almost analytically. Simple expressions are obtained for all coefficients of interest in the mass formula, to which geometrical properties of nucleon distributions are related. Surface effects will be emphasized, in particular in the description of the Giant Dipole Resonance, for which microscopic calculations are also presented.

SOME PHYSICAL QUANTITIES OF INTEREST

The precise form of the Skyrme functional is given in App. A. It is composed of a sum of powers of the nucleon densities (volume part) plus gradient corrections important at the surface. In order to characterize the volume part it is convenient to consider the equation of state in nuclear matter and to use an expansion around saturation (see fig. 1).

The volume part of the functional is separated, for small neutron excess, into an isoscalar part $h(\rho)$ and isovector one by expanding up to second order in $\rho_n-\rho_p$.

$$H(\rho_n,\rho_p) = h(\rho) + \rho\ \varepsilon_\delta(\rho)\ \left(\frac{\rho_n - \rho_p}{\rho}\right)^2 . \qquad (1)$$

The isoscalar part will be written as

$$h(\rho)= \sum_s B_s \rho^s = \rho\left[\lambda_{nm} + \frac{K}{18}\left(\left(\frac{\delta\rho}{\rho_{nm}}\right)^2 + c\left(\frac{\delta\rho}{\rho_{nm}}\right)^3\right)\right] \qquad (2)$$

$$K \equiv K_{nm} = 9\ \rho_{nm} \frac{d^2(e)}{d\ \rho_{nm}^2} \qquad (3)$$

$$c = \frac{3}{K_{nm}} \rho_{nm}^2 \frac{d^3(\rho e)}{d\rho_{nm}^3} - 1 \qquad (4)$$

$$e = \frac{h(\rho)}{\rho} \qquad \delta\rho = \rho - \rho_{nm} .$$

The third order term is necessary for the following reason: in a finite nucleus the surface tension produces a compression in the volume ; the resulting change in the central density, compared to the saturation density, is given by :

$$\frac{\delta\rho}{\rho_{nm}} = 6 \frac{\varepsilon_s}{K_{nm}} A^{-1/3} \qquad (5)$$

where ε_s denotes the surface energy. Consequently the resistance of the nucleus to compression (e.g. during a monopole vibration) will not be characterized by the asymptotic value K_{nm} but rather by $K_{nm} + \delta K$ where δK is given by

$$\delta K = (K + 9\rho_{nm}^2 \frac{\partial^3(\rho e)}{\delta \rho_{nm}^3}) \frac{\delta\rho}{\rho_{nm}} . \qquad (6)$$

Now it has been shown in ref. 12 that for all commonly used interactions, for which the cubic approximation of the equation of state is good, the following relation holds:

$$\rho_{nm}^2 \frac{d^3(\rho e)}{d\rho_{nm}^3} \simeq \frac{K_{nm}}{2} - 45 \text{ [MeV]} \qquad (7)$$

so that with $\varepsilon_s \simeq 20$ MeV and $K_{nm} \simeq 250$ MeV one finds:

$\delta K \simeq$ 79 MeV in ^{208}Pb

$\simeq$ 104 MeV in ^{90}Zr.

One sees that these variations, of the order of 40 %, are far from negligible.

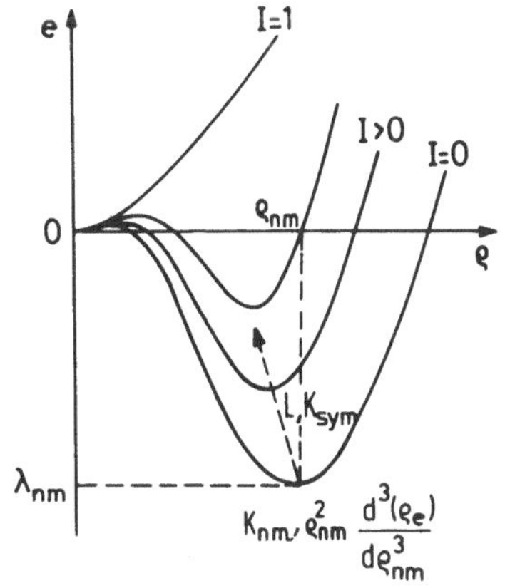

Fig. 1 Typical nuclear matter curves for different asymmetries $I = (N-Z)/A$. ρ_{nm} and λ_{nm} are well determined experimentally. K_{nm} and the anharmonicity of energy per particle near saturation are coupled in the description of the GMR; the present consensus is $K_{nm} = 220 \pm 20$ MeV. Little is known about L and K_{sym} which characterize the change in saturation density with neutron excess.

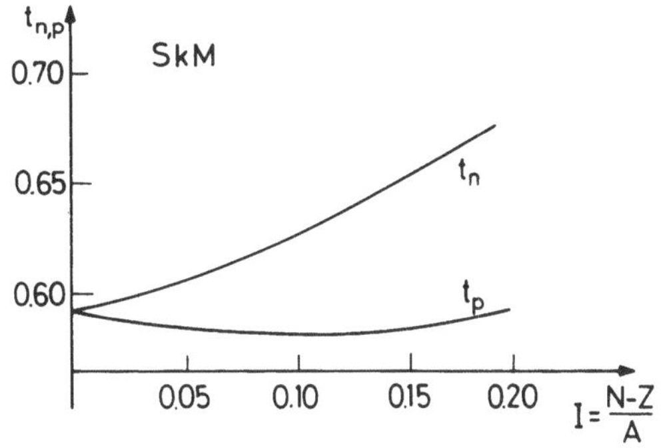

Fig. 2 Variation of the surface thickness with increasing neutron excess. The interaction used is SkM.

Concerning the isovector part, we shall consider an expansion up to second order :

$$\varepsilon_\delta(\rho) = J + \frac{L}{3}\left(\frac{\delta\rho}{\rho_{nm}}\right) + \frac{K_{sym}}{18}\left(\frac{\delta\rho}{\rho_{nm}}\right)^2 . \tag{8}$$

The coefficient J denotes the volume symmetry energy ; in the case of a neutron excess the saturation density changes and for small excess this change is characterized by L ; the coefficient K_{sym} allows for a better description of the symmetry potential at all densities.

Expressions for these various coeffecients in terms of the parameters of the Skyrme interaction are collected in App. A.

ON THE DENSITY PROFILE

Nuclear densities are often parametrized by a Fermi distribution. In doing so, one assumes that the surface is symmetrical around inflexion point, in other word that there is no skewness in the distribution. There is no reason a priori for such a feature: indeed the internal part of the surface is determined by the properties of the nuclear matter curve around saturation (partly at least) whereas the external part is related to the last filled orbits, i.e. to the Fermi energy. This question can be analyzed very simply within the EDF.

If one uses the cubic expansion eq. 2, the Euler equation corresponding to an semi-infinite system with equal number of neutrons and protons can be integrated analytically. However the solution has a rather complicated expression (giving in particular the coordinate x as a function of the density instead of the opposite) and we shall not give it here. In the asymptotic regions, though, one can invert the solution, which writes:

far inside : $\rho(x) \underset{x\to -\infty}{\sim} \rho_{nm}\,(1-c_1 \exp \frac{x}{a})$

c_1/c_2 constants

far outside : $\rho(x) \underset{x\to +\infty}{\sim} \rho_{nm}\, c_2 \exp(- \nu\frac{x}{a})$

$$\text{where } a = \left[\frac{18}{K_{nm}}\left(\frac{h^2}{2m}\beta + d\rho_{nm}\right)\right]^{1/2} \tag{9}$$

$$b = \frac{1}{1 + \frac{\hbar^2}{2m} \frac{\beta}{d\rho_{nm}}} \tag{10}$$

$$\nu = [\frac{1 - c}{1 - b}]^{1/2} . \tag{11}$$

The above behaviour suggests to represent the density by the ν-th power of a Fermi distribution, hereafter labelled F-ν, rather than by a simple Fermi distribution:

$$\rho(x) = \rho_{nm} f^{\nu} \equiv \frac{\rho_{nm}}{(1+\exp\frac{x}{a})^{\nu}} . \tag{12}$$

Indeed, one has

$$\rho(x) \underset{x\to -\infty}{\sim} \rho_{nm}(1-\nu\exp \frac{x}{a}) , \quad \rho(x) \underset{x\to +\infty}{\sim} \rho_{nm}\exp(- \nu\frac{x}{a}) .$$

We then define an internal diffuseness a_{in} and an external one a_{out} by :

$$\begin{aligned} a_{in} &\equiv a \\ a_{out} &= \frac{a}{\nu} . \end{aligned} \tag{13}$$

The value of ν thus characterizes the skewness of the nuclear surface. For such a distribution, the distance between the points where the density has decreased by 10% and 90% with respect to the interior density is given by:

$$t = a \ln \frac{10^{1/\nu}-1}{(\frac{10}{9})^{1/\nu} - 1} . \tag{14}$$

Notice that the external diffuseness a_{out} should be compared to the result obtained from the exact solution of the Euler equation (without using the cubic approximation) in the extreme outer part of the surface where the potential is completely negligible :

$$\rho(x) \sim \exp - \sqrt{\frac{2m}{\hbar^2} \frac{\lambda_{nm}}{\beta}}\, r . \tag{15}$$

This corresponds to an external diffuseness given by

$$a'_{out} = [- \frac{\hbar^2}{2m} \frac{\beta}{\lambda_{nm}}]^{1/2} . \tag{16}$$

Indeed a_{out} and a'_{out} are found to agree very well (which is just a test of the validity of the cubic approximation). Of course the values of a_{out} do not differ very much for various interactions, as it depends on the Fermi energywhich is rather well determined. Hence the skewness depends essentially on a_{in}, and in particular on the value of K_{nm} : The smaller the value of K_{nm}, the larger the value of ν. The value of the Weiszacker coefficient plays also a role. The semi-classical value $\beta = 1/36$ leads to $\nu \sim 4$ but it is known to underestimate the external diffuseness. It has been shown that an emperical value of $\beta \simeq 1/9$ leads to density profiles in good agreement with HF calculations. The corresponding values of ν are in the range (2-2.5) depending on the value of K_{nm}.

In finite systems the skewness is smaller : a'_{out} increases as the Fermi energy decreases (in absolue value) and one can see, using eqs. 2,3, and 5, that a_{in} gets smaller as the resistance of the bulk to compression increases. The resulting change in the value of ν is found to be $\sim -1.A^{-1/3}$, i.e. a change of -0.5 when going from semi-infinite medium to a medium nucleus. Hence the average skewness should be in the range (1.5-2) if one neglects the effect of the neutron excess and that of the Coulomb interaction. The role of the neutron excess will be analyzed below; the effect of the Coulomb interaction is difficult to discuss analytically; self-consistent calculations show that it has little effect on the skewness of the proton distribution and does not contradict the above analysis.

Let us now turn to the discussion of the neutron excess. In the case of a two-component system, defined by a relative neutron excess $I = (N - Z)/A$, one deals with a set of two coupled Euler equations. In the semi-infinite case, one can show, by looking at the asymptotic region in the interior -where one can linearize the equations- that both densities decrease exponentially with the same decaying constant; that of the total density. In other words the internal diffuseness, to first order in I, is the same for protons and neutrons. Now in the outer part of the surface, one knows that the external diffuseness is determined by the Fermi energies:

$$\begin{aligned} \lambda_n &= \lambda_{nm} + JI(2-I) \\ \lambda_p &= \lambda_{nm} - JI(2+I) \end{aligned} \tag{17}$$

and one has

$$a^q_{out} = [- \frac{h^2}{2m} \frac{\beta}{\lambda_q}]^{1/2} \qquad q = n,p . \tag{18}$$

Moreover there is a shift of the neutron distribution toward the proton distribution, due to the strong attraction between both kinds of particles at high densities. The following conclusions result:

i) the skewness of the proton distribution increases with neutron excess while the opposite is true for the neutrons.

ii)the formation of the neutron skin (the difference between the equivalent sharp radii for neutrons and protons[1])is the consequence of two contradictory tendencies: in the outside region the Fermi energies, fixed by the value of the neutron excess, force both distributions apart whereas in the interior the large value of the volume symmetry energy produces a shift of one density toward the other.

The above considerations allow one to understand also the evolution of surface thicknesses with increasing I: the evolution will be different for neutrons and protons, because the external diffusenesses vary with opposite signs (to first order in I) while the internal ones remain the same (also to first order); fig. 2 shows the result of a variational calculation using SkM interaction: for small neutron excess, both surface thicknesses vary in opposite direction, then I^2 terms come into play and produce an increase of the common internal diffuseness. As a result, the proton surface thickness remains almost constant whereas it increases monotonically for the neutrons. Notice that the difference between the two is not small. This difference contributes to the determination of the difference between neutron and proton radii, although it has been neglected in some recent studies[14,15]. The present discussion may also be relevant in the description of isovector surface vibrations. In that case, the restoring force is roughly proportional to the surface thickness. The degree of freedom associated with the difference between surface thicknesses is then a first order correction -and not a second order one as it is the case in the description of a volume property. Finally let us notice that here also the presence of the Coulomb interaction would not alter significantly the result of the discussion. The increase of the proton Fermi energy, which should produce an increase of a^p_{out}, is in fact compensated by the effect of the repulsive barrier.

We shall now turn to the determination of various coefficient of the mass formula, where the binding energy of a spherical nucleus is written in the form of an expansion in powers of the ratio of the surface thickness (i.e. the range of the nuclear force) to the radius of the system. Due to the saturation property of the nuclear force, this ratio becomes smaller when A increases so that it is the natural expansion parameter of a so-called leptodermous system. In the following, we shall be concerned with the surface tension and the surface energy, with the surface symmetry energy and with the curvature energy. The presence of a constant term in the mass formula will also be discussed.

SURFACE TENSION

When the energy density of a system (N = Z) is described by an energy density which has the form

$$\mathcal{H}(\rho,\rho') = h(\rho) + g(\rho)\rho'^2 \tag{19}$$

where h (ρ) and g (ρ) are functions of ρ only, the surface tension can be obtained using an elegant method due to Wilets[16]. Consider first the Euler equation

$$\frac{h}{\rho} - \frac{dg}{d\rho}\rho'^2 - 2g\rho'' = \lambda . \tag{20}$$

It can be integrated once and yields

$$h(\rho) - g(\rho)\rho'^2 = \lambda\rho . \tag{21}$$

Now the surface tension is defined by

$$\sigma = \int[h(\rho) + g(\rho)\rho'^2 - \lambda\rho]dx . \tag{22}$$

Eq. 21 expresses the fact that the contribution of the volume term to σ is equal to that of the gradients terms, so that one can write:

$$\sigma = 2\int[h(\rho) - \lambda\rho]dx = 2\int g(\rho)\rho'^2 dx. \tag{23}$$

Using eq. 21 again, one can integrate over ρ instead of x

$$\sigma = 2\int_0^{\rho_{nm}} [g(\rho)\ (h(\rho) - \lambda\rho)]^{1/2}\ d\rho . \tag{24}$$

Notice that one does not have to solve the Euler equation to calculate σ.

Although one has in eq. 24 an exact expression (within the formalism used here) it is interesting to derive an approximate expression using a method which will be useful also in the calculation of the curvature energy.

One first notices that an integral of the form

$$I_\nu = \int_0^\infty \frac{4\pi\, r^2 dr}{(1+\exp\frac{r-R}{a})^\nu} \tag{25}$$

where ν is a positive number, can be written as[17]

$$I_\nu = 4\pi \frac{R^3}{3}[1+3\eta_\nu^{(o)}\frac{a}{R} + 6\eta_\nu^{(1)}\frac{a^2}{R^2} + 3\eta_\nu^{(2)}\frac{a^3}{R^3}] + \sigma(e^{-\frac{R}{a}})$$

where (26)

$$\eta_\nu^{(k)} = (-)^k \int_0^\infty [\frac{1+(-)^k e^{-u\nu}}{(1+e^{-u})^\nu} - 1]\, u^k du \ .$$

This theorem, which generalizes the Sommerfeld Lemma to non integer powers of ν, allows one to use distributions of the form eq. 12 in order to calculate the coefficients of the leptodermous expansion. For example one gets for σ (without spin-orbit)

$$\sigma = a \sum_s B_s \rho^s_{nm}(\eta^{(o)}_{s\nu} - \eta^{(o)}_\nu) + \nu\frac{\rho_{nm}}{a}[\frac{\hbar^2}{2m}\frac{\beta}{\nu+1} + d\rho_{nm}\frac{1}{2(2\nu+1)}] . \tag{27}$$

For sake of simplicity, we shall from now on write the expressions obtained when one constraints $\nu = 1$. Although this should not be done when discussing geometrical properties, as we have seen, it is not a bad approximation when one deals with integrated quantities as long as one uses a variational procedure. In that case, σ reduces to (spin-orbit is included here)

$$\sigma = a \sum_s B_s \rho^s_{nm}\eta_s^{(o)} + \frac{\rho_{nm}}{2a}[\frac{\hbar^2}{2m}\beta + \frac{1}{3}d\rho_{nm} + V_{so}\rho^2_{nm}(\frac{1}{6} - \frac{1}{10}k\rho_{nm})] \ . \tag{28}$$

Minimizing with respect to a gives :

$$a = [\frac{18}{K_{nm}} \frac{\frac{\hbar^2}{2m}\beta + \frac{1}{3}d\rho_{nm} + V_{so}\rho^2_{nm}(\frac{1}{6} - \frac{1}{10}k\rho_{nm})}{(1 - \frac{2}{3}c)}]^{1/2} \tag{29}$$

where we have made use of the cubic approximation. Then σ is given by

$$\sigma = \frac{\rho_{nm}}{a} [\frac{\hbar^2}{2m} \beta + \frac{1}{3} d\rho_{nm} + V_{so} \rho_{nm}^2 (\frac{1}{6} - \frac{1}{10} k\rho_{nm})] . \tag{30}$$

From eq. 30 the surface energy is simply given by

$$\varepsilon_s = 4\pi r_{nm}^2 \sigma, \qquad r_{nm} = (3/4\pi\rho_{nm})^{1/3} . \tag{31}$$

SURFACE SYMMETRY ENERGY

This quantity is more interesting as it is not so well determined from experimental data and difficult to extract directly from HF calculations.

In the case of a neutron excess, there are(at least) two ways of generalizing eq. 22 :

i) as in the case N = Z, one can consider the difference between the toal energy and the energy of a system having the same density distribution but with a constant energy density equal to that of the uniform medium found in the interior region, i.e.

$$\sigma_o(I) = \int \{\mathcal{H} - (\frac{\mathcal{H}}{\rho})_o \rho \} dx, \tag{32}$$

ii) one can also consider the following expression

$$\sigma_\lambda(I) = \int \{\mathcal{H} - \lambda_n \rho_n - \lambda_p \rho_p \} dx \tag{33}$$

where one substracts from the total energy the energy of a system of neutrons and protons having all an energy equal to the separation energy of the actual system, λ_n and λ_p .

Both expressions are not at all equivalent and indeed it can be shown that if one expands each one for small neutron excess I

$$\sigma_o(I) = \sigma + \sigma_s^o I^2 , \qquad \sigma_\lambda(I) = \sigma + \sigma_s^\lambda I^2 \tag{34}$$

one has

$$\sigma_s^\lambda = - \sigma_s^o . \tag{35}$$

This can be shown within the Droplet Model as well as in the EDF. It can also be shown that using eq. 33 one gets directly the coefficient of the $I^2 A^{2/3}$ term in the mass formula, i.e. the so-called surface symmetry coefficient.

Without going into details, let us mention briefly that the surface symmetry coefficient is the sum of two terms: i) a genuine surface term, which correspond to eq. 32 and which is positive and ii) a negative term, twice as large as the first one, coming from the volume. The origin of the first one is simple: one looses energy by separating neutrons from protons in the surface. The second one comes from the fact that, in doing so, the local asymmetry in the bulk gets smaller than the average one, i.e. I, and this represents a gain in energy, which happens to be twice as large (this result is completely general) as the loss in the surface region. It is somewhat a surprise that in order to calculate the sum of these contributions, one can go directly to semi-infinite medium and use eq. 33 although in this situation the interplay between volume and surface described above cannot take place: the asymmetry in the bulk is indeed equal to the average one.

It is convenient to use eq. 33 rather than eq. 32 because one can make use of the (coupled) Euler equations. Like in the isoscalar case, one can show that the contribution of the volume type terms in the functional to the surface tension is equal to that of the gradient terms (as long as the dependence is quadratic). The isoscalar part of the functional does not contribute to the surface symmetry coefficient (due to the so-called $\dot{\sigma} = 0$ theorem[1]) and one gets :

$$\sigma_\lambda(I) = \sigma + 2I^2 \int\rho [\, \varepsilon_\delta(\rho) \frac{\delta}{I^2} - 2\, J \frac{\delta}{I} + J\,]dx \qquad (36)$$

where δ denotes the local asymmetry $(\rho_n - \rho_p)/\rho$ and where ρ is the total density of symmetric matter. Eq. 36 is simple but not yet tractable since we do not know δ. Notice that the degree of freedom associated with the different diffuseness for neutrons and protons is trivially taken into account by the factor of 2 in front of the integral. A simpler but still accurate expression is obtained by just setting $\delta = I$ in eq. 36, leading to[18]

$$\sigma_\lambda(I) = \sigma + 2 \int\rho\ [J - \varepsilon_\delta(\rho)]\ dx \quad . \qquad (37)$$

Eq. 37 gives a very simple physical interpretation of σ_s^λ : it is the average, over the surface region, of the defect of symmetry energy relative to the volume symmetry energy J. Now one can use the expanded form for $\varepsilon_\delta(\rho)$ and assume a Fermi shape for the isoscalar density. Finally one obtains

$$\varepsilon_{ss} = 4\pi r_{nm}^2 \sigma_s^\lambda = -\frac{2a}{r_{nm}} \left(L - \frac{K_{sym}}{12} \right) \tag{38}$$

where a is the surface diffuseness of the distribution. This very simple expression allows one to recover the results of completely variational calculations within 20% for all Skyrme interactions, the worst accuracy being obatined for interactions having negative L's, i.e. for which the expansion eq. 8 may not be accurate.

Table 1 shows results obtained using eq. 38 for various parametrization of the Skyrme functional (a term 2 $\varepsilon_s L/K_{nm}$ has to be added which accounts for the change in the saturation density with neutron excess).The rather large spreading of the numbers reflects the fact already mentioned in the introduction that similar fits to ground state masses can be obtained with very different sets of volume and surface symmetry energies. Notice however that SkM* is the only interaction which gives the correct fission barrier of ^{240}Pu and that the corresponding value of the surface symmetry coefficient agrees fairly well with the one obtained through a recent analysis of experimental masses and fission barrier heights using the macroscopic-microscopic model.

Table 1

Values of volume symmetry energies and surface symmetry coefficients for various Skyrme forces.

	SII	SIII	SIV	SV	SVI	SkM	SkM*
J (MeV)	34.1	28.2	31.2	32.8	26.9	30.7	30.0
ε_s^s (MeV)	-54	-30	-59	-76	-23	-53	-53

Finally let us mention that it is possible to show that within the EDF the Droplet Model theory of the neutron skin is valid; this theory relates the neutron skin τ to the surface symmetry energy in the following way

$$\tau = \frac{2}{3} r_{nm} \frac{|\varepsilon_{ss}|}{J} I . \tag{39}$$

Now τ is given by

$$\tau = \int [\frac{\rho_n}{\rho_n^0} - \frac{\rho_p}{\rho_p^0}] \, dx \, . \qquad (40)$$

Comparing with eq. 39 leads immediately to another simple expression for ε_{ss}

$$\varepsilon_{ss} = 4\pi r_{nm}^{r} \, J \int [\frac{\delta}{I} - 1] \, \rho dx \qquad (41)$$

and another interpretation of ε_{ss} : it is the symmetry energy of nuclear matter weighted by the average of the excess of local asymmetry ($\frac{\delta}{I}$ – 1) ρ (the function δ varies from I far inside to 1 in the outside).

CURVATURE ENERGY

The choice usually made for the curvature energy coefficient ε_c in the recent macroscopic approaches to nuclear masses is 0[2,3]. In the folding model of ref. 3, ε_c = 0 exactly. The fit to the masses is compatible with no $A^{1/3}$ term (such a term could well be accounted for by a slight reajustment of the surface energy)and it seems that the fit to the fission barriers of light nuclei militates against its presence: the fission barrier heights predicted in the mass region A > 100 using a model with a sharp surface are too high by several MeV[19] because it overestimates the surface energy of saddle configurations exhibiting pronounced necking; the presence of a positive curvature energy would increase the disagreement. However the existence of a diffuse surface together with the finite range of the interaction (mocked up by the gradient terms in the Skyrme functional) will reduce the surface energy in the neck region; whether this reduction is sufficient to allow for a curvature energy remains to be seen, as no calculation of fission barries in light nuclei is available at the moment.

The calculation of the curvature energy can be easily done by picking up the $(a/R)^2$ – term in the expansion of the total energy (see eq. 26). One gets

$$\varepsilon_c = \frac{K_{nm}}{6} (1-c) \frac{a^2}{r_{nm}^2} + \frac{3}{r_{nm}^2} [\frac{\hbar^2}{2m} \beta - \frac{m^*}{m} \frac{V_{so}}{12} \rho_{nm}^2] \, . \qquad (42)$$

The values obtained for the different Skyrme interactions are shown in Table 2. They lie in the range (10–14) MeV. One could have expected a larger range for a quantity which is not fitted to any experimental data. However one sees from eq. that ε_c is mainly determined by the product $K_{nm}a$, which can be written (taking ν = 1 and c = 0 for simplicity)

$$K_{nm}a^2 = \frac{18a\sigma}{\rho_{nm}} \tag{43}$$

so that ε_c can be given the following approximate form

$$\varepsilon_c \simeq \frac{3a\sigma}{\rho_{nm} r_{nm}^2} + \frac{3}{r_{nm}^2} \left[\frac{\hbar^2}{2m}\beta - \frac{m^*}{m} \frac{V_{so}}{12} \rho_{nm}^2 \right] . \tag{44}$$

This explains why ε_c does not vary much from one interaction to another. Notice that this conclusion is not "Skyrme dependent". It is grounded only on the assumption of an energy density functional, which is a fairly general assumption, so that eq. 44

Table 2

Values of the curvature energy coefficient for various Skyrme interactions (in MeV).

	SII	SIII	SIV	SV	SVI	SkM	SkM*
ε_c	11.2	10.3	12.6	14.5	9.6	12.8	13.0

should give a reasonable estimate also for finite range interactions such as Gogny's or Campi-Sprung's interaction G-0. Indication that it should be so is given in ref. 20 where one recovers, in a macroscopic calculation of the potential energy landscape of deformed ^{240}Pu, the smooth trend of a microscopic calculation using Gogny's interaction if one incorporates a curvature energy of ~10 MeV.

HIGHER ORDER TERMS

From eq. 26 one sees that the expansion of the energy of the system contains a constant term and an exponential term.

The constant term is made up of 4 terms: a genuine one which corresponds to the $(a/R)^3$ - term in eq. 26 plus 3 terms coming from the volume, the surface and the curvature due to the squeezing of the bulk by the surface tension. In order to calculate these contributions one has to go to the next order in the evaluation of the central compression (see eq. 5) but it can be shown that indeed the leptodermous expansion breaks down at that order: the correction to eq. 5 is rather a rational fraction in $A^{-1/3}$ which cannot be expanded. It is then more convenient to evaluate the total constant term by fitting a series of energies calculated for nuclei without Coulomb interaction, knowing the values of the volume, surface and curvature energies. One then finds that the constant term is in the range (-10,-20) MeV, the larger (absolute) value being obtained for interactions having low incompressibility.

Finally let us briefly discuss the exponential term in eq. 26. A simple prescription has been proposed in ref. 21 to evaluate it. The idea is just to ask that the energy per particle goes to zero when the number of particle goes to zero. The simplest form for the total energy of the nucleus is then

$$E = \varepsilon_v A + \varepsilon_s A^{2/3} + \varepsilon_c A^{1/3} + \varepsilon_o + (\lambda_2 A^{2/3} + \lambda_1 A^{1/3} + \lambda_o) e^{-\Lambda A^{1/3}} \qquad (45)$$

and by expanding the exponential one determines the coefficients $\lambda_o, \lambda_1, \lambda_2, \Lambda$. One gets the following equations

$$\begin{aligned} \lambda_o &= - \varepsilon_o \\ \lambda_1 &= - (\varepsilon_c + \Lambda\varepsilon_o) \\ \lambda_2 &= - (\varepsilon_s + \Lambda\varepsilon_c + \frac{\Lambda^2}{2} \varepsilon_o) \end{aligned} \qquad (46)$$

where Λ is solution of

$$\Lambda^3 \frac{\varepsilon_o}{6} + \Lambda^2 \frac{\varepsilon_c}{2} + \Lambda\varepsilon_s + \varepsilon_v = 0 \ . \qquad (47)$$

It is interesting to notice that the value obtained for Λ compares nicely with that of Krappe, Nix and Sierk although derived in a different context. Whether a parametrisation such as eq. 45 could improve the fit to ground state masses and fission barriers remains to be seen.

SEMI-CLASSICAL CALCULATIONS OF NUCLEAR LEVEL DENSITIES

The large number of nuclear excited states observed at a few MeV excitation energy has long ago suggested that a statistical approach should be appropriate. This program was initiated by Bethe in the 30's; assuming a constant single particle density of states g, one derives for the density of n-particle states

$$\rho(E_x) = \frac{\sqrt{\pi}}{12} a^{-1/4} E_x^{-5/4} \exp(2\sqrt{aE_x}) \tag{48}$$

where E_x denotes the excitation energy and $\underline{a}$ represents the so-calles level density parameter. Fig. 3a shows the dependence of $\underline{a}$ on mass number and exhibits strong shell effects. An important progress has been achieved in the past decade in the study of these shell effects; instead of considering a constant g, the authors of ref. 23 start from a Fourier expansion of the shell fluctuations in the single particle level density $G(\varepsilon)$

$$G(\varepsilon) = g(\varepsilon) + \delta g(\varepsilon) = g(\varepsilon) + \sum_m g_m \cos(m\omega\varepsilon - \phi_m) \tag{49}$$

where ω is a parameter characteristic of the wavelength of shell oscillations and therefore of the major shell spacing. By keeping only one term in the Fourier expansion, it is still possible to derive simple analytical expressions for the entropy of the system and for the excitation energy in terms of the thermodynamic temperature of the nucleus; the following form is obtained,

$$S = 2\, a_{LDM} T + \frac{\Delta g_s}{T} \left[\frac{\pi^2 \omega^2 T^2 \cosh\pi\omega T}{\sinh^2 \pi\omega T} - \frac{\pi\omega T}{\sinh\pi\omega T}\right] \tag{50}$$

$$E_x = a_{LDM} T^2 + \Delta g_s \left[\frac{\pi^2 \omega^2 T^2 \cosh\pi\omega T}{\sinh^2 \pi\omega T} - 1\right] \tag{51}$$

where Δg_s represents the ground state shell correction.
Using this model in a systematic of neutron resonance spacings, one gets the curve reproduced in fig. 3b, which justifies the validity of the model : all shell effects have been removed and one is left with a macroscopic part to which semi-classical calculations should be tested. In particular a rather detailed and interesting feature could be even pinpointed in ref. 23, namely the fact that when one considers a series of isotopes, a_{LDM} shows a minimum for the β-stable element. Another question of interest is the determination of the variation of surface energy with temperature: from the data plotted in fig. 3b a negative temperature dependence was found, whereas model calculations using a Hill-Wheeler box as well as a harmonic oscillator predict an increase of surface energy (or a decrease of surface free-energy) with temperature[24]. We shall now address these questions in the context of the Thermal Thomas Fermi model (TTF) which allows one to derive a simple expression for a_{LDM}[25].

The equilibrium state at a given temperature T is calculated by minimizing the free energy of the nucleus F=H-TS, where the entropy S is approximated by the non-interacting Fermi gas expression

$$S = - \Sigma n_\alpha \ln n_\alpha + (1-n_\alpha)\ln(1-n_\alpha) \ , \quad n_\alpha = [1+\exp \frac{\varepsilon_\alpha - \mu}{T}]^{-1} \ .$$

In the Thomas Fermi approximation, the discrete sums over states are replaced by integrals over momentum and one gets for example for the matter density and the kinetic energy density a parametric representation

$$\rho = \frac{1}{2\pi^2} \left(\frac{2m^* T}{\hbar^2} \right)^{3/2} J_{1/2}(\eta), \quad \tau = \frac{1}{2\pi^2} \left(\frac{2m^* T}{\hbar^2} \right)^{5/2} J_{3/2}(\eta),$$

$$J_\nu(\eta) = \int_0^\infty \frac{x^\nu \, dx}{1+e^{x-\eta}} \ .$$

The Fermi integrals can be expanded for low temperature (one should have in mind that this expansion is certainly not valid in the outside region where the local Fermi momentum goes to zero so that the temperature cannot be considered as small) and one gets

$$E(T) = E_0 + a_{LDM} T^2 \ , \quad F(T) = E_0 - a_{LDM} T^2$$

where E_0 is the ground state energy and a_{LDM} is given by

$$a_{LDM} = \frac{\pi^2}{4} \sum_{n,\rho} \int \rho_q \frac{2m^*_q}{\hbar^2 k^2_{F_q}} \, d\vec{r} \ , \quad k_{F_q} = [3\pi^2 \rho_q]^{1/3} \quad q = n,\rho. \quad (52)$$

Notice that a_{LDM} is expressed in function of the equilibrium nucleon densities at T = 0, which makes it easy to calculate: the change in ρ appears only at the order T^4 in the expansion of the free energy. Consequently the effective interaction appears explicitely through the effective mass only (of course it plays an indirect role in determining the equilibrium density profiles).

Next, assuming a Fermi shape for the nucleon density, we can derive eq. 52 an $A^{1/3}$ - expansion of a_{LDM}, using the coefficients $\eta_\nu(k)$ defined in eq. 26.

Writing

$$a = a_v A + a_s A^{2/3} + a_c A^{1/3} \tag{53}$$

one gets (we shall here denote the surface diffuseness by d)

$$a_v = \frac{\pi^2}{4} \frac{2m^*}{h^2 k_F^2} \tag{54}$$

$$a_s = \frac{m}{m^*} a_v \left[\frac{3d}{r_{nm}} (\eta_{1/3}^{(o)} - (1 - \frac{m^*}{m}) \eta_{4/3}^{(o)}) - \frac{6\varepsilon_s}{K_{nm}} (1 - \frac{1}{3} \frac{m^*}{m}) \right] . \tag{55}$$

An expression can also be derived for a_c but we shall not write it in full length here due to the fact that, besides the leading term which writes

$$a_c^L = \frac{m}{m^*} a_v \frac{6d^2}{r_{nm}^2} \left[\eta_{1/3}^{(1)} - \eta_1^{(1)} - (1 - \frac{m^*}{m}) (\eta_{4/3}^{(1)} - \eta_1^{(1)}) \right] \tag{56}$$

a number of small corrections, arising from the central compression of the nucleus, have to be added to a_c^L . Taking some typical values $m^*/m = 0.8$, $k_F = 1.35$, $\varepsilon_s = 20$ MeV, $d = 0.55$ fm, $K_{nm} = 200$ MeV, one has

$$a_v = 0.052 \text{ MeV}^{-1}$$

$$a_s = 0.22 \text{ MeV}^{-1}$$

$$a_c^L = =.67 \text{ MeV}^{-1} .$$

Hence the temperature dependence of both the surface and the curvature energy coefficient is positive, and large (compared to the volume contribution). For example for A = 216 one has

$$a_{LDM} = 0.052\text{x}216 + 0.22\text{x}36 + 0.67\text{x}6 = 11.23 + 7.92 + 4.02$$

$$= 23.2 \text{ MeV}^{-1} .$$

Although surface and curvature corrections appear to contribute as much as the volume term, the value of a_{LDM} is still too small compared to the experimental value $a_{LDM} = 30 \text{ MeV}^{-1}$. Self-consistent calculations including Coulomb and asymmetry effects show that the discrepancy of 25% cannot be removed. Clearly a physical effect is missing, which will be discussed below. Let us first discuss the isospin behaviour of a_{LDM} in a series of isotopes.

We have mentioned above that a_{LDM} exhibits a minimum for the β-stable isotope. This minimum is also obtained numerically when one feeds eq. 52 with the self-consistent ETF ground state densities, although it is less pronounced. The mechanism is the following: eq. 55 shows that the contribution of each type of particle to the surface part of a_{LDM} is proportional to its surface thickness, which is different for neutrons and protons in the case of a neutron excess. We have seen (see above the discussion concerning the geometrical properties of nuclear densities) that the difference in surface thickness came from different behaviour of the external part of the surface, i.e. to the difference in separation energies. The contribution of the neutrons increases with asymmetry while it decreases for the protons. These opposite behaviours produce a minimum in the total. It is interesting to note that the semi-classical interpretation of this fact involves a rather detailed feature of the nucleon densities, indeed not established experimentally, namely the evolution of the nuclear surface thickness with asymmetry.

We now come to the discrepancy between calculated and empirical values of the level density parameter. Indeed this discrepancy is even greater than mentioned because the present approach overestimates the values of a_{LDM} due to the breakdown of the low-temperature expansion in the external part of the surface. Strutinsky type of calculations as well as calculations making use of the Wigner-Kirkwood expansion instead of the ETF functional give values smaller by 15 to 20% than those presented here. Hence the discrepancy between calculated and empirical values is more likely to be of the order of 40%. Several directions have been investigated recently in order to overcome this difficulty.[26,27]

A first attempt is concerned with the discussion of the effective mass. In the HF scheme, the effective mass, which plays a central role here, arises from the non-locality of the potential. The smaller the value of m, the smaller the corresponding density of particles states; a value of m in the range (0.6,0.8) leads to a reasonable agreement with experiment concerning the deep states but not in the vicinity of the Fermi level; this is due to the fact that the energy dependence of m^* , obtained in optical model analysis, is neglected in the HF method. The coupling of particle states with collective ones, calculated in the RPA, produces such an energy dependence of m^* which raises its value up to near the Fermi level. It has been proposed to incorporate this feature in the effective interaction used in HF calculations by modifying the r-dependence of m^* so that it exhibits a bump in the geometrical surface region: in a semi-classical picture, the Fermi level corresponds to the geometrical surface, in the sense that the nuclear

surface is built up mainly by states near the Fermi level. In this procedure the desired enhancement of the level density parameter is obtained through a very large increase of the surface term a_s, as a_v is of course non altered (a_c is also increased). One of the consequences pointed out in ref. 26 concerns the temperature dependence of the fission barrier: a rapid evaluation shows that the fission barrier in 240 Pu is reduced by a factor of ~ 2 at a temperature of T = 1.5 MeV, i.e. at an excitation energy of ~ 70 MeV. Let us mention however that a serious objection can be raised against this line of thought[28] : the energy dependence of the effective mass is clearly an effect lying beyond HF. Then the states calculated when coupling particle states with collective states do not have occupation number 1, they are fragmented, and so the condition of applicability of Bethe's statistical model are not longer met: one does not know how to calculate the n-particle density of states.

An alternative idea is presently being investigated[29]. In the analysis of experimental data summerized in fig. 3b, deformed and spherical nuclei have been analyzed in the same way. However in the case of deformed systems, when the excitation energy increases, there is a release of macroscopic deformation energy together with the disappearance of shell effects. This macroscopic deformation energy is now able to go into excitation energy, so the corresponding level density parameter extracted when taking into account this mechanism should be smaller, i.e. closer to the calculated values.

Let us end this section by mentioning the relevance of the present discussion to astrophysical applications: the equation of state of hot dense matter utilized in calculations of supernovae collapses. The extremely large number of excited states provides a way of storing energy during the collapse. Variations of 20 to 40% in the level density parameter, which appears in an exponential, might influence the amount of energy available in the shock wave which is thought to be responsible for the explosion of the star.

GIANT RESONANCES

Electric Giant Resonances in nuclei are of collective character so that one can expect shell effects to be of minor importance in their description: the smooth variation of the peak energies with mass number supports this idea. This fact calls for description of the vibrations in terms of macroscopic variables (densities, displacements, currents...) which are taken as basic variables of the theory in the different fluid-dynamical and semi-classical approaches. Simplifying the dynamics, one can hope to gain more physical insight in the properties of the effective interaction: nuclear incompressibilities in the case of the Giant Monopole Resonance (GMR), symmetry energy coefficients in the case of the Giant Dipole Resonance (GDR)... When the collective state exhaust a large fraction of the Energy Weighted Sum Rule (EWSR) the knowledge of a few moments of the strength function suffices to extract the physical information[30]. The self-consistent HF-RPA scheme is here particularly well adapted since it preserves the sum rules, as was shown by Thouless. The strength distribution corresponding to a given excitation operator Q is given by

$$S(E) = \sum_n |\langle n|Q|0\rangle|^2 \, \delta(E-E_n) \tag{57}$$

The m_k moment of S(E) is defined as

$$m_k = \int E^k S(E)dE = \sum_n |\langle n|Q|0\rangle|^2 \, E_n^k \tag{58}$$

where the sum is taken over all excited states $|n\rangle$. It can be easily checked that

$$m_1 = \sum_n |\langle n|Q|0\rangle|^2 E_n = \langle 0|[Q,[H,Q]]|0\rangle \tag{59}$$

where $|0\rangle$ denotes the exact ground state of the system. Now the Thouless theorem writes[31]

$$\left(\sum_n |\langle n|Q|0\rangle|^2 E_n\right)_{RPA} = {}_{HF}\langle 0|[Q,[H,Q]]|0\rangle_{HF} \tag{60}$$

where the sum on the left hand-side is taken over all RPA states and where $|0\rangle_{HF}$ in the right hand-side stands for the HF (i.e. uncorrelated) ground state. Of course from eq. 60 other relations between RPA sum rules and HF ground state properties can be generated by simply replacing the operator Q by the commutator [H,Q]. As the effective interaction is fitted to ground state properties, Thouless theorem gives confidence in the calculation of RPA sum rules with the <u>same</u> interaction. Thus one has a <u>unified and self-consistent framework</u> to discuss both static and dynamical properties. Some of these moments can be calculated semi-classically and this method has been extensively used in the description of the GMR[11,12] and to a lesser extend the GDR[13,32,35]. The extension of the method to finite temperatures is being explored now and will give informations, with a minimum of computational efforts, on Giant Resonances built on excited states: displacement of peak energies and evolution of escape width with energy. We shall restrict ourselves here to a discussion of the GDR in relation with the symmetry energy coefficients. Besides the dipole polarizability, we shall discuss the (unphysical) fragmentation of the strength obtained in RPA calculations with commonly used interactions, even when giving satisfactory results concerning the moments; we shall see that the transparency of the semi-classical approach allows one to give argument that this wrong feature is related to erroneous values of the surface symmetry coefficient.

SURFACE EFFECTS IN THE DIPOLE POLARIZABILITY

The static dipole polarizability p is related to the moment m_{-1} and to the integrated cross-section by

$$p = 2m_{-1} = 4\pi^2 \frac{\hbar c}{e^2} \sigma_{-2} = 4\pi^2 \frac{\hbar c}{e^2} \int \frac{\sigma(\omega)}{\omega^2} d\omega \quad . \tag{61}$$

It is a rather well known quantity over the mass table. Due to the weighting factor ω^{-2} in eq. 61 it is not sensitive to the high energy part of the cross-section, i.e. to short range correlations. Hence it is a good quantity to which effective interactions should be tested.

The role of the diffuse nuclear surface on p can be identified by looking at the general trend of the quantity $\sigma_{-2} A^{-5/3}$ over the mass table. A constant density model leads to Migdal's formula involving the volume symmetry coefficient J :

$$\sigma^M_{-2} = 2\pi^2 \frac{e^2}{\hbar c} A \frac{\langle r^2 \rangle}{24J} = 2\pi^2 \frac{e^2}{\hbar c} \frac{r_{nm}^2}{40J} A^{5/3} \qquad r_{nm}=1.15\text{fm}. \tag{62}$$

Fig. 4 shows the experimental points, plotted versus $A^{-1/3}$ in order to visualize more clearly finite size effects. The error bar in the nuclear matter value is obtained by considering 2 extreme values of J, namely J = 28 MeV and J = 37 MeV (this last value is favoured by recent mass formula fits). One sees that surface effects are indeed important: the experimental value in the region of medium and heavy nuclei, almost constant = $(2.9\pm0.2)\mu b\ MeV^{-1}$ is ~ 50% greater than the largest estimate of the Migdal value obtained using a low value of J. In light systems the experimental value of $\sigma_{-2} A^{-5/3}$ gets larger by a factor of 2 or 3.

This trend has been confirmed by RPA calculations[33]; semi-classical and hydro-dynamical investigations have analyzed in more detail the dependence of σ_{-2} on the isospin properties of the effective interaction. The EDF used in ref. 13 gives results in remarkable agreement with the RPA calculations of ref. 33. The polarizability is obtained through a constrained calculation, i.e. one minimizes the energy of the system under the constraint of a dipole operator D_z. The transition density is obtained as the

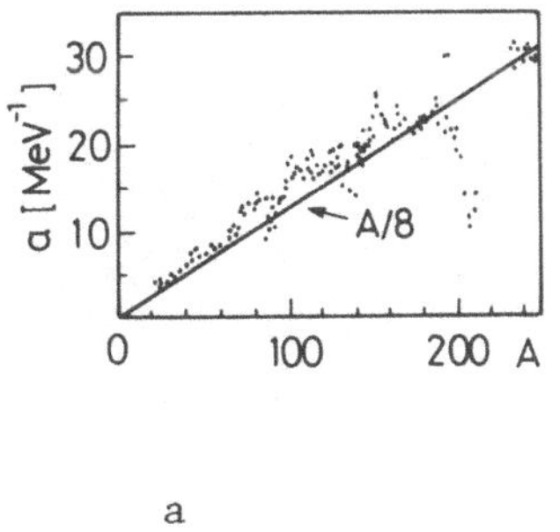

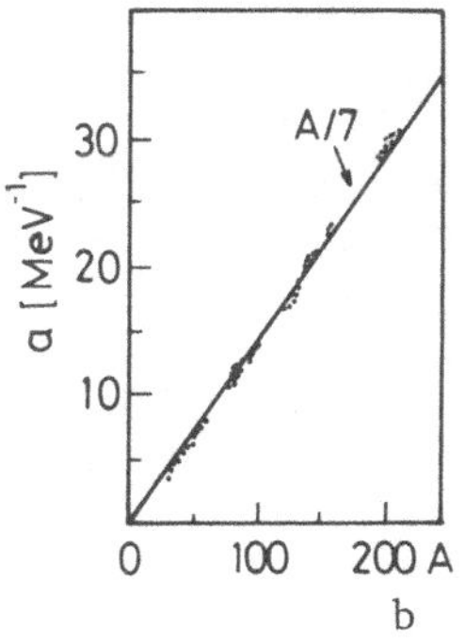

Fig. 3 Plot of the level density parameter as a function of mass number. (3a from Ref. 22, 3b from Ref. 23).

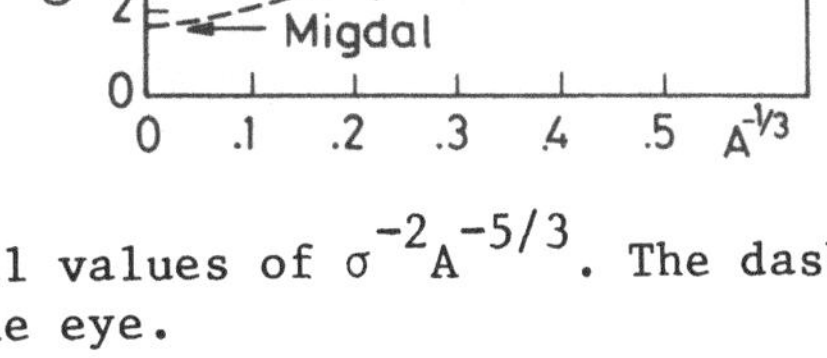

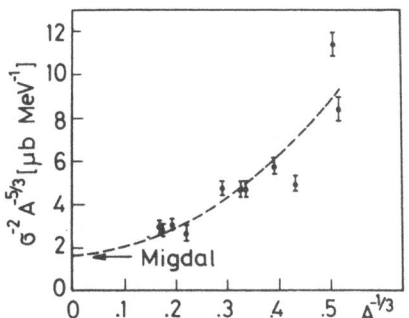

Fig. 4 Experimental values of $\sigma^{-2}A^{-5/3}$. The dashed line is drawn to guide the eye.

solution of a differential equation and the method is fully self-consistent. Fig. 5 shows the result in the case of calcium and lead. One has a volume type of deformation corresponding to the Steinwedel-Jensens model, but of course with a diffuse surface.

In order to investigate the role of J and ε_{ss}, one can change separately both quantities, by changing the symmetry potential $\varepsilon_\delta(\rho)$ as a function of density. Surprisingly one finds that similar agreement with experiment is obtained with very different couples of values of J and ε_{ss}. The expression which can be derived within the Droplet Model helps understanding this feature; surface effects on σ_{-2} appear through the ratio of surface to volume symmetry coefficient [32,34]

$$\sigma_{-2} = \sigma_{-2}^{M}\left(1 + \frac{15}{4}\frac{J}{Q}A^{-1/3}\right) = \sigma_{-2}^{M}\left(1 + \frac{5}{3}\frac{\varepsilon_{ss}}{J}A^{-1/3}\right) \tag{63}$$

where Q is the so-called stiffness parameter. This relation, while describing correctly the qualitative behaviour of σ_{-2} with mass number, should not be used as such to analyze experimental data: besides Coulomb effects which are neglected and which in fact lower the values of σ_{-2} by ~ 10%, fig. 4 shows that $A^{-2/3}$ corrections are indeed not negligible if one wants to consider the whole mass table. On the other hand if one restricts oneself to heavy systems where one expects curvature corrections to be small, eq. 63 gives only a correlation between J and ε_{ss}/J : similar values of σ_{-2} are obtained with low or high values of both J and ε_{ss}/J. Our conclusion is that the semi-classical calculations of the polarizability is in good agreement with experiment for a large class of effective interactions; little constraint can be extracted concerning J and ε_{ss} separately.

OTHER MOMENTS OF THE STRENGTH FUNCTION

Besides m_{-1}, the moments m_1 and m_3 can be calculated in an EDF, and used to characterize the energy of the resonance by considerung the ratios

$$E_1 = \sqrt{\frac{m_1}{m_1}} \qquad E_3 = \sqrt{\frac{m_3}{m_1}} \tag{64}$$

(there is no theoretical expression for the peak energy). The m_1 moment is given by

$$m_1 = \frac{\hbar^2}{m}\,\frac{NZ}{A}\,(1+\kappa) \tag{65}$$

where κ is the so-called enhancement factor, related to the exchange component of the interaction (=velocity dependent part of the force in the case of a Skyrme interaction). The m_3 moment brings some new information; it can be calculated by performing a dipole scaling on the ground state density, which corresponds to the Goldhaber-Teller model, i.e. to a vibration in the surface; it is thus sensitive to the symmetry potential at low density, hence to the surface symmetry coefficient, as can be seen from the approximate expression derived in ref. 35

$$E_3 = \left[\frac{2}{3} \frac{\hbar^2}{m} \frac{\int [\frac{\rho'}{\rho} (1+2\chi\rho)]^2 \rho \varepsilon_\delta(\rho) d\vec{r}}{A (1 + \kappa)} \right]^{1/2} , \quad \chi = \frac{\kappa}{\rho_o} \tag{66}$$

Notice the weighting factor ρ'^2 in the numerator. Identifying E_3 with the peak energy then leads to values of $\varepsilon_{ss}/J \simeq 2$.

We mentioned above that the analysis in terms of moments of the strength could be considered sufficient if the collective state exhausted almost all the EWSR. However RPA calculations of the strength have up to now always shown an unphysical fragmentation which has to be understood. We shall now turn to this discussion.

RPA CALCULATIONS OF THE DIPOLE STRENGTH

Fig. 6 shows the strength calculated in ^{208}Pb with SkM interaction together with the experimental curve. This force gives satisfactory results concerning the moments, yet the photoabsorption cross section exhibits a fragmentation which is not seen experimentally. Similar results have been obtained by various authors: the strength shows mainly two concentrations in energy, separated by 4 to 5 MeV, most of the strength (~ 2/3) being concentrated in the lower peak, which corresponds more or less to the experimental one[36].

The presence of the unphysical peak raises the following alternative: either its origin lies in some particularity of the particle-hole spectrum and it will disappear when more complex configuration (2p-2h) are included in the calculation; or the fragmentation is linked to a macroscopic aspect and indeed reveals some erroneous macroscopic property of the interaction, in which case it will survive even after including some damping mechanism. In the first case, one should not be preoccupied by the presence of some structure in the strength. However calculations of the damping, although becoming available, are still too preliminary to answer the quantitative question of how much fragmentation can be tolerated at the RPA level. Besides, fluid-dynamical calculations[37,38] give a fragmentation of the dipole strength, which is interpreted[38,39] as resulting from the coupling of transverse to longitudinal com-

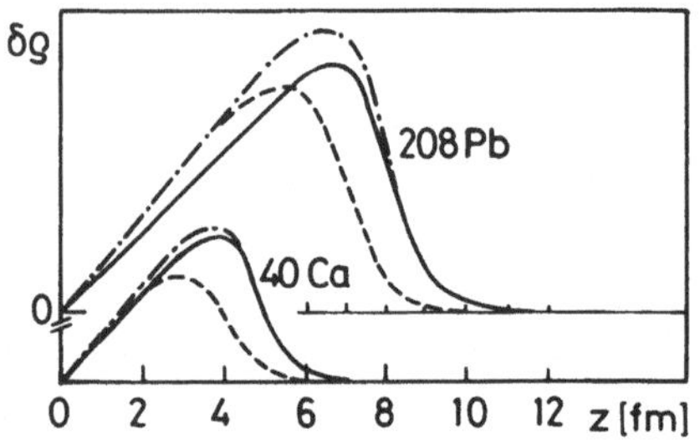

Fig. 5 Transition densities (full lines: with Coulomb, dashed-dotted: without Coulomb, dotted: Migdal). SkM force.

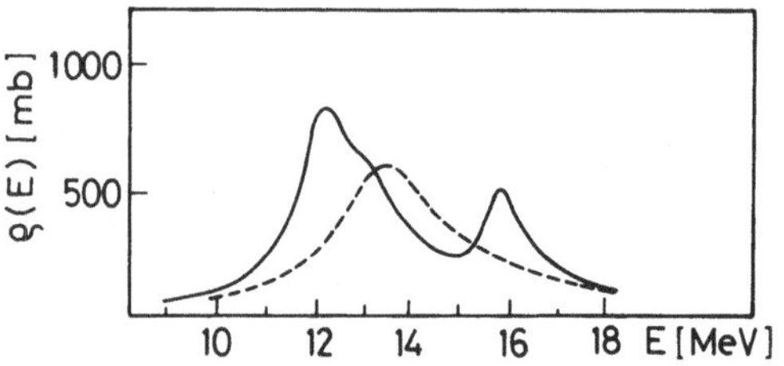

Fig. 6 RPA photoabsorption cross-section calculated in ^{208}Pb with SkM interaction using an averaging interval of 1 MeV. Dashed line: experimental curve.

ponents in the solutions of the equations of motion. It is then worth exploring the second branch of the alternative.

In order to do so we have made different sets of RPA calculations in the following model conditions:

* no velocity dependence in the Skyrme functional.

* N = Z = 70 nucleus (no Coulomb, no spin-orbit) : the unperturbed cross-section plotted in fig. 7 is thus independent of the isospin properties of the interaction.

Fig. 8 shows the results obtained for the strength when, fixing the value of J = 37 MeV, one changes the surface symmetry energy ε_{ss}, by changing the shape of the curve $\varepsilon_\delta(\rho)$ as a function of ρ. One sees that the unphysical peak disappears for large values of ε_{ss}. Indeed one could have predicted that going to large values of ε_{ss} would help, by considering eq. 66. In order to bring E_3 closer to E_1 (if there were one single collective state both quantities would be identical) one has to lower the values of $\varepsilon_\delta(\rho)$ at low densities; in doing so one increases the defect of repulsion (compared to J) in the surface region, which amounts to increasing the surface symmetry coefficient.

What could not be predicted by the semi-classical approach was whether the upper peak would come closer to the lower one or whether the strength would be transferred from the higher peak to the lower one, which is actually what happens. Notice that in fig.8 the lower peak is shifted down by 2 MeV; however in a more realistic case in order to preserve the masses one has to increase J if one increases ε_{ss}. Then the shift of the lower peak is reduced. This can be seen in fig. 9, where are plotted the RPA strength functions for ^{208}Pb calculated with 2 different interactions giving the same binding energy (Coulomb and spin-orbit are now included). The first one is the original SkM force, the second one has exactly the same isoscalar properties but J has been increased to 37 MeV and ε_{ss} to -140 MeV (these values are close to the Droplet Model values). The result is the same as in the model case: the unphysical peak is no longer present in this second case, i.e. for a ratio $\varepsilon_{ss}/J \simeq 4$. Only a very weak structure is seen at ~ 15 MeV whereas in the case of SkM, the upper peak at ~ 16 MeV exhaust ~ 30% of the EWSR.

Now that we have shown that the fragmentation of the RPA strength is related to the isospin properties of the interaction, we want to stress that our aim is not to give a definite value for the ratio ε_{ss}/J. As long as one does not know how much structure can be washed out by coupling to 2p-2h configurations, one cannot give a precise answer, although there is probably too much fragmentation with the interactions commonly used. Besides, the position of the peak for the modified SkM interaction of fig. 9 is too low compared to experiment; hence m_{-1} is too large. Other physical

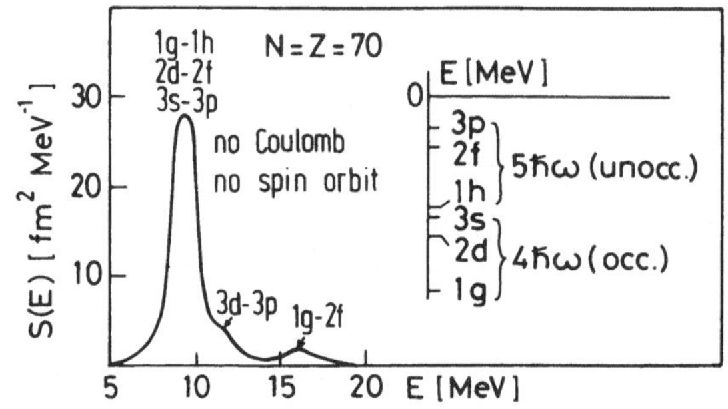

Fig. 7 Model calculation of the RPA dipole strength for N=Z=70. (unperturbed strength function, no Coulomb, no spin-orbit).

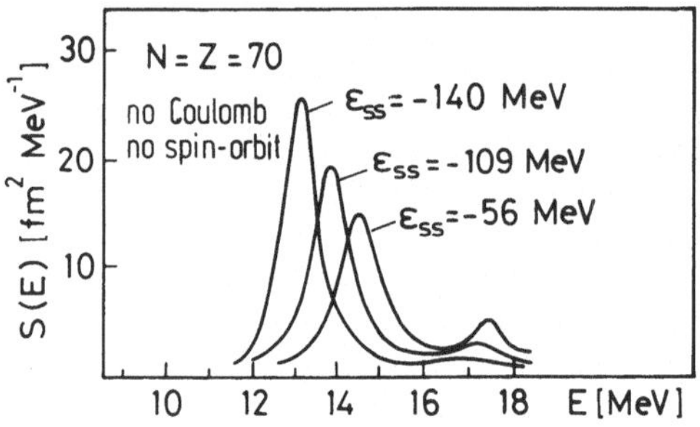

Fig. 8 Model calculation of the RPA dipole strength for N=Z=70. (effect of the surface symmetry coefficient on the fragmentation of the strength).

phenonema where isospin properties are involved have also to be taken into account simultaneously: the neutron skin thickness increases with increasing surface symmetry coefficient, the proton r.m.s. radius in ^{48}Ca decreases and the fission barrier heights decrease. Such a study lies beyond the scope of the present lectures, where we just want to show a possible mechanism by which the unphysical feature of the strength can be corrected.

We show in fig. 10 the transition densities $\delta\rho(r,E)$ as functions of r for different energies. As expected the amplitude of the vibration is maximum at the peak energies; the dashed lines locate the nuclear surface. One sees that the lower peak is of volume type while the upper one is a surface vibration. Notice the change in $\delta\rho(r,E)$ around 12 MeV when one increases ε_{ss}: a large bump appears now at the surface, while the amplitude of the vibration around 15 MeV has been significantly reduced (we have checked this aspect in the model calculation also).

The corresponding currents are displayed in fig. 11 in the case of the force with the large value of ε_{ss}. The vertical axis is the z-axis and only one fourth of the nucleus is considered. At low energy the flow lines tend to be parallel to the surface except for large z where the large value of ε_{ss} allows for a crossing of the nuclear surface: in that case one does not loose much energy by separating the neutrons from the protons at the surface. When going to higher energies, the flow lines tend to become parallel to the z-axis, i.e. to be more of Goldhaber and Teller type, but surprisingly this happens at the minimum of the strength function. Around 15 to 16 MeV a vortex appears for small values of z in the surface region: this was unexpected as one would have rather predicted in this region a Goldhaber-type of vibration which is irrotational, whereas one has in fact large transverse components in the vortex. The results obtained here do not confirm those obtained in the fluid-dynamical calculation of ref. 38, although a fragmentation of the strength is obtained in both cases. But in ref. 38 the lower peak gets more strength than the upper one and exhibits vorticity. This disagreement might be linked to the treatment of the collective kinetic energy, which is assumed to keep its classical form in the fluid-dynamical approach even in the calculation of the transverse modes. This question deserves clearly further study[39].

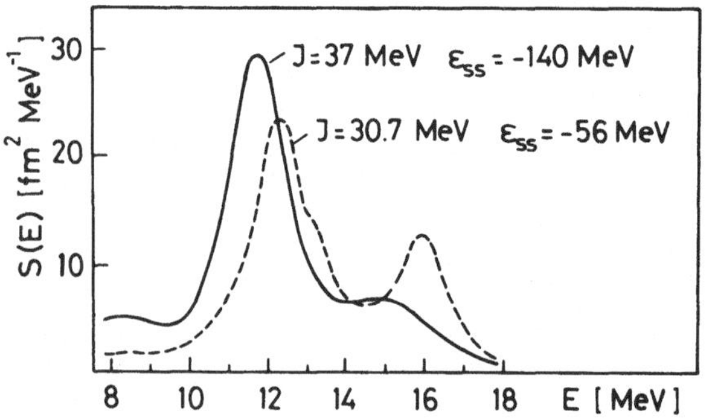

Fig. 9 RPA strength functions in ^{208}Pb calculated with SkM and a modified force having J=37 MeV and ε_{ss} = −140 MeV.

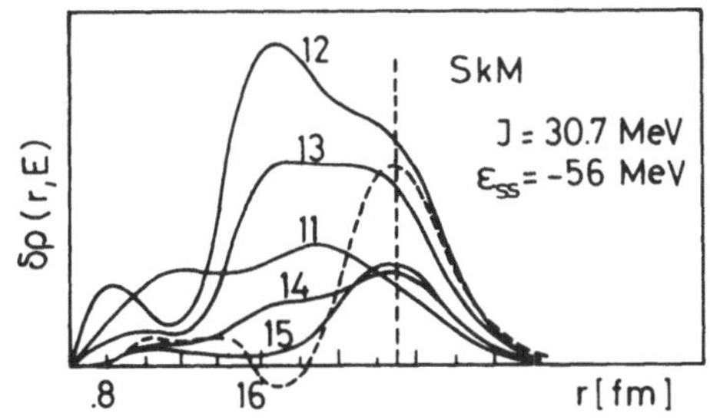

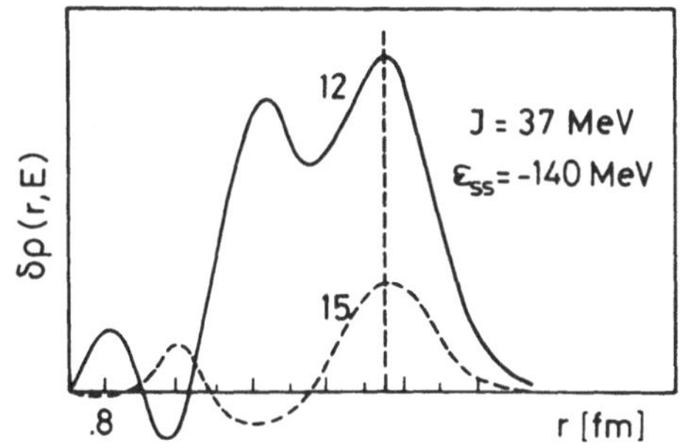

Fig. 10 Transition densities for different energies in ^{208}Pb. The dashed vertical line locates the nuclear surface.
top: SkM interaction
bottom: modified force indicated in Fig. 9.

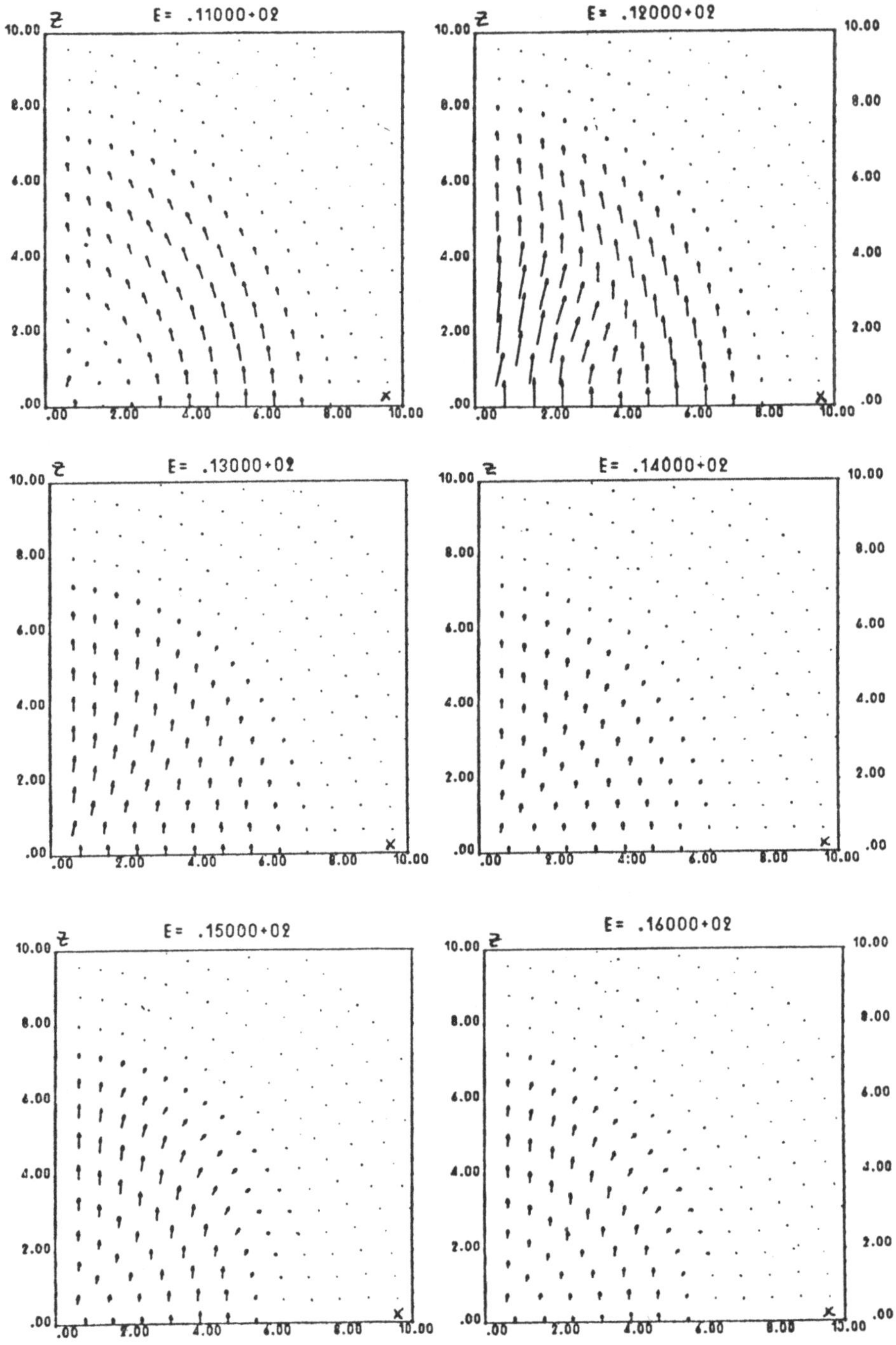

Fig. 11 plot of the current for different energies. Notice the vortex in the xy plane for energies around 15-16 Mev.

CONCLUSION

The choice made in these lectures has been a practical one: to show that a large variatyof situations can be understood in simple terms using the EDF based on the Extended Thomas Fermi approximation. In particular much computational effort can be saved by using analytical formulae which exhibit the main macroscopic aspects of effective interactions to be used in more microscopic approaches. This can be useful in the process of improving existing interactions by incorporating more experimental information. For example, it would be interesting to calculate the fission barrier of light systems (between say mass 100 and mass 150) in a semi-classical approximation in order to check whether or not the rather high value of the curvature energy obtained with Skyrme type interactions is contradictory with experiment. Another question being presently investigated is the description of Giant Resonances built on excited states[42]. The semi-classical may well be the only tractable approach, due to the numerical difficulties encountered in generalizing the RPA to finite temperature. The moment approach will help in particular in discussing whether such vibrations should be considered at constant temperature or rather at constant entropy[43].

Much of the work presented here results from an earnest collaboration with O. Bohigas and H. Krivine during the past few years. The discussion of the GDR has been developed also with C. Scmit and Nguyen Van Giai, whose code was used in the RPA calculations. It is also a pleasure to acknowledge numerous and fruitful discussions with M. Brack, G. Eckart, R. Hasse, G. Holzwarth, J. Martorell, P. Schuck and S. Stringari, especially during the period of the annual Workshop on Semi-Classical Methods in Nuclear Physics, initiated in 1977 at the ILL, Grenoble. Last but not least, thanks to R. Dreizler for organizing this school during which, instead of confronting different methods on a given subject, one has had the opportunity to measure the variety of subjects discussed in the same framework.

APPENDIX A.

The Skyrme interaction is used together with the following form for the kinetic energy density

$$\tau_q = \frac{\hbar^2}{2m}\left(\alpha\rho_q^{5/3} + \beta\frac{(\nabla\rho)^2}{\rho_q} + \gamma\Delta\rho_q\right) \quad q = n,p \quad \alpha = \frac{3}{5}(3\pi^2)^{2/3}$$

with $\beta = 1/18$ and $\gamma = 1/3$. The value of β is adjusted to reproduce the HF value of the surface energy for SIII interaction in the semi-infinite calculation of ref. 40 but a somewhat smaller value agrees better with the results of Brack and col.[41]

The total energy density takes the form

$$\mathcal{H} = \Sigma \frac{\hbar^2}{2m_q^*} \alpha\rho_q^{5/3} + \frac{t_o}{2}[(1+\frac{x_o}{2})\rho^2 - (\frac{1}{2} + x_o)(\rho_n^2 + \rho_p^2)]$$

$$+ \frac{1}{12} t_3\rho^\sigma [(1 + \frac{x_3}{2})\rho^2 - (\frac{1}{2} + x_3)(\rho_n^2 + \rho_p^2)]$$

$$+ d_1\rho'^2_n \rho'^2_p + d_2\rho'_n \rho'_p + d_3(\rho_p \frac{\rho'^2_n}{\rho_n} + \rho_n \frac{\rho'^2_p}{\rho_p})$$

$$+ \frac{\hbar^2}{2m} \beta (\frac{\rho'^2_n}{\rho_n} + \frac{\rho'^2_p}{\rho_p}) + 4V_{so} \Sigma_q \frac{m^*_q}{m} \rho_q\rho'^2_q ,$$

the effective mass is given by

$$\frac{m}{m^*_q} = 1+ \frac{2m}{\hbar^2}\frac{1}{4} [(t_1(1+ \frac{x_1}{2}) + t_2(1+ \frac{x_2}{2}))\rho$$

$$+ (t_2(\frac{1}{2} + x_2) - t_1(\frac{1}{2} + x_1))\rho_q], q = n,p$$

and V_{so} is related to the strength of the spin-orbit interaction W_o

$$V_{so} = - \frac{9}{32} (W_o)^2 \frac{2m}{h^2} .$$

The coefficients d_1, d_2 and d_3 are given by

$$d_1 = \frac{3}{32} [t_1(1-x_1)-t_2(1+x_2)]+ \frac{1}{8}[3t_2(1+x_2)+t_1(1-x_1)](\beta-\gamma)$$

$$d_2 = \frac{1}{8} [3t_1(1+ \frac{x_1}{2})-t_2(1+ \frac{x_2}{2})] - \frac{\gamma}{2}[t_1(1+ \frac{x_1}{2})+t_2(1+ \frac{x_2}{2})]$$

$$d_3 = \frac{1}{4} [t_1(1+ \frac{x_1}{2}) +t_2(1+ \frac{x_2}{2})] \beta .$$

In the case where $\rho_n = \rho_p$, reduces to

$$\mathcal{H} = \frac{\hbar^2}{2m}\frac{3}{5}\left(\frac{3\pi^2}{2}\right)^{2/3}\rho^{5/3} + \frac{3}{8}t_o\rho^2 + \frac{1}{16}t_3\rho^{2+\sigma}$$

$$+ d\rho'^2 + \frac{h^2}{2m}\beta\frac{\rho'^2}{\rho} + V_{so}\frac{m^*}{m}\rho\rho'^2$$

$$d = \frac{d_1}{2} + \frac{d_2}{4} + \frac{d_3}{2} = \frac{1}{64}[9t_1 - t_2(5+4x_2)]$$

$$+\frac{(\beta-\gamma)}{16}[3t_1 + t_2(5+4x_2)]$$

We collect now the expressions for the various quantities used in eqs. 2 and 8

$$K_{nm} = -2\frac{\hbar^2}{2m}\alpha\rho_{nm}^{2/3}\left[1 - \frac{5}{16}\frac{2m}{\hbar^2}3t_1 + t_2(5+4x_2))\rho_{nm}\right]$$

$$+ \frac{9}{16}(1+\sigma)\sigma t_3\rho_{nm}^{1+\sigma}$$

$$\rho_{nm}^2\frac{d^3(\rho e)}{d\rho_{nm}^3} = -\frac{10}{27}\frac{\hbar^2}{2m}\alpha\rho_{nm}^{2/3}\left[1 - \frac{1}{2}\frac{2m}{\hbar^2}(3t_1 + t_2(5+4x_2))\rho_{nm}\right]$$

$$+ \frac{(2+\sigma)(1+\sigma)}{16}\sigma t_3\rho_{nm}^{1+\sigma}$$

$$J = \frac{5}{9}\frac{\hbar^2}{2m}\alpha\rho_{nm}^{2/3} - \frac{t_o}{4}(x_o + \frac{1}{2})\rho_{nm} - \frac{t_3}{24}(x_3 + \frac{1}{2})\rho_{nm}^{1+\sigma}$$

$$+ \frac{5}{72}[t_2(4+5x_2) - 3t_1x_1]\alpha\rho_{nm}^{5/3}$$

$$L = \frac{10}{9}\frac{\hbar^2}{2m}\alpha\rho_{nm}^{2/3} - \frac{3}{4}t_o(x_o+\frac{1}{2})\rho_{nm} - (1+\sigma)\frac{t_3}{8}(x_3+\frac{1}{2})\rho_{nm}^{1+\sigma}$$

$$+ \frac{25}{72}[t_2(4+5x_2) - 3t_1x_1]\alpha\rho_{nm}^{5/3}$$

$$K_{sym} = -\frac{10}{9}\frac{\hbar^2}{2m}\alpha\rho_{nm}^{2/3} - 3\frac{(1+\sigma)}{8}\sigma t_3(x_3+\frac{1}{2})\rho_{nm}^{1+\sigma}$$

$$+ \frac{50}{72}[t_2(4+5x_2) - 3t_1x_1]\alpha\rho_{nm}^{5/3} \quad .$$

REFERENCES

1. W.D. Myers and W.J. Swiatecki, Ann. of Phys. 55:395(1969).
2. W.D. Myers, Droplet Model of Atomic Nuclei,Plenum, NY (1977).
3. H.J. Krappe, J.R. Nix and A.J. Sierk, Phys. Rev. C20:992(1979).
4. V.M. Strutinsky, Nucl. Phys. A95:420(1967); 122:1(1968).
5. T.H.R. Skyrme, Phil. Mag. 1:1043(1956)
   D. Vautherin and D.M. Brink, Phys. Lett. 32B:149(1970)
   M. Beiner, H. Flocard, N. Van Giai and P. Quentin, Nucl. Phys. A238:29 (1975).
6. J. Decharge and D. Gogny, Phys. Rev. C21:1568(1980).
7. M. Brack, Lecture notes to this school and refs. therein.
8. O. Bohigas, X. Campi, H. Krivine and J. Treiner, Phys. Lett. 64:381(1976).
9. Y.H. Chu, B.K. Jennings and M. Brack, Phys. Lett.68B:407(1977).
10. H. Krivine and J. Treiner, Phys. Lett. 88B:212(1979).
11. J.P. Blaizot and B. Grammeticos, Nucl. Phys. A335:115(1981).
12. J. Treiner, H. Krivine, O. Bohigas and J. Martorell, Nucl. Phys. A371:253(1981).
13. H. Krivine, C. Schmit and J. Treiner, Phys. Lett. 112B:281(1982).
14. W.D. Myers and W.J. Swiatecki, Nucl. Phys. A336:267(1980).
15. S. Stringari and E. Lipparine, Phys. Lett. 117B:141(1982).
16. L. Wilets, Phys. Rev. 101:1805(1956).
17. H. Krivine and J. Treiner, J. Math. Phys. 22:2484(1981).
18. H. Krivine and J. Treiner, Phys. Lett. 124B:127(1983).
19. M. Dahlinger, D. Vermeulen and K.H. Scmidt, Nucl. Phys. A376:94(1982).
20. J. Treiner, R.W. Hasse and P. Schuck, J. Physique L-733(1983).
21. B. Grammaticos, Ann. of Phys. 126:450(1980).

22. A. Gilbert and A.G.B. Cameron, Can. J. Phys. 43:1446(1965).
23. S.K. Kataria, V.S. Ramamurthy and S.S. Kapoor, Phys. Rev. C18:549(1978).
24. J. Treiner, Workshop on Semi-Classical Methods in Nuclear Physics, ILL Grenoble(1981).
25. M. Barranco and J. Treiner, Nucl. Phys. A351:269(1981).
26. X. Campi and S. Stringari, Z. Phys.A309:239(1983).
27. M. Prakash, J. Wambach and Z.Y. Ma,Stony Brook preprint.
28. J. Martorell, private communication.
29. S.K. Kataria, private communication.
30. O. Bohigas, A.M. Lane and J. Martorell, Phys. Rep. 51:267(1979).
31. D.J. Thouless, The Quantum Mechanics of Many Body Systems,NY (1972)
32. E. Lipparini and S. Stringari, Phys. Lett.112B:281(1982).
33. O. Bohigas, N. van Giai and D. Vautherin, Phys. Lett.102B:105 (1981).
34. J. Meyer, P. Quentin and B.K. Jennings, Nucl. Phys. A385:269 (1982).
35. H. Krivine, J. Treiner and O. Bohigas, Nucl. Phys. A336:155 (1980).
36. G.F. Bertsch and S.F. Tsai, Phys. Rep. 18:125(1975).
K.F. Liu and G.E. Brown, Nucl. Phys.A265:385(1976).
J.P. Blaizot, These d'Etat, Universite de Paris(1977).
N. Auerbach and A. Klein, Nucl. Phys. A395:77(1983).
N. van Giai, private communication.
37. G. Holzwarth and G. Eckart, Z. Phys.A284:291(1978).
38. H. Koch, G. Eckardt, B. Schwesinger and G. Holzwarth, Nucl. Phys.A373(1982).
39. S. Stringari, Ann. of Phys. 151:35(1983).
40. M. Farine, J. Cote and J.M. Pearson, Nucl. Phys. A338:86(1980).
41. M. Brack, private communication.
42. J. Meyer, P. Quentin and M. Brack, Phys. Lett. 133B:279(1983).
43. M. Barranco, S. Marcos and J. Treiner, in preparation.

ON CHARGE SHARING IN DIATOMIC QUASIMOLECULES

Jörg Eichler*

Bereich Kern- und Strahlenphysik, Hahn-Meitner-Institut für Kernforschung Berlin, D- 1000 Berlin 39, West Germany

If a diatomic molecule is pulled apart into two atomic fragments, the electronic charge is shared between the separating atoms in a way that is closely related to the nature of the chemical bond which is broken. Whereas in chemistry one usually deals with neutral sytems and hence with very few competing stable fragmentation channels, the situation is quite different in slow collisions of highly charged projectile ions with neutral target atoms. Here, with increasing net positive charge, an increasing number of stable fragmentation channels becomes accessible. Owing to the present development of sources for slow highly charged ions, charge sharing becomes an increasingly more important degree of freedom.

The simplest way to describe charge sharing for well separated systems is to specify the numbers N_A, N_B of electrons attached to systems A and B, respectively. It has recently been noticed by Perdew et al.[1] that with this description one encounters an apparent difficulty: Since for separated neutral atoms the system energy $E_{AB} = E_A + E_B$ is simply the sum of the atomic energies, any variation of electron density that shifts $\delta N_B > 0$ electrons from atom A (with the higher chemical potential μ_A) to atom B (with the lower chemical potential μ_B) will lower the system energy and hence appears to lead to a better variational ground state. Perdew et al.[1] have resolved the resulting "paradox of fractional charges" by showing that the open system in question has to be described as a statistical mixture of pure states, each having an integral charge number. The associated curve of E_{AB}

* Also at Fachbereich Physik, Freie Universität Berlin, D-1000 Berlin 39, West Germany

versus $N_A = N-N_B$ is a series of straight-line segments linking the values for integral asymptotic charge fragmentations. The energy then minimizes nonanalytically at $N_A = Z_A$, $N_B = Z_B$ which corresponds to the proper dissociation limit. The discontinuity of the slope $\mu = \partial E/\partial N$ at integral N for a single atom finds a natural interpretation in the jump of $|\mu|$ from the ionization potential to the electron affinity.

Nevertheless, one may feel that discontinuities in the slope of the energy curve remain an unsatisfactory feature of this particular choice for describing charge fragmentation channels. Furthermore, it is not easily possible to apply this concept to finite internuclear separations.

An alternative approach for describing charge fragmentation channels has been introduced[2,3] by the present author a few years ago in the context of charge-changing ion-atom collisions. This concept not only allows for an extension to finite internuclear separations but also lends itself easily for the computation of complete potential energy surfaces[3,4]. The basic difference compared to Ref. 1 is that electrons are not just counted but are weighted with their average distance from the midplane between the nuclei. From the outset, this leads to a meaningful continuous variable without the need to consider fractional charges.

In practice, the variable, termed "dipole deformation" or "charge asymmetry", is defined in the following way[2-4]: Let us choose the z axis along the internuclear line directed from A to B and with the origin at the midpoint between the nuclei. Then for large internuclear distances R, the quantity $2\langle\hat{z}_i\rangle/R$ is +1 or -1, if the ith electron is centered about nucleus A or B, respectively. Hence for $R\to\infty$, this quantity "counts" electrons, but for finite R deviations from ±1 reflects distortions of electron orbitals from their positions centered around the nuclei A or B. For the complete diatomic system, we define the operator for dipole deformation as

$$\hat{\zeta} = -\sum_i 2\hat{z}_i/R + Z_B - Z_A, \qquad (1)$$

where the nuclear charges Z_A, Z_B have been introduced in order to ensure that, asymptotically, the expectation value $\zeta = \langle\hat{\zeta}\rangle = k = k_B - k_A$ equals the charge state difference of the separated systems. Obviously, if an electron is to be removed from A and added to B (or vice versa) the process has to be initiated by a continuous change of the dipole moment.

A given asymptotic charge state difference k or, more generally, a given charge asymmetry ζ can be imposed upon a system by minimizing the constrained energy density functional $E^C[n]$ for a fixed external potential determined by Z_A, Z_B, R. This functional is obtained by adding the constraint $\int \hat{\zeta}\, n(\vec{r}) d^3r = \zeta$ with a Lagrange multiplier λ to the conventional[5,6] energy functional $E[n]$. Within the class of number-conserving variations of the density $n(\vec{r})$, the minimum requirement is hence expressed as

$$\delta E^c[n] = \delta\{ E[n] - \lambda \int \zeta\, n(\vec{r}) d^3r \} = 0. \tag{2}$$

For the prescribed dipole deformation ζ, the ground state energy of the system is then

$$E_\zeta = E_\zeta^c + \lambda_\zeta \int \zeta\, n_\zeta(\vec{r}) d^3r, \tag{3}$$

and the associated ground state density is $n_\zeta(\vec{r})$. The Lagrange multiplier

$$\lambda_\zeta = \partial E_\zeta / \partial \zeta \tag{4}$$

is interpreted as the slope of the energy curve with respect to ζ. Instead of the linear constraint in Eq.(2), one may also use other differentiable constraint functions[3]. In particular, nonlinear constraints are indispensible whenever there is no one-to-one correspondence between ζ and the slope of the curve. The energy surface $E(R,\zeta)$ as a function of the "collective" variables R and ζ is, however, independent of the method of its construction.

Since the slope $\lambda = F_c R/2$ is proportional to a constraint force F_c (or a constraining electric field), the second derivative of the energy with respect to ζ is simply the inverse longitudinal polarizability of the molecule

$$\alpha = [\,\partial^2 E(R,\zeta)/\partial\zeta^2]^{-1}\,(R/2)^2. \tag{5}$$

For a large but finite internuclear separation R, such that the subsystems do not overlap, the system energy $E_{AB}=E_A+E_B$ is easy to visualize: For each fragmentation ζ_0, this energy* has a local minimum as shown in the schematic illustration of Fig. 1. A variation of ζ in the vicinity of the stationary point ζ_0 describes a polarization of both separated atoms and thus raises the energy which hence may be expanded as

$$E(R,\zeta) = E^c(R,\zeta_0) + 1/2\,(\alpha_A+\alpha_B)^{-1}\,(R/2)^2\,(\zeta-\zeta_0)^2 + \ldots, \tag{6}$$

---

* Since for a given value of the continuous variable ζ the eigenvalue equation corresponding to the variational prescription of Eq. (2) defines a unique stationary eigenstate, the densities determined by Eq.(2) correspond to pure states and not to statistical mixtures or ensembles.

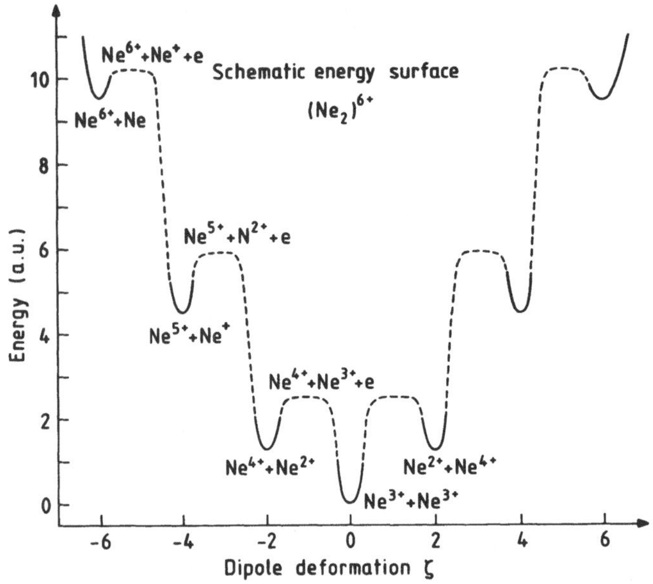

Fig. 1: Schematic potential energy surface for the system $(Ne_2)^{6+}$ as a function of the dipole deformation at large but finite internuclear separations R. Those parts of the curve shown as solid lines result as an expansion around the local minima at integral charge fragmentations, the dashed lines indicate the expected behavior between the minima. The positions of the minima and maxima reflect the correct system energies $E_{AB} = E_A + E_B$ with respect to the lowest minimum.

where use has been made of Eq.(5). With increasing distortion $|\zeta-\zeta_0|$, the energy must increase further until the ionization limit of one of the subsystems is reached (see Fig. 1). The charge state difference is then changed by one full unit from ζ_0. A further increase of $|\zeta-\zeta_0|$ will again lower the energy because the detached electron is then trapped by the other atom, such that $|\zeta-\zeta_0| = 2$.

Clearly, when R is also varied, the minima of Fig. 1 develop into valleys. Then Eq. (6) shows that for given polarizabilities α_A, α_B the valleys become narrower in ζ space (not in z space) with increasing R. This is a simple consequence of the definition (1) of $\hat{\zeta}$.

Energy surfaces for a number of neutral diatomic quasimolecules have been calculated[3,4] using a constrained Hartree-Fock method. As an example, Fig. 2 shows a contour map for the system HF. One clearly notes a valley for the atomic fragmentation H-F corresponding to $\zeta = 0$. The ionic valley for H^+-F^- is higher in energy and, owing to the long-range Coulomb force, has not yet reached the asymptotic value $\zeta = 2$ at R<8 a.u. The molecular bound state appears as a minimum at $R_e = 1.733$ a.u. and $\zeta_e = 0.882$. In fact, the quantity ζ_e is just twice the "fractional ionic character" of the molecular bond[7] and its

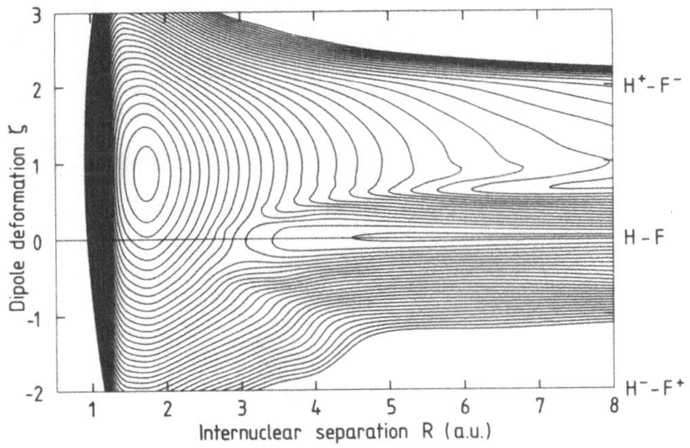

Fig. 2: Contour map of the potential energy surface (E, ζ) for HF, see Ref. 4. The lowest contour line around the equilibrium point at E_e = -100.0703 a.u. corresponds to the energy -100.06 a.u. Successive lines are separated by ΔE = 0.02 a.u.

rather large value clearly indicates a polar molecule. The energy surfaces E(R, ζ) thus at once exhibit the polar or covalent character of the molecule as well as its polarizability and its fragmentation channels.

In conclusion I would like to point out some further features of the constrained variational approach[2-4] discussed here: (i) The approach allows one to control the dissociation property of a molecule by simply imposing the appropriate dipole deformation. It thus remedies (e.g. in Hartree-Fock theory) a well-known defect of the unconstrained treatment. (ii) The same method with an obvious modification of the variable ζ can be applied if one of the atoms is replaced by a surface. (iii) It is possible to visualize and to calculate[8] elastic as well as charge-changing ion-atom collisions as a classical motion following trajectories in the potential energy surface. (iv) An extension to other properties of the molecule or to polyatomic molecules can be achieved by choosing a constraint operator suited to the geometry.

References

1. J.P. Perdew, R.G. Parr, M. Levy, and J.L. Balduz, Jr., Phys. Rev. Lett. 49, 1691 (1982)
2. J. Eichler, Phys. Rev. Lett. 40, 1560 (1978),
3. J. Eichler, Phys. Rev. Lett. 46, 1619 (1981)
4. J. Eichler and T.S. Ho, Z. Phys. A 311, 19 (1983)
5. P. Hohenberg and W. Kohn, Phys. Rev. 136, B 864 (1964)
6. W. Kohn and L.J. Sham, Phys. Rev. 140, A 1133 (1965)
7. R. McWeeny, Coulson's Valence, 3rd Edition (Oxford University Press, 1979), p. 153 ff.
8. J. Eichler and T.S. Ho, Z. Physik A 311, 29 (1983)

GRADIENT EXPANSIONS AND QUANTUM MECHANICAL EXTENSIONS OF THE CLASSICAL PHASE SPACE

Byron K. Jennings

TRIUMF
4004 Wesbrook Mall
Vancouver, B.C., Canada V6T 2A3

INTRODUCTION

One common feature of most derivations of the gradient expansion for the kinetic energy density is the use of a mock phase space in intermediate steps. By mock phase space I mean a space parametrized by two coordinates p and r which in the classical limit reduce to the classical phase space. Of course a real phase space can not be defined quantum mechanically since p and r can not be specified simultaneously. Also there are many different mock phase spaces of which two in particular are frequently used. We will discuss these two in detail later. I will begin by giving a brief derivation of the Thomas-Fermi relation between the kinetic energy density and the spatial density to motivate the use of phase space and then indicate how this must be generalized. I then will discuss the various quantum mechanical generalizations of the phase space and show using the Wigner transform as an example one method of obtaining an $\hbar$ expansion. More details can be found in a review article by N.L. Balazs and myself.[1]

MOCK PHASE SPACES

In the limit when $\hbar$ is small we can write the phase space density for fermions as

$$f_{cl}(p,r) = \Theta\left(\lambda - \frac{p^2}{2m} - V(r)\right) \tag{1}$$

where V(r) is a one-body potential. The density and kinetic energy density can be written as

$$\rho(r) = \int d^3p\ f_{cl}(p,r) \sim \left(\lambda - V(r)\right)^{3/2} \tag{2}$$

and

$$\tau(r) = \int d^3p \, \frac{p^2}{2m} f_{cl}(p,r) \sim \big(\lambda - V(r)\big)^{5/2} \tag{3}$$

respectively. From this the well known Thomas-Fermi result $\tau(r) \sim \rho(r)^{5/3}$ follows immediately. To extend this procedure to obtain the gradient expansion it is necessary to perform the following steps: 1) Define f(p,r) quantum mechanically; 2) Expand f(p,r) in powers of $\hbar$; 3) Compute $\rho\big(\lambda-V(r)\big)$ and $\tau\big(\lambda-V(r)\big)$ term by term in $\hbar$; and 4) Express τ as a functional of ρ by eliminating V.

Let me now discuss the first step, i.e. defining f(p,r). Quantum mechanically a probability is found by taking a trace. Thus the probability for finding a particle at a point r is

$$\Pr(r) = \mathrm{Tr}\ \delta(r-\hat{r})\hat{\rho} = \langle r|\hat{\rho}|r\rangle \tag{4}$$

where $\hat{\rho}$ is one body density matrix and the ^ denotes operators. Similarly for the momentum distribution

$$\Pr(p) = \mathrm{Tr}\ \delta(p-\hat{p})\hat{\rho} = \langle p|\hat{\rho}|p\rangle \tag{5}$$

The most naive method of attempting to define a joint probability is then

$$f_s(p,r) = \mathrm{Tr}\ \delta(p-\hat{p})\,\delta(r-\hat{r})\hat{\rho} = \langle r|\hat{\rho}|p\rangle\langle p|r\rangle \tag{6}$$

and most other putative quantum mechanical phase space densities are simple generalizations of this obtained through different orderings of the $\hat{p}$ and $\hat{r}$ operators. When $\hat{\rho} = \theta(\lambda-\hat{H})$ the $f_s(p,r)$ defined by Eq. (6) is just the function used by E.K.U. Gross[2] in his lecture on the Kirzhnits method.[3] He wrote the density as

$$\rho \propto \int d^3p\ \langle r|\,\theta(\lambda-\hat{H})\,|p\rangle\langle p|r\rangle \tag{7}$$

and did an $\hbar$ expansion on the integrand. In the Wigner-Kirkwood expansion[4] one works with the Laplace transform of $f_s(p,r)$ (see Ref. 1 for a discussion of the Wigner-Kirkwood expansion from this point of view). Note that the $f_s(p,r)$ defined by Eq. (6) is <u>not</u> the Wigner distribution function.

The delta functions in Eq. (6) can be written through the Fourier transform. Thus we have

$$\begin{aligned}\hat{\Delta}_s(p,r) &= \delta(p-\hat{p})\,\delta(r-\hat{r})\\ &= \frac{1}{(2\pi)^6}\int d^3u\,d^3v\ \exp\big(iu\cdot(p-\hat{p})+iv\cdot(r-\hat{r})\big)\\ &= \frac{1}{(2\pi)^6}\int d^3u\,d^3v\ \exp\big(iu\cdot(p-\hat{p})+iv\cdot(r-\hat{r})\big)\exp\Big(\frac{i\hbar u\cdot v}{2}\Big)\,.\end{aligned} \tag{8}$$

In the last line the phase $i\hbar u\cdot v/2$ has come from combining the two exponentials. We can define a new $\hat{\Delta}$ by dropping the phase. Thus we have

$$\hat{\Delta}_w = \frac{1}{(2\pi)^6} \int d^3u d^3v \exp\big(iu\cdot(p-\hat{p})+iv\cdot(r-\hat{r})\big) \; . \tag{9}$$

This new $\hat{\Delta}$ differs from $\hat{\Delta}_s$ just in the ordering of the $\hat{p}$ and $\hat{r}$ operators. It is Hermitian and symmetric in $\hat{p}$ and $\hat{r}$. We can now define a new distribution function

$$f_w(p,r) = \mathrm{Tr}\,\hat{\Delta}_w(p,r)\hat{\rho} \; . \tag{10}$$

This is the Wigner distribution function. Given either $f_w(p,r)$ or $f_s(p,r)$ the full off diagonal density matrix may be easily reconstructed. While $f_s(p,r)$ is in general complex $f_w(p,r)$ is real although it can be negative. Thus neither f_s or f_w is a true probability. In many cases $f_s(p,r)$ can be more easily calculated than $f_w(p,r)$. For example the lack of symmetry between p and r in $f_s(p,r)$ is exploited[1] in deriving the Wigner-Kirkwood expansion.[4,1]

The two f's are related by

$$f_w(p,r) = \exp \frac{i\hbar\nabla_p\cdot\nabla_r}{2} \; f_s(p,r) \; . \tag{11}$$

From this equation it should not be concluded that $f_w(p,r)$ and $f_s(p,r)$ are equal when $\hbar\to 0$ since both may be rapidly oscillating so the gradients in effect go like $1/\hbar$. However, if there is averaging the two will become the same.

For any operator $\hat{\Theta}$ it is possible to define a function

$$O_s(p,r) = \mathrm{Tr}\; \hat{\Delta}_s(p,r)\hat{\Theta} \tag{12}$$

and thus with each operator associate a function of p and r. (One can proceed similarly with $\hat{\Delta}_w$.) The point of this is that all of quantum mechanics can be recast in terms of these associated functions rather than the original operators.[1]

I have by no means exhausted the possible choices for $\hat{\Delta}$. For almost every ordering it is possible to generate a new $\hat{\Delta}$. Thus there is a tremendous freedom which can be exploited in the solution of individual problems.

THE $\hbar$ EXPANSION

In this section I will present a method of obtaining the $\hbar$ expansion first given by Voros.[5] In this method the density matrix $\hat{\rho} = \Theta(\lambda-\hat{H})$ is formally expanded about a c-number. Thus we have

$$\hat{\rho} = \Theta(E-\hat{H}) = \Theta\left(E-\mathcal{H}+(\mathcal{H}-\hat{H})\right)$$
$$= \Theta(E-\mathcal{H}) + (\mathcal{H}-\hat{H}) \frac{\partial}{\partial E} \Theta(E-\mathcal{H})$$
$$+ \frac{1}{2} (\mathcal{H}-\hat{H})^2 \frac{\partial^2}{\partial E^2} \Theta(E-\mathcal{H}) + \ldots \tag{13}$$

The Wigner transform is now taken term by term. Thus

$$f_w(p,r) = \mathrm{Tr}\hat{\Delta}_w\, \Theta(E-\hat{H})$$
$$= \Theta(E-\mathcal{H})\ \mathrm{Tr}\hat{\Delta} + \delta(E-\mathcal{H})\ \mathrm{Tr}\{\hat{\Delta}(\mathcal{H}-\hat{H})\}$$
$$+ \frac{1}{2}\ \delta'(E-\mathcal{H})\ \mathrm{Tr}\{\hat{\Delta}(\mathcal{H}-\hat{H})^2\} + \ldots \tag{14}$$

I could equally well have used $\hat{\Delta}_s$ and $f_s(p,r)$ rather than $\hat{\Delta}_w$ and $f_s(p,r)$. The latter was chosen simply for the sake of definiteness. To evaluate Eq. (14) it is only necessary to calculate $\mathrm{Tr}\ \hat{\Delta}(\hat{H})^N$ for various values of N. The $\hbar$ corrections arise because

$$(\mathrm{Tr}\ \hat{\Delta}\hat{H})^N \neq \mathrm{Tr}\ \hat{\Delta}(\hat{H})^N . \tag{15}$$

Now we must decide how to choose $\mathcal{H}$. One possibility is $\mathcal{H} = \mathrm{Tr}\ \hat{\Delta}\hat{H}$. Thus a different $\mathcal{H}$ is chosen for each point in the p,r plane i.e. $\mathcal{H}$ is a function of p and r. For the normal Hamiltonian $\hat{H} = \hat{p}^2/2m + V(\hat{r})$, $\mathcal{H}$ is independent of $\hbar$ and the $\hbar$ expansion of $f_w(p,r)$ is obtained by rearranging terms in Eq. (14). Thus we have

$$f_w(p,r) = \Theta\left(E-\mathcal{H}(p,r)\right) + \hbar^2\left\{-\delta'(E-\mathcal{H})\ \frac{\nabla^2 V}{8m}\right.$$
$$\left. + \delta''(E-\mathcal{H})\left[\frac{(\nabla V)^2}{24m} + \frac{(p\cdot\nabla)^2 V}{24m^2}\right]\right\}$$
$$+ \Theta(\hbar^4) . \tag{16}$$

Note that in Eq. (16) there is no term proportional to $\hbar$ and $f_w(p,r)$ is real. If we had used $f_s(p,r)$ instead there would have been an imaginary term proportional to $\hbar$. The above procedure is easily extended to more complex Hamiltonians.[6]

The rest of the procedure in obtaining the gradient expansion for the kinetic energy is rather tedious and not particularly enlightening so I will not go through it here. It is discussed in Refs. 1, 2, 3 and 6.

ACKNOWLEDGEMENTS

I would like to thank the organizers of this conference for asking me to give this contribution and the Natural Sciences and Engineering Research Council of Canada for financial support.

REFERENCES

1. N.L. Balazs and B.K. Jennings, TRIUMF preprint TRI-PP-83-55 (submitted to Physics Reports).
2. E.K.U. Gross, this proceedings.
3. D.A. Kirzhnits, Field Theoretical Methods in Many-Body Systems, Pergamon, Oxford (1967).
4. J.G. Kirkwood, Phys. Rev. 44:31 (1933).
5. A. Voros, Doctoral thesis, ORSAY (1977), unpublished.
6. B. Grammaticos and A. Voros, Ann. Phys. (N.Y.) 123:359 (1979).

POSTER SESSION

A. List of Posters

G.E.W. Bauer
Hahn-Meitner-Institut, Berlin
Electronic densities in the HKS formalism

P. Blaha and J. Redinger
Institut für Technische Elektrochemie, Wien
Electron densities in metals and insulators
(a test of the validity of the LDA)

E. Engel
Institut für Theoretische Physik, Frankfurt/M
Gradient corrections to the relativistic exchange potential

S.R. Gadre
University of Poona, Pune
A density functional model for atoms in momentum space

P. Hedegard
Institute of Physics, Aarhus
Auger electron energy shifts for s-d transition elements

A. Henne
Institut für Theoretische Physik, Frankfurt/M
Time-dependent TF description of ion-atom scattering

D. Kolb
Institut für Theoretische Physik, Kassel
Energy conserving density matrix expansion

P. Malzacher
Institut für Theoretische Physik, Frankfurt/M
Photoabsorption of TFDW-atoms

F.A. Parpia
Department of Physics, Notre Dame
The relativistic, time-dependent LDA

M. Puska
Laboratory of Physics, Helsinki
Atoms inbedded in an electron gas

A. Scheidemann
Institut für Theoretische Physik, Frankfurt/M
A parametrisation of the TFDW screening function

M.J. Stott
Department of Physics, Queens University, Kingston
Developments in quasi-atom theory

A. Toepfer
Institut für Theoretische Physik, Frankfurt/M
Density functional description of effective screening
in ion-atom collisions

G. Zumbach
Institute de Physique thèorique, Ecole Polytechnique, Lausanne
a. A new approach to the calculation of density functional
b. Density theory for the N-particle ground state

POSTER SESSION

B. Abstracts of Posters

DENSITY-FUNCTIONAL EXCHANGE-CORRELATION POTENTIALS AND ORBITAL EIGENVALUES FOR LIGHT ATOMS

C.-O. Almbladh and A.C. Pedroza

Department of Theoretical Physics
University of Lund, Sölvegatan 14A
S - 223 62 Lund, Sweden

Using accurate correlated wavefunctions calculated earlier by Bunge and by Larsson, we have constructed the Hohenberg-Kohn-Sham density functionals and exchange-correlation (ground-state) potentials, and obtained orbital energy eigenvalues for a number of light atoms by in principle exact numerical algorithms. While the uppermost occupied density-functional eigenvalue always gives an exact excitation energy as has been shown earlier, we find that eigenvalues for deeper shells lie above the corresponding excitation energy. We have compared our essentially exact density-functional (DF) results with those obtained in the local-density (LD) approximation. We find that the LD theory approximates the exchange-correlation energy rather well, but that it gives larger errors in the exchange-correlation potential and in the DF orbital eigenvalues. In all cases we have found that the LD error in the orbital eigenvalue is larger that the difference between the true DF eigenvalue and the corresponding exact excitation energy. Possible implications of these results for solid state work are briefly discussed.

REFERENCES

C.-O. Almbladh and A.C. Pedroza, Phys. Rev. A29:2322 (1984).

ELECTRONEGATIVITY OF ATOMS, POSITIVE IONS AND FRACTIONALLY CHARGED ATOMS IN THE DENSITY FUNCTIONAL THEORY

L.C. Balbas[1] and J.A. Alonso[2]

Departamento de Optica[1]
Departamento de Fisica Teórica[2]
Universidad de Valladolid
Valladolid, Spain

The electronegativity ϕ of an atom or ion with nuclear charge Z and N_0 electrons (identified with the negative of the chemical potential μ of the Density Functional Theory (DFT))

$$\phi(Z,N_0) = -\left[\frac{\partial E(Z,N)}{\partial N} \right]_{N=N_0} = -\mu \;, \tag{1}$$

has been studied using the following approximate energy functional

$$E(Z,n) = \frac{1}{2} \iint \frac{\rho(\vec{r})\rho(\vec{r}\,')}{|\vec{r} - \vec{r}\,'|} d\vec{r}\, d\vec{r}\,' + \int V_c(\vec{r})\rho(\vec{r})d\vec{r}$$

$$+\int \left[\frac{(\nabla\rho(\vec{r}))^2}{8\rho(\vec{r})} + \frac{3}{10}(3\pi^2)^{2/3}\rho(\vec{r})^{5/3} \right] d\vec{r} + \int \left[-\frac{3}{4}\left(\frac{3}{\pi}\right)^{1/3}\rho(\vec{r})^{4/3} - \frac{0.44\,\rho(\vec{r})}{7.8 + \left(\frac{4\pi\,\rho(\vec{r})}{3}\right)^{-1/3}} \right] d\vec{r} \;. \tag{2}$$

Only the n valence electrons have been taken into account and the effect of the ionic core was simulated by an empty core pseudopotential $V_c(\vec{r})$.

From (2), plus the normalization condition for $\rho(\vec{r})$, the following equation is obtained for the chemical potential

$$\mu = V + \frac{1}{2}(3\pi^2)^{2/3}\rho^{2/3} + \frac{(\nabla\rho)^2}{8\rho^2} - \frac{\nabla^2\rho}{4\rho} - \left(\frac{3\rho}{\pi}\right)^{1/3}$$

$$- \frac{3.43+0.36\,\rho^{-1/3}}{[7.8+(\frac{3}{4\pi\rho})^{1/3}]^2} \quad , \qquad (3)$$

where $V(\vec{r}) = V_c(\vec{r}) + \int \frac{\rho(\vec{r}\,')}{|\vec{r} - \vec{r}\,'|} d\vec{r}\,'$, is the total electrostatic potential. Eqn. (4) is then solved coupled with Poisson's equation

$$\nabla^2 V(\vec{r}) = -4\pi\rho(\vec{r}), \qquad (4)$$

which guarantes the selfconsistency between the electron density and the electrostatic potential.

We have used a set of core radii, R_c, fitted by requiring that the calculated valence electron energy $E(Z,n_a)$ of the neutral atom equals the experimental energy, that is, the sum of the n_a first ionization potentials (n_a is the number of valence electrons in the neutral atom).

Calculations have been performed for atoms and ions of the Lithium, Berillium, Boron, Carbon, Nitrogen, Oxygen and Fluorine groups. The relation between Mulliken's definition of electronegativity

$$\phi_M(Z,N_o) = \frac{1}{2}[E(Z,N_o-1) - E(Z,N_o+1)] \quad , \qquad (5)$$

and the exact electronegativity (eqn.(1)), has been studied[1] using the present DFT, that is, evaluating $\phi_M(Z,N_o)$ ($N_o < n_a-1$) from the energies $E(Z,N_o-1)$ and $E(Z,N_o+1)$ obtained from eqns. (2-4). The relative deviation $(\phi_M-\phi)/\phi$ decreases as the degree of ionization increases. An extrapolation of this trend suggests significant deviations (perhaps larger than 10 per cent) between ϕ and ϕ_M for the neutral atoms[2].

From the electronegativities of atoms and ions calculated in several isoelectronic series, the electronegativities of atoms and ions with fractional nuclear charge (Quark atoms) have been obtained by quadratic interpolation[3].

REFERENCES

1. L.C. Balbás, E.Las Heras,J.A. Alonso, Z. Phys. A 305:31 (1982).
2. L.C. Balbás, J. Alonso, E. Las Heras, Molec. Phys. 48:981 (1983).
3. L.C. Balbás, J.A. Alonso, L.M. del Rio, Z. Phys. A 312:95 (1983).

LOCAL EXCHANGE-CORRELATION APPROXIMATIONS

AND FIRST-ROW MOLECULAR DISSOCIATION ENERGIES

A.D. Becke

Department of Chemistry, Dalhousie University
Halifax, Nova Scotia
Canada B3H 4J3

Recently, we have reported the results of accurate Hartree-Fock-Slater calculations on selected first-row diatomic molecules which show very encouraging agreement with experiment*. In the worst cases, however, the HFS dissociation energies were found to overestimate the experimental values. Therefore, we have examined several refinements of the Hartree-Fock-Slater theory and their effects on molecular bond lenghts, dissociation energies, and vibrational frequencies. Among them, gradient corrections to the HFS exchange energy and also some local correlation approximations based on recent parametrizations of electron gas data are considered. We find that a local exchange-correlation approximation with gradient corrections gives dissociation energies in significantly better agreement with experiment than the Hartree-Fock-Slater approximation.

---

* A.D. Becke, J. Chem. Phys. 76:6037(1982).

A SEMIQUANTAL APPROACH TO ATOMIC GROUND STATE ENERGIES AND DENSITIES

M. Brack, Regensburg University, Regensburg, W-Germany
R.K. Bhaduri, McMaster University, Hamilton, Ont., Canada
P. Schuck, I.S.N., Grenoble, France

We have recently proposed[1] a parameter-free semiquantal approach to atomic binding energies, starting from the single-particle Bloch density matrix in the form[2] (C_0 is the plane wave solution)

$$C(r,r';\beta) = C_0(r,r';\beta)\cdot\exp\left[-\beta U(r,r';\beta)\right], \tag{1}$$

from which ground-state properties are derived in the usual way.[3] For electrons in a Coulomb potential $V(r)=-\alpha/r$, one can expand U in a perturbation series $U=U_1+U_2+\ldots$, where $U_n \propto \alpha^n$. U_1 is known analytically[1] for r=r' and finite at r=0, thus avoiding the well-known singularity of TF theory. U_2 can be obtained in a gradient expansion in terms of U_1.[1] Adding the e-e-potential to U_1, we reproduce without any adjustable parameter the exact atomic HF energies within 3% and obtain at the same time good densities which fulfill the Kato theorem.[4]

We want to report here that for a screened Coulomb potential of Yukawa form, $V(r)=-\alpha\exp(-\mu r)/r$, the exact $U_1(r,r;\beta)$ can also be given:

$$U_1(r,r;\beta) = -\alpha\sqrt{\pi/2\beta}\exp(\beta\mu^2/8)\left\{\left[\exp(\mu r)-\exp(-\mu r)\right] + 2\exp(-\mu r)\,\mathrm{erf}(\mu\sqrt{\beta/8}) + \exp(-\mu r)\,\mathrm{erf}(\sqrt{2/\beta}\,r-\mu\sqrt{\beta/8}) - \exp(\mu r)\,\mathrm{erf}(\sqrt{2/\beta}\,r+\mu\sqrt{\beta/8})\right\}/\mu r. \tag{2}$$

This allows to approximate the full HF potential by a superposition of Yukawas: $V(r)=\Sigma_i\alpha_i\exp(-\mu_i r)/r$. The full U_1 is then just the corresponding sum of U_1 in eq.(2). Calculating U_2 as in ref.[1] necessitates a trivial numerical integration over β. The total energy of the atom is then obtained variationally by minimizing with respect to the α_i,μ_i. This approach is straightforwardly applied to multicenter molecules.

REFERENCES

1. R.K.Bhaduri, M.Brack, H.Gräf and P.Schuck, J.Physique 41:L347(1980)
2. see N.H.March, Phys.Lett. 64A:185(1977) and refs. quoted therein
3. see, e.g., M.Brack, these Proceedings
4. T.Kato, Commun. Pure Appl. Math. 10:151(1951); see also E.Steiner, J.Chem.Phys. 39:2365(1963)

CALCULATION OF CORE EXCHANGE POLARIZATION FOR FREE ALKALI ATOMS THROUGH QUASI-RELATIVISTIC LOCAL DENSITY APPROXIMATIONS

H. Chermette

Institut de Physique Nucléaire (et IN2P3)
Université Claude Bernard Lyon I
69622 - Villeurbanne Cedex (France)

Core exchange effects in free atoms have been examined through the determination of the spin density at the nucleus of (^2S) Li, Na, Rb and Cs atoms. The use of relativistic corrections and of different local spin density approximations is evaluated in both restricted and spin polarized self-consistent-field wave functions. Simultaneously the incidence of relativistic corrections on the electronic density at the nucleus is investigated.

APPLICATION OF A SIMPLIFIED SELF-INTERACTION CORRECTION TO INSULATORS AND TRANSITION METALS

M.R. Norman and J.P. Perdew

Dept. of Physics and Quantum Theory Group
Tulane University
New Orleans, La. 70118

A size-consistent simplified self-interaction correction (SSIC)[1] is applied to solids. The band gaps of the insulators Neon and Sodium Chloride with the inclusion of SSIC become 20.2eV (experimentally 21.4eV) and 9.2eV (experimentally 9.0eV) compared to the local density values of 11.5eV and 5.6eV respectively.[2] Thus we find a large improvement over the local density results. When applied to the transition metals Copper and Zinc, it is found to be necessary to screen the SSIC to obtain reasonable results since the hole left in a photoemission experiment suffers a large relaxation shift due to metallic screening. An approximate screened calculation results in d band widths and d band positions that are a large improvement over their local density counterparts as compared to photoemission data.[3] We conclude from this work that in the initial state before photoemission, the d states in Copper are delocalized, but those in Zinc are localized below the 4s-4p band. In the final state, we interpret the d holes in both cases as semi-localized, screened holes propagating through the lattice.

REFERENCES

1. J.P. Perdew and M.R. Norman, Phys. Rev. B 26: 5445 (1982).
2. M.R. Norman and J.P. Perdew, Phys. Rev. B 28: 2135 (1983).
3. M.R. Norman, submitted, Phys. Rev. B.

A KOHN-SHAM FUNCTIONAL FOR ATOMS

M.R. Norman and D.D. Koelling

Material Sciences Division
Argonne National Laboratory
Argonne, Ill. 60439

Recently, Talman and Shadwick[1] proposed a method for obtaining an optimized, local, effective potential from the Hartree-Fock Hamiltonian and applied this technique to atoms. We apply the same technique[2], but to the self-interaction corrected local spin density (SIC-LSD) functional.[3] This yields a state-independent potential which is self-interaction free, and thus should be a good approximation to the exact Kohn-Sham functional. Applications to atoms yield potentials similar to those of Talman.The highest occupied eigenvalues agree with those of SIC-LSD and thus are a good approximation to the ionization potentials. The core eigenvalues are higher than SIC-LSD, though, because of state mixing in the Talman equations. The lowest unoccupied eigenvalues (a measure of the electron affinity) are too low since these states should have zero SIC, but the effective potential has SIC built into it. To conclude, an energy (state)-independent potential cannot yield all the eigenvalues to agree with experimental removal energies. As a consequence, the band gap in insulators will not be given by the exact Kohn-Sham density functional.

REFERENCES

1. J.D. Talman and W.F. Shadwick, Phys. Rev. A 14: 36 (1976).
2. M.R. Norman and D.D. Koelling, unpublished.
3. J.P. Perdew and A. Zunger, Phys. Rev. B 23: 5048 (1981).

VARIATIONAL APPROACH TO NUCLEAR FLUID DYNAMICS
ZERO SOUND IN FINITE FERMI SYSTEMS WITH INCLUSION
OF LOCAL EQUILIBRIUM DEFORMATIONS

João P. da Providência
Departamento de Fisica
Universidade de Coimbra
3000 Coimbra, Portugal

A variational derivation of a fluid-dynamical formalism, which is based on a single determinant as variational function and includes the possibility of transverse flow, is presented. Therefore the explicit specification of the time-odd part of the operator i F generating the collective modes has to go beyond the local approximation χ while the time-even part is restricted to the generalized scaling form. We allow this operator to act on a state $|\Phi_f\rangle$ which already includes local equilibrium deformation and is therefore associated to a density matrix

$$f_f = \Theta(\lambda - \frac{p^2}{2m} - U_o - W(\vec{r}))$$

where U_o is the ground state selfconsistent potential and $W(\vec{r})$ is a small quantity.

When we restrict our approach to the classical limit the equilibrium density is a step function, $\rho_o(r) = \rho_o(0)\Theta(R-r)$ where $\rho_o(0)$ is the infinite matter equilibrium density and R is the nuclear radius.

In the harmonic approximation the distribution function may be written in the classical limit as

$$f = f_f + \{F, f_f\} + \frac{1}{2}\{F, \{F, f_o\}\}$$

where

$$F = \chi(\vec{r},t) + p_\alpha s_\alpha(\vec{r},t) + \frac{1}{2} p_\alpha p_\beta \phi_{\alpha\beta}(\vec{r},t)$$

and f_o is the ground state distribution function.

The density is

$$\rho = g \int \frac{d^3p}{(2\pi\hbar)^3} f \simeq \rho_f + \vec{\nabla}\cdot(\rho_o \vec{s})$$

and the current is

$$j_\alpha = g \int \frac{d^3p}{(2\pi\hbar)^3} f \frac{p_\alpha}{m} = \rho_o\{\partial_\alpha \chi + \frac{p_F^2}{5} \left(\frac{1}{2} \partial_\alpha \phi_{\beta\beta} + \partial_\beta \phi_{\alpha\beta}\right)\}$$

where the boundary condition $x_\alpha \phi_{\alpha\beta}\Big|_{r=R} = 0$ has been imposed at the nuclear surface in order to avoid a pure surface current.

The quantum mechanical Lagrangian $L = i\hbar \langle\Phi|\dot{\Phi}\rangle - \langle\Phi|H|\Phi\rangle$ leads in the classical limit to the Lagrangian

$$L^{(2)} = \int d^3r\{-\rho_f^{(1)} (\dot{\chi} + \frac{p_F^2}{6} \dot{\phi}_{\alpha\alpha}) + \rho_o s_\alpha [\partial_\alpha \dot{\chi} + \frac{p_F^2}{5} (\frac{1}{2} \partial_\alpha \phi_{\beta\beta} + \partial_\beta \phi_{\alpha\beta})]$$

$$- \frac{\rho_o}{2m} [(\vec{\nabla}\chi)\cdot(\vec{\nabla}\chi) + \frac{2}{5} p_F^2 (\partial_\alpha \chi) (\frac{1}{2} \partial_\alpha \phi_{\beta\beta} + \partial_\beta \phi_{\alpha\beta})$$

$$+ \frac{p_F^4}{35} ((\frac{1}{2}\partial_\alpha \phi_{\beta\beta} + \partial_\beta \phi_{\alpha\beta})^2 + \frac{1}{6}(\partial_\alpha \phi_{\beta\gamma} + \partial_\beta \phi_{\gamma\alpha} + \partial_\gamma \phi_{\alpha\beta})^2)]\} - E^{(2)}[\rho_f^{(1)}, \vec{s}]$$

with $E^{(2)}[\rho_f^{(1)}, \vec{s}] = \int d^3r\{(\frac{\rho_o p_F^2}{6m} + \sum_\sigma a_\sigma \frac{\sigma(\sigma-1)}{2} \rho_o^\sigma)(\rho_f^{(1)} + \rho_o \vec{\nabla}\cdot\vec{s})$

$$+ \frac{p_F^2 \rho_o}{10m} [- \frac{2}{3} (\vec{\nabla}\cdot\vec{s})^2 + \frac{1}{2} (\partial_i s_j + \partial_j s_i)^2]\}$$

and $\rho_f^{(1)} = \rho_f - \rho_o$. The Euler-Lagrange equations corresponding to this Lagrangian describe collective oscillations of the system.

DENSITY FUNCTIONAL CALCULATION OF THE GROUND-STATE PROPERTIES OF GRAPHITE *

N.A.W. Holzwarth
Department of Physics
Wake Forest University
Winston-Salem
North Carolina 27106

Steven G. Louie
Department of Physics
University of California
Berkeley
California 94720

Sohrab Rabii
University of Pennsylvania
The Moore School of Electrical Engineering
Philadelphia, Pennsylvania 19104

The energy-band structure of graphite is calculated using an *ab-initio* self-consistent pseudopotential technique[1], within the density functional theory[2]. The local density approximation is used to represent the exchange-correlation potential[3]and the wave functions are represented by a mixed-basis set consisting of plane waves and localized orbitals[4]. The calculated valence-charge density is in excellent agreement with that obtained from high-resolution data on x-ray diffraction from graphite[5]. In particular, our calculation accurately reproduces the double-humped structure in the C-C bond, observed in the x-ray diffraction. The energy band structure is also in good agreement with previous calculations as well as with experimental results.

* This work was supported by National science Foundation Materials Research Grant No. DMR79-23647, Army Research Office Contract DAAD-29-77-C400, and National Science Foundation Grant No.DMR78-22465. One of us (S.G.L.) would like to acknowledge support from a Sloan Foundation Fellowship.

References:

1. D.R. Hamann, M. Schlüter and C. Chiang, Phys. Rev. Lett. 43:1494 (1979).
2. P. Hohenberg and W. Kohn, Phys. Rev. 136:B864 (1964); W. Kohn and L.J. Sham, ibid. 140:A1133 (1965).
3. L. Hedin and B.J. Lundqvist, J. Phys. C4:2064 (1971).
4. S.G. Louie, K.M. Ho and M.L. Cohen, Phys. Rev. B19:1774 (1979).
5. R. Chen, P. Trucano and R.F. Stewart, Acta Crystallogr. Sec. A 33:823 (1979).

LSD CORRELATION IN ATOMS AND MOLECULES:

IONIZATIONS AND EXCITATIONS WITH SPIN FLIP

A. Savin, H. Stoll and H. Preuss

Institut für Theoretische Chemie, Universität Stuttgart
Pfaffenwaldring 55, D-7ooo Stuttgart 8o, West Germany

The Hartree-Fock method predicts an incorrect ground state for the CH molecule. This error is removed by including correlation through local spin-density functionals. The contributions of the correlation energy to the ionization potentials and to the energies of excitation with spin flip are correctly reproduced with LSD also for other systems (with six, seven and eight electrons).

THOMAS-FERMI-DIRAC-LIKE DESCRIPTION OF ELECTRON CORRELATION

M. Tönhardt and A.M.K. Müller

Institut für Theoretische Physik
Technische Universität
D-3300 Braunschweig

Institut für Mathematische Physik

1. TWO-PARTICLE DENSITY FOR INTERACTING SYSTEMS

An interacting N-Electron system is described by the one-body density matrix d_1 and the two-body density n_2. For non-interacting systems, it is known from Hartree-Fock theory that

$$n_2(x_1,x_2) = n(x_1)\cdot n(x_2) - d_1(x_2;x_1)\cdot d_1(x_1;x_2) \tag{1}$$

with occupation numbers $\nu_i=1$ or $\nu_i=0$ in

$$d_1(x_1';x_1) = \Sigma\ \nu_i\chi_i(x_1')\chi_i(x_1), \tag{2}$$

χ_i being the natural orbitals. We extend the expression (1) by defining the "square root of d_1"

$$D_1(x_1';x_1) := \Sigma\ \sqrt{\nu_i}\ \chi_i(x_1')\chi_i(x_1) \tag{3}$$

and we write

$$\tilde{n}_2(x_1,x_2) = n(x_1)\cdot n(x_2) - D_1(x_2;x_1)\cdot D_1(x_1;x_2). \tag{4}$$

Now the normalization condition for the two-body density holds for broken occupation numbers, too. However, the Pauli principle is violated, since $\tilde{n}_2(x_1,x_1) \leq 0$.
The N-representability is destroyed: there is no lower bound for the ground state energy; the approximate value may be lower than the exact value. We hope that $\tilde{n}_2$ and n_2 don't differ significantly, so we calculate expectation values with $\tilde{n}_2 \simeq n_2$.

2. DENSITY DEPENDENCE OF THE ENERGY FUNCTIONAL E

(For the following, summation over spin coordinates has been carried out).

The natural orbitals are eliminated by connecting d_1 and D_1 with a one-body Hamiltonian $H_V = T + V$, V local, that has the natural orbitals as eigenfunctions, but eigenvalues ε_i. Choose a real function f(s), so that $f(\varepsilon_i) = \nu_i$. Then,

$$f(H_V(\vec{r}))\ \chi_i(\vec{r}) = f(\varepsilon_i)\ \chi_i(\vec{r}) = \nu_i \chi_i(\vec{r}) \tag{5}$$

and, by using the closure relation,

$$\begin{aligned} d_1(\vec{r}\,';\vec{r}) &= \Sigma\ f(H_V)\ \chi_i(\vec{r}\,')\ \chi_i(\vec{r}) = f(H_V)\ \delta(\vec{r}\,'-\vec{r}) \\ D_1(\vec{r}\,';\vec{r}) &= \Sigma\sqrt{f(H_V)}\ \chi_i(\vec{r}\,')\ \chi_i(\vec{r}) = \sqrt{f(H_V)}\ \delta\vec{r}\,'-\vec{r}) \end{aligned} \tag{6}$$

are now functionals of f and V, which have to be optimized to get E minimal. For simplifying the evaluation, the statistical approximation [T,V] = 0 is used.

3. OPTIMIZATION

Subject to the constraint that n[f,V] equals a given $n(\vec{r})$ which conserves particle number, we get back the results of the Hartree-Fock case, if we treat non-interacting systems. The energy functional is that of Thomas-Fermi-Dirac theory.
The requirement that this shall be included in the treatment of interacting systems, suggests the form of fig. 1 for f.

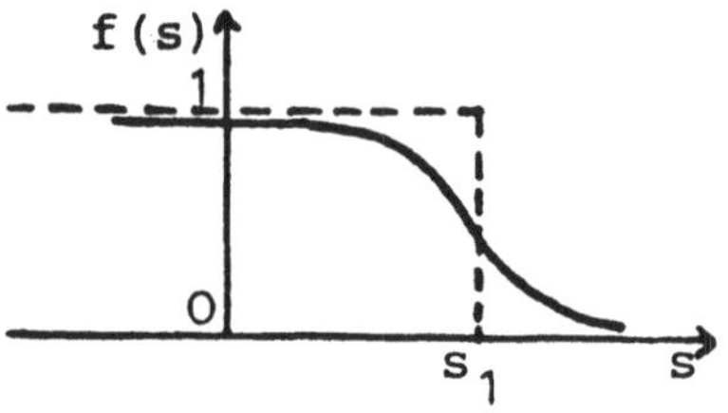

fig. 1

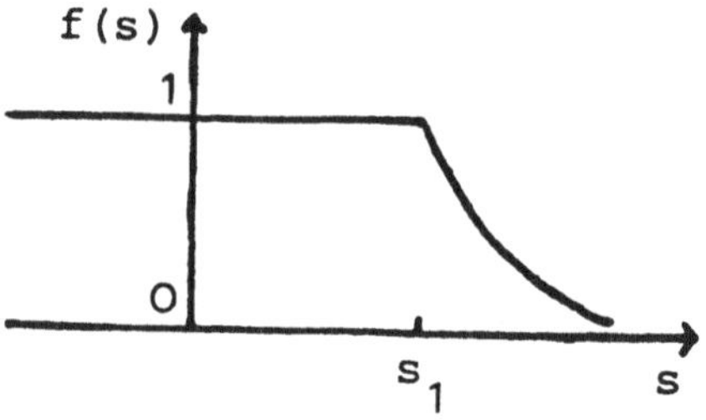

fig. 2

Optimization of V ensures already that the energy may be written as a density functional. For high particle densities, we can show that the optimal f(s) is that of fig. 2. Variation of the particle density in the optimized energy functional E[n] gives a Thomas-Fermi-Dirac like equation, the constants in the expressions of the kinetic and exchange energy now being functionals of the particle density.

4. RESULTS

For high particle densities, the correlation energy is given by

$$E_{corr} = -\frac{e^2}{a_o}\frac{3}{2\pi^2} \cdot N. \qquad (7)$$

These values may be added to those of SCF-calculations. They are compared with the results of Tong and Sham [Phys. Rev. 144, 1-4(1966)].

Atom	Correlation energy [Ry]		Total energy [Ry]		
	Tong	TFD-like	Experiment	Tong	TFD-like
He	-0.20	-0.61	-5.8	-5.7	-6.1
Ne	-1.36	-3.04	-257.9	-256.3	-258.0
Ar	-2.68	-5.47	-1055.2	-1051.7	-1054.5
Kr	-6.26	-10.94	?	-5499.9	-5504.6
Li	-0.31	-0.91	-15.0	-14.7	-15.3
Na	-1.51	-3.34	-324.5	-322.8	-324.6
K	-2.83	-5.78	-1200.0	-1196.2	-1199.2
Rb	-6.43	-11.25	?	-5872.3	-5876.9
O	-1.01	-2.43	-150.0	-148.9	-150.3

INDEX

Action integral, 82, 89, 385
atomic properties, 106, 197
atomic scattering, 106, 114

Bag model, 323
Banach space, 40
band gap, 7, 219, 297ff
band structure, 252, 259, 266, 272, 297
BCS approximation, 334, 430
binding energies, 170
    of molecules, 189
    of nuclei, 332, 355, 427
Bloch density, 329, 333
Bloch equation, 431
Boltzmann equation, 402
bond length, 192
Born-Oppenheimer approximation, 160, 163, 168
bounds
    for excited states, 22ff
    lower for Coulomb repulsion, 71
    lower for kinetic energy, 70
    upper for groundstate energy, 13, 50
    upper for kinetic energy, 70
    variational principle for upper, 72
breathing mode, nuclear, 374, 428
Breit interaction, 120
Brueckner
    G-matrix, 332, 368

Cascade model, 314, 402
charge asymmetry, diatomic, 498
chemical potential, 143, 146ff, 225, 233, 236, 265, 271, 276ff
chemisorption, 252
circulant orbitals, 151
collision term, 403
commutator expansion, 96
compressibility, atomic, 153, 278
compressibility, nuclear, 460
concave envelope, 48, 67
concavity, 41, 67
condensation energy, nuclear, 431
configuration interaction, 177, 201
confinement, 317ff
constrained search method, 3, 7 12ff, 26, 152, 167, 171, 225, 278
continous tangent functional, 49ff
convex envelope, 48, 67
convexity, 15, 34, 43ff, 55ff, 67, 270
correlation diagram, diatomic, 111ff
correlation energy, 177ff, 182
Coulomb hole, 195
coupling constant integration, 181, 246, 282
curvature energy, nuclear, 471ff

Degeneracy problem, 5, 16, 44
density
    atomic radial, 107, 191
    current, 85, 165, 383, 394, 440, 487

magnetisation, 4, 165, 169
nuclear, 349ff, 354ff, 428, 462
one particle, 2, 11, 32
one particle, time dependent, 83
spin orbit, 334, 346
density matrix
one particle, 16, 37, 57, 73, 94, 145,385,420,450,504
one particle, relativistic, 120, 125
two particle, 74, 94, 181, 198
N-particle, 38, 57, 65
density of states
coulombic, 235, 244, 257
hadronic, 309ff, 318, 324
nuclear, 310, 371,440ff,473
diatomic systems, 105, 109, 145, 200, 497
discontinuity, of exchange-correlation potential, 7, 227, 270, 272, 294
dissociation energy, 191, 268, 273, 301

Electron affinity, 144, 186, 221, 267, 297
electron negativity, 143, 277
electron removal energies, 295
equation of state, 326, 408
exchange-correlation energy, 4, 13, 211, 222, 245
exchange-correlation functional, 3, 8, 59, 90, 93, 148, 170ff, 179, 181, 234, 239, 266, 285
exchange-correlation potential, 4, 169, 211ff, 234, 284
exchange-correlation hole,242,283
exchange energy functional, 97, 102, 292
exchange energy functional, relativistic, 127, 131
exchange hole, 213
excitation energies
see Kohn-Sham orbital energies
excited states, density functional theory for, 19, 22, 65ff
excited states, variational principle for, 66ff
extended Thomas-Fermi theory, *see* Thomas-Fermi theory

Fermi-Amaldi correction, 13
Fermi energy, 212, 223, 309,462
Fermi energy, local, 95, 127,442
Fermi function, 368, 422, 425, 463
Fermi momentum, local, 99, 128, 161
Fermi surface, 238, 432
Fermi temperature, 160
finite temperature density functional theory, 367, 429,474
see also Hohenberg-Kohn-Mermin theory
fireball model, 314
fission barrier, 332, 359, 373, 429
fluid dynamical model, *see* hydrodynamical model
fractional occupation numbers, 16, 150, 220, 224, 267ff, 497
free energy,277, 283ff, 368, 370

Giant resonances, nuclear, 332, 381ff, 390, 459, 478ff
gradient corrections, 4, 8, 193, 287, 433
gradient expansion, 95ff, 142 234, 342, 503
gradient expansion, relativistic, 120, 127
gradient expansion, for finite temperature systems, 371
grand canonical potential, 166, 224, 276ff
groundstate energy functional, 3, 7, 12, 16, 40, 46, 94, 234, 382, 408, 499
groundstate energy functional, N-dependence of, 55
groundstate energy functional, relativistic Foldy-Wouthuysen limit, 118

Hadrons, 309
hardness, chemical
see compressibility
Hartree-Fock-Bogoliubov theory, 429
Hartree-Fock theory, 21, 91, 94, 106, 151, 177, 199, 217, 288, 295, 332, 334, 367, 387, 405, 425, 452, 458, 477, 500
heavy ion collisions, 314, 401, 410
high density limit, of density functional theory, 160
high temperature limit, of density functional theory, 161
Hohenberg-Kohn-Mermin theory, 150, 267, 275ff
Hohenberg-Kohn theorem, 2, 11, 13, 20, 31, 42, 173, 179, 212, 265, 331, 336
Hohenberg-Kohn theorem, for time-dependent systems, 83ff
Hohenberg-Kohn theorem, generalised, 165
homogeneous electron gas, 162, 195
hydrodynamic model, 88, 382, 387, 402ff, 450

Inhomogeneity correction, 85, 109, 115, 458
interatomic potential, 202
interatomic potential surface, 500
ionisation energy, 7, 26, 170, 187, 270
ionisation potential, 56, 144, 184, 212, 221, 222, 267, 297, 498

Jellium model, 237ff, 253

Kinetic energy functional, 13, 33, 56, 59ff, 97, 290, 334, 341, 382, 385, 504
Kinetic energy functional, exchange correction, 103
kinetic energy functional, relativistic, 116
Kirzhnits method, 95, 490
Kohn-Sham equations, 3, 12, 16, 18, 104, 181, 211, 234, 241, 250, 258, 265, 284, 336
Kohn-Sham equations, for relativistic systems, 168
Kohn-Sham equations for time dependent systems, 90
Kohn-Sham-Mermin theory, 222, 280
Kohn-Sham orbital energies, 187, 188, 217, 219, 271, 294

Landau parameter, 388
Landau-Vlasov equation, 392, 452
Laplace transform, 97, 319, 339, 433, 504
lattice model, 250, 257, 270
Legendre transform functional, 21, 31, 43, 56, 148
liquid drop model, nuclear, 338, 362ff, 469, 475
liquid drop model, at finite temperature, 368, 475
local density approximation, 4, 6, 8, 14, 162, 170, 181, 186, 209, 217, 234, 266, 272, 334, 442ff
local density approximation, at finite temperature, 172
local equilibrium, 404
local spin density approximation, 179, 195

Mass formula, nuclear, 362, 417
mean free path, 401, 404
mock phase space, 503
moment of inertia, nuclear, 435ff

Neutron excess, 458
N-representability problem, 74, 167, 179
nuclear matter, 335, 412, 459

Optimum potential, 288

Pair correlation energy, 184ff
partition function, 165
plasma, 161
polarisability, electronic, 149, 199, 213, 275
polarisability, nuclear dipole, 480
pressure, 149, 327
pressure tensor, 389, 406

Quark-gluon plasma, 312, 317, 329, 412
quark model, 312

Random phase approximation, 381, 391, 445, 479, 483ff
reciprocal relations, 148
relativistic density functional theory, 115ff, 161ff
removal energies, nuclear, 390
renormalisation, 164
resummation technique, partial, 366, 424
rotational motion, nuclear, 434

Scaling approximation, hydrodynamic, 384, 393, 419
scaling properties, of density functionals, 144
screening charge, 249
self interaction correction, 182, 287, 295
semicontinuity, 43, 47ff, 53, 69
shell effects, 337, 419
Skyrme interaction, 332, 334ff, 342, 352, 426, 459
sound velocity, 394
spin density functional theory, 161, 179, 213, 287
square density model, nuclear, 388
strength function, nuclear, 396, 482
structure function, nuclear, 446
Strutinsky method, 332, 337, 418, 423, 458
sum rule, for exchange correlation hole, 283
sum rule, for nuclear giant resonances, 384, 448, 472
surface charge density, 251, 256
surface energy, 237, 245ff
surface, nuclear, 352, 421, 462ff
surface symmetry energy, nuclear, 468
symmetry specifications, 20

Tassie model, 390, 446
thermodynamic density functional theory, 163
see also Hohenberg-Kohn-Mermin theory
Thomas-Fermi model, 2, 7, 31, 73, 100, 104, 142, 234, 239, 331, 340, 382, 419, 423, 433, 448, 458, 503
Thomas-Fermi model, at finite temperature, 368
Thomas-Fermi model, relativistic, 115ff
Thomas-Fermi model, time dependent, 106
Thomas-Fermi theory, extended, 74, 102, 142, 240, 333, 338, 382, 419, 449, 459
time dependent density functional theory, 7, 82ff, 154
transition density, nuclear, 396, 480, 487
turning point problem, 100, 133, 340, 366

Uniform back ground model, *see* jellium model

Vacuum polarisability, field theoretical, 320
variational principle,
for excited states, 66ff
for free energy, 166
for grand canonical potential, 225, 276, 278
for ground state energy, 2, 16, 40, 73, 104, 331, 348, 382
for time dependent systems, 82, 385

velocity field, 381, 383, 406
vibrations, nuclear collective,
381, 450
virial theorem, 14
viscosity, 406
Vlasov equation,
see Landau-Vlasov equation
v-representability, 2, 5, 12, 15,
24, 46, 74, 91, 167

Wave vector analysis, 248
Wigner distribution function,
382, 391, 420, 503, 505
Wigner-Kirkwood expansion, 338ff,
364, 419, 477, 504ff
Wigner transform, 386, 391, 420,
425, 433, 450, 506
Woods-Saxon potential, 338, 422
work function, 234ff, 250, 266

x-α method, 14, 102, 145